BARRON'S

AP*

PHYSICS B

5TH EDITION

Jonathan S. Wolf, M.A., Ed.M.
Physics Teacher
Scarsdale High School
Scarsdale, New York

About the author:

Jonathan Wolf has been teaching high school physics for more than twenty-eight years. He has also been an Adjunct Assistant Professor of Astronomy at Hofstra University. He has published over forty professional papers in the fields of Astronomy, Physics, and Physics Education and served for more than ten years as Assistant Editor for *The Science Teachers Bulletin* published by the Science Teachers Association of New York State (STANYS). In addition to being the author of Barron's *AP Physics B*, he is also the author of Barron's *College Review: Physics* and co-author of Barron's *SAT Subject Test in Physics*.

Credit lines:
Figures 5.6, 6.6, 7.1, 8.2, 8.5, 10.4, 12.1, 14.2, 15.1, 15.4, 17.1, 19.10, 20.1, 20.2, 23.6, 24.1, and 24.5 are from *Physics the Easy Way*, Third Edition, by Robert L. Lehrman.

Figure 24.6 is from *Let's Review: Physics* by Miriam A. Lazar and Albert S. Tarendash.

All inquiries should be addressed to:
Barron's Educational Series, Inc.
250 Wireless Boulevard
Hauppauge, New York 11788
http://www.barronseduc.com

ISBN (book only): 978-0-7641-4568-1
ISBN (with CD-ROM): 978-1-4380-7037-7

ISSN: 2156-9487 (book)
2156-9495 (book and CD-ROM)

PRINTED IN THE UNITED STATES OF AMERICA
9 8 7 6 5 4 3 2 1

Mixed Sources
Product group from well-managed forests and other controlled sources

Cert no. SW-COC-002507
www.fsc.org
© 1996 Forest Stewardship Council

Contents

Preface

In this review book, you will be taught several strategies for making your learning and studying more efficient. These are skills that will always be useful and that you can apply across the curriculum for many different subjects.

Each topical review chapter contains multiple-choice questions that differ in variety, style, and level of difficulty. Their purpose is to provide you with a balanced set of questions that test your level of understanding of the review material. Some questions may be easier or more difficult than the actual AP Physics B exam questions because, unlike the actual test, which covers a broad range of subjects, each chapter in this book deals with a specific topic. In addition to the multiple-choice questions there are free-response problems that also vary in difficulty. Full solutions and explanations follow the questions. Additional problem-solving strategies are also provided.

Before the review chapters, there is a diagnostic examination designed to measure your initial level of understanding or to use for practice. At the end of the book, two additional sample examinations are provided. Each examination is fully explained with solutions and guidelines.

This new fifth edition includes many changes. Included with some books is a CD-ROM, which has two more practice exams. Each exam has a self-assessment scoring guide as well as a guide for score improvement. I would like to thank a student, Alex Ramek, for helping me with the score improvement guide and helping me from a student's perspective. Additionally, more sample problems have been included in each of the chapters, and the practice tests have been updated. Each content chapter review begins with a listing of key concepts and ends with a brief summary of main ideas. My colleagues Robert Draper, Patricia Jablonowski, and Joseph Vaughan have been very helpful with insight and ideas for improvement.

I am grateful to Linda Turner, Senior Editor at Barron's Educational Series, for all her ideas and suggestions, and to my wife, Karen, and my daughters, Marissa and Ilana, for their understanding, love, and support.

Scarsdale, New York
July 2010

Jonathan S. Wolf

Introduction to Advanced Placement Physics B

CHAPTER 1

KEY CONCEPTS

- About the Course
- About the Exam
- How to Use This Book

1.1 ABOUT THE COURSE

Advanced Placement Physics B is a one-year, algebra-based survey course in college-level physics. If you have registered for this course, you should cover many different topics to prepare for the AP Physics B exam given in May of each year.

The curriculum guidelines for this course have been established by the College Board. You should consult their website *www.collegeboard.com* to obtain further information. Your teacher and guidance counselors will advise you on registration procedures. As a survey course in physics, you will explore motion, forces, energy, heat, waves, light, optics, electricity, magnetism, and modern physics. You can use many different textbooks, and the College Board's website lists several commonly used ones.

If this is your first physics class, you should be prepared to devote at least 30–45 minutes of work per night to keep up with the material. The types of problems you will be solving will vary in depth and detail. Learning how to solve problems in physics can be tricky, especially when they are not the traditional "plug and chug." I hope this book will help you not only prepare for the examination in May but also help you during the year.

If this is your second physics course, then you have already been exposed to some of the major concepts of physics. You may have taken conceptual physics or project physics classes. These first courses may or may not have been algebra based. However, they did expose you to basic concepts early in your high school career. In that case, this year you will be exposed to some new concepts as well as a more in-depth look at some old concepts.

In either case, be assured that physics is a fundamental science that teaches you about how the universe works. These are exciting times for physics and engineering. Learning how to become an effective problem solver is a lifelong skill that is highly valued.

1.2 ABOUT THE EXAM

The AP Physics B exam is designed by the Educational Testing Service (ETS) in Princeton, New Jersey. The exam is 3 hours long and consists of 2 parts that are 90 minutes each.

- Part I: 70 multiple-choice questions. No calculators or formulas are allowed for this part, but you will be given a table of information.
- Part II: 6–8 free-response problems of varying lengths and score weightings. You will be provided with a table of formulas, and you may use approved calculators for this part. Consult the College Board's website for further details.

Each part of the exam is worth 50 percent of your grade. Within Part II, though, each question may not be worth the same amount of points. For the multiple-choice questions, you will be graded on the number of correct answers.

Your final grade on the examination is based on a weighted curve that varies from year to year and is established by ETS after the exam is given. Final grades range from 1–5, with 5 being the highest. We will discuss test-taking strategies and study skills in the next chapter.

Since ETS copyrights their examinations, all practice examinations in review books are necessarily simulated. This means knowing exactly how you will do on an actual exam is difficult. However, if you study hard, practice, and use all of your skills, this book can help you to improve your score. This book contains three practice exams. If you purchased the edition with the CD-ROM, two additional exams are provided. All exams have answers fully explained and offer suggestions for self-assessment and score improvement. The general topics and approximate distribution of material for the AP Physics B exam are shown in the following table.

TABLE 1.1

Test Topics and Distribution

Content Area	Approximate Percentage
I. Newtonian Mechanics	35%
Kinematics (which includes vectors, vector algebra, coordinate systems, displacement, velocity, acceleration, one-dimensional motion, and projectile motion)	7%
Newton's laws (which includes static and dynamic equilibrium, dynamics on one body, accelerated systems of two bodies, and friction)	9%
Work, energy, and power (which includes the work-energy theorem, work on a body, kinetic energy, gravitational and elastic potential energy, conservation of energy, and power)	5%
Linear momentum (which includes impulse, conservation of momentum, and collisions of systems of particles)	4%
Circular motion and rotation (which includes torque, uniform circular motion, and rotational equilibrium)	4%
Oscillations and gravitation (which includes simple harmonic motion, mass on a spring, simple pendulum, Newton's law of gravitation, and circular orbits)	6%

TABLE 1.1

Test Topics and Distribution *(continued)*

Content Area	Approximate Percentage
II. Fluid Mechanics and Thermal Physics	15%
Fluid mechanics (which includes hydrostatic pressure, buoyancy, fluid flow, and Bernoulli's equations)	6%
Temperature and Heat (which includes mechanical equivalent of heat, heat transfer, and linear expansions)	2%
Kinetic theory and thermodynamics (which includes ideal gases; kinetic model; first law of thermodynamics—pV diagrams; second law of thermodynamics—heat engines)	7%
III. Electricity and Magnetism	25%
Electrostatics (includes static charges, Coulomb's law, electric fields, and electric potential)	5%
Conductors and capacitors (includes capacitance, parallel plates, and electrostatics with conductors)	4%
Electric circuits (includes direct current, Ohm's law, resistance, resistivity, power, simple circuits, series circuits, parallel circuits, combination circuits, and capacitors in steady state)	7%
Magnetic fields (includes forces on moving charges, forces on wires in external magnetic fields, and fields of long wires)	4%
Electromagnetism (includes induction, induced EMF, Faraday's law, and Lenz's law)	5%
IV. Waves and Optics	15%
Wave motion (includes traveling waves, sound, superposition, and standing waves)	5%
Geometric optics (includes light, reflection, refraction, mirrors, and lenses)	5%
Physical optics (includes diffraction, interference, dispersion, and electromagnetic radiation)	5%
V. Atomic and Nuclear Physics	10%
Atomic physics and quantum effects (includes photons, the photoelectric effect, the Compton effect, X rays, matter waves, wave-particle duality, and atomic energy levels)	7%
Nuclear physics (includes conservation of mass number and charge, and mass-energy equivalence)	3%

1.3 HOW TO USE THIS BOOK

If you are using this book throughout the year, study each chapter as the topics are covered in class. At the end of each chapter are a series of multiple-choice questions and free-response problems. Since you will be reviewing specific content material, some of the multiple-choice questions may require you to use a calculator. Remember that on the actual AP exam, you cannot use a calculator for the multiple-choice questions. All of the multiple-choice questions in this book reflect the level of difficulty found on the exam.

The free-response questions at the end of the chapters review the content material. You may find that some of these questions are a bit difficult. On the practice exams, the free-response questions simulate the level of difficulty on the actual exam. As you proceed during the year, doing the end-of-chapter questions will help you to understand the material, help you in the classroom, and also help you prepare for the actual AP exam. When March arrives (see Chapter 2 for a timeline schedule), you can begin to do the diagnostic and practice exams. These are cross-indexed back to the appropriate content areas.

If you are using this book just before the AP exam, take the diagnostic test, assess what content areas you need to review, and then go back to those appropriate areas. You can then work through the appropriate content review followed by the remaining practice exams.

Chapter Summary

- AP Physics B is an algebra-based survey course in college physics.
- The AP Physics B exam is given in May of each year.
- The AP Physics B exam consists of 2 parts that are 90 minutes each. Part I consists of 70 multiple-choice questions. No calculators or formulas are allowed. Part II consists of 6–8 free-response problems. Calculators are allowed, and a formula sheet is provided.
- Contact the College Board at *www.collegeboard.com* for registration and exam information.
- Make sure all of your registration information is complete, and make a plan for studying during the year and before the examination.

CHAPTER 2

Study Skills and Tips

KEY CONCEPTS

- Units
- Relationships and Review of Mathematics
- Tips For Answering Multiple-Choice Questions
- Tips For Solving Free-Response Questions
- Study Skills and Scheduling Your Review

2.1 UNITS

TIP

Make sure you set up a review schedule.

Preparing for an AP exam takes time and planning. In fact, your preparation should begin in September when you start the class. As mentioned in the last chapter, if you are using this review book during the year, the content review chapters should parallel what you are covering in class. If you are using this review book a few weeks prior to the exam in May, your strategy needs to change. The review material should help you refresh your memory as you work on the practice exams. In either case, you should have a plan.

In this chapter, we will look at study skills and tips for helping you do well on the Physics B exam. One of the most important things to remember is that most physical quantities have units associated with them. You must memorize units since you can be asked questions about them in the multiple-choice section. In the free-response questions, you must include all units when using equations, making substitutions, and writing final answers.

TIP

Make sure you memorize all units. Be sure to include them with all calculations and final answers.

A list of standard fundamental (SI) units as well as a list of some derived units are shown in the following two tables. As you work through the different chapters, make a note (on index cards, for example) of each unit.

Table 2.1

Fundamental SI Units Used in Physics

Quantity	Unit Name	Abbreviation
Length	Meter	m
Mass	Kilogram	kg
Time	Second	s
Electric current	Ampere	A
Temperature	Kelvin	K
Amount of substance	Mole	mol

Table 2.2

Some Derived SI Units Used in Physics

Quantity	Unit Name	Abbreviation	Expression in Other SI Units
Area			m^2
Velocity			m/s
Acceleration			m/s^2
Force	Newton	N	$kg \cdot m/s^2$
Momentum			$kg \cdot m/s$
Impulse			$N \cdot s = kg \cdot m/s$
Spring constant		N/m	kg/s^2
Frequency	Hertz	Hz	s^{-1}
Pressure	Pascal	Pa	$N/m^2 = kg/(m \cdot s^2)$
Work, energy	Joule	J	$N \cdot m = kg \cdot m^2/s^2$
Power	Watt	W	$J/s = kg \cdot m^2/s^3$
Electric charge	Coulomb	C	$A \cdot s$
Electric field		N/C	$kg \cdot m/(A \cdot s^3)$
Electric potential	Volt	V	$J/C = kg \cdot m^2/(A \cdot s^3)$
Resistance	Ohm	Ω	$V/A = kg \cdot m^2/(A^2 \cdot s^3)$
Capacitance	Farad	F	$C/V = A^2 \cdot s^4/(kg \cdot m^2)$
Magnetic flux	Weber	Wb	$V \cdot s = kg \cdot m^2/(A \cdot s^2)$
Magnetic flux density	Tesla	T	$Wb/m^2 = N/(A \cdot m) = kg/(A \cdot s^2)$
Inductance	Henry	H	$Wb/A = kg \cdot m^2/(A^2 \cdot s^2)$

2.2 RELATIONSHIPS AND REVIEW OF MATHEMATICS

Reminder

These relationships are also useful for analyzing data to answer laboratory-based questions. A laboratory-based question is usually on the exam. See the Appendix for a review of graphing skills and data analysis techniques.

Since AP Physics B is an algebra-based course, the Appendix reviews some essential aspects of algebra. In physics, we often discuss how quantities vary using proportional relationships. Four special relationships are commonly used. You can review them in more detail by referring to Appendix A. You should memorize these relationships.

- Direct relationship—This is usually represented by the algebraic formula $y = kx$, where k is a constant. This is the equation of a straight line, starting from the origin. An example of this relationship is Newton's Second Law of Motion, $\mathbf{a} = \frac{\mathbf{F}_{net}}{m}$, which states that the acceleration of a body is directly proportional to the net force applied (see Chapter 7).

- Inverse relationship—This is usually represented by the algebraic formula $y = \frac{k}{x}$. This is the equation of a hyperbola. An example of this relationship can be seen in a different version of Newton's Second Law, $\mathbf{F}_{net} = m\mathbf{a}$. In this version, if a constant net force is applied to a body, the mass and acceleration are inversely proportional to each other. Some special relationships, such as gravitation and static electrical forces, are known as inverse square law relationships. The forces are inversely proportional to the square of the distances between the two bodies (see Chapters 12 and 15).
- Squared (quadratic) relationship—This is usually represented by the algebraic formula $y = kx^2$ and is the equation of a parabola starting from the origin. An example of this relationship can be seen in the relationship between the displacement and uniform acceleration of a mass from rest $\mathbf{d} = \frac{1}{2}\,\mathbf{a}t^2$ (see Chapter 5).
- Square root relationship—This is usually represented by the algebraic formula $y = k\sqrt{x}$ and is the equation of a "sideways" parabola. This relationship can be seen in the relationship between the period of a simple pendulum and its length, $T = 2\pi\sqrt{L/\mathbf{g}}$ (see Chapter 11).
- As you review your material, you should know each of these relationships and their associated graphs (see Appendix A for more details).

2.3 TIPS FOR ANSWERING MULTIPLE-CHOICE QUESTIONS

Without a doubt, multiple-choice questions can be tricky. The AP Physics B exam asks 70 multiple-choice questions. These can range from a simple recall of information to questions about units, graphs, proportional relationships, formula manipulations, and simple calculations (without a calculator). The questions cover all areas of the course. An approximate distribution of concepts was presented in Chapter 1.

One tip to remember is that there is no penalty for wrong answers. This means that you may want to try to answer all questions. Instead of randomly guessing, however, you can improve your chances of getting a correct answer if you can eliminate at least two answer choices. Guess intelligently.

When you read a multiple-choice question, try to get to the essential aspects. You have 90 minutes for this part, so do not waste too much time per question. Try to eliminate two or three choices. If a formula is needed, you may try to use approximations (or simple multiplication and division). For example, the magnitude of the acceleration due to gravity (**g**) can be approximated as 10 m/s^2. You can also use estimations or order of magnitude approximations to see if answers make sense.

Remember, no formulas or calculators are allowed for this part. However, you are supplied with a table of information. As you work on the multiple-choice questions in the practice exams, look for distractors. These are choices that may look reasonable but are incorrect. For example, if the question is expecting you to divide to get an answer, the distractor may be an answer obtained by multiplying. Watch out for quadratics (such as centripetal force) or inverse squares (such as gravitation).

If you cannot recall some information, perhaps another similar question will cue you as to what you need to know. (You may work on only one part of the exam at a time.) When you read the question, try to link it to the overall general topic, such as kinematics, dynamics, electricity. Then narrow down the specific area and the associated formula. Finally, you must know which quantities are vectors and which quantities are scalars (see Chapter 4).

Each multiple-choice question in the practice exams is cross-indexed with the general topic area of physics to guide you on your review. As you work on the exams and check your answers, you can easily go back to the topic area to review. At the start of your review, you may want to work on the multiple-choice questions untimed for the diagnostic and first practice exam. A few days before the exam (see the timeline schedule later in this chapter), you should do the last practice exam timed (90 minutes).

2.4 TIPS FOR SOLVING FREE-RESPONSE QUESTIONS

The AP Physics B exam includes 6–8 free-response questions. You have 90 minutes for this section. You may use an approved calculator. (Check the College Board's website for details.) A formula sheet is provided. One of the first things you may notice is that you are not given every formula you ever learned. Some teachers may let you use a formula sheet on their classroom exams, and some teachers may require you to memorize formulas. Even if you get to use a formula sheet on a classroom exam, you should memorize derivations and variations of formulas.

Since you are not given specific formulas for some concepts, you should begin learning how these formulas are derived starting at the beginning of the year. For example, you are not given the specific formulas for projectile motion problems since these are easily derived from the standard kinematics equations. If you begin reviewing a few weeks before the AP exam, you may want to make index cards of formulas to help you to memorize them.

For the free-response questions, each question may be worth a different amount. In fact, each subsection may be worth a different amount. However, each part of the exam is worth 50 percent of your grade to determine your "raw score." As previously discussed, the curve for the exam changes from year to year.

You must read the entire question carefully before you begin. Make sure you know where the formulas and constants can be found on the supplied tables. Also, make sure that you have a working calculator with extra batteries.

TIP 

Make sure you show all of your work on Part II. Include all formulas, substitutions with units, and general concepts used. Remember to label all diagrams. Communicate with the grader!

As you begin to solve the problem, make sure that you write down the general concept being used, for example, conservation of mechanical energy or conservation of energy. Then, you must write down the equations you are using. For example, if the problem requires you to use conservation of mechanical energy (potential and kinetic energies), write out those equations:

$$\text{Initial total mechanical energy} = \text{Final total mechanical energy}$$
$$m\mathbf{g}h_i + \frac{1}{2}\,m\mathbf{v}^2{}_i = m\mathbf{g}h_f + \frac{1}{2}\,m\mathbf{v}^2{}_f$$

When you are making substitutions, you must include the units! For example, if you are calculating net forces on a mass (such as a 2 kg mass that has an acceleration of 4 m/s²), you must write as neatly as possible:

$$\Sigma\mathbf{F} = \mathbf{F}_{\text{net}} = m\mathbf{a} = (2\text{ kg})(4\text{ m/s}^2) = 8\text{ N}$$

Include all relevant information. Communicate with the grader by showing him/her that you understand what the question is asking. You may want to make a few sketches or write down your thoughts in an attempt to find the correct solution path. If a written response is requested, make sure that you write neatly and answer the question in full sentences.

Sometimes the question refers to a lab experiment typically performed in class or simulated data is given. In that case, you may be asked to make a graph (refer to Appendix A). Make sure the graph is labeled correctly (with axes labeled and units clearly marked), points plotted as accurately as possible, and best-fit lines or curves used. Do not connect the dots. Always use the best-fit line for calculating slopes. Make sure you include your units when calculating slopes. Always show all of your work.

If you are drawing vectors, make sure the arrowheads are clearly visible. For angles, there is some room for variation. However, make sure you use your protractor correctly.

TIP

Make sure you have pencils, pens, a calculator, extra batteries, a metric ruler, and a protractor with you for the exam!

Since angles are measured in degrees, be sure your calculator is in the correct mode. If scientific notion is used, make sure you know how to input the numbers into your calculator correctly. Remember, each calculator is different.

If you are asked to draw a free-body diagram (see Chapter 7), make sure you include only actual applied forces. Do not include component forces. Centripetal force is not an applied force and should not be included on a free-body diagram.

What do you do if you are not sure how to solve a problem? Follow these 10 tips.

1. Make sure you understand the general concepts involved, and write them down.
2. Write down all appropriate equations.
3. Try to see how this problem may be similar to one you may have solved before.
4. Make sure you know which information is relevant and which information is irrelevant to what is being asked.
5. Rephrase the question in your mind. Maybe the question is worded in a way that is different from what you are used to.
6. Draw a sketch of the situation if one is not provided.
7. Write out what you think is the best way to solve the problem. This sometimes triggers or cues a solution.
8. Use numbers or estimations if the solution is strictly algebraic manipulation, such as deriving a formula in terms of given quantities or constants.
9. Relax. Sometimes if you move on to another problem, take a deep breath, close your eyes, and just relax for a moment, the tension and anxiety may go away and allow you to continue.
10. Do not leave anything out. Unlike on the multiple-choice questions, you need to show all of your work to earn credit.

2.5 STUDY SKILLS AND SCHEDULING YOUR REVIEW

Preparing for any Advanced Placement exam takes practice and time. Effective studying involves managing your time so that you efficiently review the material. Do not cram a few days before the exam. Getting a good night's sleep before the exam and having a good breakfast the day of the exam is a better use of your time than "pulling an all-nighter." Working in a study group is a good idea. Using index cards to make your own flash cards of key concepts, units, and formulas can also be helpful.

When you study, try to work in a well-lighted, quiet environment, when you are well rested. Studying late at night when you are exhausted is not an effective use of your time. Although some memorization may be necessary, physics is best learned (and studied) by actively solving problems. Remember, if you are using this book during the year, working through the chapter problems as you cover each topic in class, memorizing the units, and familiarizing yourself with the formulas at that time will make your studying easier in the days before the exam.

If you are using this book in the weeks before the exam, make sure you are already familiar with most (if not all) of the units, equations, and topics to be covered. You can either use the chapter review for a quick overview and practice or dive right in to the diagnostic exam. You do not need to take the diagnostic test under timed conditions. See how you do, and then review the concepts for those questions that you got wrong. You can use the end-of-chapter questions (some of which may be more difficult than the actual AP exam) to test your grasp of specific topics and then work on the remaining practice exams.

Setting up a workable study schedule is also vital to success. Each person's needs are different. The following schedule is just one example of an effective plan.

Table 2.3

Test Prep Schedule

September 1–April 15	As the year progresses, make sure you memorize units and are comfortable with formulas. If you are using this book during the year, do end-of-chapter problems as they are covered in class. Make sure you register for the exam, following school procedures, and refer to the College Board's website for details: *www.collegeboard.com*
Four weeks before the exam	Most topics should be covered by now in class. If you are using this book for the first time, begin reviewing concepts and doing the end-of-chapter problems. Begin reviewing units and formulas. Devote at least 30 minutes each day to studying.
Three weeks before the exam	Start working on the diagnostic exam. Go back and review topics that you are unsure of or feel that you answered incorrectly.
Two weeks before the exam	Begin working on practice exams. The CD-ROM edition of this book has two additional practice exams. Continue to review old concepts.
One week before the exam	Do the remaining practice exams timed. Make sure you are comfortable with the exam format and know what to expect. Review any remaining topics and units
The day before the exam	Pack up your registration materials, pens, pencils, calculator, extra batteries, metric ruler, and a protractor. Put them by the door, ready to go. Get a good night's sleep.
The day of the exam	Have a good breakfast. Make sure you take all the items you prepared the night before. Relax!

Chapter Summary

- Make sure you set up a manageable study schedule well in advance of the exam.
- Make sure you memorize all units and are familiar with the exam format.
- Multiple-choice questions do not have a penalty for wrong answers, so do not skip any. If you are unsure of the answer, try to eliminate as many choices as you can, and then guess!
- Do not leave any question out on the free-response part! Show all of your work. Write down all fundamental concepts, write all equations used, and include units for all substitutions and in your final answer.
- Read each question carefully. Write your answers clearly. On the multiple-choice questions, make sure you have a #2 pencil and bubble in all information carefully. Write out short-answer questions in full sentences. Clearly label graphs with units and use best-fit lines or curves.
- Try to relax and do all of the practice exams. Work on the chapter questions to review concepts as needed.
- Get a good night's sleep before the exam.
- On the day of the exam, bring all registration materials with you, as well as pens, pencils, calculators, extra batteries, a metric ruler, and a protractor.

Relax and Good Luck!

Answer Sheet

DIAGNOSTIC TEST

1 Ⓐ Ⓑ Ⓒ Ⓓ Ⓔ
2 Ⓐ Ⓑ Ⓒ Ⓓ Ⓔ
3 Ⓐ Ⓑ Ⓒ Ⓓ Ⓔ
4 Ⓐ Ⓑ Ⓒ Ⓓ Ⓔ
5 Ⓐ Ⓑ Ⓒ Ⓓ Ⓔ
6 Ⓐ Ⓑ Ⓒ Ⓓ Ⓔ
7 Ⓐ Ⓑ Ⓒ Ⓓ Ⓔ
8 Ⓐ Ⓑ Ⓒ Ⓓ Ⓔ
9 Ⓐ Ⓑ Ⓒ Ⓓ Ⓔ
10 Ⓐ Ⓑ Ⓒ Ⓓ Ⓔ
11 Ⓐ Ⓑ Ⓒ Ⓓ Ⓔ
12 Ⓐ Ⓑ Ⓒ Ⓓ Ⓔ
13 Ⓐ Ⓑ Ⓒ Ⓓ Ⓔ
14 Ⓐ Ⓑ Ⓒ Ⓓ Ⓔ
15 Ⓐ Ⓑ Ⓒ Ⓓ Ⓔ
16 Ⓐ Ⓑ Ⓒ Ⓓ Ⓔ
17 Ⓐ Ⓑ Ⓒ Ⓓ Ⓔ
18 Ⓐ Ⓑ Ⓒ Ⓓ Ⓔ
19 Ⓐ Ⓑ Ⓒ Ⓓ Ⓔ
20 Ⓐ Ⓑ Ⓒ Ⓓ Ⓔ
21 Ⓐ Ⓑ Ⓒ Ⓓ Ⓔ
22 Ⓐ Ⓑ Ⓒ Ⓓ Ⓔ
23 Ⓐ Ⓑ Ⓒ Ⓓ Ⓔ
24 Ⓐ Ⓑ Ⓒ Ⓓ Ⓔ
25 Ⓐ Ⓑ Ⓒ Ⓓ Ⓔ
26 Ⓐ Ⓑ Ⓒ Ⓓ Ⓔ
27 Ⓐ Ⓑ Ⓒ Ⓓ Ⓔ
28 Ⓐ Ⓑ Ⓒ Ⓓ Ⓔ
29 Ⓐ Ⓑ Ⓒ Ⓓ Ⓔ
30 Ⓐ Ⓑ Ⓒ Ⓓ Ⓔ
31 Ⓐ Ⓑ Ⓒ Ⓓ Ⓔ
32 Ⓐ Ⓑ Ⓒ Ⓓ Ⓔ
33 Ⓐ Ⓑ Ⓒ Ⓓ Ⓔ
34 Ⓐ Ⓑ Ⓒ Ⓓ Ⓔ
35 Ⓐ Ⓑ Ⓒ Ⓓ Ⓔ
36 Ⓐ Ⓑ Ⓒ Ⓓ Ⓔ
37 Ⓐ Ⓑ Ⓒ Ⓓ Ⓔ
38 Ⓐ Ⓑ Ⓒ Ⓓ Ⓔ
39 Ⓐ Ⓑ Ⓒ Ⓓ Ⓔ
40 Ⓐ Ⓑ Ⓒ Ⓓ Ⓔ
41 Ⓐ Ⓑ Ⓒ Ⓓ Ⓔ
42 Ⓐ Ⓑ Ⓒ Ⓓ Ⓔ
43 Ⓐ Ⓑ Ⓒ Ⓓ Ⓔ
44 Ⓐ Ⓑ Ⓒ Ⓓ Ⓔ
45 Ⓐ Ⓑ Ⓒ Ⓓ Ⓔ
46 Ⓐ Ⓑ Ⓒ Ⓓ Ⓔ
47 Ⓐ Ⓑ Ⓒ Ⓓ Ⓔ
48 Ⓐ Ⓑ Ⓒ Ⓓ Ⓔ
49 Ⓐ Ⓑ Ⓒ Ⓓ Ⓔ
50 Ⓐ Ⓑ Ⓒ Ⓓ Ⓔ
51 Ⓐ Ⓑ Ⓒ Ⓓ Ⓔ
52 Ⓐ Ⓑ Ⓒ Ⓓ Ⓔ
53 Ⓐ Ⓑ Ⓒ Ⓓ Ⓔ
54 Ⓐ Ⓑ Ⓒ Ⓓ Ⓔ
55 Ⓐ Ⓑ Ⓒ Ⓓ Ⓔ
56 Ⓐ Ⓑ Ⓒ Ⓓ Ⓔ
57 Ⓐ Ⓑ Ⓒ Ⓓ Ⓔ
58 Ⓐ Ⓑ Ⓒ Ⓓ Ⓔ
59 Ⓐ Ⓑ Ⓒ Ⓓ Ⓔ
60 Ⓐ Ⓑ Ⓒ Ⓓ Ⓔ
61 Ⓐ Ⓑ Ⓒ Ⓓ Ⓔ
62 Ⓐ Ⓑ Ⓒ Ⓓ Ⓔ
63 Ⓐ Ⓑ Ⓒ Ⓓ Ⓔ
64 Ⓐ Ⓑ Ⓒ Ⓓ Ⓔ
65 Ⓐ Ⓑ Ⓒ Ⓓ Ⓔ
66 Ⓐ Ⓑ Ⓒ Ⓓ Ⓔ
67 Ⓐ Ⓑ Ⓒ Ⓓ Ⓔ
68 Ⓐ Ⓑ Ⓒ Ⓓ Ⓔ
69 Ⓐ Ⓑ Ⓒ Ⓓ Ⓔ
70 Ⓐ Ⓑ Ⓒ Ⓓ Ⓔ

CHAPTER 3

Diagnostic Test

The purpose of a diagnostic examination is to assess your beginning level of understanding before reviewing the topical material. Since ETS prohibits use of its examinations in review books, practice examinations are necessarily simulated. In some cases, the questions may be slightly more difficult or easier than those on an actual exam. Additionally, since the awarding of points for Section II—Free-Response varies from year to year, as do the cutoffs for the various AP grades, it is difficult to predict what your grade on the actual exam would be from your performance on a practice test. However, as the saying goes, "Practice makes perfect," and taking the practice examinations will almost certainly improve your performance on the actual examination. After taking the Diagnostic Test, go over the solutions and explanations that follow and then plan your further studies accordingly, using the self-assessment guide.

Topical review of material from the AP Physics B curriculum begins with Chapter 4 and continues through Chapter 24. Following the topical review chapters are two more practice examinations, also with fully explained solutions, a glossary, and a mathematical appendix. A cumulative index of material appears at the end of the book. Note: Vectors are represented in boldface type.

The Table of Information may be used throughout the exam; however, the Formula Sheet may be used with Section II only.

Good luck!

For this examination the following conventions hold:

I. All frames of reference are assumed to be inertial unless otherwise indicated.
II. Electrical current will follow the direction of a positive charge (conventional current).
III. For any isolated charge, the potential at infinity is taken to be equal to zero at an infinite distance from the charge.
IV. The work (W) done on a thermodynamic system is defined as a positive quantity.

Table of Information

Useful Constants

1 atomic mass unit	$1\ u = 1.66 \times 10^{-27}$ kg
Rest mass of the proton	$m_p = 1.67 \times 10^{-27}$ kg
Rest mass of the neutron	$m_n = 1.67 \times 10^{-27}$ kg
Rest mass of the electron	$m_e = 9.11 \times 10^{-31}$ kg
Magnitude of the electron charge	$e = 1.60 \times 10^{-19}$ C
Avogadro's number	$N_0 = 6.02 \times 10^{23}$ per mol
Universal gas constant	$R = 8.32$ J/(mol · K)
Boltzmann's constant	$k_B = 1.38 \times 10^{-23}$ J/K
Speed of light	$c = 3 \times 10^8$ m/s
Planck's constant	$h = 6.63 \times 10^{-34}$ J · s $= 4.14 \times 10^{-15}$ eV · s
1 electron volt	$1\ \text{eV} = 1.6 \times 10^{-19}$ J
Vacuum permittivity	$\varepsilon_0 = 8.85 \times 10^{-12}\ C^2/N \cdot m^2$
Coulomb's law constant	$k = (1/4)\pi\varepsilon_0 = 9 \times 10^9\ N \cdot m^2/C^2$
Vacuum permeability	$\mu_0 = 4\pi \times 10^{-7}$ Wb/(A · m)
Magnetic constant	$k' = k/c^2 = \mu_0/4\pi = 10^{-7}$ Wb/(A · m)
Acceleration due to gravity at Earth's surface	$\mathbf{g} = 9.8\ m/s^2$
Universal gravitational constant	$G = 6.67 \times 10^{-11}\ m^3/(kg \cdot s^2)$
1 atmosphere pressure	$1\ \text{atm} = 1.0 \times 10^5\ N/m^2 = 1.0 \times 10^5$ Pa
1 nanometer	$1\ \text{nm} = 1.0 \times 10^{-9}$ m

Unit Symbols

meter, m	mole, mol	watt, W	farad, F
kilogram, kg	hertz, Hz	coulomb, C	tesla, T
second, s	newton, N	volt, V	degree Celsius, °C
ampere, A	pascal, Pa	ohm, Ω	electron volt, eV
Kelvin, K	joule, J	henry, H	

Prefixes

Factor	Prefix	Symbol
10^9	giga	G
10^6	mega	M
10^3	kilo	k
10^{-2}	centi	c
10^{-3}	milli	m
10^{-6}	micro	μ
10^{-9}	nano	n
10^{-12}	pico	p

Values of Trigonometric Functions for Common Angles

θ	0°	30°	37°	45°	53°	60°	90°
sin θ	0	$\frac{1}{2}$	$\frac{3}{5}$	$\frac{\sqrt{2}}{2}$	$\frac{4}{5}$	$\frac{\sqrt{3}}{2}$	1
cos θ	1	$\frac{\sqrt{3}}{2}$	$\frac{4}{5}$	$\frac{\sqrt{2}}{2}$	$\frac{3}{5}$	$\frac{1}{2}$	0
tan θ	0	$\frac{\sqrt{3}}{3}$	$\frac{3}{4}$	1	$\frac{4}{3}$	$\sqrt{3}$	∞

Diagnostic Test

Section I

MULTIPLE-CHOICE QUESTIONS

70 QUESTIONS

90 MINUTES

50 PERCENT OF TOTAL GRADE

Directions: For each of the questions or incomplete statements below there are five choices. In each case select the best answer or completion and fill in the corresponding oval on the answer sheet. You may not use a calculator for this part.

1. Which of the following quantities is *not* a vector?

   (A) Momentum
   (B) Displacement
   (C) Acceleration
   (D) Work
   (E) Impulse

2. A 0.05-kg ball is thrown upward from the ground with an initial velocity of 30 m/s. At its maximum height, the magnitude of the ball's acceleration is approximately

   (A) 0 m/s^2
   (B) 10 m/s^2
   (C) 30 m/s^2
   (D) 45 m/s^2
   (E) Not enough information is given.

3. A projectile is launched at an angle θ to the horizontal with an initial velocity **v**. In the absence of any air resistance, which of the following statements is correct?

   (A) The horizontal velocity increases and then decreases during the flight.
   (B) The horizontal velocity remains constant.
   (C) The vertical velocity remains constant.
   (D) The horizontal velocity decreases and then increases during the flight.
   (E) At the projectile's maximum height, the acceleration vector is zero.

4. Which two graphs represent one-dimensional uniformly accelerated motion?

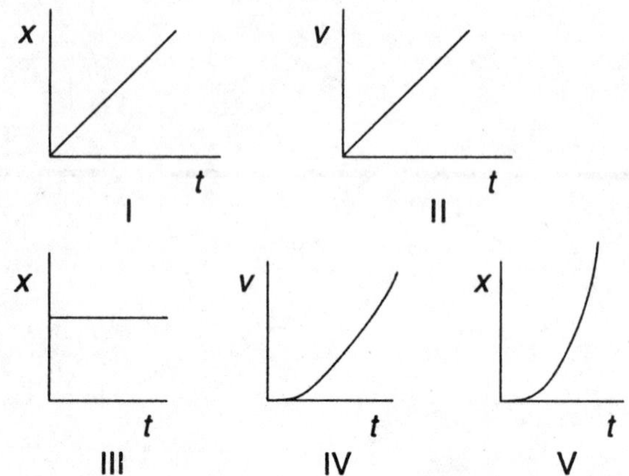

(A) I and IV
(B) II and III
(C) I and V
(D) IV and V
(E) II and V

5. A rock is thrown horizontally off of the top of a building that is 7.5 m above the ground. It is observed that the rock lands 22 m away from the base of the building. Which of the following would increase the time it takes the rock to reach the ground?

I. Increasing the height
II. Increasing the initial horizontal velocity
III. Increasing the mass of the rock

(A) I only
(B) II only
(C) III only
(D) I and II
(E) II and III

Questions 6 and 7 are based on the following information:

A student swings a 0.035-kg rubber stopper attached to a string in a horizontal circle over her head. The length of the string is 0.6 m. The stopper is observed to complete 10 revolutions in 11.7 s.

6. What is the magnitude of the period of revolution?

(A) 0.85 s
(B) 1.17 s
(C) 3.22 s
(D) 0.65 s
(E) 11.7 s

7. Which of the following expressions represents the relationship between the centripetal force **F** acting on the mass and its kinetic energy K?

(A) K/r
(B) $K^2/2r$
(C) $2K/r$
(D) K/r^2
(E) $K/2r$

8. Which of the following statements is correct about a projectile in flight (near Earth)?

(A) Its acceleration increases during its flight.
(B) Its acceleration decreases during its flight.
(C) Its acceleration decreases, then increases, during its flight.
(D) Its acceleration increases, then decreases, during its flight.
(E) Its acceleration remains constant.

9. A 15-newton force is applied to a mass M that is adjacent to a wall, as shown. If the mass is 2 kilograms, the force that the wall exerts on the mass is equal to

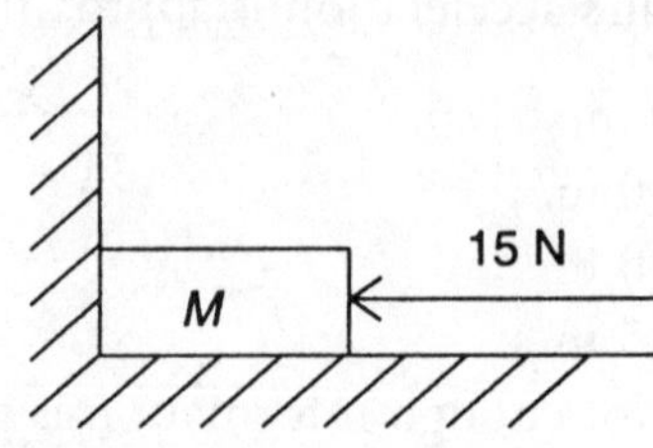

(A) 0 N
(B) 2 N
(C) 15 N
(D) 19.6 N
(E) 30 N

10. In the diagram, a force **F** is applied to a mass M at an angle θ to the horizontal. The mass is moving along a flat, smooth horizontal surface. What is the magnitude of the normal force?

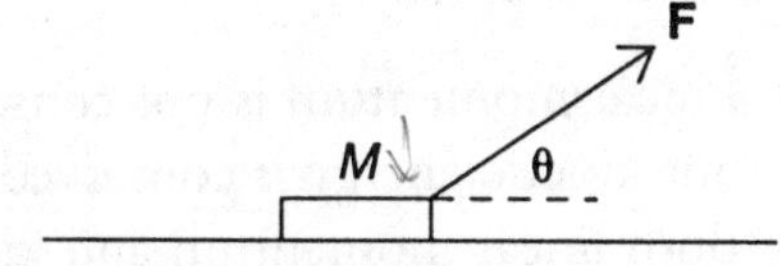

(A) $M\mathbf{g} - \mathbf{F}\sin\theta$
(B) $M\mathbf{g} + \mathbf{F}\sin\theta$
(C) $M\mathbf{g}/\mathbf{F}\sin\theta$
(D) $\mathbf{F}\sin\theta/M\mathbf{g}$
(E) $M\mathbf{g}$

11. In the situation shown, a 2-kilogram mass is attached by a light string over a frictionless pulley to a 5-kilogram mass hanging below. The 2-kilogram mass rests on a frictionless surface. If the system is released, what will be the approximate acceleration of both masses?

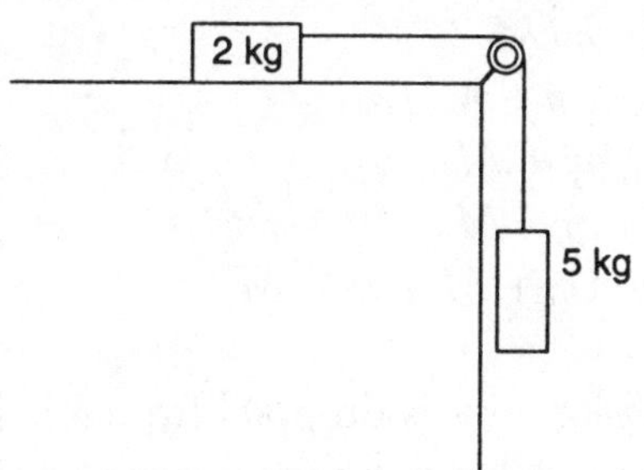

(A) 7 m/s^2
(B) 25 m/s^2
(C) 12 m/s^2
(D) 15 m/s^2
(E) 20 m/s^2

12. A mass M rests on top of a frictionless inclined plane. Which of the following statements is correct about the normal force acting on the mass as the angle of elevation increases?

(A) The normal force increases.
(B) The normal force decreases.
(C) The normal force increases, then decreases.
(D) The normal force decreases, then increases.
(E) The normal force remains constant.

13. A projectile is launched with a velocity **v** and angle θ from level ground. If air resistance is neglected, which of the following graphs corresponds to the relationship between the horizontal component of the velocity and time?

(A)

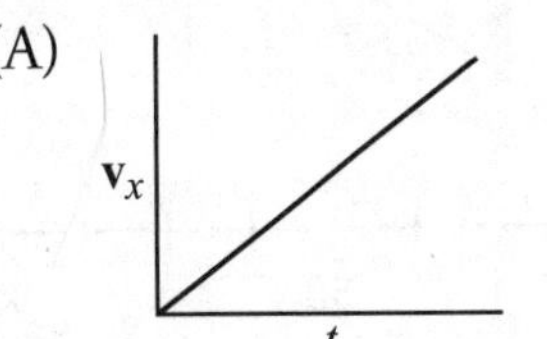

(B)

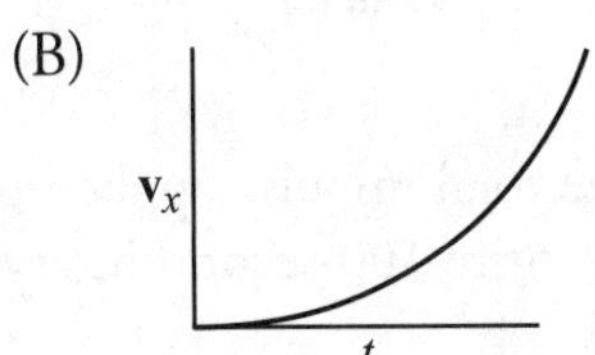

(C)

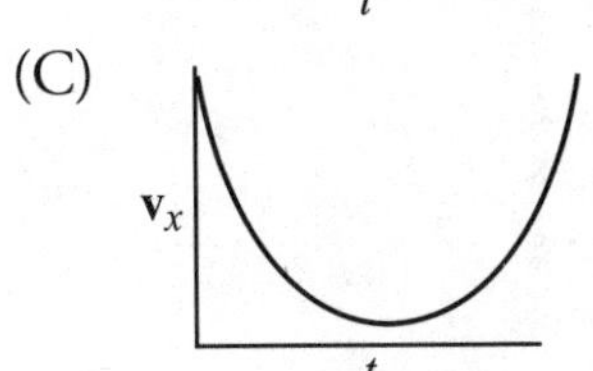

(D)

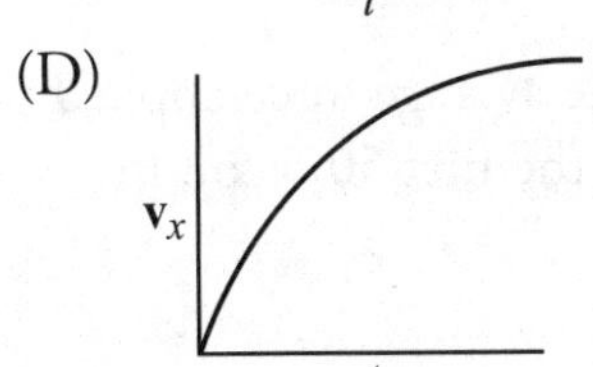

(E)

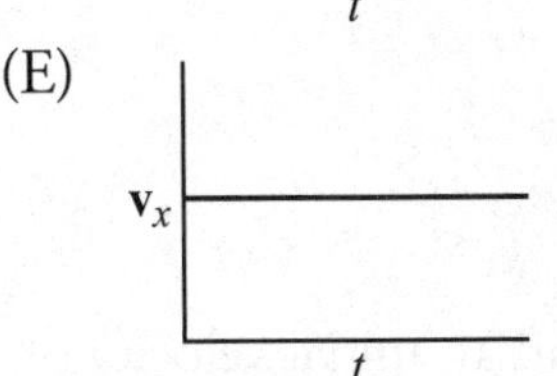

14. A crate weighing 15 N is moving along a rough horizontal surface with a constant velocity **v**. The coefficient of kinetic friction between the crate and the surface is 0.50. What is the approximate magnitude of the force maintaining the constant velocity?

(A) 0 N
(B) 15 N
(C) 75 N
(D) 30 N
(E) 7.5 N

Questions 15–17 are based on the graph of force versus time shown below for a 15-kilogram mass.

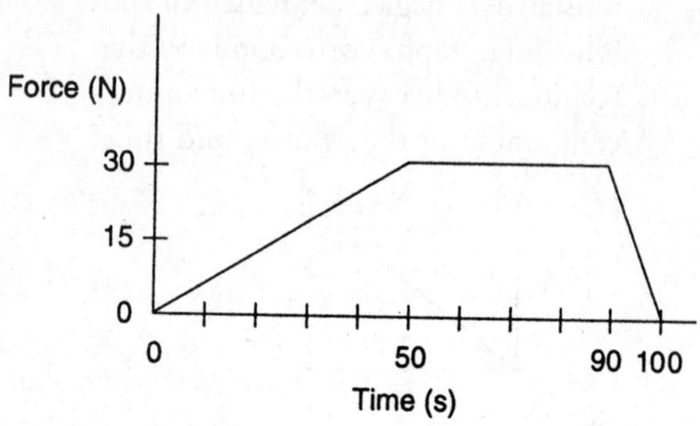

15. What was the total impulse applied to the mass for the entire 100-second interval?

    (A) 1750 N · s
    (B) 1800 N · s
    (C) 1900 N · s
    (D) 2100 N · s
    (E) 2500 N · s

16. What was the average force applied to the mass during the first 50 seconds?

    (A) 7 N
    (B) 15 N
    (C) 18 N
    (D) 21 N
    (E) 30 N

17. If the mass had an initial velocity of 4 meters per second, what was its velocity at the end of 50 seconds?

    (A) 46 m/s
    (B) 50 m/s
    (C) 54 m/s
    (D) 60 m/s
    (E) 66 m/s

18. A 10-kg object has a velocity of 2 m/s to the right. The object is struck by a 0.05-kg wad of putty moving to the left at 10 m/s. The putty sticks to the object after the collision. Which of the following statements is correct?

    (A) Linear momentum is not conserved, but kinetic energy is conserved.
    (B) Both linear momentum and kinetic energy are conserved.
    (C) Linear momentum is conserved, but kinetic energy is not conserved.
    (D) Neither linear momentum nor kinetic energy is conserved.
    (E) None of the above statements is correct.

19. A mass m is moving with a velocity $\mathbf{v}$. It collides with and sticks to a stationary mass M. Which of the following expressions represents the ratio of the initial kinetic energy to the final kinetic energy?

    (A) m/M
    (B) $(m + M)/m$
    (C) $m + M$
    (D) $m - M$
    (E) $(m + M)/(m - M)$

20. A 5-kg mass is dropped from a height of 15 m. What will the approximate velocity of the mass be when it is 10 m above the ground?

    (A) 0 m/s
    (B) 10 m/s
    (C) 20 m/s
    (D) 50 m/s
    (E) 100 m/s

21. A pulley is used to lift a 10-kilogram mass to a height of 2 meters. If 150 newtons of effort was used, the work done to overcome friction was equal to

    (A) 14 J
    (B) 52 J
    (C) 104 J
    (D) 196 J
    (E) 300 J

22. Power is to work as force is to

(A) impulse
(B) momentum
(C) velocity
(D) acceleration
(E) displacement

23. Which pair of forces could produce a resultant force of 7 newtons?

(A) 2 N, 6 N
(B) 3 N, 11 N
(C) 1 N, 5 N
(D) 4 N, 2 N
(E) 3 N, 3 N

24. A mass m is sitting on a wooden board. One end of the board is elevated until at an angle θ, measured from the horizontal, the mass begins to slide down the board. The coefficient of static friction, μ, is given by the expression

(A) $\sin\theta$
(B) $\cos\theta$
(C) $\sec\theta$
(D) $\tan\theta$
(E) $\cot\theta$

25. An object weighs 20 N in air but weighs only 12 N when fully submerged in water. What is the value of the magnitude of the buoyant force acting on the object?

(A) 8 N
(B) 42 N
(C) 12 N
(D) 30 N
(E) 0 N

26. Which of the following expressions represents the kinetic energy of a mass m that has a momentum **p**?

(A) $2m/\mathbf{p}^2$
(B) $2m\mathbf{p}^2$
(C) $2\mathbf{p}/m$
(D) $\mathbf{p}^2/2m$
(E) $2\mathbf{p}m^2$

27. The centripetal force needed to keep Earth in orbit about the Sun is provided by

(A) inertia
(B) Earth's rotation about its axis
(C) the gravitational pull of the Sun
(D) the gravitational pull of the Moon
(E) the shape of Earth

28. A spring with a force constant $k = 400$ N/m is compressed horizontally by a 10-kg mass along a rough surface as shown below. The spring is compressed 0.2 m, and the coefficient of kinetic friction $\mu = 0.10$. If the system is released, what will be the approximate kinetic energy of the mass when the spring is no longer compressed?

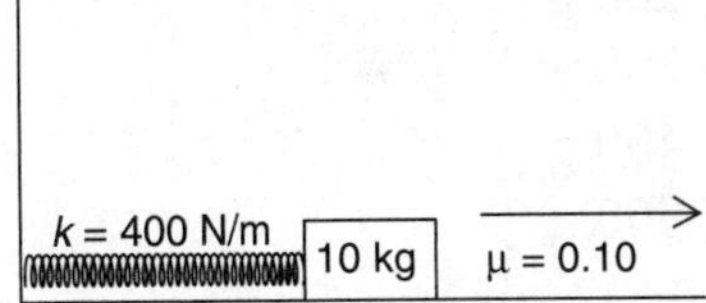

(A) 2 J
(B) 6 J
(C) 8 J
(D) 10 J
(E) 12 J

29. Which of the following expressions is equivalent to 1 watt?

(A) $kg \cdot m^2/s^3$
(B) $kg \cdot m/s$
(C) $kg \cdot m^2/s$
(D) $kg \cdot m^2/s^2$
(E) $kg \cdot m/s^2$

Diagnostic Test

30. A meter stick is balanced at the 50-centimeter mark as shown. Four masses are suspended at the positions shown. What is the magnitude, *m*, of the missing mass?

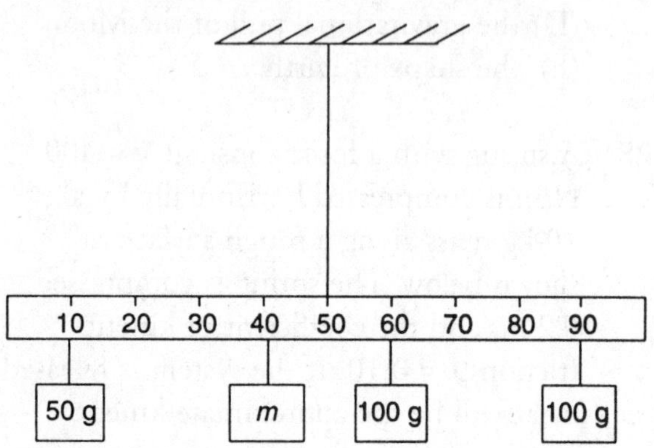

(A) 100 g
(B) 200 g
(C) 300 g
(D) 400 g
(E) 500 g

31. Materials with the highest thermal conductivities are generally

(A) gases
(B) metals
(C) plastics
(D) liquids
(E) wood

32. Heat transfer in a gas can occur by

(A) radiation only
(B) conduction only
(C) convection only
(D) conduction and convection
(E) radiation, conduction, and convection

33. In the following diagram, a wrench is being used to loosen a stubborn bolt using equal forces. Which force will produce the largest torque?

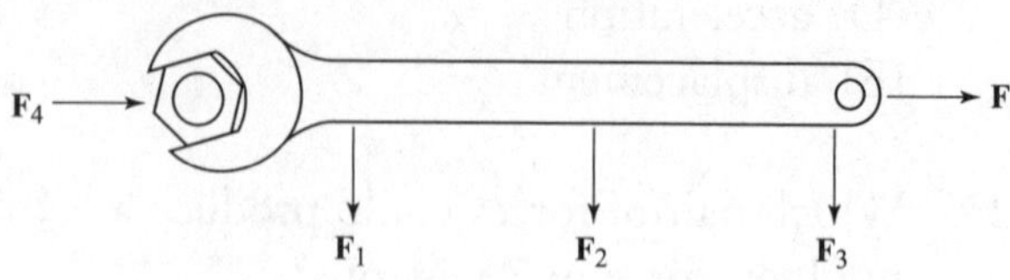

(A) $\mathbf{F}_3$
(B) $\mathbf{F}_2$
(C) $\mathbf{F}_1$
(D) $\mathbf{F}_4$
(E) $\mathbf{F}_5$

34. In an ideal gas, if both the pressure and the absolute temperature are doubled, then which of the following statements about the volume is correct?

(A) The volume is doubled.
(B) The volume remains unchanged.
(C) The volume is halved.
(D) The volume is quadrupled.
(E) The volume is quartered.

35. An insulated tank contains 2.5 cubic meters of an ideal gas under a pressure of 0.5 atmosphere. If the pressure is raised to 3 atmospheres, while the temperature remains constant, the new volume of the gas will be

(A) 2.5 m^3
(B) 5 m^3
(C) 1.25 m^3
(D) 0.42 m^3
(E) 0.37 m^3

36. Which of the following graphs represents the relationship between pressure and absolute temperature for a confined ideal gas?

(A)

(B)

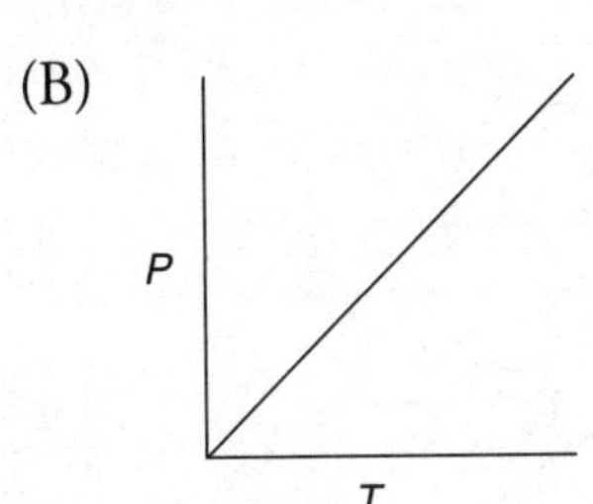

(C)

(D)

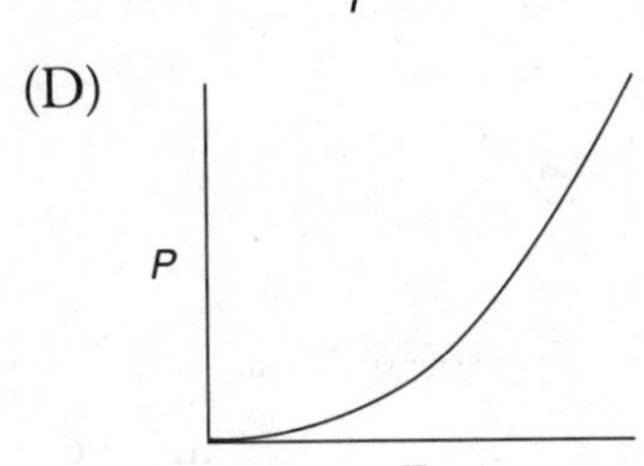

(E)

37. An ideal gas experiences an adiabatic change. Which of the following statements is correct?

(A) The temperature of the gas remains the same.
(B) No work is done to the gas.
(C) The internal energy of the gas remains the same.
(D) The entropy of the gas decreases.
(E) None of the preceding statements is correct.

38. An ideal gas is kept under constant pressure. What change in Celsius temperature must occur in order to double the volume of the gas?

(A) 100°C
(B) 373°C
(C) 273°C
(D) The required change depends on the starting temperature.
(E) The required change depends on the starting volume.

39. The first law of thermodynamics states that $\Delta Q = \Delta U + \Delta W$. Which of the following statements is correct?

(A) ΔQ is the heat supplied to the system, and ΔW is the work done to the system.
(B) ΔQ is the heat supplied to the system, and ΔW is the work done by the system.
(C) ΔQ is the heat supplied by the system, and ΔW is the work done to the system.
(D) ΔQ is the heat supplied by the system, and ΔW is the work done by the system.
(E) None of the preceding statements is correct.

40. The term "isobar" refers to a curve of constant

(A) density
(B) mass
(C) temperature
(D) volume
(E) pressure

Questions 41–43 are based on the following information:

A 10 kg mass is attached to a massless spring that has a force constant $k = 4{,}000$ N/m. It is oscillating back and forth along a horizontal frictionless surface.

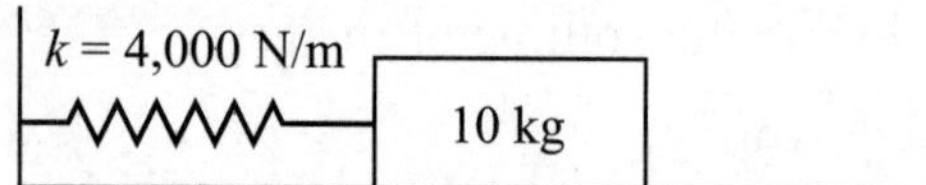

41. What is the approximate value for the period of the oscillations?

(A) $\pi/2$ s
(B) $\pi/4$ s
(C) $\pi/10$ s
(D) 2π s
(E) $\pi/5$ s

42. What is the approximate length of a simple pendulum which would have the same period?

(A) 1/20 m
(B) 1/40 m
(C) 1/10 m
(D) 1/5 m
(E) 2π m

43. Which of the following graphs represents the relationship between period and mass for the oscillating spring system described above?

(A)

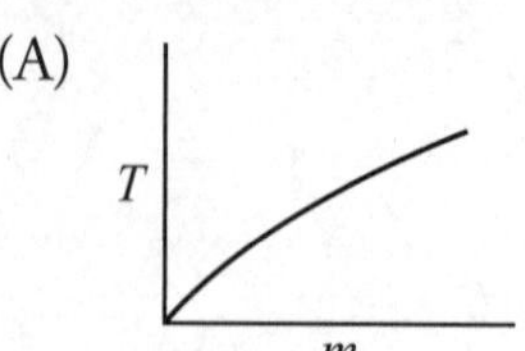

(B)

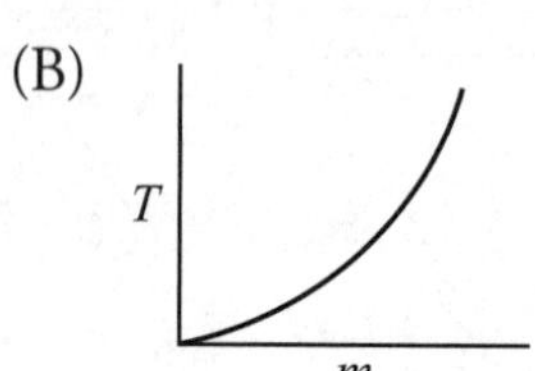

(C)

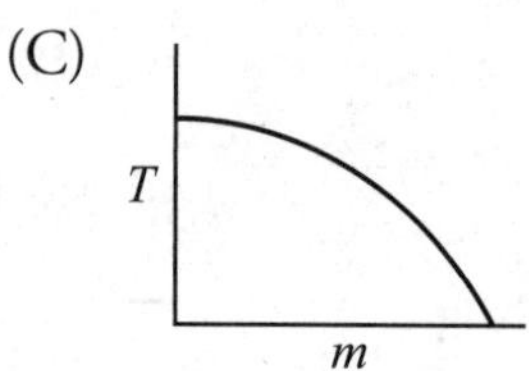

(D)

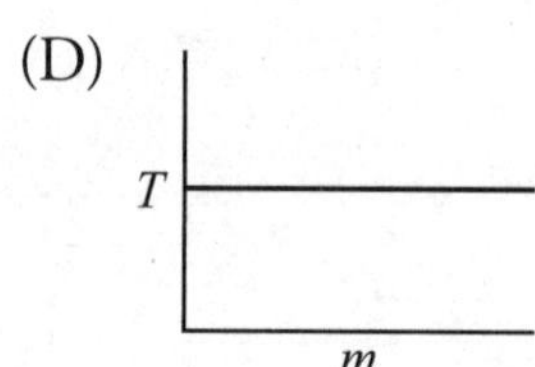

(E)

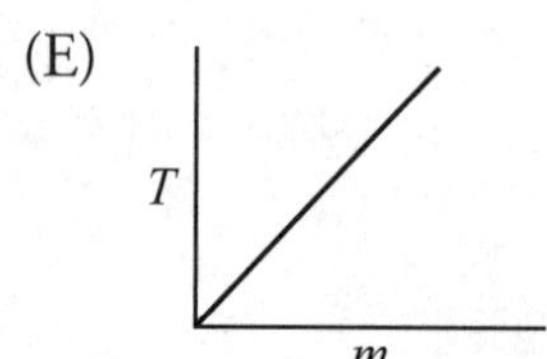

44. A taut string is vibrating in its fundamental resonance mode. If the length of the string is ℓ and the velocity of wave propagation is **v**, the frequency of the fundamental mode is

(A) $2\mathbf{v}/\ell$
(B) $2\ell\mathbf{v}$
(C) $\ell/\mathbf{v}$
(D) $2\ell/\mathbf{v}$
(E) $\mathbf{v}/2\ell$

45. If the tension of a taut string is increased, the frequency of its fundamental mode of vibration will

(A) increase
(B) decrease
(C) remain the same
(D) increase, then decrease
(E) depend on the type of material in the string

Questions 46–48 are based on the circuit shown below:

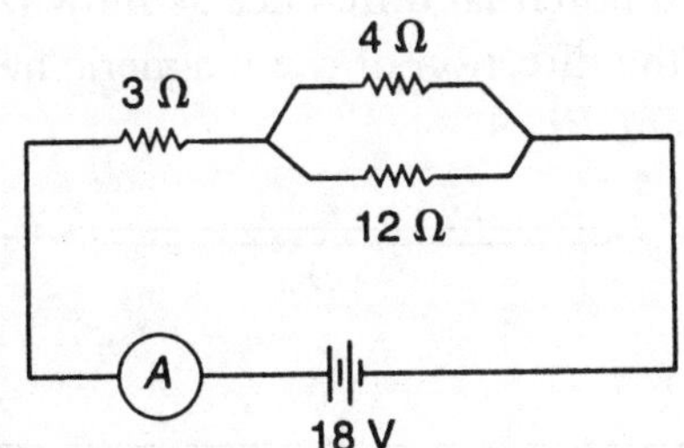

46. What is the equivalent resistance of the circuit?

(A) 3 Ω
(B) 6 Ω
(C) 9 Ω
(D) 12 Ω
(E) 16 Ω

47. What is the current measured in ammeter *A*?

(A) 1.1 A
(B) 3 A
(C) 4 A
(D) 5.2 A
(E) 6 A

48. Compared with the potential difference across the 3-Ω resistor, the potential difference across the 4-Ω resistor is

(A) greater
(B) less
(C) the same
(D) sometimes greater, sometimes less
(E) Not enough information is given to make the comparison.

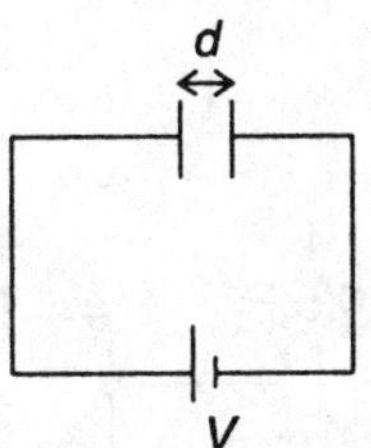

49. An electric circuit consists of two parallel plates separated by a distance *d*. The plates are connected to a potential difference *V* as shown. If the plates are separated further, which of the following statements is correct?

(A) The charge on the plates remains the same, while the electric field decreases.
(B) The charge on the plates decreases, while the electric field remains the same.
(C) The charge on the plates decreases, while the electric field decreases.
(D) The charge on the plates increases, while the electric field decreases.
(E) The charge on the plates increases, while the electric field remains the same.

50. Which two of the following electric circuits have the same equivalent capacitance?

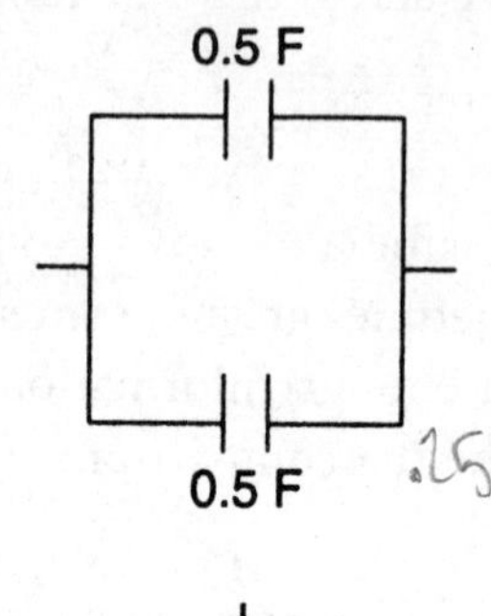

I

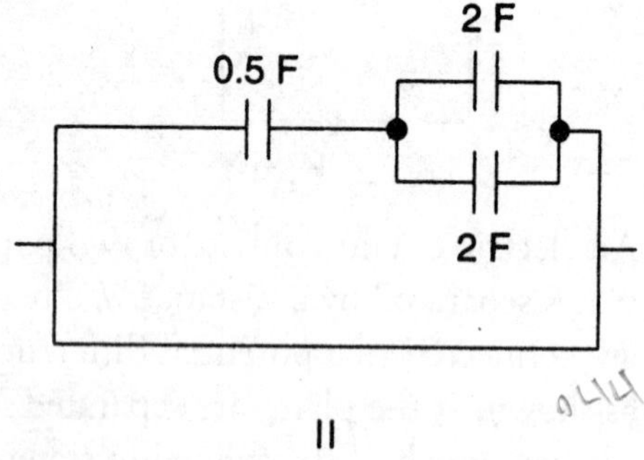

II

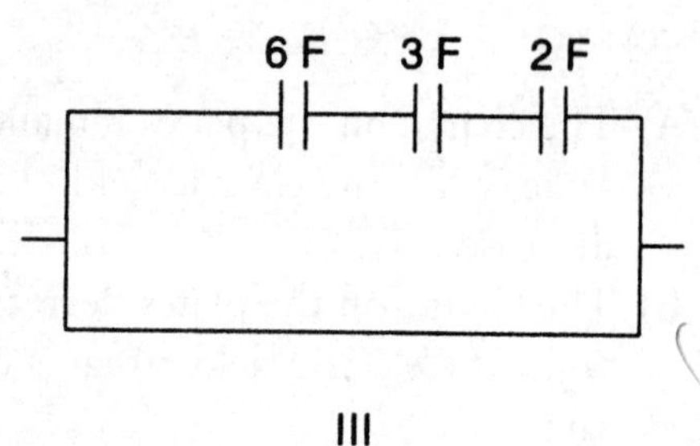

III

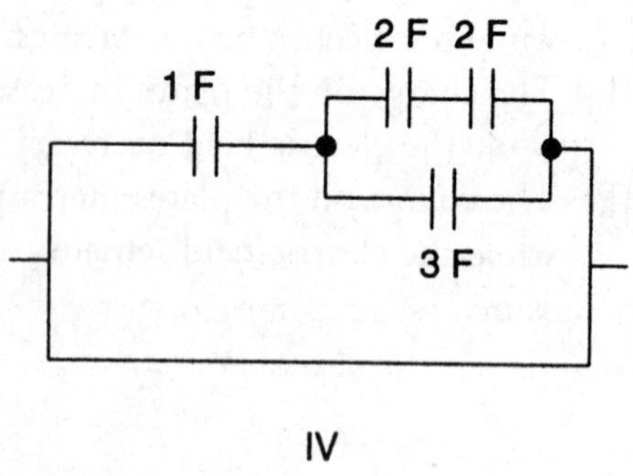

IV

(A) I and III
(B) I and IV
(C) II and III
(D) III and IV
(E) II and IV

51. A charged particle is moving through a region of crossed electric and magnetic fields that are perpendicular to each other. Which expression represents the velocity of the particle?

(A) **E**/**B**
(B) **B**/**E**
(C) **E**q/**B**
(D) **B**q/**E**
(E) **E**/**B**q

52. A cylindrical wire is connected to a circuit and potential difference as shown. What is the direction of the magnetic field at point P?

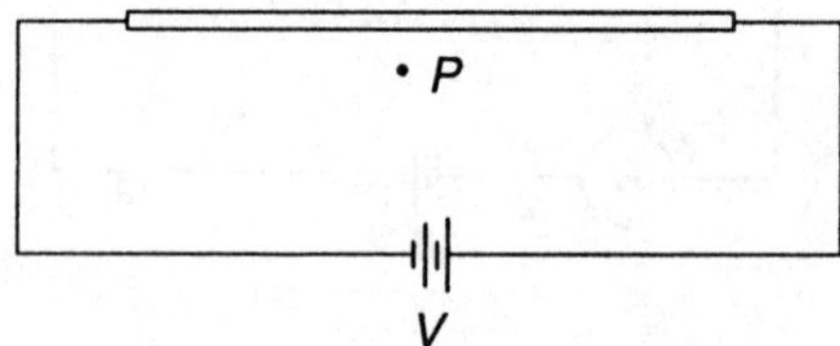

(A) To the right of the page
(B) To the left of the page
(C) Down into the plane of the page
(D) Into the page
(E) Out of the page

53. Two identical resistors in parallel have an equivalent resistance of 2 Ω. If the resistors had been connected in series, the equivalent resistance would have been equal to

(A) 8 Ω
(B) 4 Ω
(C) 16 Ω
(D) 24 Ω
(E) 2 Ω

54. The figure below shows a set of equipotential lines. The electric field has the greatest magnitude at point

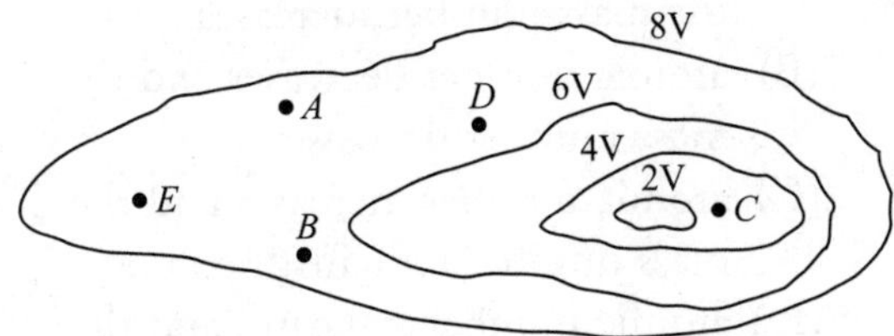

(A) A
(B) B
(C) C
(D) D
(E) E

55. An ion moves in a circular orbit with radius R, in which its velocity is always perpendicular to a constant magnetic field. If the velocity of the ion is doubled, the new radius will be equal to

(A) $\frac{R}{2}$
(B) $2R$
(C) $4R$
(D) $\frac{R}{4}$
(E) R

56. Which of the following electromagnetic waves have the shortest wavelength?

(A) Light rays
(B) Radio waves
(C) X rays
(D) Ultraviolet rays
(E) Infrared rays

57. Which of the following statements is (are) correct about transverse waves?

(A) Transverse waves can be polarized.
(B) Transverse waves can be diffracted.
(C) Microwaves are transverse waves.
(D) Vibrations are perpendicular to the propagation.
(E) All of the preceding statements are correct.

58. A light ray is incident on the boundary between two transparent media. If the velocity of light is less in medium 2 (the light is going from medium 1 to medium 2), the relative index of refraction for the two media is

(A) less than 1.0
(B) equal to 1.0
(C) greater than 1.0
(D) greater than or equal to 1.0
(E) sometimes more or less than 1.0, depending on the surface

59. A ray of monochromatic light is incident, from air, on a converging lens of focal length f. If the lens is replaced by one having a greater absolute index of refraction, the focal length will

(A) increase
(B) decrease
(C) remain the same
(D) increase or decrease, depending on the material in the lens
(E) increase or decrease, depending on the wavelength of light used

60. Total internal reflection can occur between two transparent media if their relative index of refraction is

(A) less than 1.0
(B) greater than 1.0
(C) equal to 1.0
(D) sometimes greater or less than 1.0, depending on the two media
(E) sometimes greater or less than 1.0, depending on the wavelength of light used

61. Which of the following phenomena is exclusive evidence of the wave nature of light?

(A) Reflection
(B) Refraction
(C) Diffraction
(D) Photoelectric effect
(E) Shadows

62. What is the momentum of a photon that has a wavelength of 100 nanometers?

(A) 6.63×10^{-34} kg · m/s
(B) 6.63×10^{-27} kg · m/s
(C) 6.63×10^{-41} kg · m/s
(D) 6.63×10^{-29} kg · m/s
(E) 6.63×10^{-24} kg · m/s

63. Each of the following particles is traveling with the same velocity. Which one will have the smallest de Broglie wavelength?

(A) an electron
(B) a proton
(C) a positron
(D) a neutron
(E) an alpha particle

64. In a photoelectric effect experiment, 2.7 V are required to stop the emitted electrons. The maximum kinetic energy of these electrons is equal to

(A) 2.7 eV
(B) 1.6×10^{-19} eV
(C) 3.32×10^{-19} eV
(D) 3.32 eV
(E) none of the above

65. Which particle *cannot* be accelerated by a magnetic field?

(A) a neutron
(B) a proton
(C) a positron
(D) an electron
(E) an alpha particle

66. When isotope $^{64}_{29}Cu$ decays into $^{64}_{28}Ni$, which of the following is emitted?

(A) Electron
(B) Positron
(C) Photon
(D) Proton
(E) Neutron

67. When a radioactive nucleus emits a beta particle, the

(A) atomic number remains the same and the mass number increases
(B) atomic number decreases and the mass number decreases
(C) atomic number increases and the mass number remains the same
(D) atomic number increases and the mass number decreases
(E) atomic number and the mass number both remain the same

68. The units for Planck's constant, h, are J · s, which are equivalent to the units for

(A) work times position
(B) momentum times position
(C) velocity times acceleration
(D) momentum divided by time
(E) energy divided by velocity

69. How many neutrons are in the isotope $^{191}_{77}Ir$?

(A) 77
(B) 100
(C) 114
(D) 191
(E) 268

70. The isotope $^{27}_{13}Al$ undergoes gamma decay. The decay product formed is

(A) $^{27}_{13}Al$
(B) $^{28}_{13}Al$
(C) $^{27}_{12}Mg$
(D) $^{27}_{14}Si$
(E) none of the above

STOP

This is the end of Section I. You may use any remaining time to check your work in this section.

Formula Sheet for Section II

Newtonian Mechanics

$v = v_0 + at$

$x = x_0 + v_0 t + \frac{1}{2}at^2$

$v^2 = v_0^2 + 2a(x - x_0)$

$\Sigma \mathbf{F} = \mathbf{F}_{net} = m\mathbf{a}$

$F_{fric} \le \mu N$

$a_c = \frac{v^2}{r}$

$\tau = rF \sin \theta$

$\mathbf{p} = m\mathbf{v}$

$\mathbf{J} = \mathbf{F}\Delta t = \Delta \mathbf{p}$

$K = \frac{1}{2}mv^2$

$\Delta U_g = mgh$

$W = F\Delta r \cos \theta$

$P_{avg} = \frac{W}{\Delta t}$

$P = Fv \cos \theta$

$\mathbf{F}_s = -k\mathbf{x}$

$U_s = \frac{1}{2}kx^2$

$T_s = 2\pi\sqrt{\frac{m}{k}}$

$T_p = 2\pi\sqrt{\frac{\ell}{g}}$

$T = \frac{1}{f}$

$F_G = -\frac{Gm_1m_2}{r^2}$

$U_G = -\frac{Gm_1m_2}{r}$

a = acceleration
F = force
f = frequency
h = height
J = impulse
K = kinetic energy
k = spring constant
ℓ = length
m = mass
N = normal force
P = power
p = momentum
r = radius or distance
T = period
t = time
U = potential energy
v = velocity or speed
W = work done on a system
x = position
μ = coefficient of friction
θ = angle
τ = torque

Electricity and Magnetism

$F = \frac{1}{4\pi\epsilon_0}\frac{q_1q_2}{r^2}$

$\mathbf{E} = \frac{\mathbf{F}}{q}$

$U_E = qV = \frac{1}{4\pi\epsilon_0}\frac{q_1q_2}{r}$

$E_{avg} = -\frac{V}{d}$

$V = \frac{1}{4\pi\epsilon_0}\sum_i \frac{q_i}{r_i}$

$C = \frac{Q}{V}$

$C = \frac{\epsilon_0 A}{d}$

$U_c = \frac{1}{2}QV = \frac{1}{2}CV^2$

$I_{avg} = \frac{\Delta Q}{\Delta t}$

$R = \frac{\rho\ell}{A}$

$V = IR$

$P = IV$

$C_p = \sum_i C_i$

$\frac{1}{C_s} = \sum_i \frac{1}{C_i}$

$R_s = \sum_i R_i$

$\frac{1}{R_p} = \sum_i \frac{1}{R_i}$

$F_B = qvB \sin \theta$

$F_B = BI\ell \sin \theta$

$B = \frac{\mu_0}{2\pi}\frac{I}{r}$

$\phi_m = BA \cos \theta$

$\epsilon_{avg} = -\frac{\Delta\phi_m}{\Delta t}$

$\epsilon = B\ell v$

A = area
B = magnetic field
C = capacitance
d = distance
E = electric field
ϵ = emf
F = force
I = current
ℓ = length
P = power
Q = charge
q = point charge
R = resistance
r = distance
t = time
U = potential (stored) energy
V = electrical potential or potential difference
v = velocity or speed
ρ = resistivity
θ = angle
ϕ_m = magnetic flux

Fluid Mechanics and Thermal Physics

$P = P_0 + \rho gh$

$F_{buoy} = \rho Vg$

$A_1 \upsilon_1 = A_2 \upsilon_2$

$P + \rho gy + \frac{1}{2}\rho v^2 = \text{const.}$

$\Delta \ell = \alpha \ell_0 \Delta T$

$H = \frac{kA\Delta T}{L}$

$P = \frac{F}{A}$

$PV = nRT = Nk_BT$

$K_{avg} = \frac{3}{2}k_BT$

$v_{rms} = \sqrt{\frac{3RT}{M}} = \sqrt{\frac{3k_BT}{\mu}}$

$W = -P\Delta V$

$\Delta U = Q + W$

$e = \left|\frac{W}{Q_H}\right|$

$e_c = \frac{T_H - T_C}{T_H}$

A = area
e = efficiency
F = force
h = depth
H = rate of heat transfer
k = thermal conductivity
K_{avg} = average molecular kinetic energy
ℓ = length
L = thickness
M = molar mass
n = number of moles
N = number of molecules
P = pressure
Q = heat transferred to a system
T = temperature
U = internal energy
V = volume
υ = velocity or speed
υ_{rms} = root-mean-square velocity
W = work done on a system
y = height
α = coefficient of linear expansion
μ = mass of molecule
ρ = density

Atomic and Nuclear Physics

$E = hf = pc$

$K_{max} = hf - \phi$

$\lambda = \frac{h}{p}$

$\Delta E = (\Delta m)c^2$

E = energy
f = frequency
K = kinetic energy
m = mass
p = momentum
λ = wavelength
ϕ = work function

Waves and Optics

$\upsilon = f\lambda$

$n = \frac{c}{v}$

$n_1 \sin\theta_1 = n_2 \sin\theta_2$

$\sin\theta_c = \frac{n_2}{n_1}$

$\frac{1}{s_i} + \frac{1}{s_0} = \frac{1}{f}$

$M = \frac{h_i}{h_0} = -\frac{s_i}{s_0}$

$f = \frac{R}{2}$

$d \sin\theta = m\lambda$

$x_m \approx \frac{m\lambda L}{d}$

d = separation
f = frequency or focal length
h = height
L = distance
M = magnification
m = an integer
n = index of refraction
R = radius of curvature
s = distance
υ = speed
x = position
λ = wavelength
θ = angle

Geometry and Trigonometry

Rectangle
$A = bh$

Triangle
$A = \frac{1}{2}bh$

Circle
$A = \pi r^2$
$C = 2\pi r$

Parallelepiped
$V = \ell wh$

Cylinder
$V = \pi r^2 \ell$
$S = 2\pi r\ell + 2\pi r^2$

Sphere
$V = \frac{4}{3}\pi r^3$
$S = 4\pi r^3$

Right Triangle
$a^2 + b^2 = c^2$
$\sin\theta = \frac{a}{c}$
$\cos\theta = \frac{b}{c}$
$\tan\theta = \frac{a}{b}$

A = area
C = circumference
V = volume
S = surface area
b = base
h = height
ℓ = length
w = width
r = radius

Section II

FREE-RESPONSE QUESTIONS

6 QUESTIONS

90 MINUTES

50 PERCENT OF TOTAL GRADE

Directions: Solve the following problems. Be sure to show all work, including substitutions with units, and to explain clearly. You may use a calculator for this part.

1. (15 points) In the arrangement shown, a spring with a force constant k is attached to a vertical wall. The other end of the spring is attached to a mass m that rests on a rough horizontal surface. The coefficient of kinetic friction for the mass and the surface is μ. A light string is attached to m and passed over a frictionless pulley to an overhanging mass M. The system is initially at rest. If mass M is released, it will fall a distance h until coming to rest.

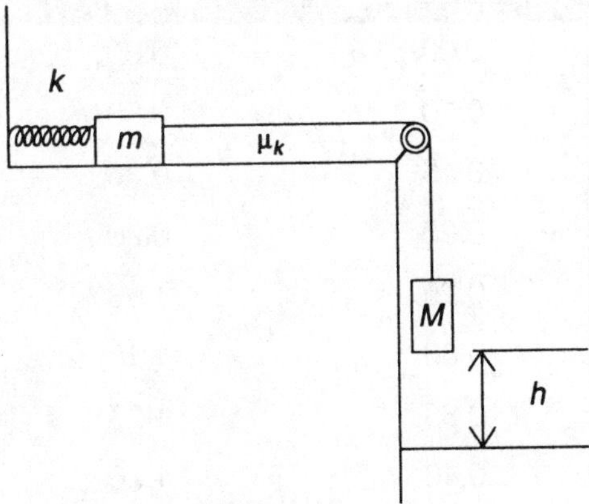

(a) Write an expression for the conservation of energy as it applies to this situation.
(b) In terms of m, M, **g**, h, and k, derive an expression for the coefficient of kinetic friction μ.
(c) If $k = 1000$ newtons per meter, $m = 0.5$ kilogram, $M = 1.5$ kilograms, and $h = 0.04$ meter, calculate the value of μ.

2. (15 points) A rigid rod with a mass of 20 kilograms and a length ℓ is attached to a wall by means of a frictionless pivot as shown. A mass of 5 kilograms is hung from the middle of the rod, and a light string is attached to the free end. The string is attached to the wall, making an angle of 25° to the horizontal. The system is in equilibrium.

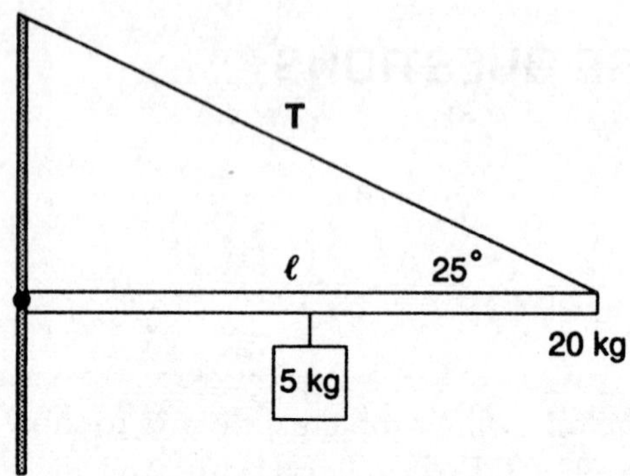

(a) Draw a free body diagram for the rod, and label all forces involved.
(b) Determine the magnitude of the tension **T**.
(c) Determine the magnitude and direction of the reaction force **R** acting at the pivot.

3. (15 points) The following data were collected by a student who measured the period of a pendulum by changing its length. A constant mass was used.

Length (m)	Period (s)
0.00	0.00
0.10	0.68
0.15	0.83
0.20	0.96
0.25	1.08
0.30	1.18
0.35	1.27
0.40	1.36
0.45	1.44
0.50	1.52

(a) Using the grid below, make a graph of length versus period for this data. Be sure to label the axes and use an appropriate scale for your values.

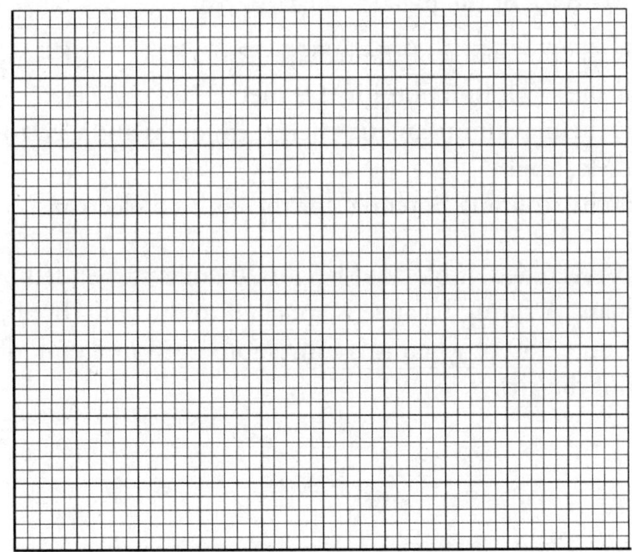

(b) Based on the shape of your graph, state the relationship between the length and period of the pendulum in words.

(c) Using the data, determine the value of the constant of proportionality for this relationship. You must describe your method for arriving at this value.

(d) Using the known equation for the period of a simple pendulum, determine the expected value for this coefficient assuming that the experiment was performed near the surface of Earth.

4. (15 points) An electron is accelerated across a 1200-volt potential difference, as shown in the diagram. Upon reaching the other side, the electron enters a downward electric field of uniform strength **E** = 1,000 newtons per coulomb. The length of the second region, containing the electric field, is 1.2 meters.

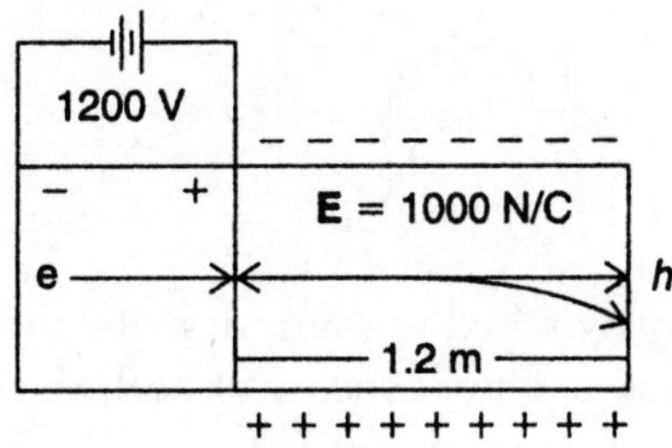

(a) How much work is done to move the electron across the potential difference region?

(b) What is the velocity of the electron as it enters the electric field region?

(c) Neglecting any gravitational effects, the electron is deflected away from its straight path. Derive an expression for the deflected distance h in terms of the mass of the electron, its charge, the length of the region ℓ, and its entrance velocity **v**. Assume constant horizontal velocity.

(d) Using the known value of each quantity, calculate the magnitude of the deflected distance h.

5. (10 points) The threshold frequency for a photoelectric metal is 2.5×10^{15} hertz. Electromagnetic radiation with a wavelength of 50 nanometers is incident on the surface of the metal.

   (a) What is the frequency of the radiation?
   (b) What is the work function for this metal?
   (c) What is the maximum kinetic energy, in joules and electron volts, for the emitted electrons?
   (d) What is the stopping potential for the emitted electrons?

6. (10 points) A rectangular block of height 20 cm is suspended by a thin wire of negligible mass. A spring balance is attached to the wire. In air, the scale reads 400 N, but only 250 N when it is fully submerged in a tank of water (neglect the density of the air).

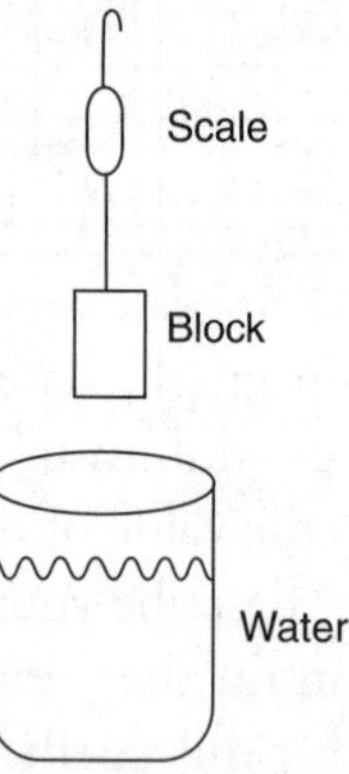

   (a) Find the density and volume of the block.
   (b) The diagram below shows a representation of the block when it is completely submerged. Draw, label, and indicate the magnitude of all of the forces acting on the block in this condition.

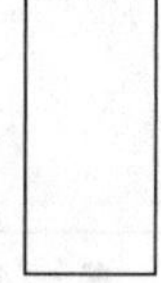

   (c) If the block is slowly raised upward, describe the changes that take place in the gravitational potential energy of both the block and the water.

This is the end of Section II. You may use any remaining time to check your work in this section.

Answer Key

SECTION I

1. D	19. B	37. E	55. B
2. B	20. B	38. D	56. C
3. B	21. C	39. A	57. E
4. E	22. B	40. E	58. C
5. A	23. A	41. C	59. B
6. B	24. D	42. B	60. A
7. C	25. A	43. A	61. C
8. E	26. D	44. E	62. B
9. C	27. C	45. A	63. E
10. A	28. B	46. B	64. A
11. A	29. A	47. B	65. A
12. B	30. C	48. C	66. B
13. E	31. B	49. C	67. C
14. E	32. E	50. A	68. B
15. D	33. A	51. A	69. C
16. B	34. B	52. E	70. A
17. C	35. D	53. A	
18. C	36. B	54. C	

ANSWER EXPLANATIONS

SECTION I

1. **(D)** Work and energy are scalar quantities. The other choices are all vectors.

2. **(B)** At maximum height, the magnitude of the ball's acceleration remains constant at $\mathbf{g} = 9.8\ \text{m/s}^2$.

3. **(B)** In the absence of any air resistance, the horizontal velocity remains constant.

4. **(E)** Graphs II and V both represent one-dimensional uniformly accelerated motion.

5. **(A)** The time required for an object to fall vertically depends only on the vertical height through which it falls.

6. **(B)** The period is equal to the number of seconds required to complete one cycle. This is equal to 11.7 s/10 cycles or 1.17 s.

7. **(C)** The centripetal force is given by $F = m\mathbf{v}^2/r$, and the kinetic energy is given by $K = (\frac{1}{2})\ m\mathbf{v}^2$. Making the correct substitution leads to $F = 2K/r$.

8. **(E)** The acceleration of the projectile, given by gravity, remains constant during its flight.

9. **(C)** Newton's third law of motion states that for every action there is an equal but opposite reaction. The reaction force by the wall is therefore equal to 15 N.

10. **(A)** If no string was attached to the mass, the normal force would be equal in magnitude to the weight, $M\mathbf{g}$. However, the upward component of the string opposes the force of gravity, and therefore we have

$$\mathbf{N} = M\mathbf{g} - \mathbf{F} \sin \theta$$

11. **(A)** To find the acceleration of both masses (since they are coupled), we set up Newton's second law of motion in both directions. The tension **T** in the string caused by the weight of the hanging mass is the same mass that accelerates that 2-kg mass to the right. In the y-direction we can write

$$\mathbf{T} - (5)(9.8) = -5\mathbf{a}$$

In the x-direction we can write

$$\mathbf{T} = 2\mathbf{a}$$

Thus, eliminating **T** from both equations, we get $\mathbf{a} = 7\ \text{m/s}^2$.

12. **(B)** On an inclined plane, the normal force is given by $M\mathbf{g} \cos \theta$. Thus, as the angle of elevation increases, the magnitude of the normal force acting on the mass will decrease.

13. **(E)** Horizontal velocity is constant for a projectile.

14. **(E)** Friction maintains the motion at a constant velocity and is equal to $\mathbf{F} = \mu\mathbf{N} = (0.5)(15 \text{ N}) = 7.5 \text{ N}$. In this case, the normal force was equal to the weight of 15 N.

15. **(D)** The total impulse applied to the mass for the entire 100-s interval is given by the total area, which equals 2,100 N · s.

16. **(B)** For the first 50 s, the impulse applied was equal to 750 N · s, determined by finding the area of that triangular segment. The average force is equal to the impulse divided by the time, 50 s, or 15 N.

17. **(C)** During the 50-s interval, the impulse was equal to 750 N · s. This is equal to the change in momentum, given by $\mathbf{m}\,\Delta\mathbf{v}$. Since the mass is equal to 15 kg, the velocity changed by 50 m/s. Since the force was directed to accelerate the mass, the new velocity is 54 m/s.

18. **(C)** In an inelastic collision, linear momentum is conserved, but kinetic energy is not conserved.

19. **(B)** The initial kinetic energy is given by

$$\text{KE}_i = \frac{1}{2}m\mathbf{v}_1^2$$

The final kinetic energy is given by

$$\text{KE}_f = \frac{1}{2}(m + M)\mathbf{v}_2^2$$

According to conservation of momentum, we can solve for the second velocity:

$$m\mathbf{v}_1 = (m + M)\mathbf{v}_2$$

Thus, the ratio of initial kinetic energy to final kinetic energy is equal to

$$\frac{m + M}{m}$$

20. **(B)** The kinetic energy gained is equal to the potential energy lost. Thus, $m\mathbf{g}h = (\frac{1}{2})m\mathbf{v}^2$. The substitutions yield a velocity of 10 m/s.

21. **(C)** The work needed to raise the 10-kg mass 2 m is equal to 196 J ($W = m\mathbf{g}h$). The actual work done is equal to 300 J. Thus, the work to overcome friction is equal to the difference between these two values: 300 J − 196 J = 104 J.

22. **(B)** Power is equal to the rate of change of work. Force is equal to the rate of change of momentum.

23. **(A)** Only the pair 2 N, 6 N could produce a resultant force of 7 N. The maximum resultant is 8 N, and the minimum resultant is 4 N. Thus, there does exist an angle at which the resultant force could equal 7 N.

24. **(D)** The mass is just beginning to slide. This means that the downward force, which is a component of gravity, is equal to the force of friction. On an incline with angle θ, the magnitude of the force of friction is given by $\mathbf{f} = \mu m\mathbf{g} \cos \theta$, while the magnitude of the component of gravity down the incline is given by $\mathbf{F} = m\mathbf{g} \sin \theta$. Thus,

$$\mu = \tan \theta$$

This angle is known as the "angle of repose."

25. **(A)** $F_B = F_g - F_{\text{apparent}} = 20\text{ N} - 10\text{ N} = 8\text{ N}$

26. **(D)** The momentum $\mathbf{p} = m\mathbf{v}$. The kinetic energy $\text{KE} = (1/2)m\mathbf{v}^2$. Thus,

$$\text{KE} = \frac{\mathbf{p}^2}{2m}$$

27. **(C)** The centripetal force needed to keep Earth in orbit about the Sun is provided by the gravitational force of attraction between Earth and the Sun.

28. **(B)** The elastic potential energy is balanced by the kinetic energy and the work done by friction. Thus,

$$(\tfrac{1}{2})kx^2 = \text{KE} + \mu m\mathbf{g}x$$

Solving for KE by making the suitable substitutions yields KE = 6 J.

29. **(A)** One watt is equal to 1 J/s. Thus, since $1\text{ J} = 1\text{ kg} \cdot \text{m}^2/\text{s}^2$, we have $1\text{ W} = 1\text{ kg} \cdot \text{m}^2/\text{s}^3$.

30. **(C)** We need to show that the torques on both sides of the balance point are equal. Since all masses would be multiplied by **g**, we can leave **g** out without any loss of generality:

$$(50)(40) + m(10) = (100)(10) + (100)(40)$$

This implies that $m = 300$ g.

31. **(B)** Metals have the highest thermal conductivities due to the presence of a large number of free electrons.

32. **(E)** Heat can be transferred in a gas by conduction, convection, and radiation.

33. **(A)** Torque = Force × Lever arm. $\mathbf{F}_3$ has the largest lever arm.

34. **(B)** In an ideal gas, we know that $\mathbf{P}V/T$ = constant. Thus, if the pressure and absolute temperature are both doubled, then the volume remains the same.

35. **(D)** At constant temperature, we use Boyle's law for an ideal gas:

$$P_1V_1 = P_2V_2$$
$$(0.5)(2.5) = (3)V_2$$
$$V_2 = 0.42\text{ m}^3$$

36. **(B)** Pressure and absolute temperature vary directly in an ideal gas. The correct choice is B.

37. **(E)** None of the given statements is correct for an adiabatic change.

38. **(D)** We need to know the starting temperature in order to answer this question.

39. **(A)** In the first law of thermodynamics, ΔQ is the heat supplied by the system while ΔW is the work done to the system.

40. **(E)** The term "isobar" refers to a curve of constant pressure.

41. **(C)** The period is given by

$$T = 2\pi\sqrt{\frac{m}{k}}$$

From this, we see that $T = \pi/10$ s.

42. **(B)** For a pendulum, $T = 2\pi\sqrt{L/\mathbf{g}}$. If we let $\mathbf{g} = 10$ m/s^2, then $L = 1/40$ m.

43. **(A)** The period is proportional to the square root of the mass.

44. **(E)** Since $\mathbf{v} = f\lambda$, then $f = \mathbf{v}/\lambda$. For a string vibrating in its fundamental mode, $\lambda = 2\ell$, so $f = \mathbf{v}/2\ell$.

45. **(A)** In a taut string, the velocity of wave propagation increases with tension. Therefore, the frequency increases as well.

46. **(B)** We first reduce the resistance in the parallel branch:

$$\frac{1}{R} = \frac{1}{4} + \frac{1}{12} \quad \text{and} \quad R = 3\ \Omega$$

This resistance gets added, in series, to the other 3-Ω resistor, making a total of 6 Ω in the circuit.

47. **(B)** Using Ohm's law, we have

$$I = \frac{V}{R} = \frac{18}{6} = 3\ \text{A}$$

48. **(C)** Since the equivalent resistance of the parallel branch is also 3 Ω, the two resistors share the same potential difference of 9 V.

49. **(C)** As the plates are separated further, both the charge on the plates and the strength of the electric field between them will decrease.

50. **(A)** The equivalent capacitance of circuits I and III is the same (1 F).

51. **(A)** If a charged particle is moving through crossed electrical and magnetic fields, then

$$\mathbf{F}_e = \mathbf{F}_m$$

$$\mathbf{E}q = \mathbf{B}q\mathbf{v}$$

$$\text{Thus, } \mathbf{v} = \frac{\mathbf{E}}{\mathbf{B}}$$

52. **(E)** According to the right-hand rule for conventional current, the magnetic field is out of the page at point *P*.

53. **(A)** If two identical resistors in parallel have an equivalent resistance of 2 Ω, then each resistor must have a resistance equal to 4 Ω. If these resistors were then connected in series, the equivalent resistance would be 8 Ω.

54. **(C)** The magnitude of the electric field can be measured in volts per meter, which is also known as a gradient. Therefore, the field is strongest where the contour lines are closest together. This occurs at point *C*.

55. **(B)** When a charged particle has a velocity which is perpendicular to a uniform magnetic field, the magnetic force sets up the centripetal force:

$$\mathbf{B}q\mathbf{v} = \frac{m\mathbf{v}^2}{r}$$

$$\mathbf{B}qr = m\mathbf{v}$$

Therefore, if the velocity is doubled, the radius is also doubled.

56. **(C)** In the electromagnetic spectrum, X rays have the shortest wavelength.

57. **(E)** All of the given statements about transverse waves are correct.

58. **(C)** If light is slowing down, the relative index of refraction for the two media is greater than 1.0.

59. **(B)** If a lens having a greater index of refraction is used, the angle of refraction will be larger, resulting in a shorter focal length.

60. **(A)** Total internal reflection can occur only if light speeds up as it goes from one medium to another. This implies that the relative index of refraction is less than 1.0.

61. **(C)** Interference and diffraction are typical wave phenomena.

62. **(B)** We use the formula

$$\mathbf{p} = \frac{h}{\lambda}$$

Thus,

$$\mathbf{p} = \frac{6.63 \times 10^{-34}}{1 \times 10^{-7}} = 6.63 \times 10^{-27} \text{ kg} \cdot \text{m/s}$$

63. **(E)** Since the alpha particle has the greatest mass, it will have the shortest de Broglie wavelength.

64. **(A)** If the stopping potential is equal to 2.7 V, then since $KE_{max} = V_o q$, in units of electron volts, the answer is 2.7 eV.

65. **(A)** Since a neutron is electrically neutral, it cannot be accelerated by a magnetic field.

66. **(B)** Since the atomic number has decreased by one, but the total number of nucleons has remained the same, the particle emitted is a positron ($_{+1}^{0}e$).

67. **(C)** In a beta decay process, a neutron transforms into a proton and an electron. Therefore, the total number of nucleons (mass number) remains the same, but the number of protons (atomic number) increases by 1.

68. **(B)** The units J · s are the same as the units of mometum times position since $J = N \cdot m = kg \cdot m^2/s^2$. Therefore $J \cdot s = kg \cdot m^2/s = (kg \cdot m/s) \cdot m$

69. **(C)** The number of neutrons in the isotope $^{191}_{77}Ir$ is equal to $A - Z = 191 - 77 = 114$.

70. **(A)** Gamma rays are photons that have neither rest mass nor charge. The isotope remains unchanged.

SOLUTIONS TO SECTION II

1. (a) (6 points) In this situation the potential energy lost by falling mass M is converted into stretching the spring and moving mass m horizontally by doing work against friction. Thus,

$$\Delta GPE = \Delta EPE + W\,(\text{friction})$$

where GPE = gravitational potential energy, and EPE = elastic potential energy.

Award 2 points for recognizing that the potential energy lost by the mass is converted into potential energy for the spring.

Award 2 points for recognizing that energy is conserved and that the difference between the two potential energies is equal to the work done against friction.

Award 2 points for the correct written statement that

$$\Delta GPE = \Delta EPE + W\,(\text{friction})$$

(b) (6 points) We write out each of the specified terms in part (a):

$$\frac{1}{2}kh^2 = M\mathbf{g}h + \mu m\mathbf{g}h$$

$$\mu = \frac{(1/2)kh - M\mathbf{g}}{m\mathbf{g}}$$

Notice that the elongation of the spring x is equal to the downward displacement h.

Award 1 point each for the correct equations for the gravitational potential energy, the elastic potential energy, and the work done against friction.

Award 2 points for the correct symbolic equation.

Award 1 point for the final correct expression for μ.

(c) (3 points) We substitute the given values to get:

$$\mu = \frac{(0.5)(1000\ \text{N/m})(0.04\ \text{m}) - (1.5\ \text{kg})(9.8\ \text{m/s}^2)}{(0.5\ \text{kg})(9.8\ \text{m/s}^2)} = 1.08$$

Award 2 points for the correct substitution with units.

Award 1 point for the correct answer.

2. (a) (3 points) The free-body diagram is shown below:

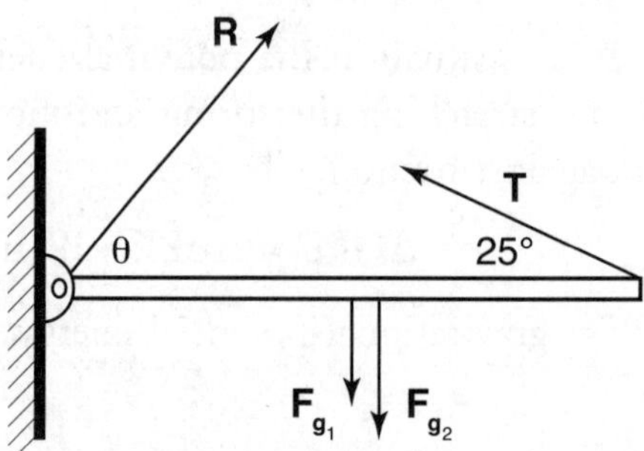

Note: Figure is not drawn to scale.

We take the action of the weight of the rod to be at its center of mass ($\frac{\ell}{2}$).

The weight of the 5-kg mass is acting at the center of the rod as well. This makes for a combined weight $\mathbf{F_g}$.

Award 1 point for each correctly drawn and labeled force.

(b) (6 points) To find the tension **T**, we choose the equilibrium point to be about the pivot. The reaction force **R** does not contribute to any torques about this point. The upward component of the tension, **T** sin θ, is a counterclockwise torque acting at length ℓ. The total weight

$$\mathbf{F_g} = (25\ \text{kg})(9.8\ \text{m/s}^2) = 245\ \text{N}$$

acts at distance $\frac{\ell}{2}$ and is a clockwise torque. Thus:

$$\Sigma\tau = 0$$

$$(\mathbf{T}\sin\theta)\ell = \frac{\mathbf{F_g}\ell}{2}$$

$$\mathbf{T}\sin 25 = \frac{245\text{ N}}{2}$$

$$\mathbf{T} = 290\text{ N}$$

Award 1 point for recognizing that the sum of all torques is equal to zero. Award 2 points for the correct calculation of the weight (with substitutions and units).

Award 1 point for the correct expression for the net torque.

Award 2 points for the correct answer showing full substitutions with units.

(c) (6 points) To find the reaction force, note that **R** has perpendicular components R_x and R_y. Now, consider the torques about the far right edge of the rod. The tension **T** and $\mathbf{R}_x$ do not contribute to any torques about this axis. However, the total weight $\mathbf{F_g}$ acts at a distance of $\ell/2$ from the right edge, and component force $\mathbf{R}_y$ acts at a distance ℓ. In equilibrium, the sum of these torques must be equal to zero:

$$\Sigma\tau = 0$$

$$\mathbf{R}_y\ell = \mathbf{F_g}\frac{\ell}{2}$$

$$\mathbf{R}_y = \frac{245\text{ N}}{2} = 122.5\text{ N}$$

To find $\mathbf{R}_x$, we note that, in equilibrium, the sum of all forces must equal zero as well:

$$\Sigma\mathbf{F} = 0$$

This means that $\mathbf{R}_x$ must balance the horizontal component of the tension, which equals $\mathbf{T}\cos\theta$.

$$\mathbf{R}_x = (290\text{ N})\cos 25 = 262.83\text{ N}$$

The magnitude of the reaction force is given by the Pythagorean theorem:

$$\mathbf{R} = \sqrt{(122.5\text{ N})^2 + (262.83\text{ N})^2} = 290\text{ N}$$

The direction of **R** is given by:

$$\tan\theta = \frac{\mathbf{R}_y}{\mathbf{R}_x} = \frac{122.5\text{ N}}{262.83\text{ N}} = 0.466$$

$$\theta = 25°$$

Award 1 point for again recognizing that the sum of all torques is equal to zero.

Award 1 point for recognizing that the sum of all forces is equal to zero.

Award 1 point for the correct expression for the horizontal reaction force.

Award 1 point for the correct expression for the vertical reaction force.

Award 1 point for the correct answer (with units) for the net reaction force.

Award 1 point for the correct answer for the direction angle.

3. (a) (6 points)

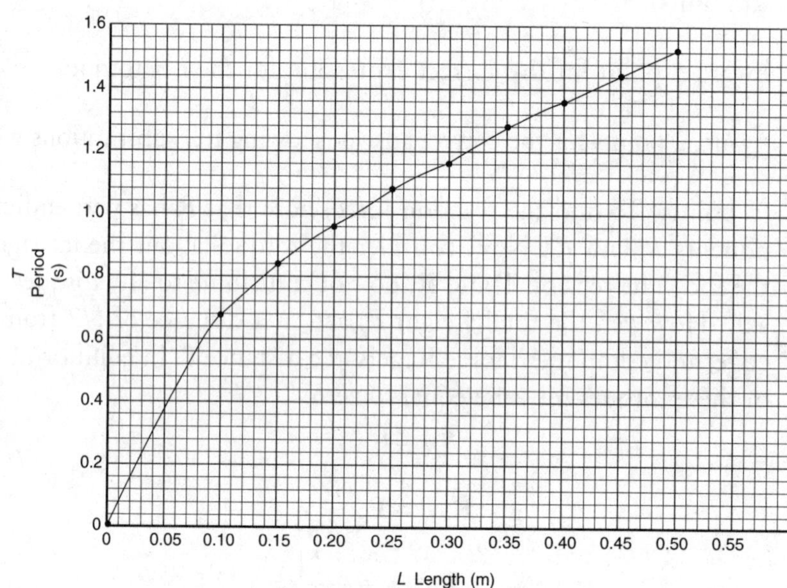

Award 2 points for correctly labeling the axes.

Award 2 points for correctly plotting the points.

Award 2 points for the best fit curve through the points.

(b) (2 points) The period of the pendulum varies directly with the square root of its length.

Award 2 points for the correct statement about the relationship.

(c) (5 points) Given that this is a square root relationship, we can find the constant of proportionality using the relationship.

$$T = k\sqrt{L}, \text{ where } T = \text{period and } L = \text{length}$$

Using several data sets, we can substitute pairs of data and find an average value for k:

(i) (0.1, 0.68) gives $k = 2.15$.
(ii) (0.2, 0.96) gives $k = 2.14$.
(iii) (0.3, 1.18) gives $k = 2.15$.
(iv) (0.4, 1.36) gives $k = 2.15$.

Thus, the average value for the experimental value of $k = 2.1475 \approx 2.15$.

Award 1 point for a correct empirical expression that $T = k\sqrt{L}$.

Award 3 points for demonstrating several data sets indicating various values of k. Award 1 point for showing the average value of k.

(d) (2 points) The known equation for the period of a simple pendulum is

$$T = 2\pi\sqrt{\frac{L}{g}}, \text{ where } g = 9.8 \text{ m/s}^2 \text{ (near Earth's surface)}$$

Thus, $T = 2.007\sqrt{L}$.

Award 2 points for showing that using the correct equation for the period of a pendulum, the formula can be written as $T = 2.007\sqrt{L}$.

4. (a) (3 points) The work done is equal to the product of the electron's charge and the potential difference:

$$W = eV = (1.6 \times 10^{-19} \text{ C})(1200 \text{ J/C}) = 1.92 \times 10^{-16} \text{ J}$$

Award 1 point for recognizing that the work done on the electron is equal to the product of the voltage and the charge. Award 2 points for the correct answer with substitutions and units.

(b) (3 points) The work done is equal to the change in the kinetic energy. Thus:

$$\mathbf{v} = \sqrt{\frac{2KE}{m}} = \sqrt{\frac{(2)(1.92 \times 10^{-19} \text{ J})}{(9.1 \times 10^{-31} \text{ kg})}} = 2.05 \times 10^{7} \text{ m/s}$$

Award 1 point for recognizing that the work done is equal to the change in kinetic energy of the electron. Award 1 point for writing the correct expression for kinetic energy. Award 1 point for a correct answer for the velocity with substitutions and units.

(c) (6 points) Since the electron enters the electric field with an assumed constant horizontal velocity, the downward acceleration will cause a parabolic projectile path. Thus, since the electric field is uniform, we know that $\mathbf{F} = e\mathbf{E}$ (where e is the magnitude of the electron charge). Also, we know that $\mathbf{F} = m\mathbf{a}$; thus $\mathbf{a} = e\mathbf{E}/m$. Now, with a constant horizontal velocity $\mathbf{v}$ and a distance ℓ, the time within the electric field is given by $\ell/\mathbf{v}$. This is the same time needed to fall the deflected distance h:

$$h = \frac{1}{2}\mathbf{a}t^2 = \frac{e\mathbf{E}\ell^2}{2m\mathbf{v}^2}$$

Award 1 point for recognizing that $\mathbf{F} = \mathbf{E}q$. Award 1 point for the expression $\mathbf{F} = m\mathbf{a}$. Award 1 point for recognizing that for distance dropped, $h = \frac{1}{2}\mathbf{a}t^2$. Award 3 points for a correct derivation of h in terms of the given variables.

(d) (3 points) Using the known values, we obtain

$$h = \frac{e\mathbf{E}\ell^2}{2m\mathbf{v}^2} = \frac{(1.6 \times 10^{-19}\ \text{C})(1000\ \text{N/C})(1.2\ \text{m})^2}{(2)(9.1 \times 10^{-31}\ \text{kg})(2.05 \times 10^7\ \text{m/s})^2} = 0.30\ \text{m}$$

Award 2 points for showing a correct substitution with units. Award 1 point for the correct answer with units for the distance.

5. (a) (2 points) The frequency of the radiation photons is given by

$$f = \frac{c}{\lambda} = \frac{3 \times 10^8\ \text{m/s}}{5 \times 10^{-8}\ \text{m}} = 6.0 \times 10^{15}\ \text{Hz}$$

Award 1 point for the correct equation $\mathbf{c} = f\lambda$. Award 1 point for the correct answer with units.

(b) (3 points) The work function is equal to the product of the threshold frequency and Planck's constant:

$$W_o = hf_o = (6.63 \times 10^{-34}\ \text{J} \cdot \text{s})(2.5 \times 10^{15}\ \text{s}^{-1}) = 1.6575 \times 10^{-18}\text{J}$$

Award 1 point for the correct equation for the work function. Award 2 points for showing the correct answer with substitutions and units.

(c) (3 points) The maximum kinetic energy is equal to the difference between the energy of one photon and the work function. Therefore:

$$\begin{aligned} \text{KE}_{\text{max}} &= E_p - W_o = (3.978 \times 10^{-18}\ \text{J}) - (1.6575 \times 10^{-18}\ \text{J}) \\ &= 2.32 \times 10^{-18}\ \text{J} = 14.5\ \text{eV} \end{aligned}$$

Award 1 point for the correct equation for the maximum kinetic energy. Award 2 points for the correct answer showing substitutions and units.

(d) (2 points) Since 1 eV is the energy given to one electron accelerated through 1 V, the stopping potential is equal to 14.5 V in this case.

Award 1 point for recognizing that in units of electron volts, the numerical answer for the stopping potential is the same number as the maximum kinetic energy. Award 1 point for the correct answer of 14.5 V.

6. (a) (6 points) If we let ρ = the average density of the block and let ρ' = the density of water, the weight of the block in air is given by

$$\mathbf{F_g} = \rho\mathbf{g}V, \text{ where } V = \text{volume of the block}$$

In water, the block weighs less because of the buoyancy of the water:

$$\mathbf{F}'_{\mathbf{g}} = (\rho - \rho')\mathbf{g}V, \text{ where } \rho' = 1000 \text{ kg/m}^3$$

Thus, the ratio of weights is given by

$$\frac{\rho}{(\rho - \rho')} = \frac{400 \text{ N}}{250 \text{ N}} = 1.6$$

Solving for the block's density, we obtain

$$\rho = 2666.7 \text{ kg/m}^3$$

In air, the block weighs 400 N, which means its mass is given by

$$M = \frac{\mathbf{F}_{\mathbf{g}}}{\mathbf{g}} = 40.8 \text{ kg}$$

Since density is the ratio of mass to volume, we easily obtain

$$V = \frac{m}{\rho} = 0.015 \text{ m}^3$$

Award 1 point for the expression for the weight $\mathbf{F}_{\mathbf{g}} = \rho\mathbf{g}V$. Award 1 point for the expression $\mathbf{F}'_{\mathbf{g}} = (\rho - \rho')\mathbf{g}V$. Award 1 point for showing that the ratio of weights is equal to 1.6. Award 1 point for the correct density of the block (with units). Award 1 point for the correct mass of the block (with units). Award 1 point for the correct volume of the block (with units).

(b) (3 points)

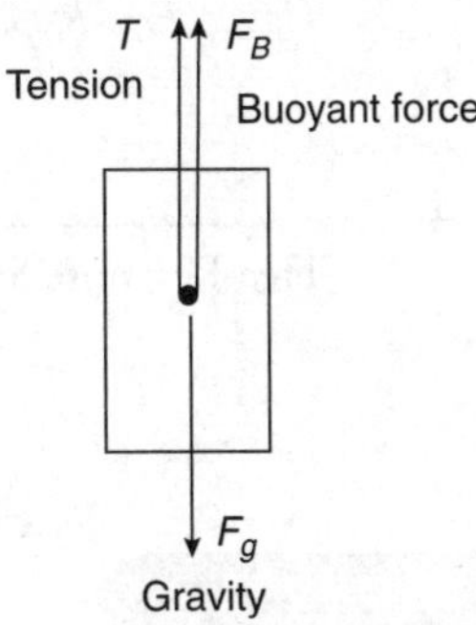

There are three forces acting on the block, the gravitational weight equal to 400 N, the upward buoyant force equal to 400 N – 250 N = 150 N, and the upward tension in the wire which must also be equal to 250 N since the block is in static equilibrium.

Award 2 points for either a correct drawing of the forces or correct explanation of equilibrium in this situation. Award 1 point for the correct answer (with units).

(c) (1 point) As the block is raised up, its gravitational potential energy increases. The water level consequently decreases, which lowers its gravitational potential energy.

Award 1 point for the correct statement about the change in level.

Test Analysis
Diagnostic Test

Section I: Multiple-Choice

Number correct (out of 70) = ____________

Raw Score = number correct × .714 = ____________
Multiple-Choice Score

Section II: Free-Response

Question 1 = ____________
(out of 15)

Question 2 = ____________
(out of 15)

Question 3 = ____________
(out of 15)

Question 4 = ____________
(out of 15)

Question 5 = ____________
(out of 10)

Question 6 = ____________
(out of 10)

Raw Score: ____________ × .625 = ____________
Free-Response Score

Final Score

____________ + ____________ = ____________

Multiple-Choice Score | Free-Response Score | Final Score (rounded to the nearest whole number)

Final Score Range	AP Score*
81–100	5
61–80	4
51–60	3
41–50	2
0–40	1

*The score range corresponding to each grade varies from exam to exam and is approximate.

SELF-ASSESSMENT GUIDE

How well did you do? Remember that on the actual AP Physics B exam, grading and scoring are based on the year and guidelines set up by the College Board for the readers who will be grading your exam. Use the results of this assessment only as a guide to further your studying and not as an absolute predictor of an AP grade.

Use the table below to help you locate the topics in the book for which you need further study.

Topic	Multiple-Choice Question Number
Motion	2, 3, 4, 5, 8, 13, 20, 41, 42, 43
Forces and Momentum	6, 7, 9, 10, 11, 12, 14, 15, 16, 17, 18, 19, 23, 24, 25, 26, 27, 28, 30, 33
Work and Energy	1, 21, 22, 29
Heat and Gases	31, 32, 34, 35, 36, 37, 38, 39, 40
Waves and Sound	44, 45, 47, 57
Light and Optics	56, 58, 59, 60, 61
Electricity and Magnetism	46, 47, 48, 49, 50, 51, 52, 53, 54, 55
Modern Physics	62, 63, 64, 65, 66, 67, 68, 69, 70

SCORE IMPROVEMENT

Not satisfied with your score? Don't worry. Here are some tips for improvement.

For the multiple-choice section:

1. Write the numbers of the questions you left blank or answered incorrectly in the first column below.
2. Go over the answers to the problems in the Answers Explained section and write the main ideas and concepts behind the problems in the second column.
3. Go back and reread the sections covering the material in the second column.
4. Retake the skipped and incorrect questions.
5. Recalculate your score to see how much you improved.

Questions	Main Idea or Concept

For the free-response section:

1. Go over the answers to the problems in the Answers Explained section, and circle any mistakes in your answers.
2. Go back and reread the sections covering the material.
3. Retake the missed free-response questions.
4. Recalculate your score to see how much you improved.

CHAPTER 4

Vectors

KEY CONCEPTS

- Coordinate Systems and Frames of Reference
- Vectors
- Addition of Vectors
- Subtraction of Vectors
- Addition Methods Using the Components of Vectors

4.1 COORDINATE SYSTEMS AND FRAMES OF REFERENCE

We begin our review of physics with the idea that all observations and measurements are made relative to a suitably chosen frame of reference. In other words, when observations are made that will be the basis of future predictions, we must be careful to note from what point of view those observations are being made. For example, if you are standing on the street and see a car driving by, you observe the car and all of its occupants moving relative to you. However, to the driver and other occupants of the car, the situation is different: they may not appear to be in motion relative to themselves; rather, you appear to be moving backward relative to them.

If you were to get into your car, drive out to meet the other car, and travel at the same speed in the same direction, right next to the first car, there would be no relative motion between the two cars. These different points of view are known as **frames of reference**, and they are very important aspects of physics.

A coordinate system within a frame of reference is defined to be a set of reference lines that intersect at an arbitrarily chosen fixed point called the **origin**. In the Cartesian coordinate system, the reference lines are three mutually perpendicular lines designated x, y, and z (see Figure 4.1). The coordinate system must provide a set of rules for locating objects within that frame of reference. In the three-dimensional Cartesian system, if we define a plane containing the x- and y-coordinates (let's say a horizontal plane), then the z-axis specifies direction up or down.

Remember

A **frame of reference** represents an observer's viewpoint and requires a coordinate system to set the origin. A frame of reference moving with a constant velocity relative to the origin is called an **inertial frame of reference.**

It is often useful to compare observations made in two different frames of reference. In the example above, the motion of the occupants of the car was reduced to zero if we transformed our coordinate system to the car moving at constant velocity. In this case we say that the car is an **inertial frame of reference**. In such a frame, it is impossible to observe whether or not the reference frame is in motion if the observers are moving with it.

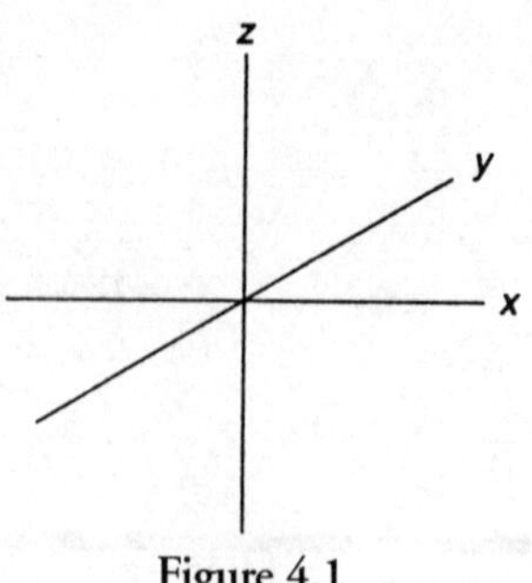

Figure 4.1

4.2 VECTORS

Vectors have both magnitude (size) and direction. **Scalars** have only magnitude.

Another way of locating the position of an object in the Cartesian system is with a directed line segment, or **vector**. If you draw an arrow, starting from the origin, to a point in space (see Figure 4.2), you have defined a position vector **R** whose magnitude is given by | **R** | and is equal to the linear distance between the origin and the point (*x*, *y*). The direction of the vector is given by the acute angle, identified by the Greek letter θ, that the arrow makes with the positive *x*-axis. Any quantity that has both magnitude and direction is called a **vector quantity**. Any quantity that has only magnitude is a **scalar quantity**. Examples of vector quantities are force, velocity, weight, and displacement. Examples of scalar quantities are mass, distance, speed, and energy.

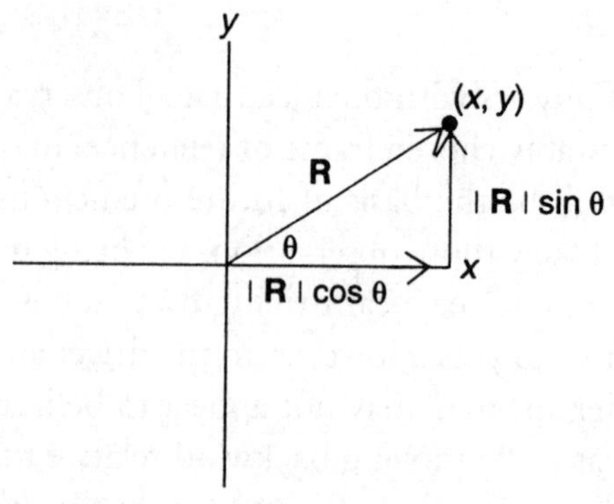

Figure 4.2

If we designate the magnitude of the vector **R** as *r*, and the direction angle is given by θ, then we have an alternative coordinate system called the **polar coordinate** system. In two dimensions, to locate the point in the Cartesian system (*x*, *y*) involves going *x* units horizontally and *y* units vertically. Additionally, we can show that (see Figure 4.2)

$$x = r\cos\theta \quad \text{and} \quad y = r\sin\theta$$

Having established this polar form for vectors, we can easily show that *r* and θ can be expressed in terms of *x* and *y* as follows:

$$r^2 = x^2 + y^2$$

$$\tan\theta = \frac{y}{x}$$

The magnitude of **R** is then given by

$$|\mathbf{R}| = r = \sqrt{x^2 + y^2}$$

and the direction angle θ is given by

$$\theta = \tan^{-1}\frac{y}{x}$$

4.3 ADDITION OF VECTORS

A. Geometric Considerations

The ability to combine vectors is a very important tool of physics. From a geometric standpoint, the "addition" of two vectors is not the same as the addition of two numbers. When we state that, given two vectors **A** and **B**, we wish to form the third vector **C** such that **A** + **B** = **C**, we must be careful to preserve the directions of the vectors relative to our chosen frame of reference.

One way to do this "addition" is by the construction of what is called a **vector diagram**. Vectors are identified geometrically by a directed arrow that has a "head" and a "tail." We will consider a series of examples. First, suppose a girl is walking from her house a distance of five blocks east and then an additional two blocks east. How can we represent these displacements vectorially? First, we must choose a suitable scale to represent the magnitude lengths of the vectors. In this case, let us just call the scale "one vector unit" or just "one unit," to be equal to one block of distance. Let us also agree that "east" is to the right (and hence "north" is directed up). A vector diagram for this set of displacements is given in Figure 4.3.

Remember

Vectors add constructively from head to tail. The vector sum is called the **resultant**. Vectors that act at the same place and at the same time are called **concurrent vectors.**

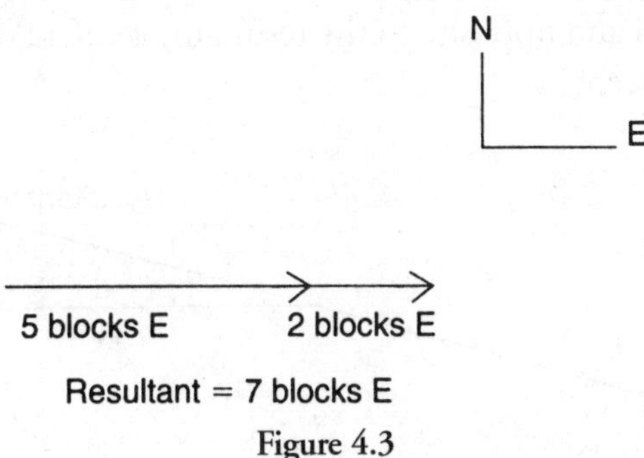

Figure 4.3

Now, for our second example, suppose the girl walks five blocks east and then two blocks northeast (that is, 45 degrees north of east). Since our intuition tells us that the final displacement will be in the general direction of north and east, we draw our vector diagram, using the same scale as before, so that the two vectors are connected head to tail. The vector diagram will look like Figure 4.4. The resultant is drawn from the tail of the first vector to the head of the second, forming a triangle (some texts use the "parallelogram" method of construction, which is equivalent). The direction of the resultant is measured from the tail of the vector inside the triangle.

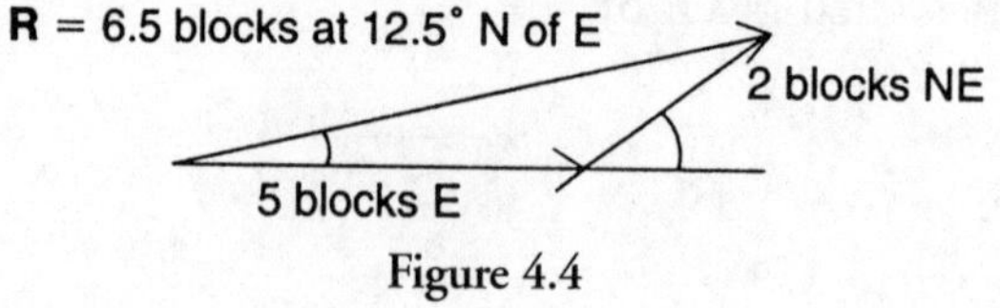

Figure 4.4

For our third example, suppose the girl walks five blocks east and then two blocks north. Figure 4.5 shows the vector diagram for this set of displacements.

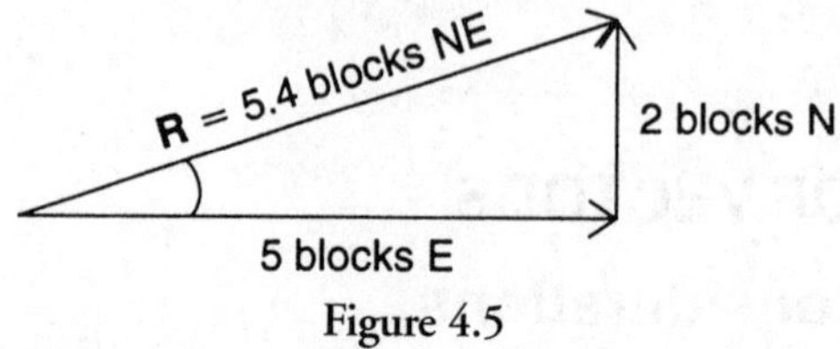

Figure 4.5

Finally, for our fourth example, suppose the girl walks five blocks east and then two blocks west. Her final displacement will be three blocks east of the starting point.

From these examples, we conclude that, as the angle between two vectors increases, the magnitude of their resultant decreases. Also, the magnitude of the resultant between two vectors is a maximum when the vectors are in the same direction (at a relative angle of 0 degree), and a minimum when they are in the opposite direction (at a relative angle of 180 degrees). It is important to remember that the vectors must be constructed head to tail and a suitable scale chosen for the system.

Note that the two vectors do not form a closed figure. The "missing link" is the resultant. This suggests an important concept. If a third vector was given that was equal in magnitude but opposite in direction to the predicted resultant, the vector sum of that resultant and the third vector would be zero. In other words, the vector sum of those three vectors would form a closed triangle of zero resultant, meaning that the girl returned to her starting point leaving a zero displacement. In physics, that third vector, equal and opposite to the resultant, is called the **equilibrant**. Figure 4.6 illustrates this concept.

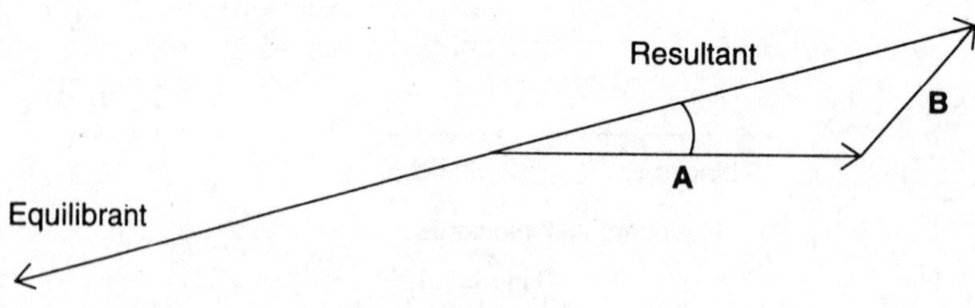

Figure 4.6

B. Algebraic Considerations

In the above examples, the resultant between two vectors was constructed using a vector diagram. The magnitude of the resultant was the measured length of the vector drawn from the tail of the first vector to the head of the second. If we sketch such a situation, we form a triangle whose sides are related by the **law of cosines** and whose angles are related by the **law of sines**.

Consider the vector triangle in Figure 4.7 with arbitrary sides *a*, *b*, and *c* and corresponding angles *A*, *B*, and *C*.

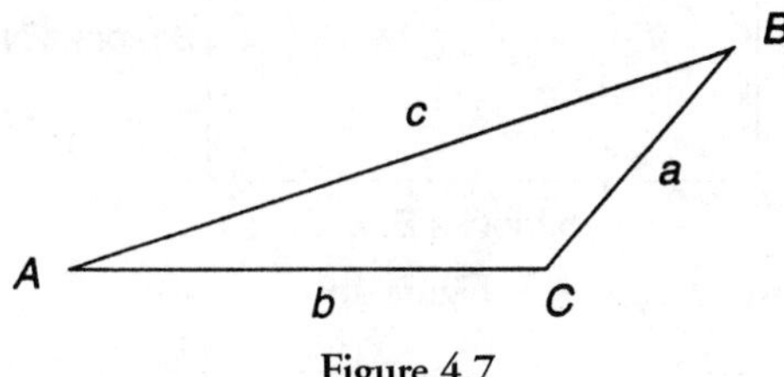

Figure 4.7

The law of cosines states that

$$c^2 = a^2 + b^2 - 2ab \cos C$$

The law of sines states that

$$\frac{a}{\sin A} = \frac{b}{\sin B} = \frac{c}{\sin C}$$

If we use the information given in our second example above, we see that in Figure 4.8 the vectors have the following magnitudes and directions:

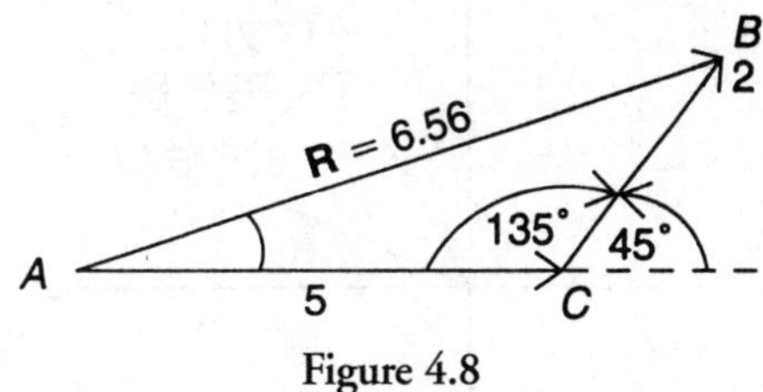

Figure 4.8

Using the law of cosines, we obtain the magnitude of the resultant in "blocks":

$$c = \sqrt{(2)^2 + (5)^2 - 2(2)(5) \cos 135} = 6.56$$

Using the law of sines, we find that the direction of the resultant is angle *A*:

$$\frac{6.56}{\sin 135} = \frac{2}{\sin A}$$

Angle *A* turns out to be equal to 12.45 degrees north of east.

C. Addition of Multiple Vectors

The discussion of the equilibrant leads to the idea of the addition of multiple vectors. As long as the vectors are constructed head to tail, multiple vectors can be added in any order. If the resultant is zero, the diagram constructed will be a closed geometric figure. This was shown in the case of three vectors forming a closed triangle, as in the example above. Thus, if the girl walks five blocks east, two blocks north, and five blocks west, the vector diagram will look like Figure 4.9, with the resultant displacement being equal to two blocks north.

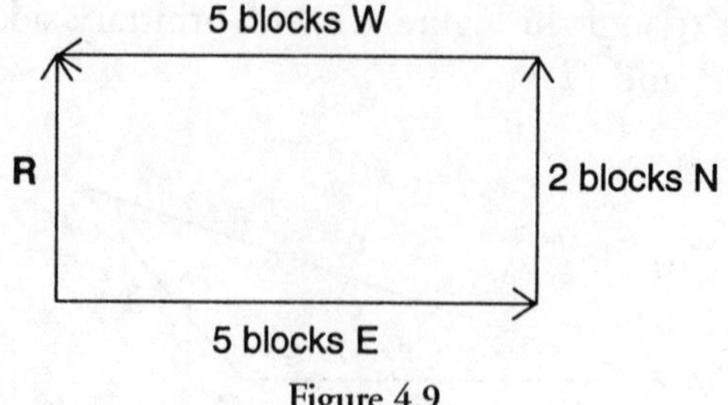

Figure 4.9

4.4 SUBTRACTION OF VECTORS

The difference between two vectors can be understood by considering the process of displacement. Suppose that, in Figure 4.10, we identify two sets of points (x, y) and (u, v) in a coordinate system as shown. Each point can be located within the coordinate system by a position vector drawn from the origin to that point. The displacement of an object from point A to point B would be represented by the vector drawn from A to B. This vector is defined to be the difference between the two position vectors and is sometimes called $\Delta\mathbf{R}$, so that $\Delta\mathbf{R} = \mathbf{R}_2 - \mathbf{R}_1$.

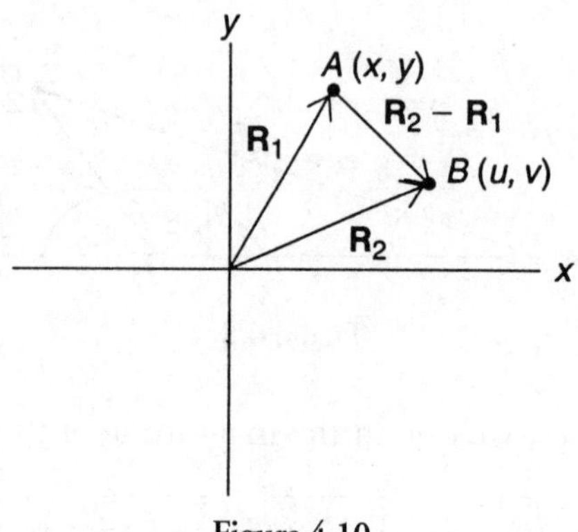

Figure 4.10

From our understanding of vector diagrams, we can see that $\mathbf{R}_1$ and $\Delta\mathbf{R}$ are connected head to tail, and thus that $\mathbf{R}_1 + \Delta\mathbf{R} = \mathbf{R}_2$!

4.5 ADDITION METHODS USING THE COMPONENTS OF VECTORS

Any single vector in space can be **resolved** into two perpendicular components in a suitably chosen coordinate system. The methods of vector resolution were discussed in Section 4.2, but we can review them here. Given a vector (sketched in Figure 4.11) representing a displacement of 100 meters northeast (that is, making a 45-degree angle with the positive x-axis), we can observe that it is composed of an x- and a y- (a horizontal and a vertical) component that can be geometrically or algebraically determined.

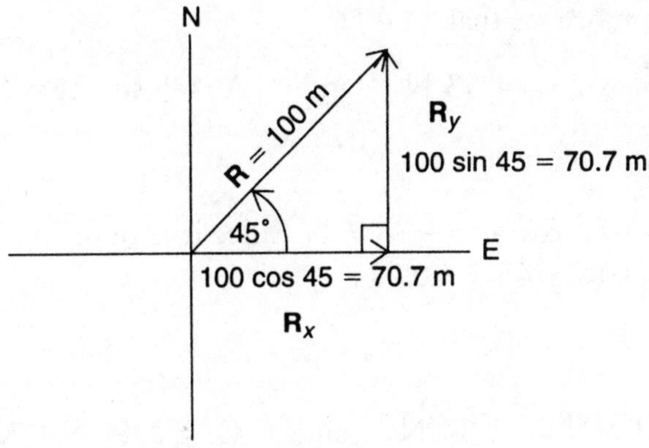

Figure 4.11

Geometrically, if we were to draw the 100-meter vector to scale at the correct angle, then projecting a perpendicular line down from the head of the given vector to the *x*-axis would construct the two perpendicular components. Algebraically, we see that if we have a given vector **R**, then, in the *x*-direction, we have

$$\mathbf{R}_x = \mathbf{R}\cos\theta \quad \text{and} \quad \mathbf{R}_y = \mathbf{R}\sin\theta$$

where the magnitude of **R** is *R*.

This method of vector resolution can be useful when adding two or more vectors. Since all vectors in the same direction add up numerically, and all vectors in opposite directions subtract numerically, therefore, if we are given two vectors, the resultant between them will be found from the addition and subtraction of the respective components. In other words, if vector **A** has components $\mathbf{A}_x$ and $\mathbf{A}_y$, and if vector **B** has components $\mathbf{B}_x$ and $\mathbf{B}_y$, the resultant vector **C** will have components $\mathbf{C}_x$ and $\mathbf{C}_y$ such that

$$\mathbf{C}_x = \mathbf{A}_x + \mathbf{B}_x \quad \text{and} \quad \mathbf{C}_y = \mathbf{A}_y + \mathbf{B}_y$$

An example of this method is shown in Figure 4.12. Suppose we have two vectors, **A** and **B**, that represent tensions in two ropes applied at the origin of the coordinate system (let's say it's a box). The resultant force between these tensions (which are vector quantities) can be found by determining the respective *x*- and *y*-components of the given vectors.

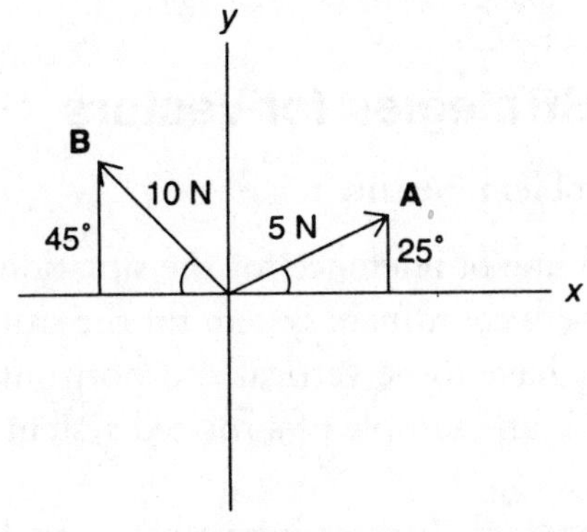

Figure 4.12

From the diagram, we see that

$$\mathbf{A}_x = 5 \cos 25 = 4.53 \text{ N} \quad \text{and} \quad \mathbf{A}_y = 5 \sin 25 = 2.11 \text{ N}$$

For the second vector, we have

$$\mathbf{B}_x = -10 \cos 45 = -7.07 \text{ N (since it is pointing left)}$$
$$\mathbf{B}_y = 10 \sin 45 = 7.07 \text{ N}$$

Therefore,

$$\mathbf{C}_x = 4.53 \text{ N} - 7.07 \text{ N} = -2.54 \text{ N} \quad \text{and} \quad \mathbf{C}_y = 2.11 \text{ N} + 7.07 \text{ N} = 9.18 \text{ N}$$

The magnitude of the resultant vector, **C**, is given by

$$|\mathbf{C}| = \sqrt{(-2.54)^2 + (9.18)^2} = 9.52 \text{ N}$$

The angle that vector **C** makes with the positive *x*-axis is about 105°.

Chapter Summary

- Vectors are quantities that have both magnitude (size) and direction.
- Scalars are quantities that have only magnitude.
- Force, displacement, and velocity are examples of vectors.
- Distance, speed, and mass are examples of scalars.
- Vectors can be "added" geometrically by constructing a parallelogram.
- The resultant of two vectors is equal to the vector obtained by "adding" them.
- If two vectors are in the same direction, their resultant is equal to their sum.
- If two vectors are in the opposite direction, their resultant is equal to their difference.
- Two or more forces are in equilibrium if their resultant is equal to zero.
- An equilibrant force is a force added to one or more forces that produces equilibrium.

Problem-Solving Strategies for Vectors

When solving a vector problem, be sure to:

1. Understand the frame of reference for the situation.
2. Select an appropriate coordinate system for the situation. Cartesian systems do not necessarily have to be vertical and horizontal (a mass sliding down an inclined plane is an example of a rotated system with the "*x*-axis" parallel to the incline).
3. Pick an appropriate scale for starting your vector diagram. For example, in a problem with displacement, a scale of 1 cm = 1 m might be appropriate.

4. Recognize the given orientation of the vectors as they correspond to geographical directions (N, E, S, W). Be sure to maintain the same directions as you draw your vector triangle or parallelogram. Use a protractor and a metric ruler.
5. Connect vectors head to tail, and connect the resultant from the tail of the first vector to the head of the last vector.
6. Identify and determine the components of the vectors. In most cases, vector problems can be simplified using components.

Practice Exercises

MULTIPLE-CHOICE

1. A vector is given by its components, $\mathbf{A}_x = 2.5$ and $\mathbf{A}_y = 7.5$. What angle does vector **A** make with the positive x-axis? *CHALLENGE

   (A) 72°
   (B) 18°
   (C) 25°
   (D) 50°
   (E) 75°

2. Which pair of vectors could produce a resultant of 35?

   (A) 15 and 15
   (B) 20 and 20
   (C) 30 and 70
   (D) 20 and 60
   (E) 20 and 70

3. A vector has a magnitude of 17 units and makes an angle of 20° with the positive x-axis. The magnitude of the horizontal component of this vector is CHALLENGE

   (A) 16 units
   (B) 4.1 units
   (C) 5.8 units
   (D) 50 units
   (E) 12 units

4. As the angle between a given vector and the horizontal axis increases from 0° to 90°, the magnitude of the vertical component of this vector

   (A) decreases
   (B) increases and then decreases
   (C) decreases and then increases
   (D) increases
   (E) remains the same

****Some questions at the end of chapters are labeled "challenge" since they may be more difficult than what you will encounter on the actual AP Physics B exam. You may need a calculator to answer these.***

CHALLENGE

5. Vector **A** has a magnitude of 10 units and makes an angle of 30° with the horizontal *x*-axis. Vector **B** has a magnitude of 25 units and makes an angle of 50° with the negative *x*-axis. What is the magnitude of the resultant between these two vectors?

(A) 20
(B) 35
(C) 15
(D) 45
(E) 25

CHALLENGE

6. Two concurrent vectors have magnitudes of 3 units and 8 units. The difference between these vectors is 8 units. The angle between these two vectors is

(A) 34°
(B) 56°
(C) 79°
(D) 113°
(E) 127°

7. Which of the following sets of displacements have equal resultants when performed in the order given?

I: 6 m east, 9 m north, 12 m west
II: 6 m north, 9 m west, 12 m east
III: 6 m east, 12 m west, 9 m north
IV: 9 m north, 6 m east, 12 m west

(A) I and IV
(B) I and II
(C) I, III, and IV
(D) I, II, IV
(E) II and IV

8. Which vector represents the direction of the two concurrent vectors shown below?

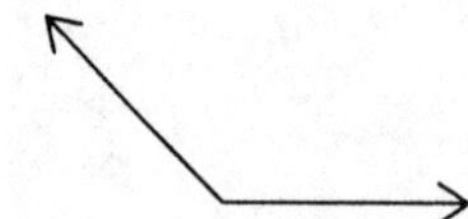

(A) 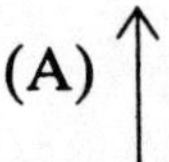 (B) 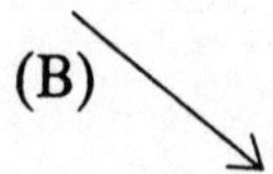 (C)  (D)  (E) 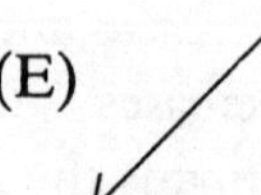

9. Three forces act concurrently on a point *P* as shown below. Which vector represents the direction of the resultant force on point *P*?

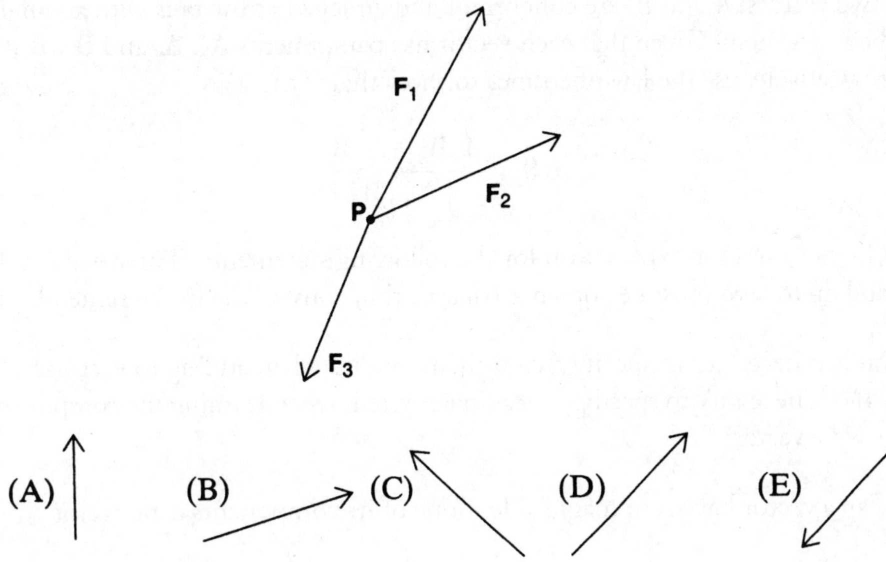

10. On a baseball field, first base is about 30 m away from home plate. A batter gets a "hit" and runs toward first base. She runs 3 m past the base and then runs back to stand on it. The magnitude of her final displacement from home plate is

    (A) 27 m
    (B) 30 m
    (C) 33 m
    (D) 36 m
    (E) 40 m

FREE-RESPONSE

1. Find the magnitude and direction of the two concurrent forces shown below using the algebraic method of components.

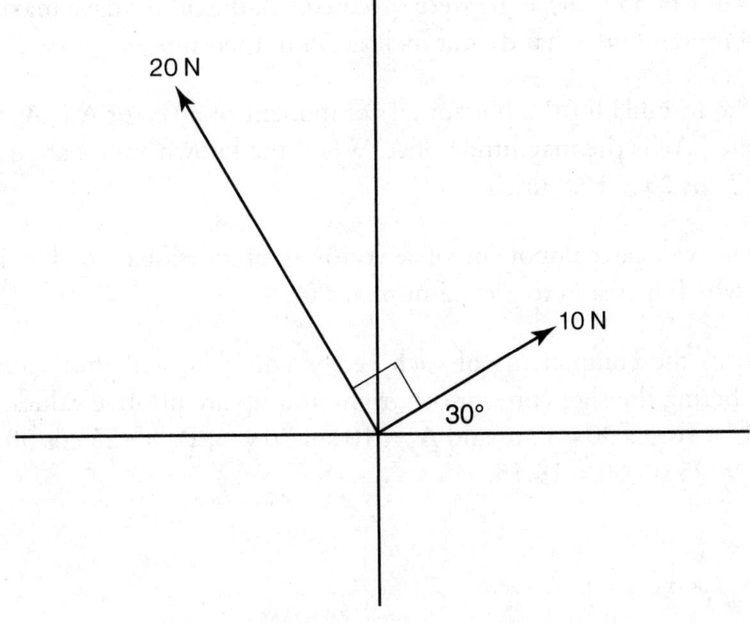

Note: Figure is not drawn to scale.

2. Two vectors, **A** and **B**, are concurrent and attached at the tails with an angle θ between them. Given that each vector has components $\mathbf{A}_x$, $\mathbf{A}_y$ and $\mathbf{B}_x$, $\mathbf{B}_y$, respectively, use the law of cosines to show that

$$\cos\theta = \frac{\mathbf{A}_x\mathbf{B}_x + \mathbf{A}_y\mathbf{B}_y}{|\mathbf{A}||\mathbf{B}|}$$

3. Give a geometric explanation for the following statement: "Three vectors that add up to zero must be coplanar (that is, they must all lie in the same plane)."

4. (a) Is it necessary to specify a coordinate system when adding to vectors?
(b) Is it necessary to specify a coordinate system when forming the components of a vector?

5. Can a vector have zero magnitude if one of its components is nonzero?

ANSWERS EXPLAINED

MULTIPLE-CHOICE PROBLEMS

1. **(A)** The angle for the vector to the positive *x*-axis is given by

$$\tan\theta = \frac{\mathbf{A}_y}{\mathbf{A}_x} = \frac{7.5}{2.5}$$

Thus $\tan\theta = 3$ and $\theta = 71.5°$ or, rounded off, 72°.

2. **(B)** Two vectors have a maximum resultant, whose magnitude is equal to the numerical sum of the vector magnitudes, when the angle between vectors is 0°. The resultant is a minimum at 180°, and the magnitude is equal to the numerical difference between the magnitudes. When this logic is used, only pair (20, 20) produce a maximum and minimum set that could include the given value of 35 if the angle were specified. All the others have maximum and minimum resultants that do not include 35 in their range.

3. **(A)** The formula for the horizontal component of a vector **A** is $\mathbf{A}_x = |\mathbf{A}|\cos\theta$, where $|\mathbf{A}|$ is the magnitude of **A**. When the known values are used, values $\mathbf{A}_x = 17\cos 20 = 15.9$ units.

4. **(D)** The vertical component of a vector is proportional to the sine of the angle, which increases to a maximum at 90°.

5. **(E)** Form the components of each vector and then add these components, remembering the sign conventions; right and up are positive values. Thus, we have $\mathbf{A}_x = 10\cos 30 = 8.66$ and $\mathbf{A}_y = 10\sin 30 = 5$, $\mathbf{B}_x = -25\cos 50 = -16.06$ and $\mathbf{B}_y = 25\sin 50 = 19.15$.

The components of the resultant are $\mathbf{C}_x = \mathbf{A}_x + \mathbf{B}_x = 8.66 - 16.06$, so $\mathbf{C}_x = -7.4$ units and $\mathbf{C}_y = \mathbf{A}_y + \mathbf{B}_y = 5 + 19.15 = 24.15$. The magnitude of this resultant vector is given by

$$|\mathbf{C}| = \sqrt{\mathbf{C}_x^2 + \mathbf{C}_y^2} = \sqrt{(7.4)^2 + (24.15)^2}$$

implying that $\mathbf{C} = 25$ units.

6. **(C)** Use the law of cosines where $a = 3$, $b = 8$, and $c = 8$, and solve for $\cos\theta$:

$$(8)^2 = (3)^2 + (8)^2 - 2(3)(8)\cos\theta$$

Solving gives $\cos\theta = 0.1875$ and $\theta = 79°$.

7. **(C)** Vectors can be added in any order. The only requirement is that the vectors have the same magnitude and direction as they are shuffled. A look at the four sets of displacements indicates that I, III, and IV consist of the same vectors listed in different orders. These three sets will produce the same resultant.

8. **(C)** Sketch the vectors head to tail as if forming a vector triangle in construction. The resultant is drawn from the tail of the first vector to the head of the second. Choose the horizontal vector as the first one; then choice C is the general direction of the resultant.

9. **(B)** Sketch the vectors in any order, as shown below. Draw the resultant from the tail of the first vector to the head of the last.

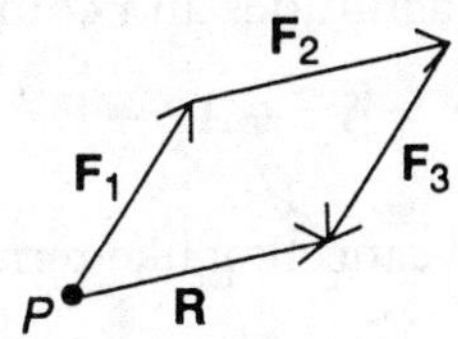

10. **(B)** The displacement is in the direction of first base and is equal in magnitude to the straight-line distance between home plate and first base (30 m).

FREE-RESPONSE PROBLEMS

1. Let $\mathbf{A} = 10$ N force and $\mathbf{B} = 20$ N force

We declare left to be negative and up to be positive.

From the diagram, the components are:

$\mathbf{A_x} = 10 \text{ N} \cos 30° = 8.66 \text{ N}$ $\mathbf{B_x} = -20 \text{ N} \cos 60° = -10 \text{ N}$

$\mathbf{A_y} = 10 \text{ N} \sin 30° = 5 \text{ N}$ $\mathbf{B_y} = 20 \text{ N} \sin 60° = 17.32 \text{ N}$

The components of the resultant are now given by the following equations. Watch your signs!

$\mathbf{R_x} = \mathbf{A_x} + \mathbf{B_x} = 8.66 \text{ N} + (-10 \text{ N}) = -1.34 \text{ N}$

$\mathbf{R_y} = \mathbf{A_y} + \mathbf{B_y} = 5 \text{ N} + 17.32 \text{ N} = 22.32 \text{ N}$

Use the Pythagorean theorem to find the magnitude of the resultant:

$$|\mathbf{R}|^2 = (-1.34\ \text{N})^2 + (22.32\ \text{N})^2$$

$$|\mathbf{R}| = 22.36\ \text{N}$$

To find the direction, notice that the resultant will lie in the second quadrant since $\mathbf{R_x}$ is negative and $\mathbf{R_y}$ is positive:

$$\tan\theta = \frac{22.32\ \text{N}}{-1.34\ \text{N}} = -16.66$$

$$\theta = 86.6° \text{ (relative to the } x\text{-axis)}$$

2. The situation is shown below. The "third" side of the vector triangle is **A** – **B**.

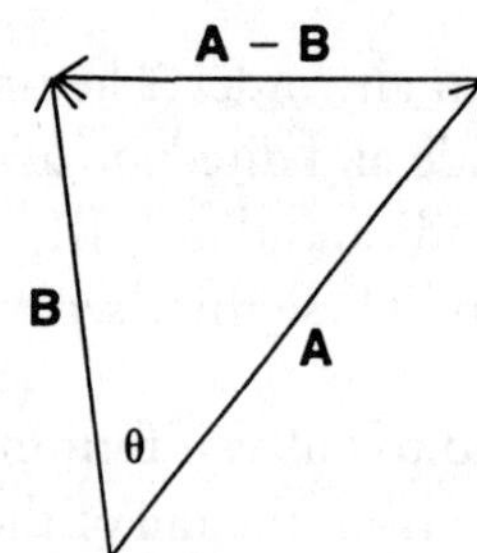

From the rules for vector subtraction, the components of **A** – **B** are $(\mathbf{A}_x - \mathbf{B}_y)$ and $(\mathbf{A}_y - \mathbf{B}_y)$. The law of cosines states that, for the given triangle,

$$|\mathbf{A} - \mathbf{B}|^2 = |\mathbf{A}|^2 + |\mathbf{B}|^2 - 2|\mathbf{A}||\mathbf{B}|\cos\theta$$

Using our definitions of magnitudes and components, we have

$$(\mathbf{A}_x - \mathbf{B}_x)^2 = (\mathbf{A}_y - \mathbf{B}_y)^2 = \mathbf{A}_x^2 + \mathbf{A}_y^2 + \mathbf{B}_x^2 + \mathbf{B}_y^2 - 2|\mathbf{A}||\mathbf{B}|\cos\theta$$

Expanding the left side and canceling like terms on both sides, we are left with

$$-2\mathbf{A}_x\mathbf{B}_x - 2\mathbf{A}_y\mathbf{B}_y = -2|\mathbf{A}||\mathbf{B}|\cos\theta$$

Again, canceling like terms on both sides leaves us with the final expression as we solve for cos θ:

$$\cos\theta = \frac{\mathbf{A}_x\mathbf{B}_x + \mathbf{A}_y\mathbf{B}_y}{|\mathbf{A}||\mathbf{B}|}$$

3. Any two vectors can be added geometrically using the parallelogram method of construction. A parallelogram is a plane figure. The resultant of two vectors is represented by the diagonal of this parallelogram. In order for three vectors to add up to zero, the third remaining vector must be equal in magnitude, but opposite in direction, to the resultant of the remaining two vectors. Hence, all three must lie in the same plane to achieve this result.

4. (a) Using the logic of problem 4, any two vectors can be added by simply constructing a parallelogram from them. They can be in any orientation and in any plane. Hence, no coordinate system is necessary.

   (b) The components of a vector, by definition, lie along the axes of a chosen coordinate system (such as the x- or y-axis). Hence, to form these components, a coordinate system must be specified.

5. The magnitude of a vector is a number obtained by taking the square root of the sum of the squares of the magnitudes of its components. If one of the components is nonzero, the magnitude of the whole vector must, by definition, be nonzero as well.

CHAPTER 5

One-Dimensional Motion

KEY CONCEPTS

- Average and Instantaneous Motion
- Uniformly Accelerated Motion
- Accelerated Motion Due to Gravity
- Graphical Analysis of Motion

5.1 INTRODUCTION

Motion involves the change in position of an object over time. When we observe an object moving, it is always with respect to a frame of reference. Since motion occurs in a particular direction, it is dependent on the choice of a coordinate system for that frame of reference. This fact leads us to the conclusion that there is a vector nature to motion that must be taken into account when we want to analyze how something moves.

In this chapter, we shall confine our discussions to motion in one dimension only. In physics, the study of motion is called **kinematics**. No hypotheses are made in kinematics concerning why something moves. Kinematics is a completely descriptive study of *how* something moves.

5.2 AVERAGE AND INSTANTANEOUS MOTION

Motion can take place either uniformly or nonuniformly. This means that, if an object is moving, relative to a frame of reference, the displacement changes, over a period of time, may be equal or unequal.

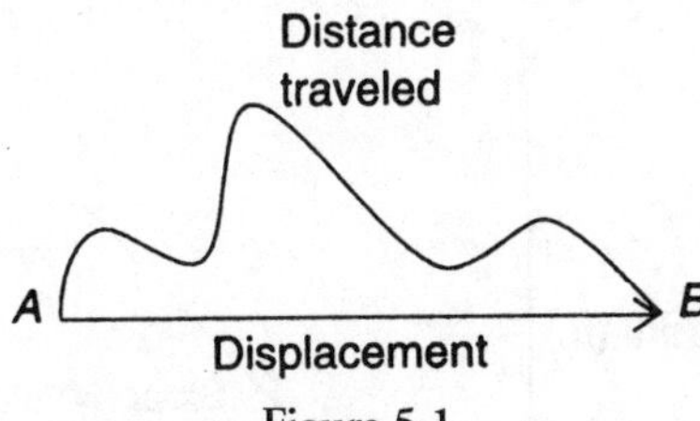

Figure 5.1

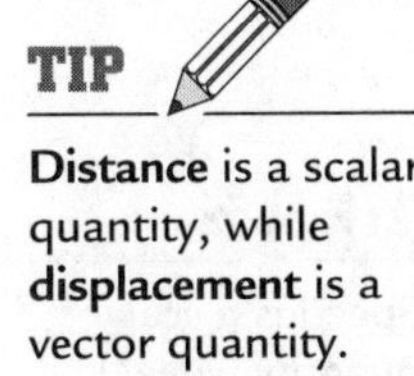

Distance is a scalar quantity, while **displacement** is a vector quantity.

If, in Figure 5.1, we consider the actual distance traveled by the object along some arbitrary path, we are dealing with a scalar quantity. The displacement vector **AB**, however, is directed along the line connecting points A and B (whether or not this is the actual route taken). Thus, when a baseball player hits a home run and runs around the bases, he or she may have traveled a distance of 360 feet (the bases are 90 feet apart), but the player's final displacement is zero (having started and ended up at the same place)!

If we are given the displacement vector of an object for a period of time Δt, we define the **average velocity**, $\bar{\mathbf{v}}$, to be equal to

$$\bar{\mathbf{v}} = \frac{\Delta \mathbf{x}}{\Delta t}$$

This is a vector quantity since directions are specified. Numerically, we think of the average speed as being the ratio of the total distance traveled to the total elapsed time. The units of velocity are meters per second (m/s).

If we are interested in the velocity at any instant in time, we can define the **instantaneous velocity** to be the velocity, **v**, as determined at any precise time period. In a car, the speedometer registers instantaneous speed, which can become velocity if we take into account the direction of motion. If the velocity is constant, the average and instantaneous velocities are equal, and we can write simply

$$\mathbf{x} = \mathbf{v}\ \Delta t$$

It is possible that the observer making the measurements of motion is likewise in motion relative to Earth. In this case, we consider the relative velocity between the two systems. If two objects, *a* and *b*, are moving in the same direction, the relative velocity between the objects is given by

$$\mathbf{v}_{rel} = \mathbf{v}_a - \mathbf{v}_b$$

In a similar way, if two objects, *a* and *b*, are approaching each other from opposite directions, the relative velocity between them is given by

$$\mathbf{v}_{rel} = \mathbf{v}_a + \mathbf{v}_b$$

5.3 UNIFORMLY ACCELERATED MOTION

If an object is moving with a constant velocity such that its position is taken to be zero when it is first observed, a graph of the expression $\mathbf{x} = \mathbf{v}t$ would represent a direct relationship between position and time (see Figure 5.2).

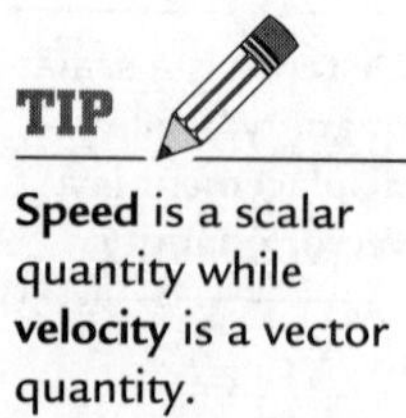

TIP

Speed is a scalar quantity while **velocity** is a vector quantity.

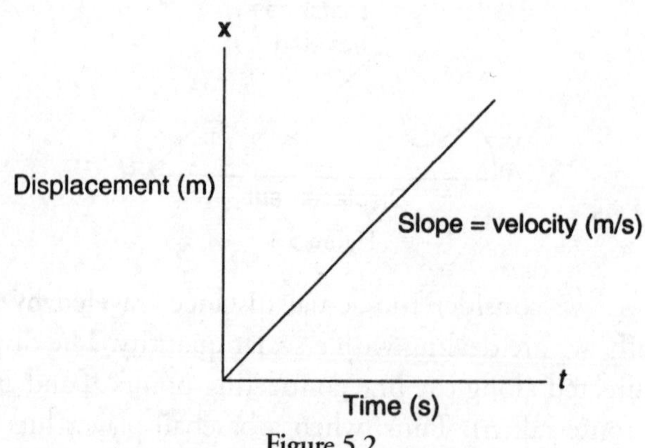

Figure 5.2

Since the line is straight, the constant slope (in which $\mathbf{v} = \Delta\mathbf{x}/\Delta t$) indicates that the velocity is constant throughout the time interval. If we were to plot velocity versus time for this motion, the graph might look like Figure 5.3.

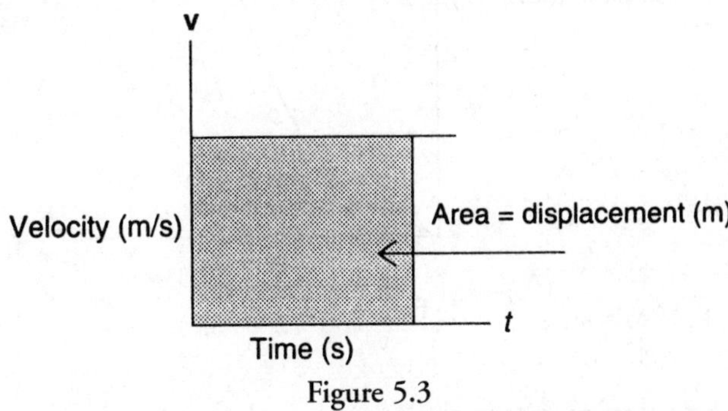

Figure 5.3

Notice that, for any time t, the area under the graph equals the displacement.

It is possible that the object is changing direction while maintaining a constant speed (uniform circular motion is an example of this situation), or that it is changing both speed and direction. In any case, if the velocity is changing, we say that the object is **accelerating.** If the velocity is changing uniformly, the object has uniform acceleration. In this case, a graph of velocity versus time would look like Figure 5.4.

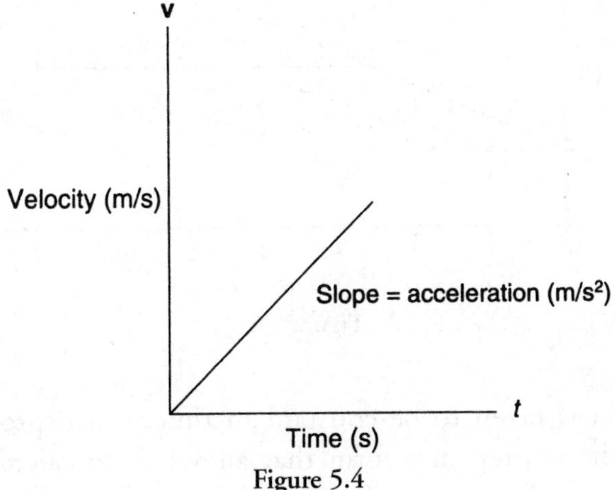

Figure 5.4

TIP

Acceleration is a vector quantity and equal to the rate of change of velocity.

The displacement from $t = 0$ to any other time is equal to the area of the triangle formed. However, between any two intermediary times, the resulting figure is a trapezoid. If we make several measurements, the displacement versus time graph for uniformly accelerated motion is a parabola starting from the origin (if we make the initial conditions that, when $t = 0$, $x = 0$, and $\mathbf{v} = 0$; see Figure 5.5).

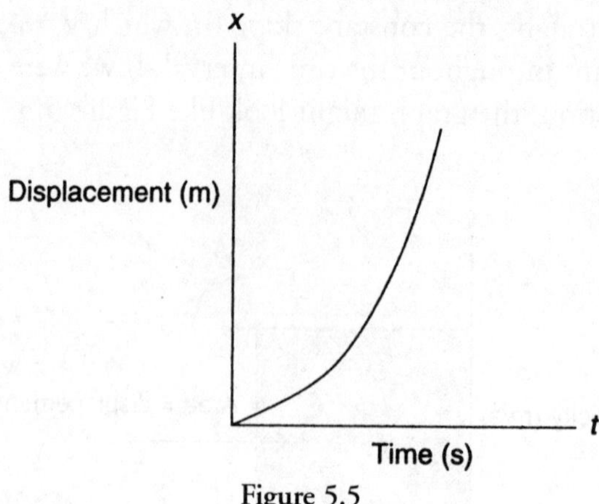

Figure 5.5

The slope of the velocity versus time graph is defined to be the average acceleration in units of meters per second squared (m/s^2). We can now write, for the average acceleration (Figure 5.6)

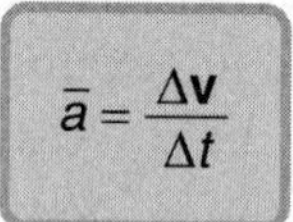

$$\bar{a} = \frac{\Delta \mathbf{v}}{\Delta t}$$

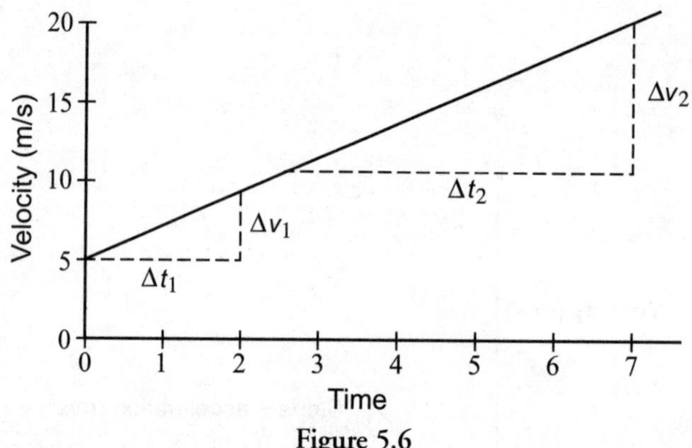

Figure 5.6

Remember

An object with zero acceleration is not necessarily at rest in a given frame of reference. The object may be moving with constant velocity.

If the acceleration is taken to be constant in time, our expression for average acceleration can be written in a form that allows us to calculate the instantaneous final velocity after a period of acceleration has taken place. In other words, $\Delta \mathbf{v} = \mathbf{a}t$ (if we start our time interval from zero). If we define $\Delta \mathbf{v}$ to be equal to the difference between an initial and a final velocity, we can arrive at the fact that

$$\mathbf{v}_f = \mathbf{v}_i + \mathbf{a}t$$

Therefore, if we plot velocity versus time for uniformly accelerated motion starting with a nonzero initial velocity, we get a graph that looks like Figure 5.7.

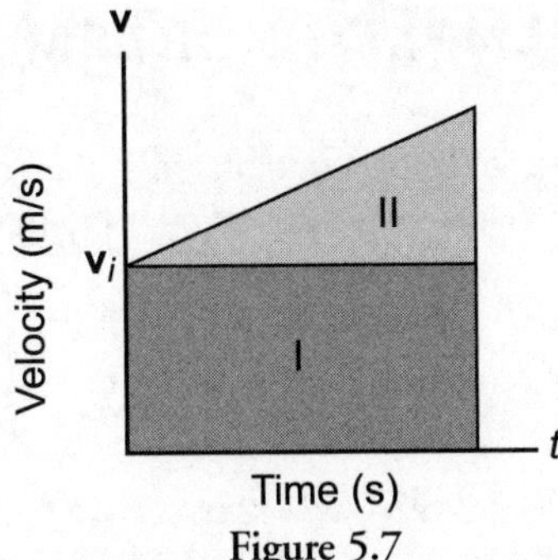

Figure 5.7

The displacement during any period of time will be equal to the total area under the graph. In this case, the total area will be the sum of two areas, one a triangle and the other a rectangle. The area of the rectangle, for some time *t*, is just $\mathbf{v}_i t$. The area of the triangle is one-half the base times the height. The "base" in this case is the time period, *t*, the "height" is the change in velocity, $\Delta\mathbf{v}$. Therefore, the area of the triangle is $(1/2)\Delta\mathbf{v}t$. If we recall the definition of $\Delta\mathbf{v}$ and the fact that $\mathbf{v}_f = \mathbf{v}_i + at$, we obtain the following formula for the distance traveled during uniformly accelerated motion starting with an initial velocity (assuming we start from the origin):

$$x = \mathbf{v}_i t + \frac{1}{2}\mathbf{a}t^2$$

This analysis suggests an alternative method of determining the average velocity of an object during uniformly accelerated motion. If we start, for example, with a velocity of 10 m/s and accelerate uniformly for 5 s at a rate of 2 m/s^2, the average velocity during that interval is the average of 10 m/s, 12 m/s, 14 m/s, 16 m/s, 18 m/s, and 20 m/s, which is just 15 m/s. Therefore, we can simply write

$$\overline{\mathbf{v}} = \frac{\mathbf{v}_i + \mathbf{v}_f}{2}$$

Occasionally, a problem in kinematics does not explicitly mention the time involved. For this reason it would be nice to have a formula for velocity that does not involve the time factor. We can derive one from all the other formulas. Since the above equation relates the average velocity to the initial and final velocities (for uniformly accelerated motion), we can write our displacement formula as

$$\mathbf{x} = \frac{\mathbf{v}_i + \mathbf{v}_f}{2}t$$

Now, since $\mathbf{v}_f = \mathbf{v}_i + \mathbf{a}t$, we can express the time as $t = (\mathbf{v}_f - \mathbf{v}_i)/\mathbf{a}$. Therefore

$$x = \left(\frac{\mathbf{v}_i + \mathbf{v}_f}{2}\right)\left(\frac{\mathbf{v}_f - \mathbf{v}_i}{\mathbf{a}}\right)$$

$$x = \left(\frac{\mathbf{v}_f^2 - \mathbf{v}_i^2}{2\mathbf{a}}\right)$$

SAMPLE PROBLEM

A particle accelerates from rest at a uniform rate of 3 m/s^2 for a distance of 200 m. How fast is the particle going at that time? How long did it take for the particle to reach that velocity?

Solution

We use the formula

$$\mathbf{v}_f^2 - \mathbf{v}_i^2 = 2\mathbf{ax}$$

Since the particle begins from rest, the initial velocity is equal to zero and

$$\mathbf{v}_f^2 = 2(3\text{m/s}^2)(200\text{ m})$$
$$\mathbf{v}_f = 34.64\text{ m/s}$$

To find the time, we use $\mathbf{v} = \mathbf{a}t$

$$t = \frac{34.64\text{ m/s}}{3\text{ m/s}^2} = 11.55\text{ s}$$

5.4 ACCELERATED MOTION DUE TO GRAVITY

Since velocity and acceleration are vector quantities, we need to consider the algebraic conventions accepted for dealing with various directions. For example, we usually agree to consider motion up or to the right as positive, and motion down or to the left as negative. In this way, negative velocity implies backward motion (since negative speed doesn't make sense). Negative acceleration, on the other hand, implies that the object is slowing down.

In Section 5.3, we considered the case of uniformly accelerated motion. A naturally occurring situation in which acceleration is constant involves motion due to gravity. Since gravity acts in the downward direction, its acceleration, common to all masses, is likewise taken to be in the downward direction. Notice that this convention may or may not imply deceleration. An object dropped from rest is accelerating as time passes, but in the downward or negative direction.

In physics, the acceleration due to gravity is generally taken to be –9.8 meter per second squared and is represented by the symbol **g**. As a vector quantity, **g** is negative and the direction of motion is reflected in either the position or the velocity equation. For example, if an object was dropped, the displacement in the –**y** direction would be given by

$$\mathbf{y} = -\frac{1}{2}\mathbf{g}t^2$$

If, however, the object was thrown down, with an initial velocity, we could write

$$\mathbf{y} = -\mathbf{v}_i t - \frac{1}{2}\mathbf{g}t^2$$

Finally, if the object was thrown upward, with some initial velocity, we would have

$$\mathbf{y} = \mathbf{v}_i t - \frac{1}{2}\mathbf{g}t^2$$

SAMPLE PROBLEM

A projectile is fired vertically upward at an initial velocity of +98 m/s. How high will it rise? How long will it take the projectile to reach that height?

Solution

We use $\mathbf{g}$ = 9.8 m/s^2 (recall its direction is negative) and note that $\mathbf{v}_{fy} = 0$ at the highest point:

$$\mathbf{v}_{fy}^2 - \mathbf{v}_{iy}^2 = -2\mathbf{g}\mathbf{y}$$
$$-(98 \text{ m/s})^2 = -2(9.8 \text{ m/s}^2)\mathbf{y}_{max}$$
$$\mathbf{y}_{max} = 490 \text{ m}$$

To find the time, we use

$$\mathbf{v} = \mathbf{a}t$$
$$t = \frac{-98 \text{ m/s}}{-9.8 \text{ m/s}^2} = 10 \text{ s}$$

5.5 GRAPHICAL ANALYSIS OF MOTION

We have seen that much information can be obtained if we consider the graphical analysis of motion. If complex changes in motion are taking place, visualization may provide a better understanding of the physics involved than algebra. The techniques of graphical analysis are as simple as slopes and areas. For example, we already know that, for uniformly accelerated motion, the graph of distance versus time is a parabola. Since the slope is changing, the instantaneous velocity can be approximated at a point **P** by finding the slope of a tangent line drawn to a given point on the curve (see Figure 5.8).

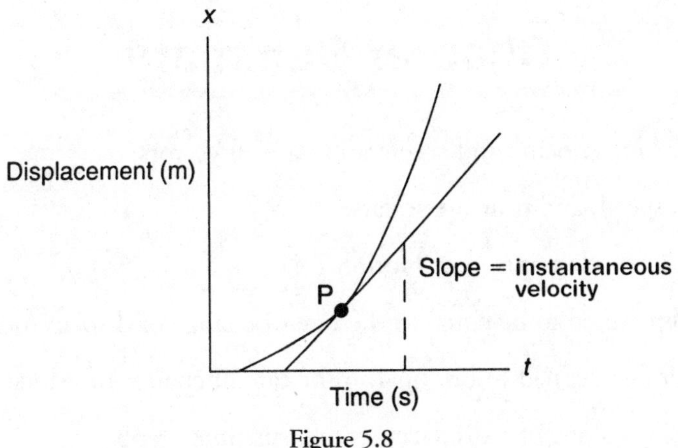

Figure 5.8

What would happen if an object accelerated from rest maintained a constant velocity for a while, and then slowed down to a stop? Using what we know about graphs of velocity and acceleration in displacement versus time, we might represent the motion as shown in Figure 5.9.

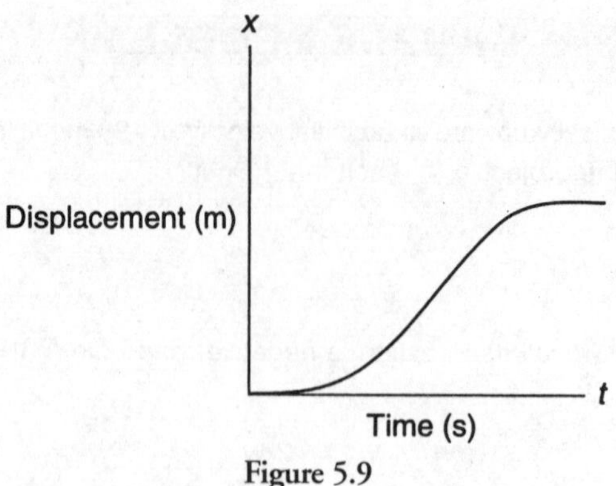

Figure 5.9

We can apply many instances of motion to graphs. In the case of changing velocity, consider the graph of velocity versus time for an object thrown upward into the air, reaching its highest point, changing direction, and then accelerating downward. This motion has a constant downward acceleration that, at first, acts to slow the object down, but later acts to speed it up. A graph of this motion is seen in Figure 5.10.

Reminder

This type of motion can also be seen with a mass oscillating while attached to a spring. See Chapter 11 for more details.

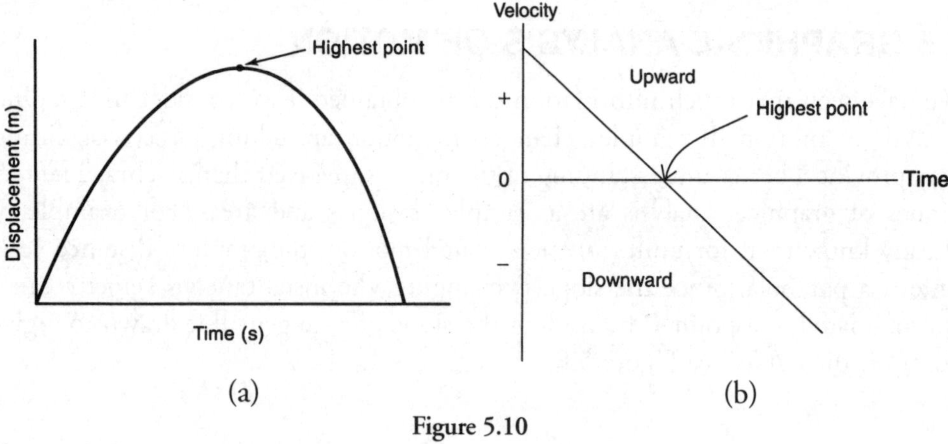

Figure 5.10

Chapter Summary

- Displacement, velocity, and acceleration are all vectors.
- Distance, speed, and time are scalars.
- Kinematics is a description of motion.
- Velocity is defined to be equal to the rate of change of displacement.
- Acceleration is defined to be equal to the rate of change of velocity.
- Velocity is the slope of a displacement versus time graph.
- Acceleration is the slope of a velocity versus time graph.
- The displacement can be obtained from a velocity versus time graph by taking the area under the graph.
- The acceleration due to gravity, near the surface of Earth, is equal to $\mathbf{g} = 9.8\ \text{m/s}^2$ and is directed downward.

Problem-Solving Strategies for One-Dimensional Motion

Since all motion is relative, it is very important to always ask yourself the following question: "From what frame of reference am I viewing the situation?" Also, whenever you solve a physics problem, be sure to consider the assumptions being made, explicitly or implicitly, about the moving object or objects. In this way, your ability to keep track of what is relevant for your solution path will be maintained. In addition, it is suggested that you:

1. Identify all the goals and givens in the problem. Recall from Chapter 2 that the goals may be explicit or implicit. If a question is based on a decision or prediction, be sure to understand the algebraic requirements necessary to reach an answer.
2. Consider the "meaningfulness" of your solution or process. Does your answer or methodology make sense?
3. Choose a proper coordinate system and remember the sign conventions for algebraically treating vector quantities. In addition, make sure you understand the nature of the concepts being discussed (for example, whether they are vectors or scalars).
4. Use proper SI units throughout your calculations, and make sure that correct units are included in your final answer. Try "dimensional analysis" to see whether the answer makes sense. Perhaps it looks too large or too small because it is expressed in the wrong units.
5. Make a sketch of the situation if one is not provided.
6. If you are interpreting a graph, be sure you understand the interrelationships among all the kinematic variables discussed (slopes, areas, etc.).
7. If you are constructing a graph, label both axes completely, choose a proper scale for each axis, and draw neatly and clearly.
8. Try different problem-solving heuristics if you get stuck on a difficult problem.

SAMPLE PROBLEM

A ball is thrown straight up into the air and, after being in the air for 9 s, is caught by a person 5 m above the ground. To what maximum height did the ball go?

Solution

The equation

$$\mathbf{y} = \mathbf{v}_i t - \frac{1}{2}\mathbf{g}t^2$$

represents the vertical position of the ball above the ground for any time t. Since the ball is thrown upward, the initial velocity is positive, while the acceleration of gravity is always directed downward (and hence is algebraically negative). Thus, we can use the above equation to find the initial velocity when $\mathbf{y}$ = 5 m and t = 9 s. Substituting these values, as well as the magnitude of $\mathbf{g}$ (9.8 m/s^2) in the equation, we get $\mathbf{v}_i$ = 44.65 m/s.

Now, to find the maximum height, we note that, when the ball rises to its maximum, its speed becomes zero since its velocity is changing direction. Since we do not know how long the ball takes to rise to its maximum height (we could determine this value if desired), we can use the formula

$$\mathbf{v}_f^2 - \mathbf{v}_i^2 = -2\mathbf{g}\mathbf{y}$$

When the final velocity equals zero, the value of **y** is equal to the maximum height. Using our answer for the initial velocity and the known value of **g**, we find that $\mathbf{y}_{max}$ = 101.7 m.

Practice Exercises

MULTIPLE-CHOICE

1. A ball is thrown upward with an initial velocity of 20 m/s. How long will the ball take to reach its maximum height?

   (A) 19.6 s
   (B) 9.8 s
   (C) 6.3 s
   (D) 3.4 s
   (E) 2.04 s

CHALLENGE

2. An airplane lands on a runway with a velocity of 150 m/s. How far will it travel until it stops if its rate of deceleration is constant at -3 m/s^2?

   (A) 525 m
   (B) 3750 m
   (C) 6235 m
   (D) 9813 m
   (E) 10,435 m

3. A ball is thrown downward from the top of a roof with a speed of 25 m/s. After 2 s, its velocity will be

   (A) 19.6 m/s
   (B) -5.4 m/s
   (C) -44.6 m/s
   (D) 44.6 m/s
   (E) -25 m/s

CHALLENGE

4. A rocket is propelled upward with an acceleration of 25 m/s^2 for 5 s. After that time, the engine is shut off, and the rocket continues to move upward. The maximum height, in meters, that the rocket will reach is

   (A) 900
   (B) 1000
   (C) 1100
   (D) 1200
   (E) 1400

Questions 5–7 refer to the velocity versus time graph shown below.

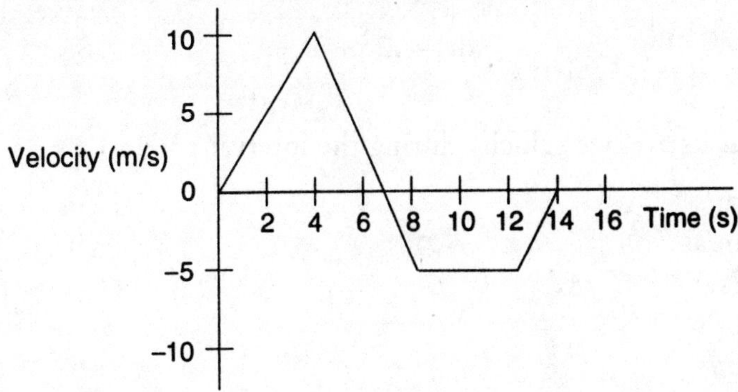

5. The total distance traveled by the object during the indicated 14 s is

   (A) 7.5 m
   (B) 25 m
   (C) 62.5 m
   (D) 77.5 m
   (E) 82.1 m

6. The total displacement of the object during the 14 s indicated is

   (A) 7.5 m
   (B) 25 m
   (C) 62.5 m
   (D) 77.5 m
   (E) 82.1 m

7. The average velocity, in meters per second, of the object is

   (A) 0
   (B) 0.5
   (C) 2.5
   (D) 4.5
   (E) 5.6

8. What is the total change in velocity for the object whose acceleration versus time graph is given below?

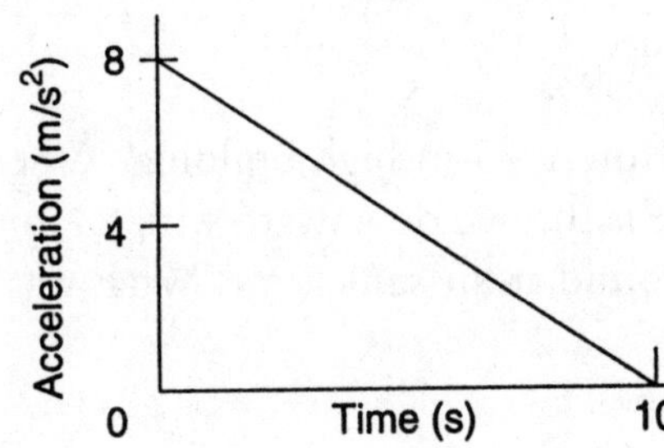

   (A) 40 m/s
   (B) −40 m/s
   (C) 80 m/s
   (D) −80 m/s
   (E) 0 m/s

CHALLENGE

9. A particle moves along the x-axis subject to the following position function:

$$x(t) = 2t^2 + 3t - 1$$

What was its average velocity during the interval $t = 0$ to $t = 3$?

(A) 2.0 m/s
(B) 3.3 m/s
(C) 5 m/s
(D) 8.3 m/s
(E) 9.0 m/s

10. An object has an initial velocity of 15 m/s. How long must it accelerate at a constant rate of 3 m/s^2 before its average velocity is equal to twice its initial velocity?

(A) 5 s
(B) 10 s
(C) 15 s
(D) 20 s
(E) 25 s

FREE-RESPONSE

1. A particle has the acceleration versus time graph shown below:

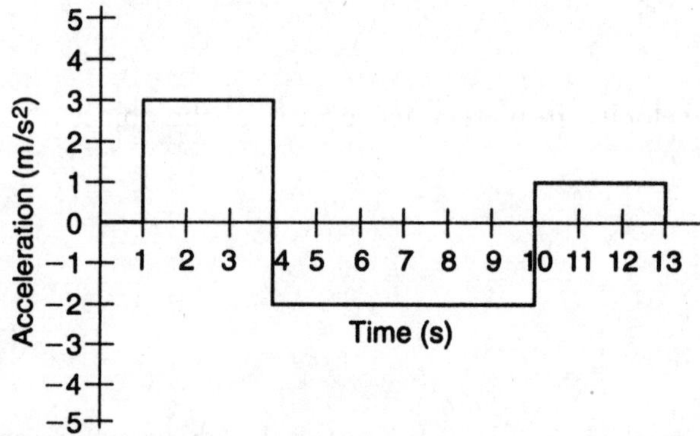

If the particle begins its motion at $t = 0$, with $v = 0$ and $x = 0$, make a graph of velocity versus time.

CHALLENGE

2. A stone is dropped from a 75-m-high building. When this stone has dropped 15 m, a second stone is thrown downward with an initial velocity such that the two stones hit the ground at the same time. What was the initial velocity of the second stone?

3. A particle is moving in one dimension along the x-axis. The position of the particle is given, for any time, by the following position function:

$$x(t) = 3t^2 + 2$$

(a) Evaluate the average velocity of the particle starting at $t = 2$ s for $\Delta t = 1, 0.5, 0.2, 0.1, 0.01, 0.001$ second.
(b) Interpret the physical meaning of your results for part (a).

4. A stone is dropped from a height h and falls the last half of its distance in 4 seconds. (a) What was the total time of fall? (b) From what height was the stone dropped? CHALLENGE

5. If the average velocity of an object is nonzero, does this mean that the instantaneous velocity of the object can never be zero? Explain.

6. A girl standing on top of a roof throws a stone into the air with a velocity $+\mathbf{v}$. She then throws a similar stone directly downward with a velocity $-\mathbf{v}$. Compare the velocities of both stones when they reach the ground.

7. Explain how it is possible for an object to have zero average velocity and still have nonzero average speed.

ANSWERS EXPLAINED

MULTIPLE-CHOICE PROBLEMS

1. **(E)** Given the initial velocity of 20 m/s, we know that the ball is decelerated by gravity at a rate of 9.8 m/s^2. Therefore, we need to know how long gravity will take to decelerate the ball to zero velocity. Clearly, from the definition of acceleration, dividing 20 m/s by 9.8 m/s^2 gives the answer: 2.04 s.

2. **(B)** We do not know the time needed to stop, but we do know that the final velocity is zero. If we use the formula

$$\mathbf{v}_f^2 - \mathbf{v}_i^2 = 2\mathbf{a}x$$

and substitute 150 m/s for the initial velocity and −3 m/s^2 for the deceleration, we get the answer: 3750 m.

3. **(C)** The initial velocity is −25 m/s downward, so $\mathbf{v} = -(25) - (9.8)(2) = -44.6$ m/s.

4. **(C)** The rocket accelerates from rest, so the first distance traveled in 5 s is given by

$$\mathbf{y}_1 = \frac{1}{2}\mathbf{a}t^2$$

Substituting the values for time (5 s) and acceleration (25 m/s^2) gives $\mathbf{y}_1 =$ 312.5 m. After 5 s of accelerating, the rocket has a velocity of 125 m/s. After this time, the engine stops and the rocket is decelerated by gravity as it continues to move upward. The time to decelerate to zero is found by dividing 125 m/s by the acceleration of gravity. Thus 125/9.8 = 12.76 s.

The distance traveled during that time is added to the first accelerated distance. The second distance is given by

$$y_2 = 125(12.76) - 4.9(12.76)^2 = 796.6 \text{ m}$$

The total distance traveled is therefore equal to 1108 m or, rounded off, 1100 m.

5. **(C)** The total distance traveled by the object in 14 s is equal to the total area under the graph. Breaking the figure up into triangles and rectangles, we find that their areas add up to 62.5 m (recall that distance is a scalar).

6. **(A)** The total displacement for the object is the sum of the positive (forward) and negative (backward) areas representing the fact that the object moves away from the origin and then back (as indicated by the velocity vs. time graph) going below the *x*-axis. Adding and subtracting the proper areas: 35 − 2.5 − 20 − 5 = 7.5 m for the final displacement.

7. **(B)** Average velocity is the change in displacement for the object over the change in time (not distance). Since from question 6 we know the displacement change to be 7.5 m in 14 s, the average velocity is therefore 0.53 m/s.

8. **(A)** The change in velocity for the object is equal to the area under the graph, which is equal to 1/2 (+8)(10) = +40 m/s.

9. **(E)** According to the formula, at $t = 0$, $x = -1$. When $t = 3$, $x = 26$. Thus $\Delta \mathbf{x} = 27$ m, and since average velocity is $\Delta \mathbf{x}/\Delta t$, the average velocity for 3 s is 9 m/s.

10. **(B)** For uniformly accelerated motion, the average velocity is

$$\overline{\mathbf{v}} = \frac{\mathbf{v}_f + \mathbf{v}_i}{2}$$

The question requires that the average velocity be equal to twice the initial velocity; thus, if the average velocity is 30 m/s, the final velocity attained must be 45 m/s. Now the question is, how long must the object accelerate from 15 m/s to achieve a speed of 45 m/s? A change in velocity of 30 m/s at a rate of 3 m/s^2 for 10 s is implied.

FREE-RESPONSE PROBLEMS

1. The acceleration graph given in the problem indicates constant acceleration from $t = 1$ to $t = 3$. Therefore, the area represents a constant change in velocity from 0 to 9 m/s. The second region corresponds to negative acceleration, which slows the object down and turns it around. We know this since the area is −12 m/s. This brings the graph from 9 m/s to −3 m/s (a change in direction).

The last area is a change of 3 m/s which brings the velocity to 0 m/s. A graph of this motion is shown below, as desired:

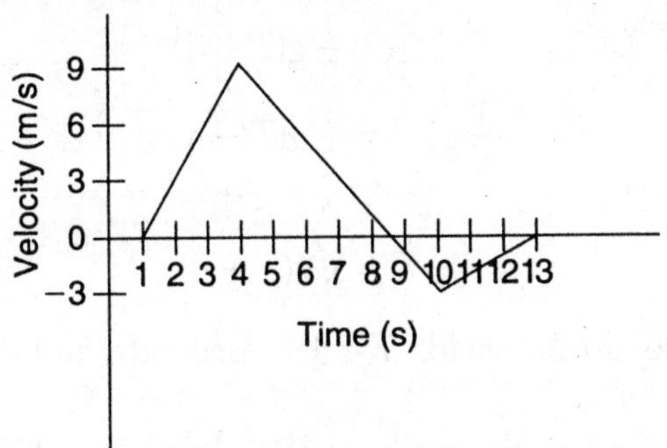

2. In this problem, we know that the first stone will be dropped from a height of 75 m, and it will be in free fall because of gravity. Therefore the time to fall is given by

$$t = \sqrt{\frac{2\mathbf{y}}{\mathbf{g}}} = 3.9 \text{ s}$$

Since the first stone is to fall 15 m before the second stone is dropped, we can likewise determine that $t = 1.75$ s to fall that distance. Thus, since the two stones must reach the ground at the same time, we know that the second stone must reach the ground after the first one has traveled for 3.9 s. Since the second stone is thrown when the distance fallen by the first stone is 15 m, we subtract 1.75 s from the total time of 3.9 s to get the second stone's duration of fall. That time is equal to 2.15 s. Since the second stone is thrown downward, it has a negative displacement, and so we can write

$$-75 = -\mathbf{v}_i(2.15) - \frac{1}{2}(9.8)(2.15)^2$$

This gives us a downward initial velocity of 24.34 m/s for the second stone.

3. (a) Using the position function, we know that when $t = 2$ s, $x = 14$ m. The average velocity is given by $\Delta x/\Delta t$. For $\Delta t = 1$ s, we find that at $t = 3$ s, $x = 29$ m. Thus $\Delta x = 15$ m and the average velocity is equal to 15 m/s. If $\Delta t = 0.5$ s, then at $t = 2.5$ s, $x = 20.75$ m and the new average velocity is equal to 13.5 m/s (6.75 m/0.5 s). We continue this procedure. For $\Delta t = 0.2$ s, the average velocity is 12.6 m/s. For $\Delta t = 0.1$ s, the average velocity is 12.3 m/s. For $\Delta t = 0.01$ s, the average velocity is 12.03 m/s. Finally, for $\Delta t = 0.001$ s, the average velocity is 12.003 m/s.
(b) It appears that as the time interval gets smaller and smaller, the average velocity approaches 12 m/s as a limit. We can say that at $t = 2$ s, the "instantaneous" velocity is approximately equal to 12 m/s.

4. Let $T =$ the total time to fall the distance h. In free fall, this is given by

$$h = \frac{1}{2}\mathbf{g}T^2$$

Now we know that the stone falls the last half of its distance in 4 seconds. This means

$$\frac{h}{2} = \frac{1}{2}\mathbf{g}(T-4)^2$$
$$\frac{1}{4}\mathbf{g}T^2 = \frac{1}{2}\mathbf{g}(T-4)^2$$
$$T^2 = 2(T-4)^2$$
$$T = \sqrt{2}(T-4)$$

Solving for the total time yields $T = 13.7$ seconds, and thus $h = 915$ meters.

5. The average velocity is the ratio of the change in the displacement to the change in the time. It is possible that the object stops for a while and then continues. Thus it is possible for a nonzero average velocity to have a zero instantaneous velocity.

6. The two stones will have the same velocity when they reach the ground. As the first stone rises, gravity decelerates it until it stops, and then it begins to fall back down. When it passes its starting point, it has the same speed but in the opposite direction. This is the same starting velocity as that of the second stone. Both stones are then accelerated through the same displacement, giving them both the same final velocity.

7. The average velocity is the total displacement divided by the total time and is a vector quantity. The average speed is equal to the total distance divided by the total time and is a scalar quantity. If an object is undergoing periodic motion, it can return to its starting point and have zero displacement in one period. This gives it zero average velocity. However, since it traveled a distance, it has a nonzero average speed.

CHAPTER 6

Two-Dimensional Motion

KEY CONCEPTS

- Relative Motion
- Horizontally Launched Projectiles
- Projectiles Launched at an Angle
- Uniform Circular Motion

6.1 RELATIVE MOTION

In Chapter 5, we reviewed the basic elements of one-dimensional rectilinear motion. In this chapter, we will consider only two-dimensional motion. In one sense, one-dimensional motion can be viewed as two-dimensional motion by a suitable transformation of coordinate systems.

For example, consider the definition of displacement from Chapter 5. If we define a coordinate system for a reference frame, the location of a point in that frame is determined by a position vector drawn from the origin of that coordinate system to the point. If the point is displaced, the one-dimensional vector drawn from point *A* to point *B* is called the **displacement vector** and is designated as **B** – **A** (see Figure 6.1).

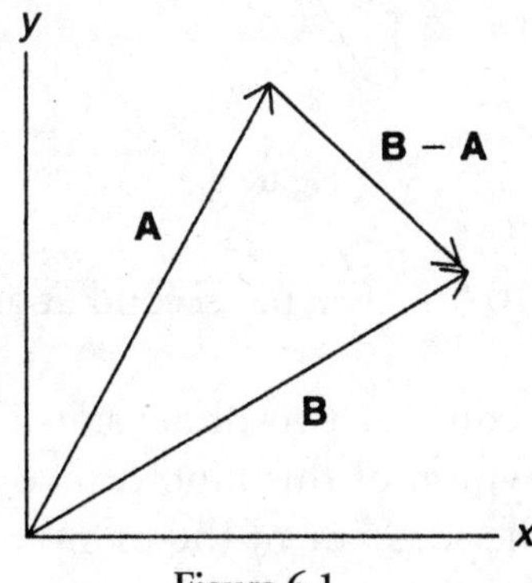

Figure 6.1

Even though the vector **B** – **A** is one dimensional, we can resolve it into two components that represent mutually perpendicular and independent simultaneous motions. An example of this type of motion can be seen when a boat trying to cross a river or an airplane meeting a crosswind is considered. In the case of the boat, its velocity, relative to the river, is based on the properties of the engine and is measured by the speedometer on board. However, to a person on the shore, its

relative velocity (or effective velocity) is different from what the speedometer in the boat may report. In Figure 6.2, we see such a situation with the river moving to the right at 4 meters per second and the boat moving upward across the page at 10 meters per second. If you like, we can call to the right "eastward" and up the page "northward."

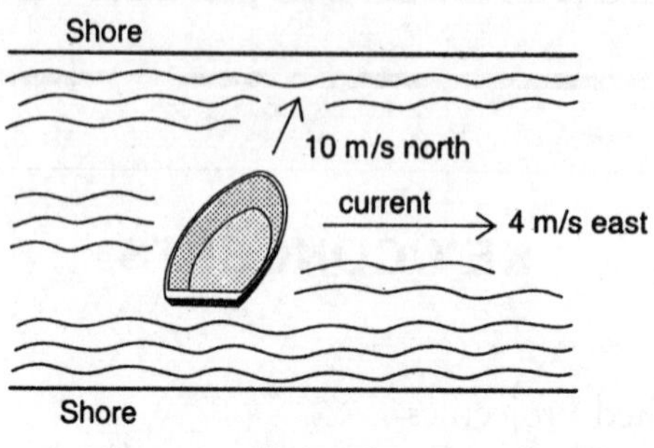

Figure 6.2

By vector methods, the resultant velocity relative to the shore is given by the Pythagorean theorem. The direction is found by means of a simple sketch (Figure 6.3) connecting the vectors head to tail to preserve the proper orientation. Numerically, we can use the tangent function or the law of sines.

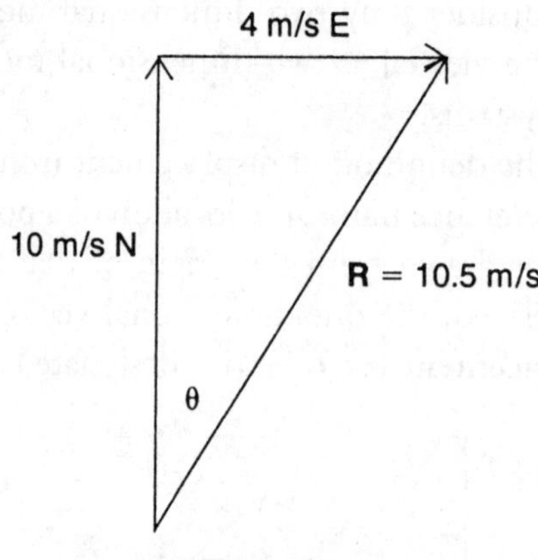

Figure 6.3

The resultant velocity is 10.5 meters per second at an angle of 22 degrees east of north.

In this chapter, we shall consider two-dimensional motion as viewed from the Earth frame of reference. Typical of this kind of two-dimensional motion is that of a projectile. Galileo Galilei was one of the pioneers in studying the mechanics in flying projectiles and the first to discover that the path of a projectile is a parabola.

6.2 HORIZONTALLY LAUNCHED PROJECTILES

If you roll a ball off a smooth table, you will observe that it does not fall straight down. With trial and error, you might observe that how far it falls will depend on how fast it is moving forward. Initially, however, the ball has no vertical velocity. The ability to "fall" is given by gravity, and the acceleration due to gravity is –9.8 meters per second squared. Since gravity acts in a direction perpendicular to the initial horizontal motion, the two motions are simultaneous and independent. Galileo demonstrated that the trajectory is a parabola.

We know that the distance fallen by a mass dropped from rest is given by the equation

$$|\mathbf{y}| = -\frac{1}{2}\mathbf{g}t^2$$

Since the ball that rolled off the table is moving horizontally, with some initial constant velocity, it covers a distance (called the **range**) of $x = |\mathbf{v}_{ix}|t$. Since the time is the same for both motions, we can first solve for the time, using the x equation, and then substitute it into the y equation. In other words,

$$t = \frac{x}{|\mathbf{v}_{ix}|}$$

and therefore

$$|\mathbf{y}| = -\frac{1}{2}\mathbf{g}\left(\frac{x}{\mathbf{v}_{ix}}\right)^2 = -\frac{\mathbf{g}x^2}{2\mathbf{v}_{ix}^{\ 2}}$$

which is of course the equation of an inverted parabola. This equation of y in terms of x is called the **trajectory** of the projectile, while the two separate equations for x and y as functions of time are called **parametric equations.** Figure 6.4 illustrates this trajectory as well as a position vector **R**, which locates a point at any given time in space.

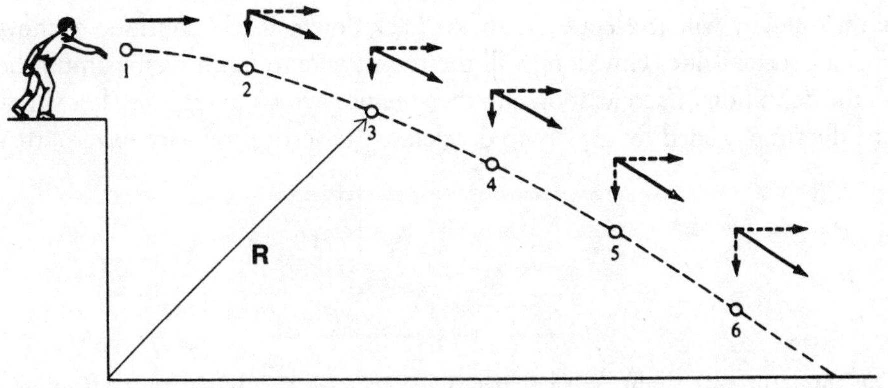

Figure 6.4

Remember

When a projectile is launched horizontally, the two component motions are independent of each other. The horizontal velocity remains constant throughout the projectile's motion.

If the height from which a projectile is launched is known, the time to fall can be calculated from the equation for free fall. For example, if the height is 49 m, the time to fall is 3.16 s. If the horizontal velocity is 10 m/s, the maximum range will be 31.6 m. We can also follow the trajectory by determining how far the object has fallen when it is 10 m away from the base. Using the trajectory formula and the velocity given, we find that the answer is –4.9 m.

SAMPLE PROBLEM

A projectile is launched horizontally from a height of 25 m, and it is observed to land 50 m from the base. What was the launch velocity?

Solution

We know that the vertical motion is independent of the horizontal motion. Thus, we can find the time that the projectile is in the air using:

$$d = \tfrac{1}{2} at^2$$

solving for *t* we get

$$t = 2.26 \text{ s}$$

Now, since $\mathbf{x} = \mathbf{v}_x t$, we see that $v_x = \dfrac{50 \text{ m}}{2.26 \text{ s}} = 22.12 \text{ m/s}$

6.3 PROJECTILES LAUNCHED AT AN ANGLE

Suppose that a rocket on the ground is launched with some initial velocity at some angle θ. The vector nature of velocity allows us to immediately write the equations for the horizontal and vertical components of initial velocity:

$$\mathbf{v}_{ix} = \mathbf{v} \cos\theta \quad \text{and} \quad \mathbf{v}_{iy} = \mathbf{v} \sin\theta$$

Since each motion is independent, we can consider the fact that, in the absence of friction, the horizontal velocity will be constant while the *y* velocity will decrease as the rocket rises. When the rocket reaches its maximum height, its vertical velocity will be zero; then gravity will accelerate the rocket back down. It will continue to move forward at a constant rate. How long will the rocket take to reach its maximum height? From the definition of acceleration and the equations in Chapter 5, we know that this will be the time needed for gravity to decelerate the vertical velocity to zero, that is,

$$t_{up} = \frac{|\mathbf{v}_{iy}|}{|\mathbf{g}|} = \frac{|\mathbf{v}_i| \sin\theta}{|\mathbf{g}|}$$

Remember

When a projectile is launched at an angle, the vertical velocity component is equal to zero at the maximum height. Additionally, for a given launch velocity, the maximum range occurs when the launch angle is equal to 45°.

The total time of flight will be just twice this time. Therefore, the range is the product of the initial horizontal velocity (which is constant) and the total time. In other words,

$$\mathbf{R} = \text{Range} = 2\mathbf{v}_{ix} t_{up} = 2\mathbf{v} \cos\theta\, t_{up}$$

Since the vertical motion is independent of the horizontal motion, the changes in vertical height are given one dimensionally as

$$\mathbf{y} = \mathbf{v}_{iy} t - \frac{1}{2}\mathbf{g}t^2$$

If we want to know the maximum height achieved, we simply use the value for the time to reach the highest point. The trajectory is seen in Figure 6.5 for a baseball being hit.

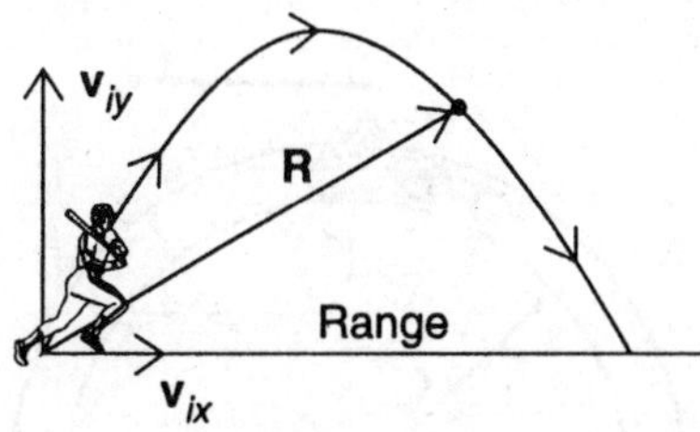

Figure 6.5

As an example, if a projectile is launched with an initial velocity of 100 m/s at an angle of 30°, the maximum range will be equal to 883.7 m. To find the maximum height, we could first find the time required to reach that height, but a little algebra will give us a formula for the maximum height independent of time. The maximum height can also be expressed as

$$\mathbf{y}_{max} = \frac{\mathbf{v}_i{}^2\mathbf{y}}{2\mathbf{g}}$$

Using the known numbers, we find that the maximum height reached is 127.55 m.

SAMPLE PROBLEM

A projectile is launched from the ground at a 40° angle with a velocity of 150 m/s. Calculate the maximum height of the projectile.

Solution

From trigonometry, we know that $\mathbf{v}_{iy} = \mathbf{v}_o \sin \theta$. Thus,

$$\mathbf{v}_{iy} = (150 \text{ m/s}) \sin (40°) = 96.42 \text{ m/s}$$

Now, at the maximum height, the vertical velocity is equal to zero, and the acceleration is **g**:

$$\mathbf{v}_{fy}{}^2 - \mathbf{v}_{iy}{}^2 = 2\mathbf{gy}$$

$$\mathbf{y}_{max} = \frac{\mathbf{v}_{iy}{}^2}{2\mathbf{g}} = \frac{-96.42 \text{ m/s}^2}{-19.6 \text{ m/s}^2} = 474.33 \text{ m}$$

6.4 UNIFORM CIRCULAR MOTION

Velocity is a vector. When velocity changes, the magnitude or direction, or both, can change. When the direction is the only quantity changing, as the result of a centrally directed deflecting force, the result is uniform circular motion. We will take up the discussion of deflecting forces in Chapter 7, but consider here an object already undergoing periodic, uniform circular motion. By this description we mean that the object maintains a constant speed as it revolves around a circle of radius *R*, in a period of time *T*. The number of revolutions per second is called the **frequency**, *f*. This is illustrated in Figure 6.6.

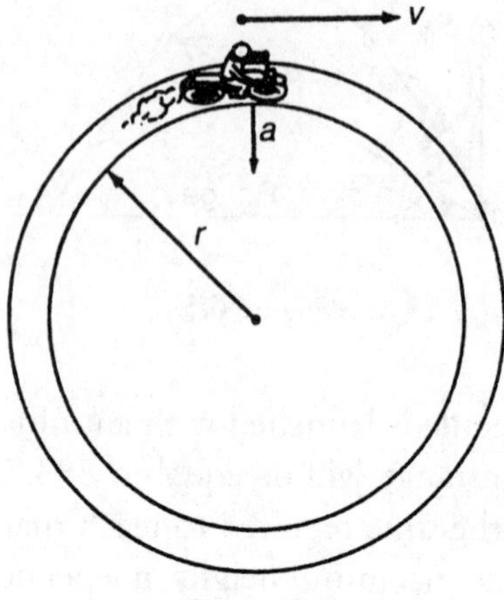

Figure 6.6

The direction of the acceleration is toward the center and is called the **centripetal acceleration**. The magnitude of the centripetal acceleration is given by two formulas:

$$\mathbf{a}_c = \frac{\mathbf{v}^2}{r} \quad \text{and} \quad \mathbf{a}_c = \frac{4\pi^2 r}{T^2}$$

where *T* is the **period** and is equal to the reciprocal of the frequency

$$T = \frac{1}{f}$$

SAMPLE PROBLEM

A 5-kg mass is undergoing uniform circular motion with a constant speed of 10 m/s in a circle of radius 2 m. Calculate the centripetal acceleration of the mass.

Solution

We use the formula

$$\mathbf{a}_c = \frac{\mathbf{v}^2}{R} = \frac{10 \text{ m/s}^2}{2 \text{ m}} = 50 \text{ m/s}^2$$

Chapter Summary

- The relative velocity between moving objects is equal to the difference between their velocities.
- When a projectile is launched, in the absence of air resistance, the horizontal motion is independent of the vertical motion.
- The only acceleration on a projectile is the downward acceleration due to gravity.
- If air resistance is present, a falling object will reach a maximum velocity called the terminal velocity.
- The vertical and horizontal components of the launch velocity can be obtained using trigonometry. The regular kinematics equations can then be used for each direction.
- If an object is moving in a circle, then the object has an acceleration acting toward the center called the centripetal acceleration.

Problem-Solving Strategies for Two-Dimensional Motion

The key to solving projectile motion problems, or any other two-dimensional problem, is to remember that the horizontal and vertical motions are independent of each other. Therefore, using the formulas from Chapter 5 for one-dimensional motion can greatly reduce the complexity of a given problem. The following guidelines should also help:

1. Draw a sketch of the situation if none is provided.
2. Determine whether the components of motion are given in the problem. If not, try to determine the components of velocity and acceleration.
3. Keep in mind that accelerations given in one dimension do not affect the other. This fact results in the curved path.
4. Remember that velocity and acceleration are vectors and that both magnitudes and directions should be specified unless otherwise noted.
5. Choose a suitable frame of reference and coordinate system for the problem.
6. Ask yourself questions about the implications of each change in motion on the whole motion of the object.
7. Remember that trajectories are expressed in terms of position variables only, and that parametric equations usually relate positions with time.

Practice Exercises

MULTIPLE-CHOICE

1. A projectile is launched at an angle of 45° with a velocity of 250 m/s. If air resistance is neglected, the magnitude of the horizontal velocity of the projectile at the time it reaches maximum altitude is equal to

   (A) 0 m/s
   (B) 175 m/s
   (C) 200 m/s
   (D) 250 m/s
   (E) 300 m/s

2. A projectile is launched horizontally with a velocity of 25 m/s from the top of a 75-m height. How many seconds will the projectile take to reach the bottom?

   (A) 15.5
   (B) 9.75
   (C) 6.31
   (D) 4.27
   (E) 3.91

3. An object is launched from the ground with an initial velocity and angle such that the maximum height achieved is equal to the total range of the projectile. The tangent of the launch angle is equal to

   (A) 1
   (B) 2
   (C) 3
   (D) 4
   (E) 5

4. At a launch angle of 45°, the range of a launched projectile is given by

   (A) $\frac{\mathbf{v}_i^2}{\mathbf{g}}$

   (B) $\frac{2\mathbf{v}_i^2}{\mathbf{g}}$

   (C) $\frac{\mathbf{v}_i^2}{2\mathbf{g}}$

   (D) $\sqrt{\frac{\mathbf{v}_i^2}{2\mathbf{g}}}$

   (E) $\frac{2\mathbf{v}_i}{\mathbf{g}}$

5. A projectile is launched at a certain angle. After 4 s, it hits the top of a building 500 m away. The height of the building is 50 m. The projectile was launched at an angle of

CHALLENGE

(A) 14°
(B) 21°
(C) 37°
(D) 76°
(E) 85°

6. A projectile is launched at a velocity of 125 m/s at an angle of 20°. A building is 200 m away. At what height above the ground will the projectile strike the building?

CHALLENGE

(A) 50.6 m
(B) 90.2 m
(C) 104.3 m
(D) 114.6 m
(E) 125.6 m

7. The operator of a boat wishes to cross a 5-km-wide river that is flowing to the east at 10 m/s. He wishes to reach the exact point on the opposite shore 15 min after starting. With what speed and in what direction should the boat travel?

CHALLENGE

(A) 11.2 m/s at 26.6° E of N
(B) 8.66 m/s at 63.4° W of N
(C) 11.4 m/s at 60.9° W of N
(D) 8.66 m/s at 26.6° E of N
(E) 5 m/s due N

8. An object is moving around a circle of radius 1.5 m at a constant velocity of 7 m/s. The frequency of the motion, in revolutions per second, is

(A) 0.24
(B) 0.53
(C) 0.67
(D) 0.74
(E) 0.98

FREE-RESPONSE

1. A ball moving horizontally 3 m/s rolls off the top of a flight of stairs. If each stair is 0.5 m high and 0.5 m wide, which stair will the ball hit? Show your work!

CHALLENGE

2. A mass attached to a string is twirled overhead in a horizontal circle of radius R every 0.3 s. The path is directly 0.5 m above the ground. When released, the mass lands 2.6 m away. What was the velocity of the mass when it was released and what was the radius of the circular path?

3. A football quarterback throws a pass to a receiver at an angle of 25 degrees to the horizontal and at an initial velocity of 25 m/s. The receiver is initially at rest 30 m from the quarterback. The instant the ball is thrown, the receiver runs at a constant velocity to catch the pass. In what direction and at what speed should he run?

4. Under what conditions can you have two-dimensional motion with a one-dimensional acceleration?

5. A car is moving in a straight line with velocity **v**. Raindrops are falling vertically downward with a constant terminal velocity **u**. At what angle does the driver think the drops are hitting the car's windshield? Explain.

ANSWERS EXPLAINED

MULTIPLE-CHOICE PROBLEMS

1. **(B)** The horizontal component of velocity remains constant in the absence of resistive forces and is equal to $\mathbf{v}_i \cos \theta$. Substituting the known numbers, we get $\mathbf{v}_x = (250)(0.7) = 175$ m/s.

2. **(E)** The time to fall is given by the free-fall formula from Chapter 5:

$$t = \sqrt{\frac{2\mathbf{y}}{\mathbf{g}}}$$

If we substitute the known numbers, we get 3.91 s for the time.

3. **(D)** Since the maximum height is equal to the range, we set these two equations equal to each other:

$$\frac{\mathbf{v}_i^2 \sin^2 \theta}{2\mathbf{g}} = \frac{\mathbf{v}_i^2 \sin 2\theta}{\mathbf{g}}$$

Recalling that $\sin 2\theta = 2 \sin \theta \cos \theta$, we solve for $\tan \theta$ and find that it equals 4.

4. **(A)** From the formula for range, we see that, at 45°, $\sin 2\theta = 1$, and so

$$R = \frac{\mathbf{v}_i^2}{\mathbf{g}}$$

5. **(A)** We know that, after 4 s, the projectile has traveled horizontally 500 m. Therefore, the horizontal velocity was a constant 125 m/s and is equal to $\mathbf{V}_i \cos \theta$. We also know that, after 4 s, the *y*-position of the projectile is 50 m. Thus we can write:

$$50 = 4\mathbf{v}_i \sin \theta - 4.9(4)^2 = 4\mathbf{v}_i \sin \theta - 78.4$$

Therefore, $\mathbf{v}_i \cos \theta = 125$, $\mathbf{v}_i \sin \theta = 32.1$, and $\tan \theta = 0.2568$, so $\theta = 14.4°$.

6. **(A)** The *x*-component of velocity is 93.97 m/s, and the *y*-component is 34.2 m/s. Since the projectile travels 200 m horizontally at 93.97 m/s, the time of flight $t = 2.128$ s. Substituting this time into our equation for the *y* position gives $y = 50.6$ m.

7. **(C)** The river is flowing at 10 m/s to the right (east) and the resultant desired velocity is 5.55 m/s up (north). Therefore, the actual velocity, relative to the river, is heading W of N. By the Pythagorean theorem, the velocity of the boat must be 11.4 m/s. The angle is given by the tangent function. In the diagram below, not drawn to scale but correct for orientation, $\tan/\theta = \dfrac{10}{5.55} = 1.8$. Therefore, $\theta = 60.9°$ W of N.

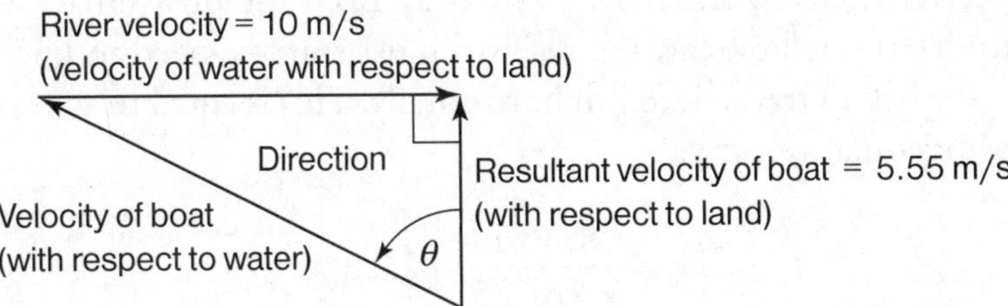

8. **(D)** The velocity is given by $\mathbf{v} = 2\pi R/T$, where T is the period, but the frequency f is just the reciprocal of the period. Thus, $\mathbf{v} = 2\pi Rf$. We substitute the known numbers to obtain $f = 0.74$ rev/s.

FREE-RESPONSE PROBLEMS

1. There are several ways to solve this problem. The correct answer is the fourth step. If we approach the problem as a projectile, then by trial and error we see that it takes 0.32 s to fall the first 0.5 m:

$$t = \sqrt{\frac{2h}{\mathbf{g}}} = \sqrt{\frac{2(0.5\text{ m})}{9.8\text{ m/s}^2}} = 0.32\text{ s}$$

During this time, it has a horizontal velocity of 3 m/s and it travels 0.96 m. This distance is beyond the first step, and we must repeat the procedure for the second step (falling 1.0 m, computing the time, and multiplying the time by 3 m/s). If we continue in this way, we see that for the fourth step the horizontal distance is less than the expected 2.0 m (which means it hits this step). There are several other alternative methods that you can come up with on your own.

2. The mass is undergoing uniform circular motion. Therefore, it has a constant speed. The velocity when it is released can be determined from the height of the mass and the horizontal range.

We know it falls vertically 0.5 m in 0.32 s (see problem 1). During this time, it moves horizontally 2.6 m. Thus, the magnitude of the horizontal initial velocity was

$$\mathbf{v} = \frac{2.6\text{ m}}{0.32\text{ s}} = 8.125\text{ m/s}$$

In circular motion, the constant velocity found above is equal to the ratio of the circumference of the circular path and the period:

$$\mathbf{v} = \frac{2\pi R}{T}$$

Substituting for the velocity and the period of 0.3 s, we find that $R = 0.388$ m.

3. The first quantity we can calculate is the football's theoretical range:

$$R = \frac{\mathbf{v}_i^2 \sin 2\theta}{\mathbf{g}} = \frac{(25)^2(\sin 50)}{\mathbf{g}} = 48.85 \text{ m}$$

The receiver needs to travel 18.85 m away from the quarterback to catch the ball. To determine how fast the receiver must run, we need to know how long it takes the ball to travel 48.85 m horizontally. This is equal to the range divided by the horizontal velocity:

$$t = \frac{48.85 \text{ m}}{\mathbf{v}_i \cos 25} = 2.16 \text{ s}$$

Therefore, to travel 18.85 m in 2.16 s, the receiver must run at 8.73 m/s.

4. Two-dimensional motion can result when the acceleration vector is not along the line of initial motion. In horizontal projectile motion, the acceleration vector due to gravity is initially perpendicular to its motion.

5. From the diagram below, we can see that, to the driver, tan $\theta = \mathbf{u}/\mathbf{v}$.

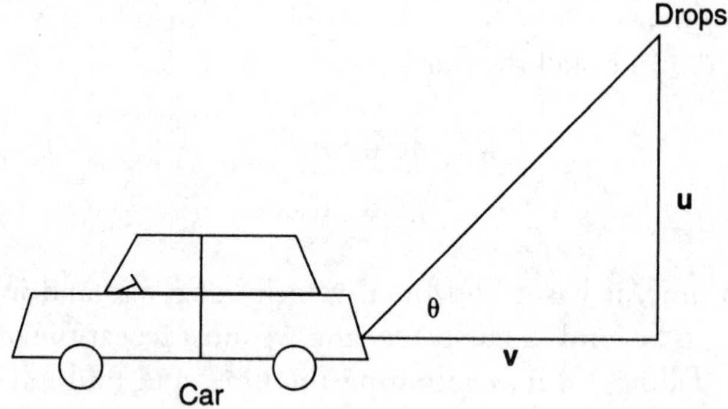

Forces and Newton's Laws of Motion

CHAPTER 7

KEY CONCEPTS

- Forces
- Newton's Laws of Motion
- Static Applications of Newton's Laws
- Dynamic Applications of Newton's Laws
- Weight Versus Mass
- Equilibrium and Net Forces
- Central Forces
- Friction

7.1 INTRODUCTION

From our discussion in Chapter 4 of frames of reference, you should be able to convince yourself that, if two objects have the same velocity, the relative velocity between them is zero, and therefore one object looks as though it is at rest with respect to the other. In fact, it would be impossible to decide whether or not such an **"inertial"** observer was moving! Therefore, accepting this fact, we state that an object that appears to be at rest in the Earth frame of reference will simply be stated to be "at rest" relative to us (the observers).

7.2 FORCES

Observations inform us that an inanimate object will not move freely of its own accord unless an interaction takes place between it and at least one other object. This interaction usually involves contact between the objects, although the gravity exerted by Earth or any other body on the object does not involve any direct contact. This type of interaction is sometimes called an **action at a distance**. Electrostatic attraction and repulsion, as well as magnetic attraction and repulsion, are other examples of this effect.

Forces are vector quantities.

When interaction occurs, we say that a **force** has been created between the objects. If an object was moving (relative to us), then the force may change the direction of the motion or it may change its speed. In other words, there will be a change in the velocity, which is a vector quantity; the magnitude and/or direction of the velocity will change. If no change occurs, we must conclude the presence of another force that resists the changes induced by the applied one. Friction is an example of an opposing force of this type.

In contrast to deflecting forces, other forces may be restorative. An example is a spring or other elastic material that has been stretched. Once the material has been elongated, it will snap back in an attempt to return to its original status.

If all the forces acting on an object produce no net change, the object is in a state of **equilibrium**. If the object is moving relative to us, we say that it is in a state of **dynamic equilibrium**. If the object is at rest relative to us, we say that it is in a state of **static equilibrium**. Figure 7.1 illustrates these ideas.

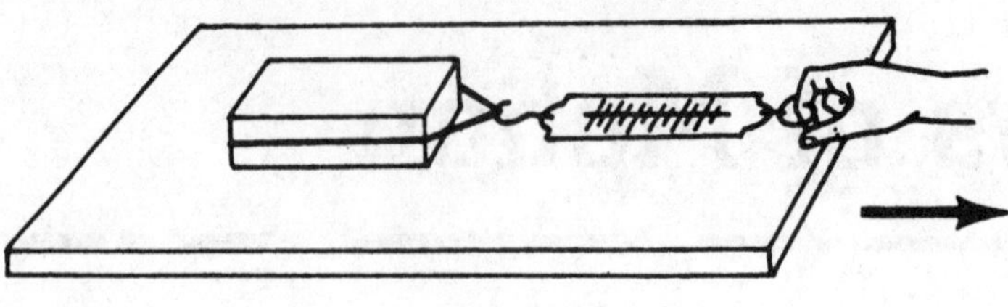

Figure 7.1

7.3 NEWTON'S LAWS OF MOTION

In 1687, at the urging of his friend Edmund Halley, Isaac Newton published his greatest work. It was titled the *Mathematical Principles of Natural Philosophy*, but is more widely known by a shortened version, *Principia*. In this book, Isaac Newton revolutionized the rational study of mechanics by the introduction of mathematical principles that all of nature was considered to obey. Using his newly developed ideas, Newton set out to explain the observations and analyses of Galileo Galilei and Johannes Kepler.

TIP 

Make sure you know all of **Newton's laws.** On an AP free-response problem, you should make a declaration like $\Sigma F = 0$ (for equilibrium) or $\Sigma F = ma$ (for accelerated systems).

The ability of an object to resist a change in its state of motion is called **inertia.** This concept is the key to Newton's first law of motion:

Every body continues in its state of rest, or of uniform motion in a straight line, unless it is compelled to change that state by forces acting on it.

In other words, an object at rest will tend to stay at rest, and an object in motion will tend to stay in motion, unless acted upon by an external force. By "rest," we of course mean the observed state of rest in a particular frame of reference. As stated above, the concept of "inertia" is taken to mean the ability of an object to resist a force attempting to change its state of motion. As we will subsequently see, this concept is covered under the new concept of **mass** (a scalar, as opposed to **weight**, a vector force).

If a mass has an unbalanced force incident upon it, the velocity of the mass is observed to change. The magnitude of this velocity change depends inversely on the amount of mass. In other words, a force directed along the direction of motion will cause a smaller mass to accelerate more than a larger mass. Newton's second law of motion expresses these observations as follows:

The change of motion is proportional to the applied force, and is in the direction which that force acts.

The acceleration produced by the force is in the same direction as the force. This does not mean that the object's direction must remain the same. Mathematically, the second law is sometimes expressed as:

$$\mathbf{F} = m\mathbf{a}$$

However, to preserve the vector nature of the forces, and the fact that by "force" we mean "net force," we can write the second law as:

$$\mathbf{F}_{net} = \Sigma\mathbf{F} = m\mathbf{a}$$

The units of force are **newtons**; 1 newton (N) is defined as the force needed to give a 1-kilogram mass an acceleration of 1 meter per second squared. Thus 1 newton equals 1 kilogram-meter per second squared. This being the case, the weight of an object is given by

$$\mathbf{F_g} = m\mathbf{g}$$

where **g** is the acceleration due to gravity and the units of weight are newtons. In the SI system of units, the kilogram is taken as the standard unit of mass and corresponds to 9.8 newtons of weight, or about 2.2 English pounds.

Newton's third law of motion is crucial for understanding the conservation laws we will discuss later. It stresses the fact that forces are the result of mutual interactions and are thus produced in pairs. The third law is usually stated as follows (Figure 7.2):

For every action, there is an equal but opposite reaction.

Figure 7.2

Think About It

Can you think of some other action-reaction pairs?

You can observe action and reaction when air is released from a balloon. The air travels in one direction as it leaves the balloon. The balloon travels in the opposite direction.

When Newton wrote his second law of motion, his concept of "quantity of motion" was defined as the product of an object's mass and velocity. In modern times, this concept is known as **momentum**; it will be discussed in Chapters 9 and 10.

SAMPLE PROBLEM

A 2-kg mass is accelerated by a net force of 20 N. What is the acceleration of the mass?

Solution

We use Newton's second law:

$$\mathbf{F}_{net} = m\mathbf{a}$$

Thus,

$$\mathbf{a} = \frac{20\text{ N}}{2\text{ kg}} = 10\text{ N/kg} = 10\text{ m/s}^2$$

SAMPLE PROBLEM

How much does a 2-kg mass weigh on the surface of Earth?

Solution

We modify Newton's second law for gravitational acceleration:

$$\mathbf{F_g} = m\mathbf{g}$$

$$\mathbf{F_g} = (2\text{ kg})(-9.8\text{ m/s}^2) = -19.6\text{ N (downward)}$$

Notice in both problems that the units for acceleration can be written as both N/kg or m/s^2. They are equivalent, and we can think of the acceleration due to gravity in terms of the force per unit mass (which is referred to as a field strength).

7.4 STATIC APPLICATIONS OF NEWTON'S LAWS

If we look more closely at Newton's second law of motion, we see an interesting implication. If the net force acting on an object is zero, the acceleration of the object will likewise be zero. Notice, however, that kinematically, zero acceleration does not imply zero motion! It simply indicates that the velocity of the object is not changing. If the object is in a state of rest and remains at rest (because of zero net forces), the object is in static equilibrium. Some interesting problems in engineering deal with the static stability of structures. Let's look at a simple example.

Place this book on a table. You will observe that it is not moving relative to you. Its state of rest is provided by the zero net force between the downward force of gravity and the upward reactive force of the table (pushing on the floor, which in turn pushes up on the table, etc.). Figure 7.3 shows this setup. The upward reactive force of the table, sometimes called the **normal force**, always acts perpendicular to the surface.

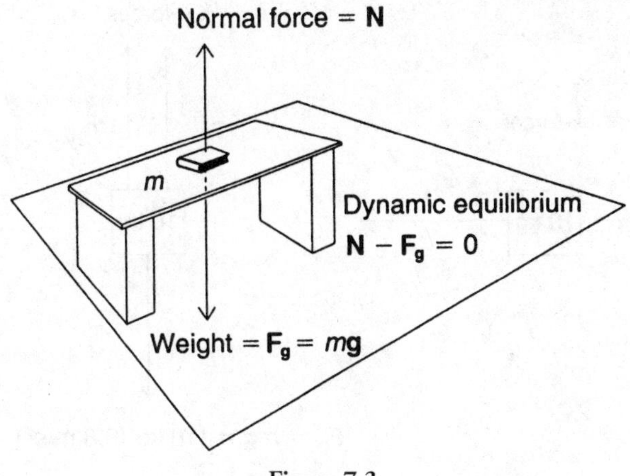

Figure 7.3

In this example, we have used both the second and third laws of motion. Remember: the second law is a vector equation, and so we must treat the sum of all vector forces in each direction separately! In this case, we have:

$$\Sigma \mathbf{F}_y = \mathbf{N} - \mathbf{F_g} = 0$$

There are no forces in the *x*-directions to be "analyzed."

A slightly more complex problem would concern a mass suspended by two strings. The tensions in the strings support the mass and are thus vector forces acting at angles. In the problem illustrated in Figure 7.4, a 10-kilogram mass is suspended by two strings, making angles of 30 and 60 degrees, respectively, to the horizontal. The question is, "What are the tensions in the two strings?"

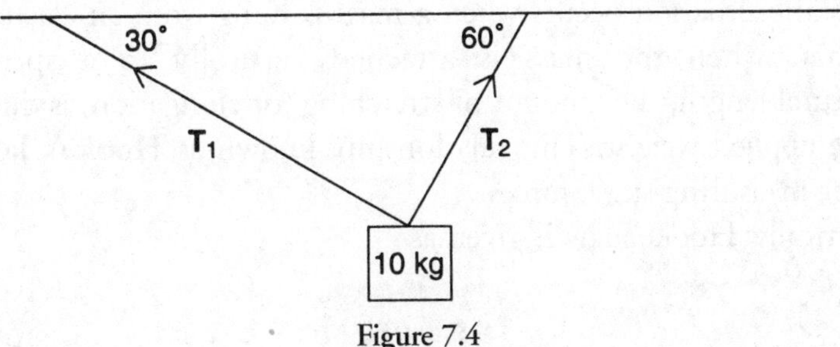

Figure 7.4

A useful heuristic for solving these types of problems is to construct what is called a "free-body diagram." In such a diagram, you "free" the body of its realistic constraints and redraw it, indicating the directions of all applied forces. You then use that diagram to set up your static equations for Newton's second law in both the *x*- and *y*-directions. The result is a system of two equations and two unknowns. In our example, a free-body diagram for both *x*- and *y*-directions would look like Figure 7.5.

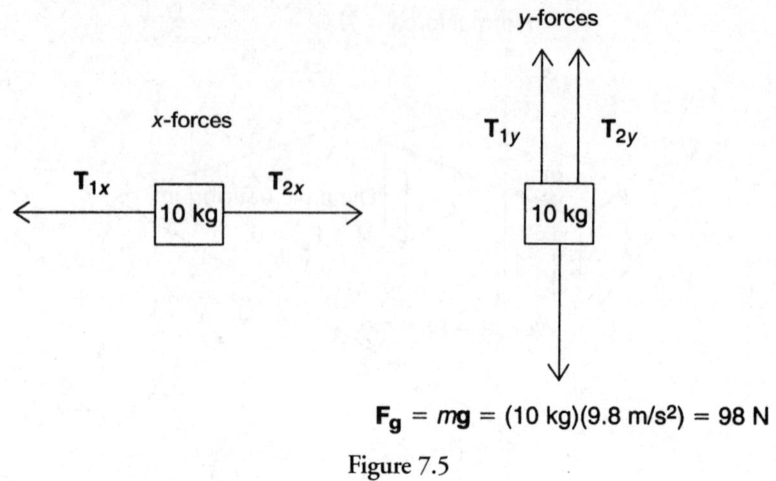

Figure 7.5

We can now set up our equations for Newton's second law, using the techniques of vector analysis reviewed in Chapter 4. We thus determine the *x*- and *y*-components of the tensions, which are given by $\mathbf{T}_{1x} = -\mathbf{T}_1 \cos 30$ and $\mathbf{T}_{2x} = \mathbf{T}_2 \cos 60$. The negative sign is used since $\mathbf{T}_{1x}$ is to the left. Thus, in the *x*-direction we write

$$\Sigma \mathbf{F}_x = 0 = \mathbf{T}_2 \cos 60 - \mathbf{T}_1 \cos 30 = 0.5\mathbf{T}_2 - 0.866\mathbf{T}_1$$

Likewise, for the *y*-direction, we see that the 10-kg mass weighs 98 N. The downward pull of gravity is compensated for by the upward pull of the two vertical components of the tensions. In other words, we can write for the *y*-direction:

$$\Sigma \mathbf{F}_y = 0 = \mathbf{T}_1 \sin 30 + \mathbf{T}_2 \sin 60 - \mathbf{F_g} = 0.5\mathbf{T}_1 + 0.866\mathbf{T}_2 - 98$$

The two simultaneous equations can then be solved for $\mathbf{T}_1$ and $\mathbf{T}_2$. Performing the algebra, we find that $\mathbf{T}_1 = 48.97$ N and $\mathbf{T}_2 = 84.87$ N.

Another static situation occurs when a mass is hung from an elastic spring. It is observed that, when the mass is attached vertically to a spring that has a certain natural length, the amount of stretching, or elongation, is directly proportional to the applied weight. This relationship, known as **Hooke's law**, supplies a technique for measuring static forces.

Mathematically, Hooke's law is given as

$$\mathbf{F} = -k\mathbf{x}$$

where k is the spring constant in units of newtons per meter (N/m), and $\mathbf{x}$ is the elongation beyond the natural length. The negative is used to indicate that the applied force is restorative so that, if allowed, the spring will accelerate back in the opposite direction. As a static situation, a given spring can be calibrated for known weights or masses, and thus used as a "scale" for indicating weight or other applied forces.

7.5 DYNAMIC APPLICATIONS OF NEWTON'S LAWS

If the net force acting on an object is not zero, Newton's second law implies the existence of an acceleration in the direction of the net force. Therefore, if a 10-kilogram mass is acted on by a 10-newton force from rest, the acceleration will be 1 meter per second squared. Heuristically, we can construct a free-body diagram for the system to analyze all forces acting in all directions and then apply the second law of motion. We must be careful, however, to choose an appropriate frame of reference and coordinate system. For example, if the mass is sliding down a frictionless incline, a natural coordinate system to use is one that is rotated in such a way that the x-axis is parallel to the incline (see Figure 7.6).

Therefore, in order to resolve the force of gravity into components parallel and perpendicular to the incline, we must first identify the relevant angle in the geometry. This is outlined in Figure 7.6, and we can see that the magnitude of the downward component of weight along the incline is given by $-m\mathbf{g} \sin \theta$, where θ is the angle of the incline. The magnitude of the normal force, perpendicular to the incline, is therefore given by $|\mathbf{N}| = |m\mathbf{g} \cos \theta|$.

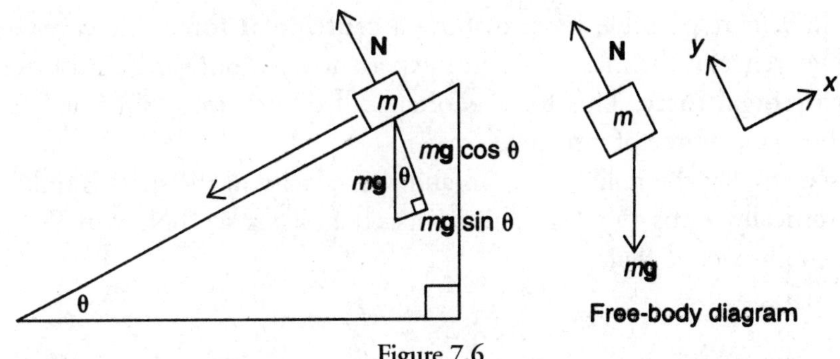

Figure 7.6

The direction of $m\mathbf{g} \cos \theta$ can be taken as inward into the incline's surface and provides the force necessary to keep the mass in contact with the surface. If friction were present (see Section 7.6), this force would be the main contributor to the frictional force, which would be directed opposite the direction of motion.

If, for example, the mass was 10 kg, the weight would be 98 N. If $\theta = 30°$, we would write for the x-forces:

$$\Sigma F_x = m\mathbf{a} = -m\mathbf{g} \sin \theta = -98 \sin 30 = 49 \text{ N}$$

and since there is no acceleration in the y-direction, we would have:

$$\Sigma \mathbf{F}_y = 0 = \mathbf{N} - m\mathbf{g} \cos \theta = \mathbf{N} - 98 \cos 30 = \mathbf{N} - 8.49$$

Thus the normal force is equal to 8.49 N.

7.6 CENTRAL FORCES

Consider a point mass moving around a circle, supported by a string making a 45-degree angle to a vertical post (the so-called **conical pendulum**). Let's analyze this situation, which is shown in Figure 7.7.

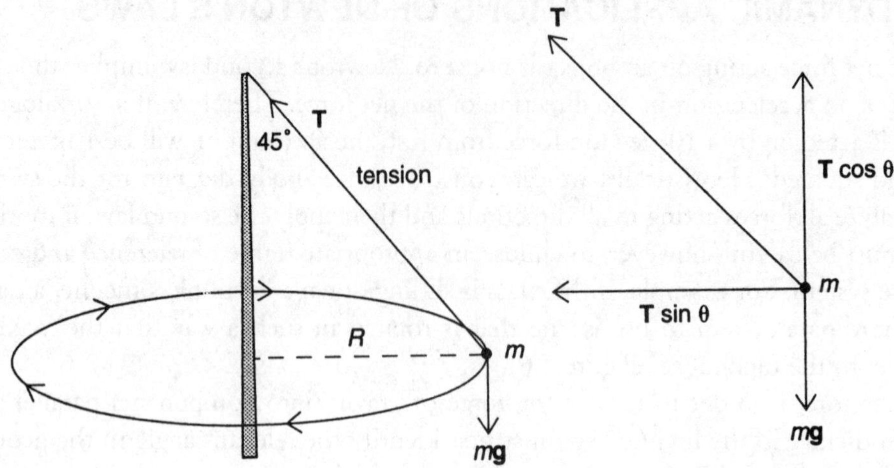

Figure 7.7

In this case, the magnitude of the weight, *mg*, is balanced by the upward component of the tension in the string, given by **T** cos θ. The inward component of tension, **T** sin θ, is responsible for providing a **centripetal force**. However, from the frame of reference of the mass, the force experienced is "outward" and is referred to as the **centrifugal force**. This force is often called "fictitious" since it is perceived only in the mass's frame of reference.

If we specify that the radius is 1.5 m and the velocity of the mass is unknown, we see that, vertically, **T** cos 45 = ***mg***. This implies that, if ***mg*** = 98 N, then **T** = 138.6 N. Horizontally, we see that

$$\mathbf{T}\sin 45 = \frac{M\mathbf{v}^2}{R}$$

where the expression for centripetal force is given by

$$\mathbf{F}_c = \frac{m\mathbf{v}^2}{R}$$

using the logic of **F** = *m***a**. Using our known information, we solve for velocity and find that **v** = 3.83 m/s. It is important to remember that the components of forces must be resolved along the principal axes of the chosen coordinate system.

7.7 FRICTION

Friction is a contact force between two surfaces that is responsible for opposing sliding motion. Even the smoothest surfaces are microscopically rough, with peaks and valleys like a mountain range. When an object is first moved, this friction plus inertia must be overcome. If a spring balance is attached to a mass and then pulled, the reading of the force scale when the mass first begins to move provides a measure of the static friction. Once the mass is moving, if we maintain a steady enough force, the velocity of movement will be constant. Thus, the acceleration will be zero, indicating that the net force is zero. The reading of the scale will then measure the kinetic friction. This reading will generally be less than the starting reading.

Observations show that the frictional force is directly related to the applied load pushing the mass into the surface. From our knowledge of forces, this fact implies that the normal force is responsible for this action. The proportionality constant linking the normal force with the frictional force is called the coefficient of friction and is symbolized by μ. There are two coefficients of friction, one for static friction and the other for kinetic friction. This linear relationship is often written as

TIP 

The coefficient of **static friction** is generally larger than the coefficient of **kinetic friction.**

$$\mathbf{f} = \mu \mathbf{N}$$

and the analysis of a situation involves identifying the normal force.

For example, in Figure 7.8 a mass is being pulled along a horizontal surface by a string, making an angle θ.

If we set up our equations of motion, we see that:

$$\Sigma \mathbf{F}_y = 0 = \mathbf{N} + \mathbf{F} \sin \theta - m\mathbf{g}$$

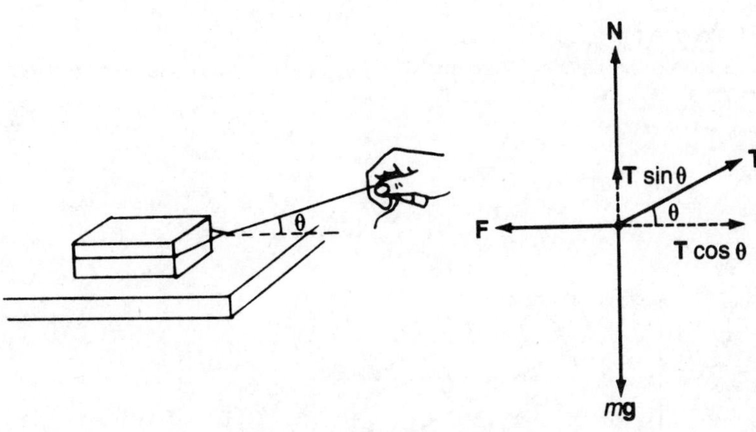

Figure 7.8

Thus the normal force is going to be less than the weight, mg, because of the upward component of the applied force. Therefore,

$$\mathbf{N} = m\mathbf{g} - \mathbf{F} \sin \theta$$

and the force of kinetic friction, given a coefficient of kinetic friction, μ_k, is written as:

$$\mathbf{f} = \mu_k \mathbf{N} = \mu_k (m\mathbf{g} - \mathbf{F} \sin \theta)$$

SAMPLE PROBLEM

A 2-kg mass is at rest at the top of an incline that makes a 40° angle with the ground. The incline is 1.5 m long and the coefficient of kinetic friction $\mu_k = 0.2$. When the mass is released, it slides down the incline with uniform acceleration.

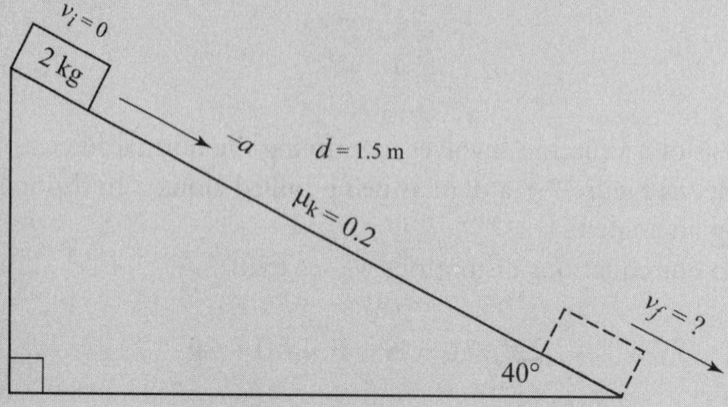

(a) Calculate the acceleration of the mass.
(b) Calculate the velocity of the mass when it reaches the bottom of the incline.

Solution

(a)

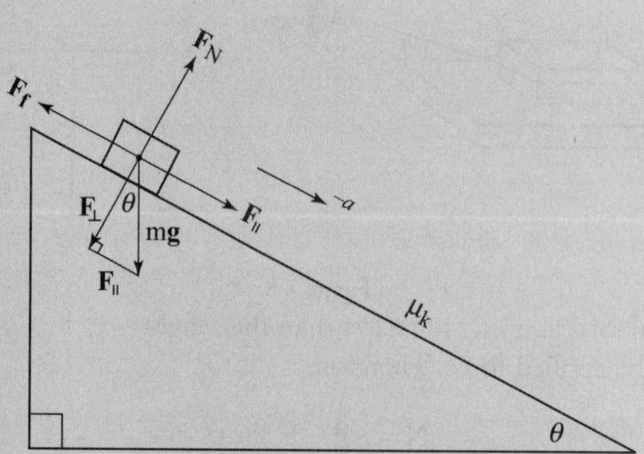

From the geometry, we see that $\mathbf{F}_\parallel = m\mathbf{g} \sin \theta$ and $\mathbf{F}_\perp = m\mathbf{g} \cos \theta = \mathbf{F}_N$.
We also know that $\mathbf{F}_f = \mu_k \mathbf{F}_N$. Thus, if we take "down the incline" as negative, then

$$\mathbf{F}_f - \mathbf{F}_\parallel = -m\mathbf{a}$$

Thus,

$$\mu_k m\mathbf{g} \cos \theta - m\mathbf{g} \sin \theta = -m\mathbf{a}$$

$$\mathbf{a} = \mathbf{g} \sin \theta - \mu_k \mathbf{g} \cos \theta = 4.8 \text{ m/s}^2$$

(b) Since the initial velocity is zero, then we have

$$\mathbf{v}_f^2 = -2\mathbf{ad} = -2\,(4.8 \text{ m/s}^2)(1.5 \text{ m})$$

and

$$\mathbf{v}_f = 3.79 \text{ m/s}$$

Chapter Summary

- Forces are pushes and pulls that can be represented by vectors.
- Inertia is the tendency of a mass to resist a force changing its state of motion in a given frame of reference.
- The normal force is a force acting perpendicular to, and away from, a surface.
- Friction is a force that opposes motion.
- Newton's three laws help us to understand dynamics, the actions of forces on masses:

  i. In the absence of an external net force, an object maintains constant velocity (law of inertia).
  ii. If a net force is acting on a mass, then the acceleration is directly proportional to the net force ($\mathbf{F}_{net} = m\mathbf{a}$).
  iii. For every action force, there is an equal but opposite reaction force.

- Weight is a force caused by the pull of Earth's gravity on a mass and is directed downward.
- An inertial frame of reference is a frame of reference moving at a constant velocity relative to Earth.
- Free-body diagrams assist in the analysis of forces by identifying and labeling all forces acting on a mass freed from the confines of the illustrated situation.
- In circular motion, the net force is directed inward toward the center and is referred to as the centripetal force. A centrifugal force is a ficticious or psuedo-force identified in an accelerated frame of reference attached to a mass undergoing circular motion.

Problem-Solving Strategies for Forces and Newton's Laws of Motion

The key to solving force problems, whether static or dynamic, is to construct the proper free-body diagram. Remember that Newton's laws operate on vectors, and so you must be sure to resolve all forces into components once a frame of reference and a coordinate system have been chosen. Therefore, you should:

1. Choose a coordinate system.
2. Make a sketch of the situation if one is not provided.
3. Construct a free-body diagram for the situation.
4. Resolve all forces into perpendicular components based on the chosen coordinate system.
5. Write Newton's second law as the sum of all forces in a given direction. If the problem involves a static situation, set the summation equal to zero. If the situation is dynamic, set the summation equal to $m\mathbf{a}$. Be sure to include only applied forces in the diagram.
6. Find the normal force. It is always perpendicular to the surface.

7. Remember that the centripetal force is always directed inward toward the circular path and parallel to the plane of the circle. Gravity is always directed vertically downward.
8. Carefully solve your algebraic equations, using the techniques for simultaneous equations.

Practice Exercises

MULTIPLE-CHOICE

CHALLENGE

1. In the situation shown below, what is the tension in string 1?

(A) 69.3 N
(B) 98 N
(C) 138.6 N
(D) 147.6 N
(E) 155 N

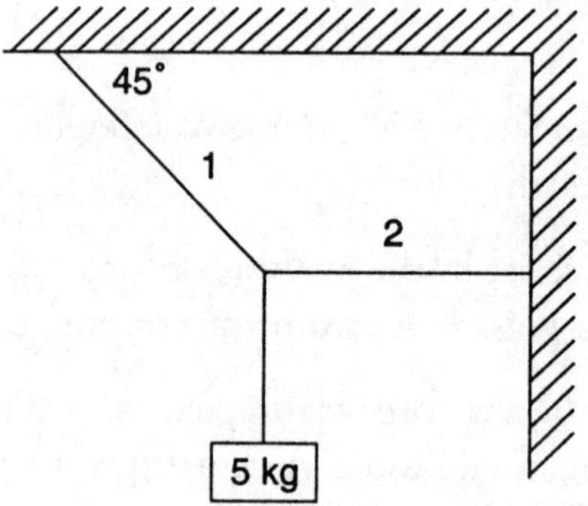

CHALLENGE

2. Two masses, M and m, are hung over a massless, frictionless pulley as shown below. If $M > m$, what is the downward acceleration of mass M?

(A) $\mathbf{g}$

(B) $\frac{(M-m)\mathbf{g}}{M+m}$

(C) $\left(\frac{M}{m}\right)\mathbf{g}$

(D) $\frac{Mm\mathbf{g}}{M+m}$

(E) $Mm\mathbf{g}$

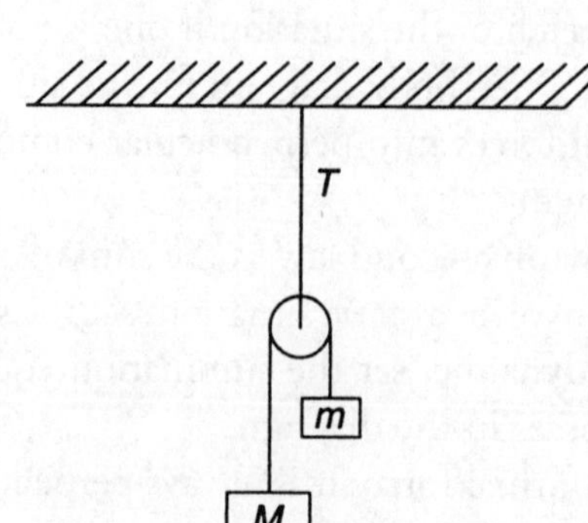

3. A 0.25-kg mass is attached to a string and swung in a vertical circle whose radius is 0.75 m. At the bottom of the circle, the mass is observed to have a speed of 10 m/s. What is the magnitude of the tension in the string at that point?

   (A) 2.45 N
   (B) 5.78 N
   (C) 22.6 N
   (D) 35.7 N
   (E) 44.7 N

4. A car and driver have a combined mass of 1000 kg. The car passes over the top of a hill that has a radius of curvature equal to 10 m. The speed of the car at that instant is 5 m/s. What is the force of the hill on the car as it passes over the top? **CHALLENGE**

   (A) 7300 N up
   (B) 7300 N down
   (C) 12,300 N up
   (D) 12,300 N down
   (E) 0 N

5. A hockey puck with a mass of 0.3 kg is sliding along ice that can be considered frictionless. The puck's velocity is 20 m/s. The puck now crosses over onto a floor that has a coefficient of kinetic friction equal to 0.35. How far will the puck travel across the floor before it stops?

   (A) 3 m
   (B) 87 m
   (C) 48 m
   (D) 92 m
   (E) 58 m

6. A spring with a stiffness constant $k = 50$ N/m has a natural length of 0.45 m. It is attached to the top of an incline that makes a 30° angle with the horizontal. The incline is 2.4 m long. A mass of 2 kg is attached to the spring, causing it to be stretched down the incline. How far down the incline does the end of the spring rest? **CHALLENGE**

   (A) 0.196 m
   (B) 0.45 m
   (C) 0.646 m
   (D) 0.835 m
   (E) 1.2 m

7. A 20-N force is pushing two blocks horizontally along a frictionless floor as shown below.

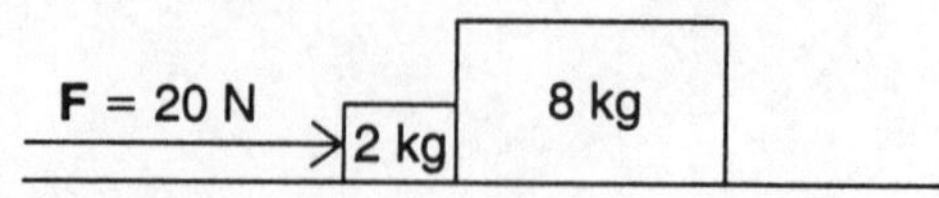

What is the force that the 8-kg mass exerts on the 2-kg mass?

(A) 4 N
(B) 8 N
(C) 16 N
(D) 20 N
(E) 24 N

CHALLENGE

8. A force of 20 N acts horizontally on a mass of 10 kg being pushed up a frictionless incline that makes a 30° angle with the horizontal, as shown below.

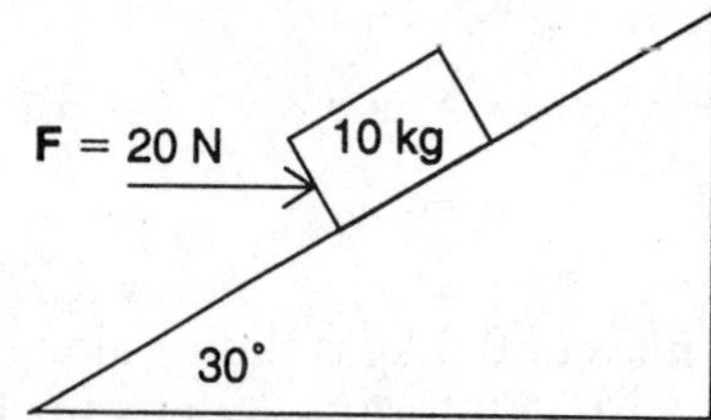

The magnitude of the acceleration of the mass up the incline is equal to

(A) 1.9 m/s^2
(B) 2.2 m/s^2
(C) 3.17 m/s^2
(D) 3.87 m/s^2
(E) 4.3 m/s^2

CHALLENGE

9. According to the diagram below, what is the tension in the connecting string if the table is frictionless?

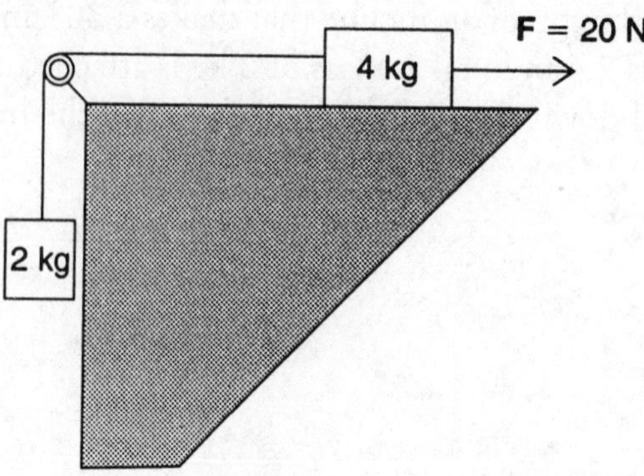

(A) 6.4 N
(B) 13 N
(C) 19.7 N
(D) 25 N
(E) 32 N

10. A mass, M, is released from rest on an incline that makes a 42° angle with the horizontal. In 3 s, the mass is observed to have gone a distance of 3 m. What is the coefficient of kinetic friction between the mass and the surface of the incline?

(A) 0.8
(B) 0.7
(C) 0.6
(D) 0.5
(E) 0.3

FREE-RESPONSE

1. A 3-kg block is placed on top of a 7-kg block as shown below.

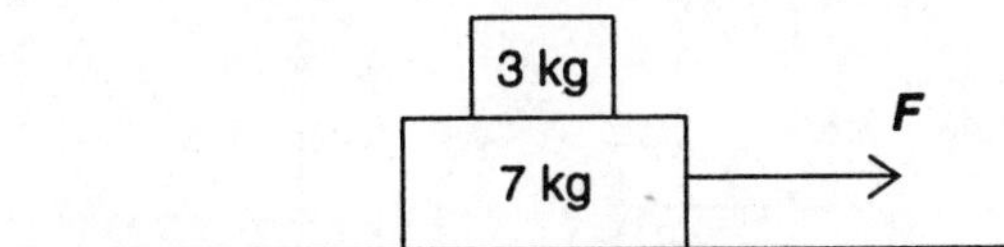

The coefficient of kinetic friction between the 7-kg block and the surface is 0.35. A horizontal force F acts on the 7-kg block.

(a) Draw a free-body diagram for each block.
(b) Calculate the magnitude of the applied force **F** necessary to maintain an acceleration of 5 m/s^2.
(c) Find the minimum coefficient of static friction necessary to prevent the 3-kg block from slipping.

2. A curved road is banked at an angle θ, such that friction is not necessary for a car to stay on the road. A 2500-kg car is traveling at a speed of 25 m/s, and the road has a radius of curvature equal to 40 m. **CHALLENGE**

(a) Draw a free-body diagram for the situation described above.
(b) Find angle θ.
(c) Calculate the magnitude of the force that the road exerts on the car.

3. The "rotor" is an amusement park ride that can be modeled as a rotating cylinder, with radius R. A person inside the rotor is held motionless against the sides of the ride as it rotates with a certain velocity. The coefficient of static friction between a person and the sides is μ. **CHALLENGE**

(a) Derive a formula for the period of rotation T, in terms of R, g, and μ.
(b) If $R = 5$ m and $\mu = 0.5$, calculate the value of the period T in seconds.
(c) Using the answer to part (b), calculate the angular velocity ω in radians per second.

4. How does the rotation of the Earth affect the apparent weight of a 1-kg mass at the equator?

5. If you overwax a floor, you can actually increase the coefficient of kinetic friction instead of lowering it. Explain how this might happen.

6. If forces occur in action-reaction pairs that are equal and opposite, how is it possible for any one force to cause an object to move?

ANSWERS EXPLAINED

MULTIPLE-CHOICE PROBLEMS

1. **(A)** We need to apply the second law for static equilibrium. This means that we must resolve the tensions into their *x*- and *y*-components. For $\mathbf{T}_1$, we have $\mathbf{T}_1 \cos 45$ and $\mathbf{T}_1 \sin 45$. In equilibrium, the sum of all the *x* forces must equal zero. This means that $\mathbf{T}_1 \cos 45 = \mathbf{T}_2$ (since $\mathbf{T}_2$ is entirely horizontal). The *y*-component of $\mathbf{T}_1$ must balance the weight $= m\mathbf{g} = 49$ N. Thus, $\mathbf{T}_1 \sin 45 = 49$ N and $\mathbf{T}_1 = 69.3$ N.

2. **(B)** The free-body diagrams for both masses, *M* and *m*, look like this:

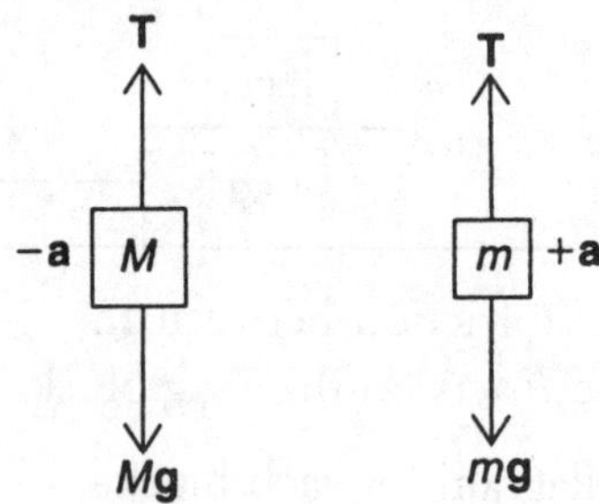

The large mass is accelerating downward, while the small mass is accelerating upward. The tension in the string is directed upward, while gravity, given by the weight, is directed downward. Using the second law for accelerated motion, we must show that $\Sigma\mathbf{F}_x = m\mathbf{a}$ and $\Sigma\mathbf{F}_y = m\mathbf{a}$ separately. Thus we have

$$\mathbf{T} - M\mathbf{g} = -M\mathbf{a} \quad \text{and} \quad \mathbf{T} - m\mathbf{g} = m\mathbf{a}$$

Eliminating the tension **T** and solving for **a** gives us $(M - m)\mathbf{g}/(M + m)$.

3. **(D)** A sketch of the situation is shown below:

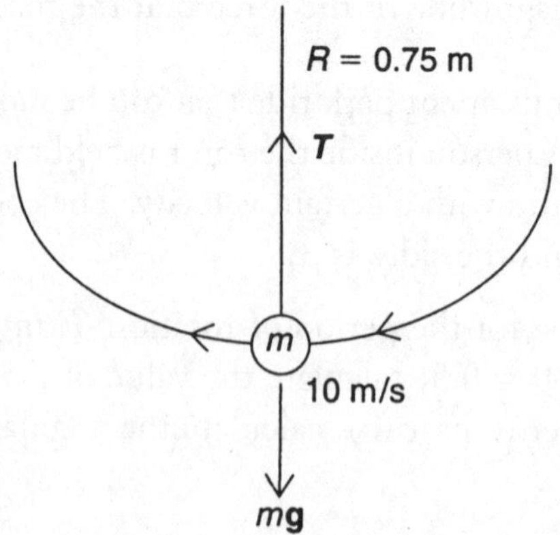

At the lowest point, the downward force of gravity is matched by the upward tension in the string. Thus we can write

$$\mathbf{T} - m\mathbf{g} = \frac{m\mathbf{v}^2}{R}$$

Substituting in the given values and solving for **T** gives us 35.7 N.

4. **(A)** As the car goes over the hill, the force that the hill exerts on the car is the normal force, **N**. This is opposed, however, by the downward force of gravity. The combination produces (as in question 3) the centripetal force $\mathbf{F}_c$. Thus, we can write

$$\mathbf{N} - m\mathbf{g} = \frac{-m\mathbf{v}^2}{R}$$

The centripetal force, in this case, is directed downward, and hence is written as negative. A sketch is shown below. Substituting in the given values and solving for **N** gives 7300 N upward.

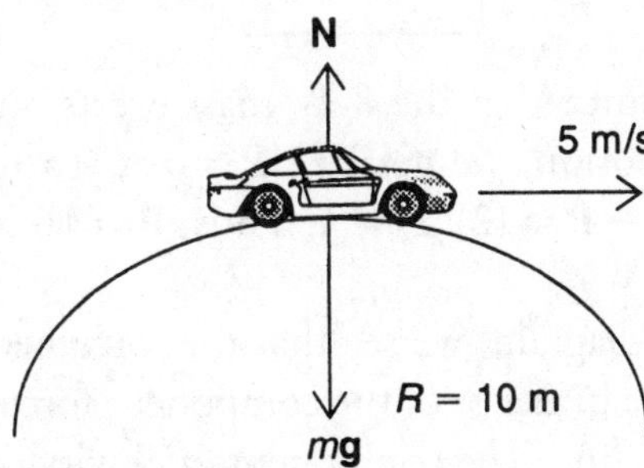

5. **(E)** With horizontal motion, the normal force is equal to the force of gravity. When the puck is moving with constant velocity, no net forces are acting on it. Thus, when friction acts to slow the puck down, the friction is the only net force. Therefore, we write

$$\mathbf{f} = \mu\mathbf{N} = \mu m\mathbf{g} = -m\mathbf{a}$$

(since the puck is decelerating). Substituting the known numbers gives a deceleration of $-3.43\ \text{m/s}^2$. Now, the final velocity will be zero; and since the time to stop is unknown, we use the following kinematic expression to solve for the stopping distance:

$$-\mathbf{v}_i^2 = 2\mathbf{ax}$$

Substituting the known numbers, we obtain approximately 58 m as the distance the puck will travel before stopping.

6. **(C)** A sketch of the situation is given below. The spring constant is k = 50 N/m.

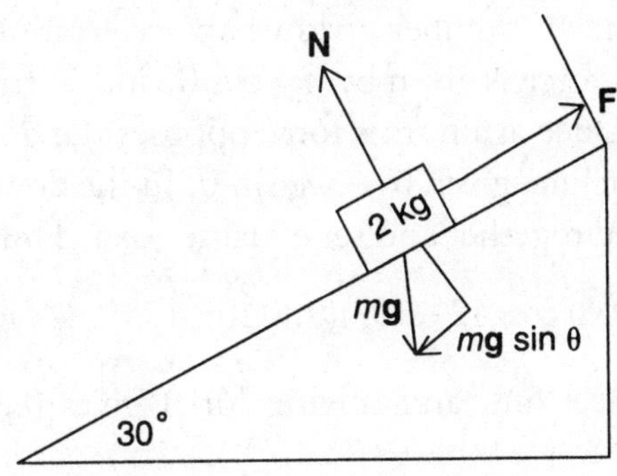

According to Hooke's law, $\mathbf{F} = k\mathbf{x}$, where $\mathbf{x}$ is the elongation in excess of the spring's natural length (here 0.45 m). The force, in this case, is provided by the component of weight parallel to the incline, which is given by $m\mathbf{g} \sin \theta$. Substituting the known numbers gives us an elongation of $\mathbf{x} = 0.196$ m. Since this must be in excess of the spring's natural length, the answer is 0.646 m.

7. **(C)** The 20-N force is pushing on a total mass of 10 kg. Thus, using $\mathbf{F} = m\mathbf{a}$, we have the acceleration of both blocks equal to 2 m/s^2. We draw a free-body diagram for the 2-kg mass as shown below.

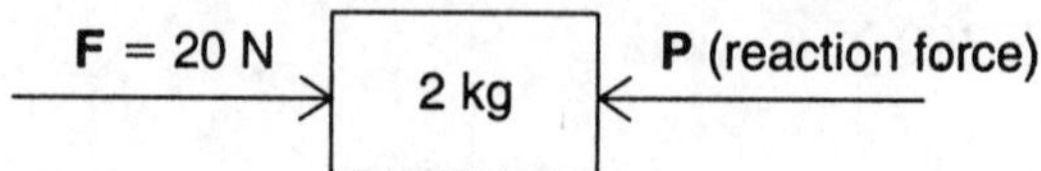

Let $\mathbf{P}$ represent the force that the 8-kg mass exerts on the 2-kg mass. Writing the second law of motion, we get $\mathbf{F} - \mathbf{P} = m\mathbf{a}$. To find $\mathbf{P}$, we substitute the known numbers; $20 - \mathbf{P} = (2)(2) = 4$. Thus, $\mathbf{P} = 16$ N.

8. **(C)** From the given diagram, we see that the force necessary to move the mass up the incline must be in excess of the component force of gravity trying to push the mass down the incline. The component of gravity down the incline is always given by $m\mathbf{g} \sin \theta$. Resolving the given force into a component parallel to the incline and a component perpendicular to the incline, we find that the force up the incline is, at the same angle, $\mathbf{F} \cos \theta$. Thus, in general we would write:

$$\mathbf{F} \cos \theta - m\mathbf{g} \sin \theta = m\mathbf{a}$$

Substituting the known numbers gives $\mathbf{a} = 3.17$ m/s^2.

9. **(C)** From the given diagram, we see that the 4-kg mass is accelerating to the right and the 2-kg mass is accelerating up. Thus we write that the sum of all forces, in each direction, equals $m\mathbf{a}$. In the x-direction we have (since the tension in the string will try to pull left), $-\mathbf{T} + 20 = 4\mathbf{a}$. In the y-direction, the tension pulls up against gravity. We therefore write $\mathbf{T} - 19.6 = 2\mathbf{a}$. Solving for $\mathbf{T}$ by eliminating $\mathbf{a}$, we get $\mathbf{T} = 19.7$ N.

10. **(A)** We know that the mass accelerates from rest uniformly in 3 s and goes a distance of 3 m. Thus, we can say that

$$d = \frac{1}{2}\mathbf{a}t^2$$

Substituting the known number gives us an acceleration of 0.67 m/s^2. On an incline, the normal force is given by $m\mathbf{g} \cos \theta$, and so friction can be expressed as $\mathbf{f} = \mu m\mathbf{g} \cos \theta$. Once again, this force opposes the downward force of gravity parallel to the incline, given by $-m\mathbf{g} \sin \theta$. In the downward direction, these two forces are added together and set equal to $-m\mathbf{a}$. Thus we write for this case

$$\mu(M)(9.8) \cos 42 - (M)(9.8) \sin 42 = -(M)(0.67)$$

The masses all cancel out, and solving for μ gives 0.8 as the coefficient of kinetic friction.

FREE-RESPONSE PROBLEMS

1. (a) A free-body diagram for each of the two blocks is given below:

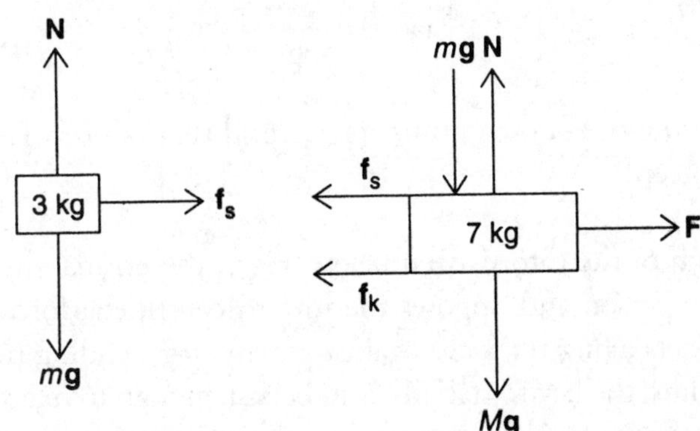

(b) The force of static friction between the two blocks is the force responsible for accelerating the 3-kg block to the right (hence the direction of friction in the free-body diagram). On the other hand, kinetic friction opposes the applied force, **F**, acting on the 7-kg block. Even so, the force **F** must accelerate both blocks combined. Thus, in the horizontal direction we can write that, for the second law of motion,

$$\mathbf{F} - \mathbf{f} = M\mathbf{a}$$

Also, since $\mathbf{f} = \mu\mathbf{N}$, we can write (since we have horizontal motion, the normal force is equal to the combined weights) $\mathbf{F} - (0.35)(10)(9.8) = (10)(5)$. Solving for **F**, we get $\mathbf{F} = 84.3$ N.

(c) Now, for the two blocks together, we know that static friction provides the force needed for the 3-kg block to accelerate at 5 m/s^2. The normal force on the "surface" is just the weight of the 3-kg block, as seen in the free-body diagram drawn in part (a). Thus $\mu(3)(9.8) = (3)(5)$. From this we get $\mu = 0.51$.

2. (a) A sketch of the situation and the free-body diagram are given below:

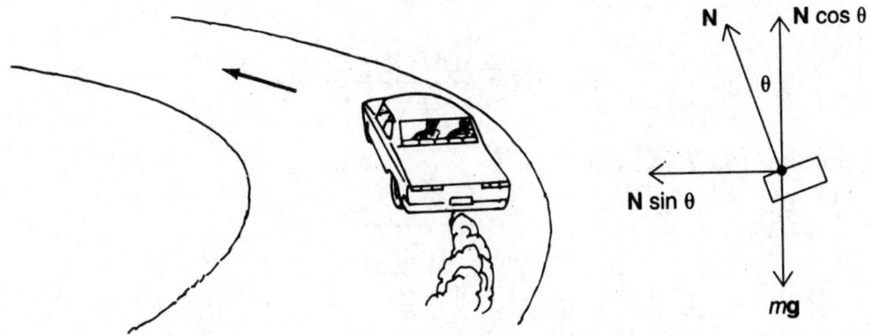

(b) In the coordinates chosen, the component of the normal force parallel to the plane of the curved road provides the centripetal acceleration. Thus we can write, in the absence of friction,

$$\mathbf{N} \sin \theta = \frac{m\mathbf{v}^2}{r}$$

We can also see that **N** cos θ = *mg*. Thus, eliminating **N** from both equations gives us

$$\tan\theta = \frac{\mathbf{v}^2}{r\mathbf{g}}$$

Substituting the known numbers, we find that tan θ = 1.59 and θ = 57.8° (rather steep).

3. (a) A diagram of the rotor is seen below. Here, the normal force is perpendicular to the person and opposes the inward centripetal force due to rotation. Friction acts along the walls against gravity, *mg*, tending to slide the person down. Thus, the key to stability is to be fast enough to maintain equilibrium.

   In the frame of reference of the person, there is an outward centrifugal force opposing the reaction force of the walls. Since **F** = *m***a**,

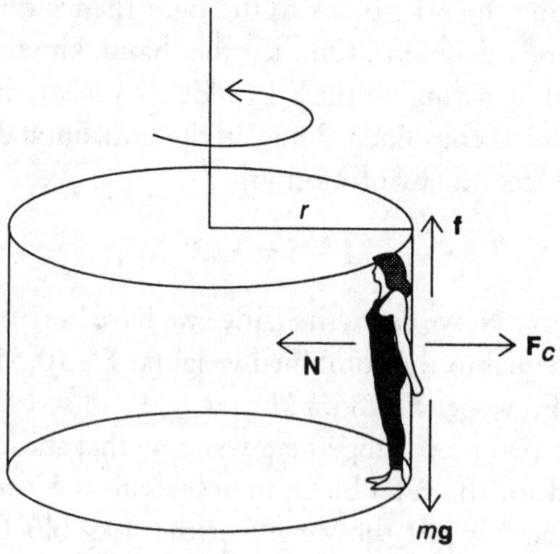

$$\mathbf{N} = \frac{4M\pi^2 r}{T^2}$$

In the vertical direction **f** = *M***g**; thus

$$\mathbf{f} = \mu\mathbf{N} = \frac{\mu M 4\pi^2 r}{T^2} = M\mathbf{g}$$

Solving for *T*, we get

$$T = \sqrt{\frac{\mu 4\pi^2 r}{\mathbf{g}}}$$

This makes sense, since if the coefficient of friction is high, the rotation rate can be small, and thus a larger period!

(b) Substituting the known numbers gives *T* = 10 s.

(c) $W = 2\frac{\pi}{T} = \frac{6.28}{10} = 0.628$ rad/s.

4. In the frame of reference of the mass, there is an apparent upward force that tends to reduce the apparent weight of the mass at the equator. This effect is very small, and only very sensitive scales can measure it.

5. Initially, the wax fills in the ridges and furrows of the floor on a microscopic level. This reduces the coefficient of kinetic friction and makes it easier for objects to slide. However, each successive wax buildup can actually overfill the ridges and then cause a sticky layer of material that increases the coefficient of kinetic friction.

6. This is a classic question and is very tricky. The answer is that the action and reaction pairs act on different objects. Thus, the applied force, if it is (or results in) a net force, can cause an object to move in a given frame of reference.

CHAPTER 8

Work and Energy

KEY CONCEPTS

- Work
- Applications of the Work Concept
- Power
- Energy
- Conservation of Energy
- Conservative and Nonconservative Forces

8.1 WORK

Imagine we have two masses, m_1 and m_2. Now suppose that a separate force acts on each mass, providing individual acceleration according to Newton's second law of motion. How can we compare the "effects" of these two forces, $\mathbf{F}_1$ and $\mathbf{F}_2$? One thing we can do is observe their ratios:

$$\frac{\mathbf{F}_1}{\mathbf{F}_2} = \frac{m_1\mathbf{a}_1}{m_2\mathbf{a}_2} \tag{1}$$

One of the arguments of the Newtonian approach would be to compare the two forces relative to the time, t_1 or t_2, each force acts. Let us assume that they act simultaneously. Then we can write

$$\frac{\mathbf{F}_1 t}{\mathbf{F}_2 t} = \frac{m_1\mathbf{a}_1 t}{m_2\mathbf{a}_2 t} = \frac{m_1\mathbf{v}_1}{m_2\mathbf{v}_2} \tag{2}$$

where we have used the expression $\mathbf{v} = \mathbf{a}t$, assuming the objects start from rest.

Another way to relate the two forces is to consider the relative displacement through which each force acts. Let us assume they act through the same displacement, $\mathbf{d}$. We can therefore write our ratio as

$$\frac{\mathbf{F}_1\mathbf{d}}{\mathbf{F}_2\mathbf{d}} = \frac{m_1\mathbf{a}_1\mathbf{d}}{m_2\mathbf{a}_2\mathbf{d}} = \frac{(1/2)m_1\mathbf{v}_1^2}{(1/2)m_2\mathbf{v}_2^2} \tag{3}$$

where, if we assume that the objects start from rest, we have used the expression $\mathbf{v}^2 = 2\mathbf{ad}$.

Comparing these two "representations" involves a comparison between vectors and scalars. The first comparison (equation 2) relates a *vector* quantity, **F***t*, expressed as the product of the force and the time (called the **impulse**). This quantity is proportional to the product of the mass and the velocity. However, in equation 3 the product of the magnitude force and distance, **Fd**, is a *scalar* quantity and is proportional to the product of the mass and the square of the velocity! This scalar quantity, **Fd**, is called the **work** done by the force.

Another way to think about this situation is to consider a graph of force versus displacement. Since an object's displacement is in the same direction as the net force applied, only the component of force acting in the direction of motion contributes to the work. Imagine that we have a constant force applied to an object. Figure 8.1 shows the graph of this relationship.

TIP

Work and energy are scalar quantities.

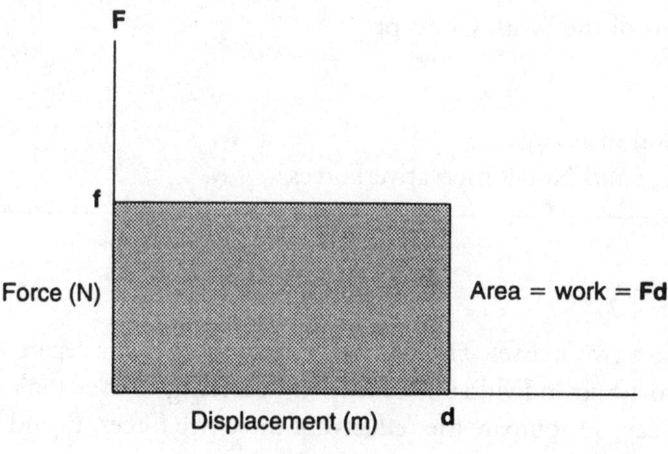

Figure 8.1

The area under the graph is equal to the work done. For variable forces, this method becomes most useful, and we can speak of the average force applied over the interval of distance displaced.

Based on equation 3, the units of work are defined to newton · meters (N · m), such that 1 newton · meter is equal to the work done by a 1-newton force displacing an object a distance of 1 meter. These units are also called **joules** (J). Using dimensional analysis, we see that, based on the definition of the newton in Chapter 7, 1 joule equals 1 kilogram · square meter per second squared. Figure 8.2 is a nice example of this idea.

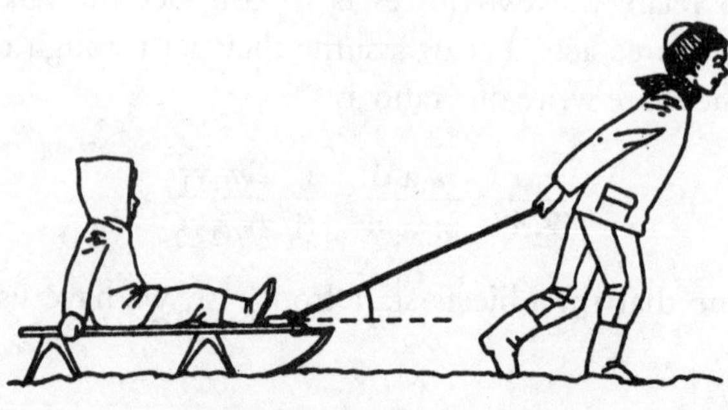

Figure 8.2

SAMPLE PROBLEM

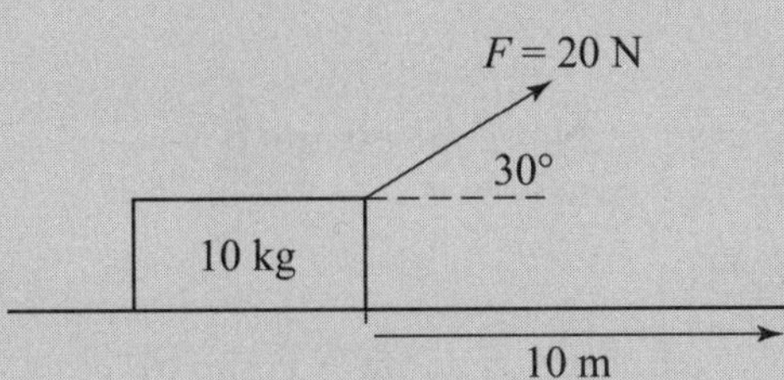

A 10-kg mass is pulled along a horizontal frictionless surface by a string making an angle of 30° as shown. If the applied force on the string is equal to 20 N, how much work is done as the mass travels a distance of 10 m?

Solution

To find the work done, we need the force to be in the same direction as the motion. This is the horizontal component of the applied force:

$$\mathbf{F}_x = \mathbf{F} \cos \theta = (20 \text{ N}) \cos (30^\circ) = 17.32 \text{ N}$$

Now,

$$W = \mathbf{F}_x\mathbf{d} = (17.32 \text{ N})(10 \text{ m}) = 173.2 \text{ J}$$

8.2 APPLICATIONS OF THE WORK CONCEPT

Let us try a more sophisticated problem. Suppose a mass of 10 kg is being pulled by a string making a 30° angle to the horizontal, as shown in Figure 8.3. The force in the string is 50 N, and the coefficient of friction between the mass and the ground, μ, is 0.2. How much work is done in displacing the mass a distance of 10 m?

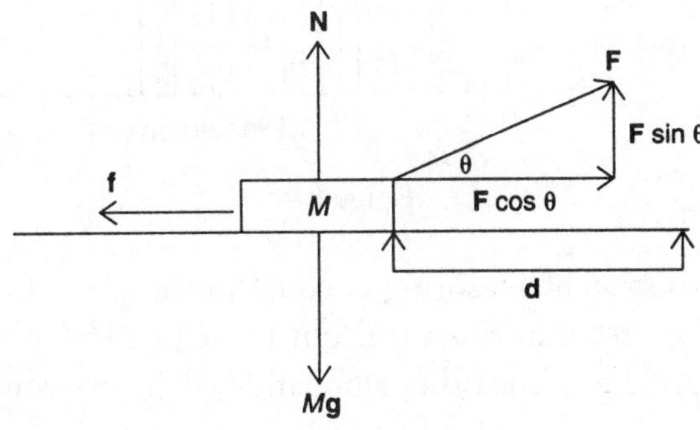

Figure 8.3

The horizontal component of the applied force is equal to **F** cos θ, and the vertical component is equal to **F** sin θ. The directions are based on the fact that the string pulls up and to the right. However, **F** cos θ is not the only horizontal force present; friction opposes the motion and is proportional to the normal force **N**. This normal force is not just the reaction force to the weight since the upward component of the string force contributes to a tendency to try and pull up the mass. Thus

$$\mathbf{N} = M\mathbf{g} - \mathbf{F} \sin \theta$$

and so the frictional force is

$$\mathbf{f} = \mu\mathbf{N} = \mu(M\mathbf{g} - \mathbf{F} \sin \theta)$$

The magnitude of the net horizontal force, which does the work, is therefore given by:

$$\mathbf{f}_{(net)} = \mathbf{F} \cos \theta - \mu(M\mathbf{g} - \mathbf{F} \sin \theta)$$

It is the product of this net force and the displacement, **d**, that evaluates the amount of work, in joules. Substituting the known numbers gives us $W = 287$ J.

As another example, consider the work done to stretch a spring. We know from Hooke's law that the elongation is directly proportional to the applied force, $\mathbf{F} = k\mathbf{x}$, where the stiffness constant, k, is in units of newtons per meter. This is not a constant force, and so the graph of force versus elongation would look as shown in Figure 8.4.

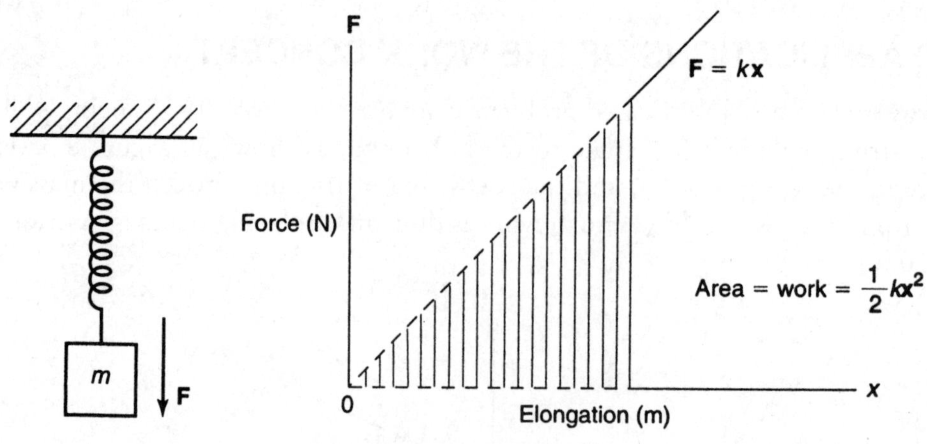

Figure 8.4

The work done to stretch the spring is equal to the area of the triangle formed. Using Hooke's law, we see that this turns out to be $(1/2)kx^2$ if we began from zero elongation. If we started at some other elongation, then we would write

$$W = \frac{1}{2}kx_f^2 - \frac{1}{2}kx_i^2$$

SAMPLE PROBLEM

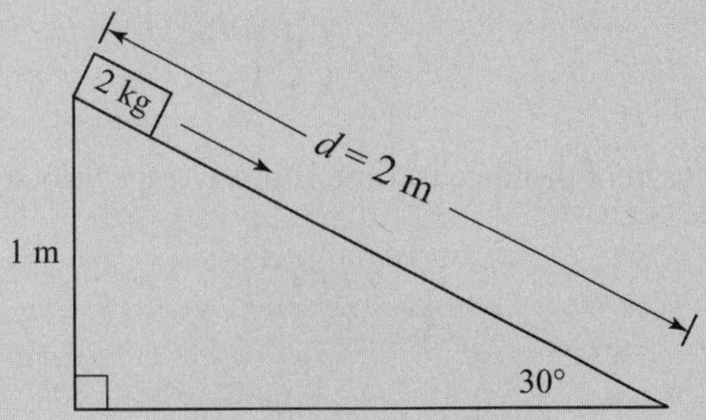

A 2-kg mass is sitting on top of a frictionless incline that is 2 m long and makes a 30° angle as shown. Compare the work done by gravity as the mass slides down the incline with the work done by gravity if the mass simply fell to the ground.

Solution

It is easy to see that if the angle of the incline is 30°, then the height of the mass is 1 m above the ground. If the mass falls to the ground, then the work done by gravity will be equal to

$$W = m\mathbf{g}h = (2\text{ kg})(9.8\text{ m/s}^2)(1\text{ m}) = 19.6\text{ J}$$

Now, if the mass slides down the incline, we know that the force down the incline is given by

$$\mathbf{F}_{\parallel} = m\mathbf{g}\sin\theta$$

Thus, the work done by gravity as the mass slides down a distance d is given by

$$W = m\mathbf{g}d\sin\theta = (2\text{ kg})(9.8\text{ m/s}^2)(2\text{ m})\sin(30°) = 19.6\text{ J}$$

We see that the work done (in the absence of friction) is the same in both cases. The interesting thing is that when the mass was sliding down the incline, the net force acting on it was less than the vertical force of gravity (its weight). However, the mass moved a greater distance such that the work done was the same. Thus, the net force down the incline was one-half of its weight, but it had to travel twice the distance. This ratio is referred to as the ideal mechanical advantage, and the incline is a simple machine, which reduces effort at the expense of distance.

8.3 POWER

It is often useful to consider the rate at which work is done. This rate, which would be expressed as the number of joules of work done per second, is called **power**. Algebraically we could write

$$\text{Power} = P = \frac{W}{t}$$

The units of power are joules per second, or **watts** (W). This relationship also implies that the work done is equal to the product of the power and the time (in seconds).

If a constant force is applied, we can relate power to the average velocity attained. Since $P = W/t$ and $W = \mathbf{Fd}$, we can write:

$$P = \mathbf{F}\left(\frac{\mathbf{d}}{t}\right)$$

Recalling that the ratio of distance to time is the average velocity, we can say also:

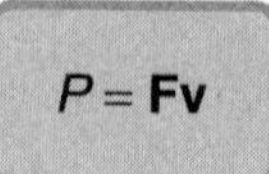

$$P = \mathbf{Fv}$$

8.4 ENERGY

While performing experiments on the dynamics of pendulum motion, Galileo discovered that, if he introduced a nail or a peg at the vertical position of a swinging pendulum, thereby interrupting the original length of the pendulum, the pendulum bob would still swing up to its original height. On the way back, the pendulum would swing away from the nail or peg, resume its normal length, and still achieve the same height. Galileo could not interpret this fact in terms of forces, but we can now attribute the observations to the law of conservation of mechanical energy. To understand this concept better, let's return to our introduction to work.

In Section 8.1, we learned that a comparison between forces and displacement leads to the fact that work is proportional to mass times the square of velocity. In fact, if we apply a constant force to a mass and write $\mathbf{F} = m\mathbf{a}$, we see something interesting:

$$\mathbf{Fd} = m\mathbf{a}\mathbf{d} = \frac{1}{2}m(\mathbf{v}_f^2 - \mathbf{v}_i^2) = \frac{1}{2}m\mathbf{v}_f^2 - \frac{1}{2}m\mathbf{v}_i^2$$

The quantity $(1/2)m\mathbf{v}^2$ is called the **kinetic energy** (KE) of the object and is defined to be the energy the object has by virtue of its being in motion relative to a frame of reference. This is crucial, since we can put ourselves into an inertial frame of reference in which the relative velocity observed is zero and the kinetic energy will be zero!

Work is therefore a measure of the change in kinetic energy; and since work implies the action of a force, we have established that forces are the agents of energy transfers. Thus we can define energy as the capacity to do work and define work as a measure of energy transferred. This relationship is known as the work-energy theorem. The units for kinetic energy are of course joules.

Another concept of energy is related to the position of an object. A mass sitting on a table has the ability to fall to the floor if allowed (the table exerts an upward force that prevents that from occurring). From our definition of energy, we can state that some type of positional energy exists within the mass. Experimentally, we know that, if we lift the mass slightly, it will land with a small velocity. However, if we lift the object very high, it will land with a large velocity. The displacement of the mass upward implies that work has been done to the object. The ability to fall is due to some kind of positional energy, which then converts itself into the kinetic energy of motion. In mechanics, this positional energy is known as **potential energy** (PE, or U in some textbooks). The units for potential energy are also joules.

If we recall the work-energy theorem, the gain in potential energy, by lifting, must be equal to the amount of work done (which we have previously shown to be equal to $m\mathbf{g}h$). Therefore, there is a connection between potential energy and kinetic energy when we have vertical displacements in a gravitational field. However, just as in the case of kinetic energy, the frame of reference for potential energy is arbitrary, and we can define the zero base level to be anywhere. What is important is only how much energy is changing (remember that work and energy are scalars!). In summary, then, we can say that

$$W = \Delta \text{KE} \quad \text{or} \quad W = \Delta \text{PE}$$

A roller coaster illustrates the conservation of energy, as seen in Figure 8.5.

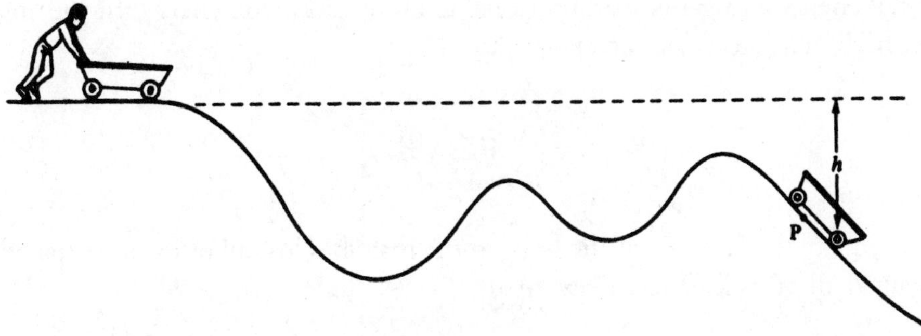

Figure 8.5

8.5 CONSERVATION OF ENERGY

Suppose (see Figure 8.6) we lift a mass m a distance h above the ground. The minimum amount of force needed to overcome gravity is equal to the weight of the mass, $m\mathbf{g}$. The work done is therefore equal to $m\mathbf{g}h$. This quantity is also accompanied by a gain in potential energy since, if we release the mass, it will fall and gain velocity (i.e., it will gain kinetic energy). If the mass is allowed to fall from height h, what will its velocity be at the bottom?

Kinematically, we know that, if a mass is dropped from a height h, the time to fall is given by

$$t = \sqrt{\frac{2h}{\mathbf{g}}}$$

Given the fact that the initial velocity is zero, the velocity at the bottom, just before impact, will be

$$\mathbf{v} = -\mathbf{g}t \quad \text{or} \quad \mathbf{v} = -\mathbf{g}\sqrt{\frac{2h}{\mathbf{g}}} = -\sqrt{2\mathbf{g}h}$$

(where the negative sign indicates the downward direction).

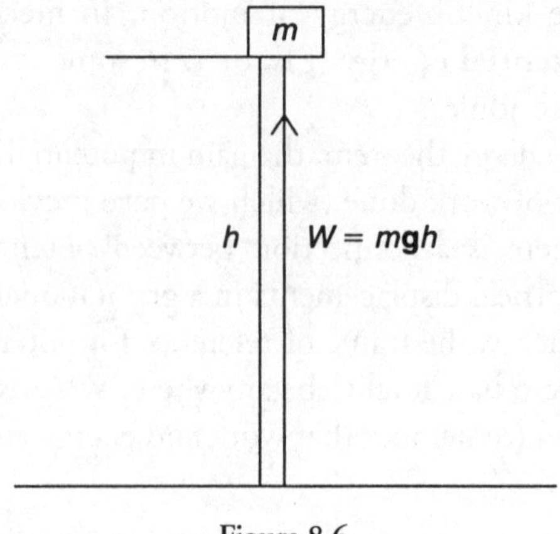

Figure 8.6

Now, since the work done to raise the mass is equal to the change (gain) in potential energy, the work done by gravity to accelerate the mass downward when released must derive from that potential energy. Since the work done through motion is equal to the change in kinetic energy, we can observe the fact that, as the mass falls, it loses potential energy and gains a proportional amount of kinetic energy. In the absence of any frictional forces, we can state that:

$$\Delta \text{PE} = \Delta \text{KE}$$

REMEMBER

In the absence of any frictional forces, the total mechanical energy before the motion is equal to the total mechanical energy after the motion.

If we equate the fact that at the bottom the mass has lost all of its potential energy and gained all of its kinetic energy, then:

$$mgh = \frac{1}{2} m\mathbf{v}^2$$

This implies, taking into account the direction of motion, that $\mathbf{v} = -\sqrt{2\mathbf{g}h}$ (as before). This discussion leads to the concept of the law of conservation of energy, which, briefly, states that evergy is never created or destroyed. In the absence of friction (which generates heat loss), the gain or loss of potential energy is balanced by a loss or gain of kinetic energy.

This conservation of energy concept applies to all mechanical systems, especially those involving springs and masses. Recall that the work done to compress (or stretch) a spring is given by $W = (1/2)kx^2$. This is also a measure of the potential energy stored in the spring. If a mass were moving on a frictionless table, at a velocity $\mathbf{v}$, toward an uncompressed spring and were to collide with it until the spring was fully compressed (and promptly absorbed all of the kinetic energy, causing the mass to stop), we could calculate the amount of compression as follows. If we assume that no energy is lost to heat (work by friction), the total energy before the interaction is contained in the kinetic energy of the mass. After the interaction, however, the energy is stored in the compressed spring as potential energy. We can therefore write that

$$\frac{1}{2} m\mathbf{v}^2 = \frac{1}{2} kx^2$$

Solving for the amount, x, of compression gives

$$x = \mathbf{v}\sqrt{\frac{m}{k}}$$

SAMPLE PROBLEM

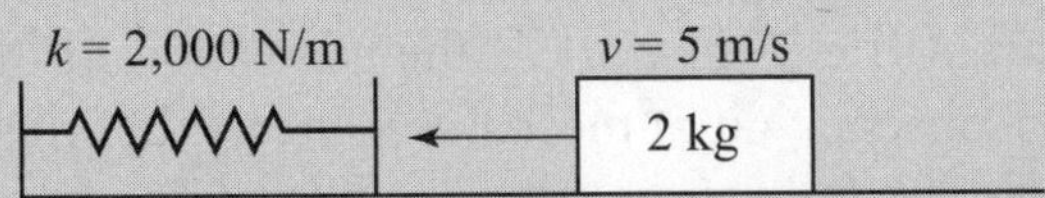

A 2-kg mass is sliding along a horizontal frictionless surface with a velocity of 5 m/s. It collides with and compresses an elastic massless spring (k = 2,000 N/m). How far will the spring have compressed when the mass comes to rest?

Solution

The mechanical energy is conserved so that the kinetic energy lost by the mass is gained by the spring as potential energy. Thus, we can write

$$\frac{1}{2}k\mathbf{x}^2 = \frac{1}{2}m\mathbf{v}^2$$

A quick substitution of the given values ($\mathbf{v}_f$ = 0 m/s) gives us that x = 0.158 m.

8.6 CONSERVATIVE AND NONCONSERVATIVE FORCES

If the work done by a force is independent of the path taken, the force is called a **conservative force.** This term implies that the mechanical energy of the system is conserved; that is, the sum of the potential and kinetic energies of the system remains constant in time. Gravity is an example of a conservative force, and so work done to or by a gravitational force (such as lifting or falling) is independent of the path taken.

As an example, consider a mass sliding down a frictionless incline of length ℓ. The magnitude of the component of gravity responsible for the acceleration down the incline is $-m\mathbf{g}\sin\theta$. Therefore, the work done by gravity is given by

$$W = -(m\mathbf{g}\ \sin\ \theta)\ell$$

As we saw in Chapter 7, in a right triangle the length ℓ is related to the height of the incline h so

$$h = \ell\ \sin\ \theta$$

TIP

The work done either by friction or to overcome friction is often converted into heat.

Therefore, the work done down the incline can be written as $W = -m\mathbf{g}h$, which is just the loss of potential energy. The work done down the incline is exactly the same as the work done if the mass fell directly down from height h. An interesting aspect of this fact is that the accelerating force down the incline is less than the gravitational force (its weight), but the displacement is larger in magnitude.

Friction is an example of a **nonconservative force.** Since friction is derived from the contact between two surfaces, any change in path affects the length of time the contact takes place and therefore increases the effect of friction. The work done by friction can be expressed as the product of the frictional force and the displacement:

$$W_f = \mathbf{f}d$$

However, an implication of frictional work is the generation of heat, which reveals itself as an apparent nonconservation of mechanical energy. In other words, the work done by any nonconservative force, in which the change in total energy, ΔE, is not equal to zero, is given by:

$$W_{nc} = E_f - E_i$$

Chapter Summary

- Work and energy are scalar quantities.
- Work is the means by which energy is transferred, and energy is the ability to do work.
- Work is calculated by the formula $W = \mathbf{F}d$ as long as the force is in the direction of motion.
- Power is the rate at which work is done (and likewise, the rate at which energy is transferred or expended).
- Potential energy is stored energy related to the position of an object in a frame of reference.
- Kinetic energy is the energy due to the motion of an object in a given frame of reference.
- Mechanical energy refers to the sum of potential and kinetic energy.
- In the absence of friction, the total mechanical energy of a system is conserved.
- In general, all energy in the universe is conserved (energy is neither created nor destroyed).

Problem-Solving Strategies for Work and Energy

1. Identify the types of energies involved in the situation.
2. Try to determine the initial energy of the system, which is usually the total energy.
3. If no friction is present, write the equations for the conservation of energy. Remember that, if a spring is involved, it has a different potential energy from gravity.

4. Remember that the work done is equal to the change in energy of the system.
5. For work problems, resolve forces into components in the direction of motion. This process involves vector analysis techniques. Identify the frame of reference involved.
6. Remember that work and energy are scalar quantities expressed in units of joules.
7. Keep in mind that, for any dynamics problem, especially one in which energy or work is not expressly stated as a concept, energy considerations may be useful in the solution.

Practice Exercises

MULTIPLE-CHOICE

1. Which of the following are the units for the spring constant, k?

   (A) kg · m^2/s^2
   (B) kg · s^2
   (C) kg · m/s
   (D) kg/s^2
   (E) kg · m^2/s

2. Which of the following is an expression for mechanical power?

   (A) $\mathbf{F}t/m$
   (B) $\mathbf{F}^2m/\mathbf{a}$
   (C) $\mathbf{F}m^2/t$
   (D) $\mathbf{F}m/t$
   (E) $\mathbf{F}^2t/m$

3. A pendulum consisting of a mass m attached to a light string of length ℓ is displaced from its rest position, making an angle θ with the vertical. It is then released and allowed to swing freely. Which of the following expressions represents the velocity of the mass when it reaches its lowest position? CHALLENGE

   (A) $\sqrt{2\mathbf{g}\ell(1-\cos\theta)}$

   (B) $\sqrt{2\mathbf{g}\ell(\tan\theta)}$

   (C) $\sqrt{2\mathbf{g}\ell(\cos\theta)}$

   (D) $\sqrt{2\mathbf{g}\ell(1-\sin\theta)}$

   (E) $\sqrt{2\mathbf{g}\ell(\cos\theta-1)}$

4. An engine maintains constant power on a conveyor belt machine. If the belt's velocity is doubled, the magnitude of its average acceleration

(A) is doubled
(B) is quartered
(C) is halved
(D) is quadrupled
(E) remains the same

CHALLENGE

5. A mass m is moving horizontally along a nearly frictionless floor with velocity $\mathbf{v}$. The mass now encounters a part of the floor that has a coefficient of kinetic friction given by μ. The total distance traveled by the mass before it is slowed by friction to a stop is given by

(A) $2\mathbf{v}^2/\mu\mathbf{g}$
(B) $\mathbf{v}^2/2\mu\mathbf{g}$
(C) $2\mu\mathbf{g}v^2$
(D) $\mu\mathbf{v}^2/2\mathbf{g}$
(E) $\mu\mathbf{vg}$

6. Two unequal masses are dropped simultaneously from the same height. The two masses will experience the same change in

(A) acceleration
(B) kinetic energy
(C) potential energy
(D) velocity
(E) momentum

7. A pendulum that consists of a 2-kg mass swings to a maximum vertical displacement of 17 cm above its rest position. At its lowest point, the kinetic energy of the mass is equal to

(A) 0.33 J
(B) 3.33 J
(C) 33.3 J
(D) 333 J
(E) 3333 J

CHALLENGE

8. A 0.3-kg mass rests on top of a spring that has been compressed by 0.04 m. Neglect any frictional effects, and consider the spring to be massless. Then, if the spring has a constant k equal to 2000 N/m, to what height will the mass rise when the system is released?

(A) 1.24 m
(B) 0.75 m
(C) 0.54 m
(D) 1.04 m
(E) 1.34 m

9. A box is pulled along a smooth floor by a force F, making an angle θ with the horizontal. As θ increases, the amount of work done to pull the box the same distance, d,

   (A) increases
   (B) increases and then decreases
   (C) remains the same
   (D) decreases and then increases
   (E) decreases

10. As the time needed to run up a flight of stairs decreases, the amount of work done against gravity

   (A) increases
   (B) decreases
   (C) remains the same
   (D) increases and then decreases
   (E) decreases and then increases

FREE-RESPONSE

1. A 0.75-kg sphere is dropped through a tall columm of liquid. When the sphere has fallen a distance of 2.0 m, it is observed to have a velocity of 2.5 m/s.

   (a) How much work was done by the frictional "viscosity" of the liquid?
   (b) What is the average force of friction during the placement of 2.0 m?

2. A 15-kg mass is attached to a massless spring by a light string that passes over a frictionless pulley as shown below. The spring has a force constant of 500 N/m and is unstretched when the mass is released. What is the velocity of the mass when it has fallen a distance of 0.3 m?

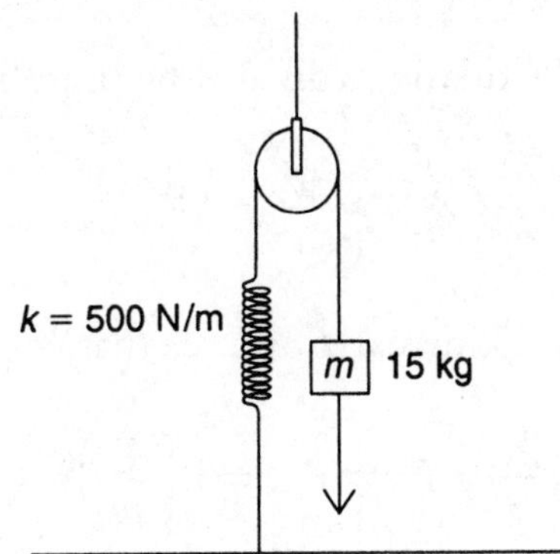

CHALLENGE

3. A 1.5-kg block is placed on an incline. The mass is connected to a massless spring by means of a light string passed over a frictionless pulley, as shown below. The spring has a force constant k equal to 100 N/m. The block is released from rest, and the spring is initially unstretched. The block moves down a distance of 16 cm before coming to rest. What is the coefficient of kinetic friction between the block and the surface of the incline?

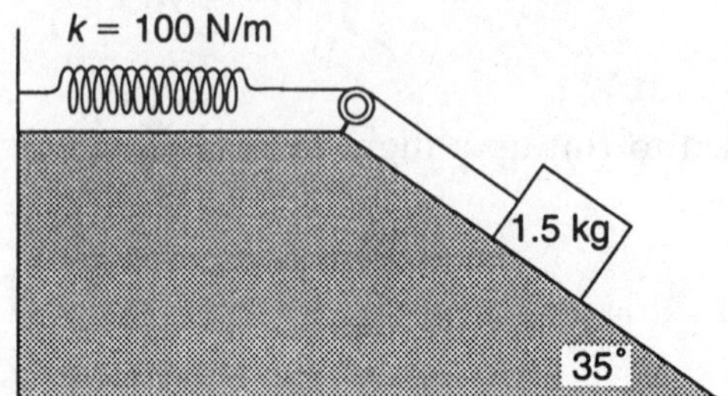

4. Explain how it might be possible for a moving object to possess, and simultaneously not possess, kinetic energy?

5. When you hold up a 10-kg mass with your arms outstretched, you get tired. However, according to physics, you have not done any work! Explain how this is possible.

ANSWERS EXPLAINED

MULTIPLE-CHOICE PROBLEMS

1. **(D)** The units for the spring or force constant are provided by Hooke's law, $\mathbf{F} = k\mathbf{x}$, and are newtons per meter. Recall that 1 N = 1 kg · m/s^2, so dividing by m gives kg/s^2.

2. **(E)** Power is equal to work done divided by time. Therefore,

$$P = \frac{W}{t} = \frac{\mathbf{Fd}}{t}$$

Using some algebra and kinematics, we see that

$$P = \frac{\mathbf{Fd}}{t} = \mathbf{Fv} = \mathbf{F}\mathbf{a}t = \mathbf{F}\left(\frac{\mathbf{F}}{m}\right)t = \frac{\mathbf{F}^2 t}{m}$$

You could also get the answer by verifying which expression has units of joules per second.

3. **(A)** Consider the sketch below of the situation:

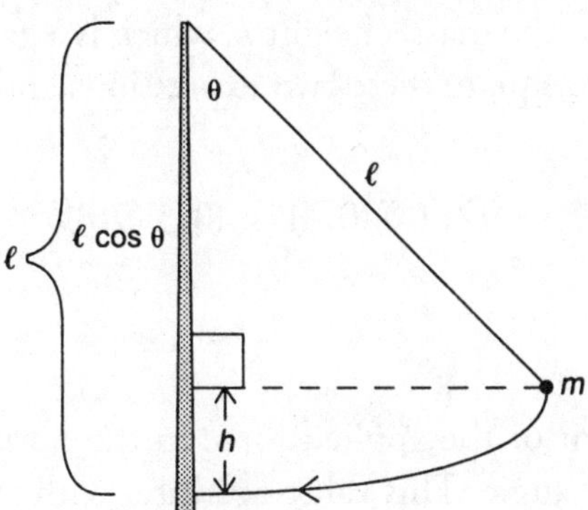

From the geometry of the sketch, note that

$$h = \ell - \ell \cos \theta = \ell(1 - \cos \theta)$$

If it is assumed that there is no friction, gravity is the only conservative force acting to do work. Therefore, ΔKE = ΔPE, and so, at the bottom,

$$\mathbf{v} = \sqrt{2\mathbf{g}h} = \sqrt{2\mathbf{g}\ell(1 - \cos \theta)}$$

4. **(C)** Power is equal to the product of the average force applied times the velocity. If the velocity is doubled, and the power is constant, the average force must be halved. Since $\mathbf{F} = m\mathbf{a}$, the average acceleration of the belt must be halved as well.

5. **(B)** The only applied force is friction, which is doing work to stop the mass. This work is being taken from the initial kinetic energy. For friction, we know that $\mathbf{f} = \mu\mathbf{N}$, and since the motion is horizontal, $\mathbf{N} = m\mathbf{g}$. Let x be the distance traveled while stopping; therefore, we can write that W_f = KE and:

$$\frac{1}{2}m\mathbf{v}^2 = \mu m\mathbf{g}x$$

Solving for x, we get:

$$x = \frac{\mathbf{v}^2}{2\mu\mathbf{g}}$$

6. **(D)** Objects dropped simultaneously from the same height have the same constant acceleration, which is the change in velocity. The unequal masses will provide for different energies, but the velocity changes will be the same.

7. **(B)** Since we have a conservative system,

$$\Delta\text{KE} = \Delta\text{PE} = (2)(9.8)(0.17) = 3.33 \text{ J}$$

Remember to change 17 cm to 0.17 m!

8. **(C)** We are dealing with a conservative system, so the initial starting energy is just the potential energy of the compressed spring. This energy supplies the work needed to raise the mass a height h, which is a gain in gravitational potential energy. Thus we equate these two expressions and solve for the height:

$$\frac{1}{2}(2000)(0.4)^2 = (0.3)(9.8)h$$

Thus $h = 0.54$ m.

9. **(E)** The component of the applied force in the horizontal direction depends on the cosine of the angle. This value decreases with increasing angle. Thus the work decreases as well.

10. **(C)** The work done is independent of the time or path taken since gravity is a conservative force. The power generated is affected by time, but the work done to run up the stairs remains the same as long as the same mass is raised to the same height.

FREE-RESPONSE PROBLEMS

1. (a) The change in potential energy is a measure of the initial energy and equals $(0.75)(9.8)(2) = 14.7$ J. After the sphere has fallen 2 m, its velocity is 5 m/s, so the kinetic energy is given by

$$\text{KE} = \left(\frac{1}{2}\right)(0.75)(5)^2 = 9.375 \text{ J}$$

The work done by friction is due to a nonconservative force that is equal to the difference between the final and initial energies:

$$E_f - E_i = 9.375 \text{ J} - 14.7 \text{ J} = -5.325 \text{ J}$$

Therefore, $W_f = -5.325$ J.

(b) The average frictional force is equal to the work done divided by the displacement of 2 m. Thus

$$\mathbf{f} = -2.67 \text{ N}$$

which of course is negative since it opposes the motion.

2. The loss of potential energy is balanced by a gain in elastic potential energy for the spring and in kinetic energy for the falling mass if we assume that the starting energy for the system is zero relative to the starting point for the mass. In the absence of friction, the displacement of the mass is equal to the elongation of the spring. Thus we can equate our energies and write:

$$0 = -m\mathbf{g}h + \frac{1}{2}kx^2 + \frac{1}{2}m\mathbf{v}^2$$

If we substitute the known numbers, we get

$$0 = -(15)(9.8)(0.3) + \left(\frac{1}{2}\right)(500)(0.3)^2 + \left(\frac{1}{2}\right)(15)(\mathbf{v}^2)$$

Solving for velocity **v** gives **v** = 1.7 m/s.

3. In this problem, the work done by gravity down the incline is affected by the work done by friction. Together, the net work is applied to stretching the spring by an amount equal to the displacement of the mass. Thus we can say

$$W_g - W_f = W_s$$

where W_g is the work done by gravity, W_f is the work done by friction, and W_s is the work done to the spring. Hence:

$$m\mathbf{g} \sin \theta d - \mu m\mathbf{g} \cos \theta d = \frac{1}{2} kx^2$$

Substituting the known numbers, we get:

$$(1.5)(9.8)(0.16) \sin 35 - \mu(1.5)(9.8)(0.16) \cos 35 = \frac{1}{2}(100)(0.16)^2$$

Solving for the coefficient of friction gives $\mu = 0.036$.

4. The motion of an object is always relative to a given frame of reference. If an object is moving relative to one frame, we can envision a second frame, moving with the object, in which it appears to be at rest. Thus, in one frame, the object possesses kinetic energy because of its relative motion. In the second frame, with the object appearing to be at rest, it does not possess kinetic energy.

5. When you hold up a mass, your arm muscles strain against the force of gravity. This requires energy from your body, which makes you feel tired.

CHAPTER 9

Impacts and Linear Momentum

KEY CONCEPTS

- Internal and External Forces
- Impact Forces and Momentum Changes
- The Law of Conservation of Linear Momentum
- Elastic and Inelastic Collisions

9.1 INTERNAL AND EXTERNAL FORCES

Consider a system of two blocks with masses m and M ($M > m$). If the blocks were to collide, the forces of impact would be equal and opposite. However, because of the different masses, the response to these forces (i.e., the changes in velocity) would not be equal. In the absence of any outside or external forces acting on the objects (such as friction or gravity), we say that the impact forces are internal.

To Newton, the "quantity of motion" discussed in his book *Principia* was the product of an object's mass and velocity. This quantity, called **linear momentum** or just **momentum**, is a vector quantity having units of kilogram · meters per second (kg · m/s). Algebraically, momentum is designated by the letter **p**, such that $\mathbf{p} = m\mathbf{v}$.

Momentum and **impulse** are both vector quantities.

To understand Newton's rationale, consider the action of trying to change the motion of a moving object. Do not confuse this with the inertia of the object; here, both the mass and the velocity are important. Consider, for example, that a truck moving at a slow 1 meter per second can still inflict a large amount of damage because of its mass. Also, a small bullet, having a mass of perhaps 1 gram or less, does incredible damage because of its high velocity. In each case (see Figure 9.1) the damage is the result of a force of impact when the object is intercepted by something else. Let's now consider the nature of impact forces.

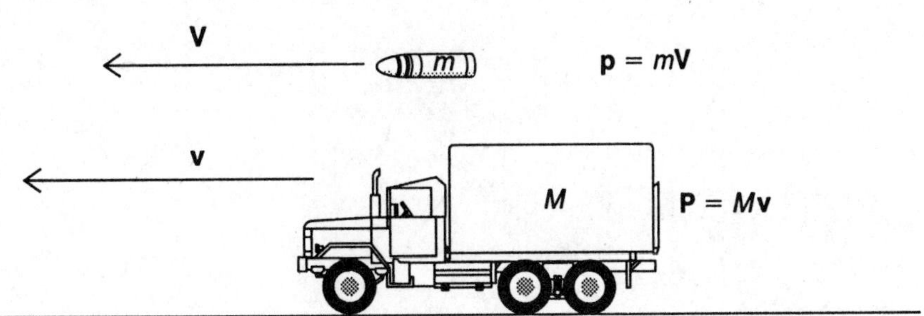

Figure 9.1 Comparison of the Momentum of a Truck with That of a Bullet

9.2 IMPACT FORCES AND MOMENTUM CHANGES

Consider a mass m moving with a velocity $\mathbf{v}$ in some frame of reference. If the mass is subjected to some external forces, then, by Newton's second law of motion, we can write that $\Sigma\mathbf{F} = m\mathbf{a}$. The vector sum of all the forces, referred to as the **net force**, is responsible for changing the velocity of the motion (in magnitude and/or direction).

If we recall the definition of acceleration as the rate of change or velocity, we can rewrite the second law of motion as

$$\mathbf{F}_{net} = m\left(\frac{\Delta\mathbf{v}}{\Delta t}\right)$$

This expression is also a vector equation and is equivalent to the second law of motion. If we make the assumption that the mass of the object is not changing, we can again rewrite the second law in the form

$$\mathbf{F}_{net} = \frac{\Delta m\mathbf{v}}{\Delta t} = \frac{\Delta\mathbf{p}}{\Delta t}$$

This expression means that the net external force acting on an object is equal to the rate of change of momentum of the object and is another alternative form of Newton's second law of motion. This **change in momentum** is a vector quantity in the same direction as the net force applied. Since the time interval is just a scalar quantity, we can multiply both sides by Δt to get

$$\mathbf{F}_{net}\ \Delta t = \Delta\mathbf{p} = m\ \Delta\mathbf{v} = m\mathbf{v}_f - m\mathbf{v}_i$$

The quantity, $\mathbf{F}_{net}\ \Delta t$, called the **impulse**, represents the effect of a force acting on a mass during a time interval Δt, and is likewise a vector quantity. From this expression, it can be stated that the impulse applied to an object is equal to the change in momentum of the object.

Another way to consider impulse is to look at a graph of force versus time for a continuously varying force (see Figure 9.2).

The area under this curve is a measure of the impulse in units of newton · seconds (N · s). Another way to view this concept is to identify the average force, $\overline{\mathbf{F}}$, such that the area of the rectangle formed by the average force is equal in area to the entire curve. This is more manageable algebraically, and we can write

$$\overline{\mathbf{F}}\ \Delta t = \Delta\mathbf{p}$$

SAMPLE PROBLEM

A 2,000-kg car is traveling at 20 m/s and stops moving over a 10-s period. What was the magnitude of the average braking force?

Solution

We know that

$$\mathbf{F} = \frac{m\Delta \mathbf{v}}{\Delta t}$$

Substituting, we obtain

$$\mathbf{F} = \frac{(2000 \text{ kg})(20 \text{ m/s})}{10 \text{ s}} = 4{,}000 \text{ N}$$

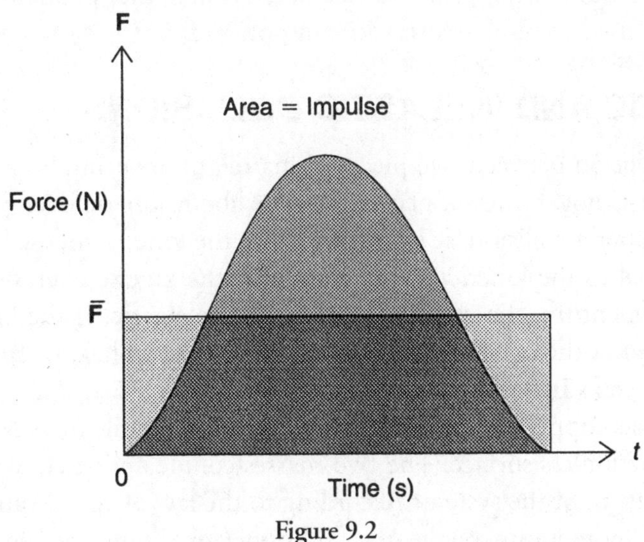

Figure 9.2

9.3 THE LAW OF CONSERVATION OF LINEAR MOMENTUM

Newton's third law of motion states that for every action there is an equal but opposite reaction. This reaction force is present whenever we have an interaction between two objects in the universe. Suppose we have two masses, m_1 and m_2, that are approaching each other along a horizontal frictionless surface. Let $\mathbf{F}_{12}$ be the force that m_1 exerts on m_2, and let $\mathbf{F}_{21}$ be the force that m_2 exerts on m_1. According to Newton's law, these forces must be equal and opposite; that is, $\mathbf{F}_{12} = -\mathbf{F}_{21}$.

Rewriting this expression as $\mathbf{F}_{12} + \mathbf{F}_{21} = 0$ leads to an interesting implication. Since each force is a measure of the rate of change of momentum for that object, we can write

$$\mathbf{F}_{12} = \frac{\Delta \mathbf{p}_1}{\Delta t} \quad \text{and} \quad \mathbf{F}_{21} = \frac{\Delta \mathbf{p}_2}{\Delta t}$$

Therefore:

$$\frac{\Delta \mathbf{p}_1}{\Delta t} + \frac{\Delta \mathbf{p}_2}{\Delta t} = 0 \quad \text{and} \quad \frac{\Delta(\mathbf{p}_1 + \mathbf{p}_2)}{\Delta t} = 0$$

The change in the sum of the momenta is therefore zero, implying that the sum of the total momentum for the system ($\mathbf{p}_1 + \mathbf{p}_2$) is a constant all the time. This conclusion is called the **law of conservation of linear momentum**, and we say simply that the momentum is conserved.

Remember

In an isolated system, the total momentum before an interaction is equal to the total momentum after the interaction.

Here is another way of writing this conservation statement in a general form for any two masses (after separating all initial and final terms):

$$m_1\mathbf{v}_{1i} + m_2\mathbf{v}_{2i} = m_1\mathbf{v}_{1f} + m_2\mathbf{v}_{2f}$$

Extension of the law of conservation of momentum to two or three dimensions involves the recognition that momentum is a vector quantity. Given two masses moving in a plane relative to a coordinate system, conservation of momentum must hold simultaneously in both the horizontal and vertical directions. These vector components of momentum can be calculated using the standard techniques of vector analysis used to resolve forces in components.

9.4 ELASTIC AND INELASTIC COLLISIONS

During any collision between two pieces of matter, momentum is always conserved. This statement is not, however, necessarily true about kinetic energy. If two masses stick together after a collision, it is observed that the kinetic energy before the collision is not equal to the kinetic energy after it. If the kinetic energy is conserved as well as the momentum, the collision is described as **elastic**. If the kinetic energy is not conserved after the collision (e.g., energy being lost to heat or friction), the collision is described as **inelastic**.

TIP

In an elastic collision, the kinetic energy of the system is conserved.

As an example, suppose that a mass m has a velocity $\mathbf{v}$ while mass M is at rest along a horizontal frictionless surface. The two masses collide and stick together. What is the final velocity, $\mathbf{u}$, of the system? According to the law of conservation of momentum, the total momentum before the collision ($m\mathbf{v}$), must be equal to the total momentum after. Since the two masses are combining, the new mass of the system is $M + m$, and the new momentum is given by $(m + M)\mathbf{u}$. Thus, we find that:

$$\mathbf{u} = \frac{m\mathbf{v}}{m + M}$$

The initial kinetic energy is $(1/2)m\mathbf{v}^2$. The final kinetic energy is given by

$$\text{KE}_f = \frac{1}{2}(m + M)\mathbf{u}^2 = \frac{1}{2}\left(\frac{m^2\mathbf{v}^2}{m + M}\right)$$

which is of course not equal to the initial kinetic energy.

If we have an elastic collision, we write both conservation laws to get two equations involving the velocities of the masses before and after the collision:

$$m_1\mathbf{v}_{1i} + m_2\mathbf{v}_{2i} = m_1\mathbf{v}_{1f} + m_2\mathbf{v}_{2f}$$

$$\frac{1}{2}m_1\mathbf{v}_{1i}^2 + \frac{1}{2}m_1\mathbf{v}_{2i}^2 = \frac{1}{2}m_1\mathbf{v}_{1f}^2 + \frac{1}{2}m_2\mathbf{v}_{2f}^2$$

If we cancel out the factor (1/2) in the second equation, and collect the expressions for each mass on each side, we can rewrite the two equations as

$$m_1(\mathbf{v}_{1i} - \mathbf{v}_{1f}) = m_2(\mathbf{v}_{2f} - \mathbf{v}_{2i})$$

$$m_1(\mathbf{v}_{1i}^2 - \mathbf{v}_{1f}^2) = m_2(\mathbf{v}_{2f}^2 - \mathbf{v}_{2i}^2)$$

The second equation is factorable and so can be simplified. Again rewriting the expressions, we get

$$m_1(\mathbf{v}_{1i} - \mathbf{v}_{1f}) = m_2(\mathbf{v}_{2f} - \mathbf{v}_{2i})$$

$$m_1(\mathbf{v}_{1i} + \mathbf{v}_{1f})(\mathbf{v}_{1i} - \mathbf{v}_{1f}) = m_2(\mathbf{v}_{2f} + \mathbf{v}_{2i})(\mathbf{v}_{2f} - \mathbf{v}_{2i})$$

If we take the ratio of the two expressions, we arrive at an interesting result after collecting terms:

$$\mathbf{v}_{1i} - \mathbf{v}_{2i} = -(\mathbf{v}_{1f} - \mathbf{v}_{2f})$$

This expression states that the relative velocity between the masses before the elastic collision is equal and opposite to the relative velocity between the masses after the elastic collision! Thus, there exists the characteristic "rebounding" observed during elastic collisions.

SAMPLE PROBLEM

A mass of 2 kg is moving at a speed of 10 m/s along a horizontal, frictionless surface. It collides, and sticks, with a 3-kg mass moving in the same direction at 5 m/s.

(a) What is the final velocity of the system after the collision?
(b) What percentage of kinetic energy was lost in the collision?

Solution

(a) Using conservation of momentum, we see that

$$m_1\mathbf{v}_1 + m_2\mathbf{v}_2 = (m_1 + m_2)\mathbf{v}_f$$

$$(2\text{ kg})(10\text{ m/s}) + (3\text{ kg})(5\text{ m/s}) = (5\text{ kg})\mathbf{v}_f$$

$$35\text{ kg}\cdot\text{m/s} = (5\text{ kg})\mathbf{v}_f$$

$$\mathbf{v}_f = 7\text{ m/s}$$

(b) The initial kinetic energy is given by

$$\frac{1}{2}m_1\mathbf{v}_1^2 + \frac{1}{2}m_2\mathbf{v}_2^2 = KE_i$$

$$\frac{1}{2}(2\text{kg})(10\text{ m/s})^2 + \frac{1}{2}(3\text{ kg})(5\text{ m/s})^2 = 137.5\text{ J}$$

The final kinetic energy is given by

$$\frac{1}{2}(m_1 + m_2)\mathbf{v}_f^2 = KE_f$$

$$\frac{1}{2}(5\text{ kg})(7\text{ m/s})^2 = 122.5\text{ J}$$

We now take the ratio of the final kinetic energy and the initial kinetic energy:

$$(122.5\text{ J}) / (137.5\text{ J}) = 0.89 \rightarrow 89\%\text{ left over KE}$$

Therefore, 11% of the initial kinetic energy was lost in the collision.

Chapter Summary

- Momentum is a vector quantity equal to $m\mathbf{v}$.
- The impulse ($\mathbf{F}\Delta t$) is found by taking the area under a graph of force versus time.
- Impulse is equal to the change in momentum ($\mathbf{F}\Delta t = m\Delta\mathbf{v}$).
- Force is equal to the rate of change of momentum.
- In an isolated system, the total momentum is conserved.
- In an elastic collision, the kinetic energy of the system is conserved.
- In an inelastic collision (typically, but not necessarily) where masses stick together, kinetic energy is lost (this energy loss typically transforms into heat).

Problem-Solving Strategies for Impacts and Linear Momentum

In any matter interaction, momentum is always conserved. Therefore, when you read a problem that does not state explicitly that momentum is involved, you can safely assume, in the case of collision or an impact, that momentum is conserved, and you should write the equations for the conservation of momentum. The kinetic energy, however, is not necessarily conserved unless the collision is elastic. In summary, you should:

1. Decide whether an impact or a collision is involved. If there is a collision, observe whether it is elastic or inelastic.
2. Remember that, if the collision is inelastic, masses will usually stick together, so be sure to determine the new combined mass.
3. If the collision is elastic, write the equations for the conservation of both momentum and kinetic energy.
4. Remember that the impulse given to a mass is equal to its change in momentum. The change in momentum is a vector quantity in the same direction as the impulse or net force.
5. Be sure to take into account algebraically any reversal of directions, and remember the sign conventions for left, right, up, and down motions.
6. Keep in mind that motions may be simpler if studied in the center-of-mass frame of reference. In such a frame, one considers the motion of the center of mass, as well as the motions of mass particles relative to the center of mass.
7. Remember that in two dimensions the center of mass will follow a smooth path after an internal explosion since the forces involved were internal and the initial momentum in that frame was zero. For example, for a projectile launched at an angle and then exploded the center of mass still follows the regular parabolic trajectory.

Practice Exercises

MULTIPLE-CHOICE

1. Which of the following expressions, where **p** represents the linear momentum of the particle, is equivalent to the kinetic energy of a moving particle?

   (A) $m\mathbf{p}^2$
   (B) $m^2/2\mathbf{p}$
   (C) $2\mathbf{p}/m$
   (D) $\mathbf{p}/2m$
   (E) $\mathbf{p}^2/2m$

2. Two carts having masses 1.5 kg and 0.7 kg, respectively, are initially at rest and are held together by a compressed massless spring. When released, the 1.5-kg cart moves to the left with a velocity of 7 m/s. What is the velocity and direction of the 0.7-kg cart?

   (A) 15 m/s right
   (B) 15 m/s left
   (C) 7 m/s left
   (D) 7 m/s right
   (E) 0 m/s

3. The product of an object's instantaneous momentum and its acceleration is equal to its

   (A) applied force
   (B) kinetic energy
   (C) power output
   (D) net force
   (E) displacement

4. A ball with a mass of 0.15 kg has a velocity of 5 m/s. It strikes a wall perpendicularly and bounces off straight back with a velocity of 3 m/s. The ball underwent a change in momentum equal to

   (A) 0.30 kg · m/s
   (B) 1.20 kg · m/s
   (C) 0.15 kg · m/s
   (D) 5 kg · m/s
   (E) 7.5 kg · m/s

5. What braking force is supplied to a 3000-kg car traveling with a velocity of 35 m/s that is stopped in 12 s?

   (A) 29,400 N
   (B) 3000 N
   (C) 8750 N
   (D) 105,000 N
   (E) 150 N

6. A 0.1-kg baseball is thrown with a velocity of 35 m/s. The batter hits it straight back with a velocity of 60 m/s. What is the magnitude of the average impulse exerted on the ball by the bat?

   (A) 3.5 N · s
   (B) 2.5 N · s
   (C) 7.5 N · s
   (D) 9.5 N · s
   (E) 12.2 N · s

7. A 1-kg object is moving with a velocity of 6 m/s to the right. It collides and sticks to a 2-kg object moving with a velocity of 3 m/s in the same direction. How much kinetic energy was lost in the collision?

   (A) 1.5 J
   (B) 2 J
   (C) 2.5 J
   (D) 3 J
   (E) 0 J

8. A 2-kg mass moving with a velocity of 7 m/s collides elastically with a 4-kg mass moving in the opposite direction at 4 m/s. The 2-kg mass reverses direction after the collision and has a new velocity of 3 m/s. What is the new velocity of the 4-kg mass?

   (A) −1 m/s
   (B) 1 m/s
   (C) 6 m/s
   (D) 4 m/s
   (E) 5 m/s

CHALLENGE

9. A mass m is attached to a massless spring with a force constant k. The mass rests on a horizontal frictionless surface. The system is compressed a distance x from the spring's initial position and then released. The momentum of the mass when the spring passes its equilibrium position is given by

   (A) $x\sqrt{mk}$

   (B) $x\sqrt{k/m}$

   (C) $x\sqrt{m/k}$

   (D) $x\sqrt{k^2m}$

   (E) xmk

10. During an inelastic collision between two balls, which of the following statements is correct?

    (A) Both momentum and kinetic energy are conserved.
    (B) Momentum is conserved, but kinetic energy is not conserved.
    (C) Momentum is not conserved, but kinetic energy is conserved.
    (D) Neither momentum nor kinetic energy is conserved.
    (E) Momentum is sometimes conserved, but kinetic energy is always conserved.

FREE-RESPONSE

1. Two blocks with masses 1 kg and 4 kg, respectively, are moving on a horizontal frictionless surface. The 1-kg block has a velocity of 12 m/s, and the 4-kg block is ahead of it, moving at 4 m/s, as shown in the diagram below. The 4-kg block has a massless spring attached to the end facing the 1-kg block. The spring has a force constant k equal to 1000 N/m. CHALLENGE

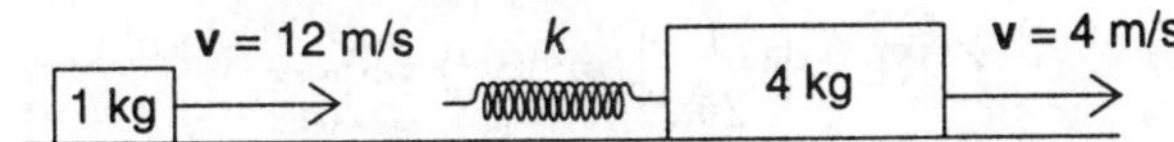

   (a) What is the maximum compression of the spring after the collision?

   (b) What are the final velocities of the blocks after the collision has taken place?

2. A 0.4-kg disk is initially at rest on a frictionless horizontal surface. It is hit by a 0.1-kg disk moving horizontally with a velocity of 4 m/s. After the collision, the 0.1-kg disk has a velocity of 2 m/s at an angle of 43° to the positive x-axis.

   (a) Determine the velocity and direction of the 0.4-kg disk after the collision.

   (b) Determine the amount of kinetic energy lost in the collision.

3. A 50-kg girl stands on a platform with wheels on a frictionless horizontal surface as shown below. The platform has a total mass of 1000 kg and is attached to a massless spring with a force constant k equal to 1000 N/m. The girl throws a 1-kg ball with an initial velocity of 35 m/s at an angle of 30° to the horizontal. CHALLENGE

   (a) What is the recoil velocity of the platform-girl system?

   (b) What is the elongation of the spring?

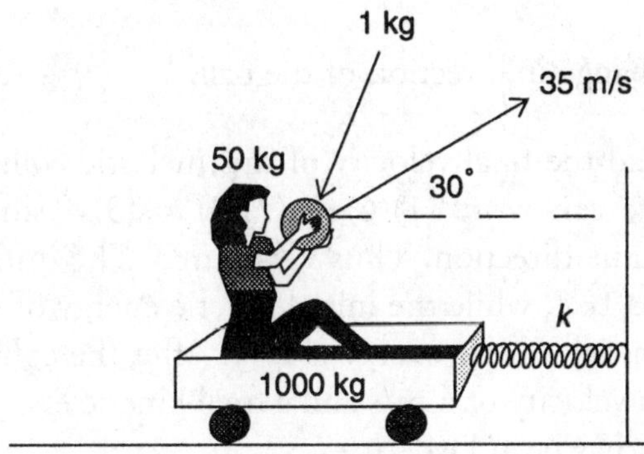

4. (a) Can an object have energy without having momentum? Explain.

   (b) Can an object have momentum without having energy? Explain.

5. Explain why there is more danger when you fall and bounce as opposed to falling without bouncing.

6. A cart of mass M is moving with a constant velocity $\mathbf{v}$ to the right. A mass m is dropped vertically onto it, and it is observed that the new velocity is less than the original velocity. Explain what has happened in terms of energy, forces, and conservation of momentum (as viewed from different frames of reference).

ANSWERS EXPLAINED

MULTIPLE-CHOICE PROBLEMS

1. **(E)** If we multiply the formula for kinetic energy by the ratio m/m, we see that the formula for kinetic energy becomes

$$\text{KE} = \left(\frac{1}{2m}\right)(m^2)(\mathbf{v}^2) = \frac{\mathbf{p}^2}{2m}$$

2. **(A)** Momentum is conserved, so $(1.5)(7) = 0.7\mathbf{v}$. Thus $\mathbf{v} = 15$ m/s. The direction is to the right since in a recoil the masses go in opposite directions.

3. **(C)** An object's instantaneous momentum times its acceleration will equal its power output in units of joules per second or watts.

4. **(B)** The change in momentum is a vector quantity. The rebound velocity is in the opposite direction, so $\Delta\mathbf{v} = 5 - (-3) = 8$ m/s. The change in momentum is

$$\Delta\mathbf{p} = (0.15)(8) = 1.20 \text{ kg} \cdot \text{m/s}$$

5. **(C)** The formula is $\mathbf{F}\,\Delta t = m\,\Delta\mathbf{v}$. Solving for the force, we get

$$\mathbf{F} = \frac{(3000)(35)}{12} = 8750 \text{ N}$$

6. **(D)** Impulse is equal to the change in momentum, which is

$$\Delta\mathbf{p} = |\ 0.1\,[-60 - 35]\ | = 9.5 \text{ N} \cdot \text{s}$$

because of the change in direction of the ball.

7. **(D)** First, we find the final velocity of this inelastic collision. Momentum is conserved, so we can write $(1)(6) + (2)(3) = (3)\mathbf{v}'$ since both objects are moving in the same direction. Thus $\mathbf{v}' = 4$ m/s. The initial kinetic energy of the 1-kg object is 18 J, while the initial kinetic energy of the 2-kg mass is 9 J. Thus the total initial kinetic energy is 27 J. After the collision, the combined 3-kg object has a velocity of 4 m/s and a final kinetic energy of 24 J. Thus, 3 J of kinetic energy has been lost.

8. **(B)** Momentum is conserved in this elastic collision but the directions are opposite, so we must be careful with negative signs. We therefore write $(2)(7) - (4)(4) = -(2)(3) + 4\mathbf{v}'$ and get $\mathbf{v}' = 1$ m/s.

9. **(A)** We set the two energy equations equal to solve for the velocity at the equilibrium position. Thus

$$\frac{1}{2}kx^2 = \frac{1}{2}m\mathbf{v}^2$$

since no gravitational potential energy is involved, and we write that the velocity is $\mathbf{v} = x\sqrt{k/m}$. Now momentum $\mathbf{p} = m\mathbf{v}$, so we multiply by m and factor the "mass" back under the radical sign, where it is squared so that we get $\mathbf{p} = x\sqrt{mk}$.

10. **(B)** Momentum is always conserved in an inelastic collision, and kinetic energy is not conserved because the objects stick together.

FREE-RESPONSE PROBLEMS

1. (a) Upon impact, the spring becomes compressed, but both blocks are still in motion. Therefore, for an instant, we have an inelastic collision, and momentum is of course conserved. Thus we can write for the moment of impact

$$m_1\mathbf{v}_{1i} + m_2\mathbf{v}_{2i} = (m_1 + m_2)\mathbf{v}_f$$

Solving for the final velocity, we get $\mathbf{v}_f = 53.6$ m/s. Using the initial values for the velocities, we find that the initial kinetic energies are 72 J and 32 J for the 1-kg and 4-kg blocks, respectively. Thus the total initial kinetic energy is 104 J. Using the final velocity of 5.6 m/s and the combined mass of 5 kg, we get a final kinetic energy of 78.4 J. The difference in kinetic energy of 25.6 J is used to compress the spring in this inelastic collision. Using the formula for the work done against a spring, $(1/2)kx^2$, we get $x = 0.05$ m for the maximum compression of the spring after the collision.

(b) After the collision has taken place, the two blocks again separate. If we treat the situation as elastic, since we assume that the work done to compress the spring will be used by the spring in rebounding, we can use the initial velocities as a "before" condition for momentum and kinetic energy, and seek to solve for the two final velocities of the blocks after the rebound has taken place. To find both velocities, we need two equations, and we use the conservation of kinetic energy in this case to assist us. Thus we write

$$m_1\mathbf{v}_{1i} + m_2\mathbf{v}_{2i} = m_1\mathbf{v}_{1f} + m_2\mathbf{v}_{2f}$$

and for the kinetic energy

$$\frac{1}{2}m_1\mathbf{v}_{1i}^2 + \frac{1}{2}m_2\mathbf{v}_{2i}^2 = \frac{1}{2}m_1\mathbf{v}_{1f}^2 + \frac{1}{2}m_2\mathbf{v}_{2f}^2$$

Substituting the known numbers, we obtain for the momentum $28 = \mathbf{v}_{1f} + 4\mathbf{v}_{2f}$, and for the kinetic energy we get $208 = \mathbf{v}_{1f}^2 + 4\mathbf{v}_{2f}^2$.

If we solve for $\mathbf{v}_{1f}$ from the momentum equation, square the result, and substitute into the kinetic energy equation, we obtain a factorable quadratic equation for $\mathbf{v}_{2f}$:

$$\mathbf{v}_{2f}^2 - 11.2\mathbf{v}_{2f} + 28.2 = (\mathbf{v}_{2f} - 7.2)(\mathbf{v}_{2f} - 4) = 0$$

Of the two choices for the second final velocity, only one provides a physically meaningful set of solutions. This is so because the first mass must rebound, and hence its final velocity must be negative. Our final answers are therefore $\mathbf{v}_{2f} = 7.2$ m/s and $\mathbf{v}_{1f} = -0.8$ m/s.

2. (a) This is a two-dimensional collision, and we consider the conservation of momentum in each direction, before and after. In the *x*-direction, we have only the initial momentum of the 0.1-kg disk with a velocity of 4 m/s. After the collision, the 0.1-kg disk has an *x*-component of momentum given by (0.1)(2)cos 43.

The 0.4-kg disk had initially zero momentum and now has some unknown velocity at some unknown angle, θ, to the *x*-axis. Let us assume that the angle is below the *x*-axis, so that the *x*-component will be positive and the *y*-component negative. If our assumption is correct, we will get a positive answer for θ. A negative answer will let us know that the assumption is incorrect. The *x*-component of final momentum for the 0.4-kg disk is given by $(0.4)\mathbf{v}_{2f}\cos\theta$, and the *y*-component by $-(0.4)\mathbf{v}_{2f}\sin\theta$. We can therefore write for the *x*-direction:

$$(0.1)(4) + 0 = (0.1)(2)\cos 43 + (0.4)\mathbf{v}_{2f}\cos\theta$$

and for the *y*-direction:

$$0 = (0.1)(2)\sin 43 - (0.4)\mathbf{v}_{2f}\sin\theta$$

Solving for the angles and velocities in each we get the following two equations:

$$\mathbf{v}_{2f}\sin\theta = 0.34 \quad \text{and} \quad \mathbf{v}_{2f}\cos\theta = 0.635$$

Taking the ratio gives $\tan\theta = 0.535$ and $\theta = 28°$. Substituting this angle gives us the final velocity for the 0.4-kg disk: 0.72 m/s.

(b) To find the loss of kinetic energy, we calculate the total initial and final kinetic energies. Using the given data, we find that the initial kinetic energy is 0.8 J. Using our final data, we get a total final kinetic energy of 0.3 J. Thus 0.5 J of kinetic energy has been lost.

3. (a) In this problem, the x-component of the velocity provides the impulse to elongate the spring using recoil. Thus we have that the velocity of the ball is 35 cos 30 = 30.31 m/s. Using conservation of momentum, we state that (1)(30.31) = (1050)$\mathbf{v}'$. Thus $\mathbf{v}'$ = 0.0289 m/s in the negative x-direction, so the recoil velocity of the platform-girl system can be expressed as $\mathbf{v}' = -0.0289$ m/s.

$$\left(\frac{1}{2}\right)(1050)(0.0289)^2 = \left(\frac{1}{2}\right)(1000)x^2$$

Solving for x gives $x = 0.0296$ m for the elongation of the spring.

4. (a) In a given frame of reference, an object can appear to be at rest and possess potential energy. In this frame, the object has zero velocity and hence its momentum is zero.

   (b) If an object has nonzero momentum, it must have nonzero velocity, which means it must have kinetic energy.

5. When an object bounces, an additional upward force (impulse) is given to it. This extra force, which provides for an additional change in momentum, can be dangerous if it is large enough.

6. First, if the mass is placed vertically onto the object, the increase in mass appears to lower the velocity since no force was acting in the direction of motion. We can also state that since the object did not have any horizontal motion, the friction between the new mass and the object must act to accelerate the mass m, and this energy comes from the moving object. From the mass M frame of reference, it appears as though the mass m is coming toward it, providing a backward force which will slow down the object.

CHAPTER 10

Torque

KEY CONCEPTS

- Parallel Forces and Moments
- Torque
- More Static Equilibrium Problems Using Forces and Torques

10.1 INTRODUCTION

In Chapter 9, we discussed the fact that, if a single force is directed toward the center of mass of an extended object, the result is a linear or translational acceleration in the same direction as the force. If, however, the force is not directed through the center of mass, then the result is a rotation about the center of mass. In addition, if the object is constrained by a fixed pivot point, the rotation will take place about that pivot (e.g., a hinge). These observations are summarized in Figure 10.1.

Torque produces rotation.

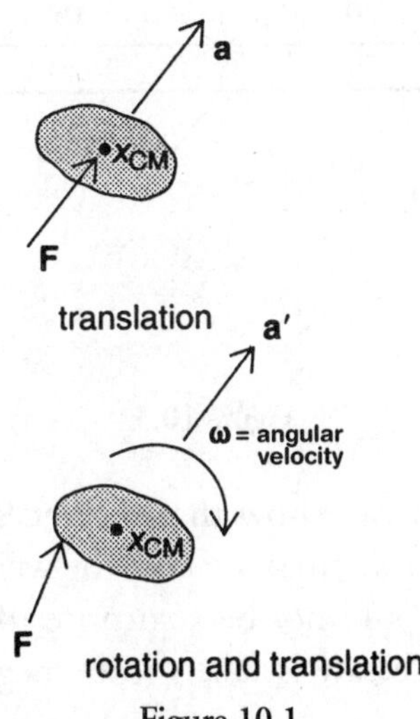

Figure 10.1

10.2 PARALLEL FORCES AND MOMENTS

If two forces are used, it is possible to prevent the rotation if these forces are parallel and are of suitable magnitudes. If the object is free to move in space, translational motion may result (see Figure 10.2).

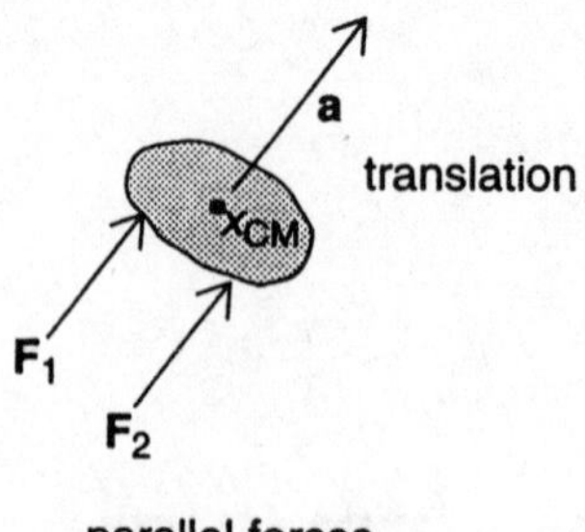

Figure 10.2

As an example of parallel forces, consider two people on a seesaw, as shown in Figure 10.3. The two people have weights $\mathbf{F}_{g1}$ and $\mathbf{F}_{g2}$, respectively, and are sitting distances d_1 and d_2 from the fixed pivot point (called the **fulcrum**). From our discussion about forces, we see that the tendency of each force (provided by gravity), is to cause the seesaw to rotate about the fulcrum. Force $\mathbf{F}_{g1}$ will tend to cause a counterclockwise rotation; force $\mathbf{F}_{g2}$, a clockwise rotation. Arbitrarily, we state that a clockwise rotation is taken as being a negative, while counterclockwise is taken as positive. What factors will influence the ability of the two people to remain in "balance"; that is, what conditions must be met so that the system remains in rotational equilibrium (it is already in a state of translational equilibrium and constrained to remain that way)?

Remember

In **static equilibrium**, the net torque on a system is equal to zero. In other words, the sum of the clockwise torques must be equal to the sum of the counter-clockwise torques.

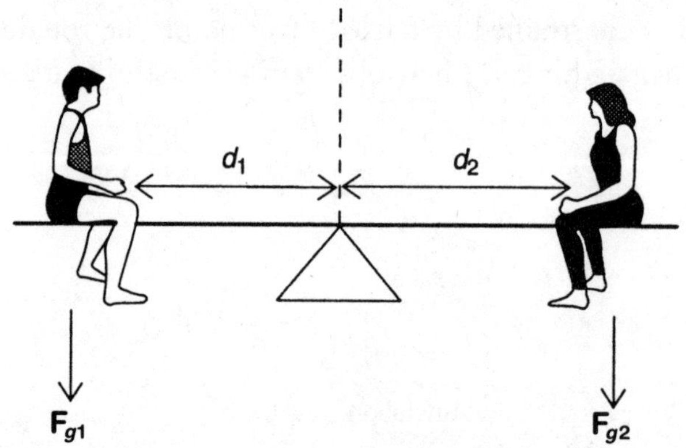

Figure 10.3

Through experiments, you can show that distances d_1 and d_2 (called **moment arm distances**) play a crucial role since we take the weights as being constant. If the two weights were equal, it should not be a surprise to learn that $d_1 = d_2$. In these examples, the weight of the seesaw is taken to be negligible (this is not a realistic scenario).

It turns out that, if $\mathbf{F}_{g2}$ is greater, that person must sit closer to the fulcrum. If equilibrium is to be maintained, the following condition must hold:

$$(\mathbf{F}_{g1})(d_1) = (\mathbf{F}_{g2})(d_2)$$

These are force and distance products, but they do not represent work in the translational sense because the two vectors are not parallel or resolvable into parallel components. In rotational motion, the product of a force and a perpendicular moment arm distance (relative to a fulcrum) is called **torque**. Let us now consider some more examples of torques and equilibrium.

SAMPLE PROBLEM

A 45-kg girl and a 65-kg boy are sitting on a see-saw in equilibrium. If the boy is sitting 0.7 m from the fulcrum, where is the girl sitting?

Solution

In equilibrium,

$$\mathbf{F}_{g1}d_1 = \mathbf{F}_{g2}d_2$$
$$(65\text{ kg})(9.8\text{ m/s}^2)(0.7\text{ m}) = (45\text{ kg})(9.8\text{ m/s}^2)d^2$$
$$d_2 = 1.01\text{ m}$$

10.3 TORQUE

When you tighten a bolt with a wrench, you apply a force to create a rotation or twist. This twisting action, called torque in physics, is represented by the Greek letter τ. Even though its units are newton · meters (N · m), you must not confuse it with translational work. In static equilibrium, the two conditions met are that the vector sum of all forces acting on the object equals zero and that the vector sum of all torques equals zero:

$$\Sigma\mathbf{F} = 0 \quad \text{and} \quad \Sigma\tau = 0$$

When we say that $\tau = \mathbf{F}d$, d is the moment arm distance from the center of mass, or the pivot point, and **F** is the force perpendicular to the vector displacement. If the force is applied at some angle θ to the object, as seen in Figure 10.4, the component of the force, perpendicular to the vector displacement out from the pivot, is taken as the force used. Algebraically this is stated as $\mathbf{F}_T = \mathbf{F}\sin\theta$

Remember

Torque is a vector quantity, but **work** is not! The direction of the torque is taken as either positive or negative, depending on whether the rotation is counterclockwise or clockwise.

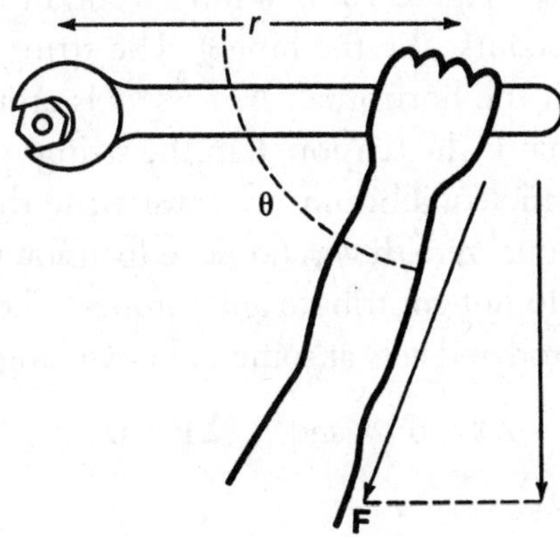

Figure 10.4

SAMPLE PROBLEM

Consider a light string wound around a frictionless and massless wheel, as shown in Figure 10.5. The free end of the string is attached to a 1.2-kg mass that is allowed to fall freely. The wheel has a radius of 0.25 m. What torque is produced?

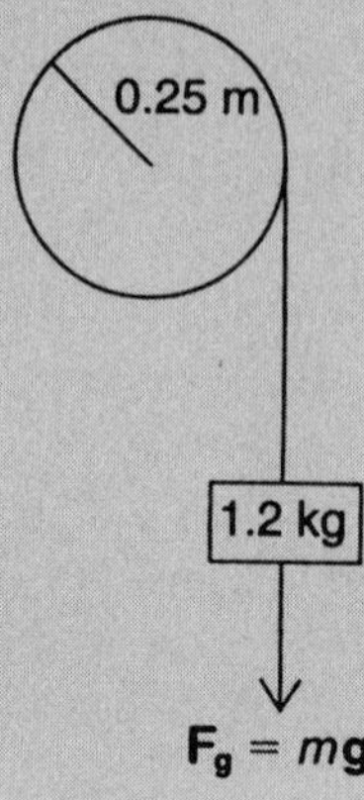

Figure 10.5

Solution

The force acting at right angles to the center of the wheel is the weight of the mass, given by

$$\mathbf{F}_g = m\mathbf{g} = (1.2)(9.8) = 11.76 \text{ N}$$

The radius serves as the moment arm distance, $d = 0.25$ m, so we can write

$$\tau = -\mathbf{F}d = (11.76)(0.25) = -2.94 \text{ N} \cdot \text{m}$$

The torque is negative since the falling weight induces a clockwise rotation in this example.

10.4 MORE STATIC EQUILIBRIUM PROBLEMS USING FORCES AND TORQUES

In the following example (see Figure 10.6), a hinged rod of mass M is attached to a wall with a string (and, of course, by the hinge). The string is considered massless, and makes an angle θ with the horizontal. A mass m is attached to the end of the rod, whose length is ℓ. What is the tension $\mathbf{T}$ in the string?

Since the system is in static equilibrium, we must write that the sum of all forces and the sum of all torques equal zero. If we choose to focus on the pivot, then all forces acting through that point do not contribute any torques. The reaction force $\mathbf{R}$ is the response of the wall to the rod and acts at some unknown angle ϕ. Thus we state that

$$\Sigma\tau = 0 \qquad \text{and} \qquad \Sigma\mathbf{F} = 0$$

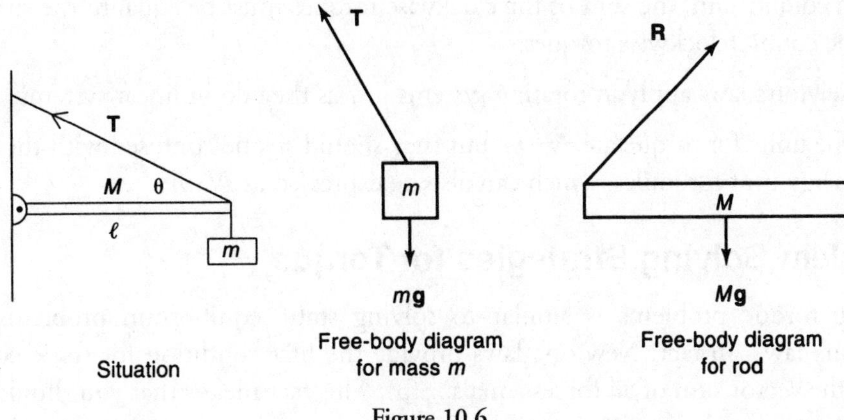

Figure 10.6

For the forces, we see from the free-body diagram in Figure 10.6 that in the *x*-direction

$$\Sigma \mathbf{F}_x = 0 = \mathbf{R} \cos \phi - \mathbf{T} \cos \theta$$

and in the *y*-direction

$$\Sigma \mathbf{F}_y = 0 = \mathbf{R} \sin \phi + \mathbf{T} \sin \theta - m\mathbf{g} - M\mathbf{g}$$

Since $\mathbf{T} \sin \theta = m\mathbf{g}$, it follows that $\mathbf{R} \sin \phi = M\mathbf{g}$ by direct substitution. Hence, angle ϕ is also equal to zero, and the reaction force is directed horizontally along the rod.

For the torques, we see that the component of tension perpendicular to the rod contributes a counterclockwise torque, while the hanging weight contributes a clockwise torque:

$$\Sigma \tau = 0 = (\mathbf{T} \sin \theta)\ell - m\mathbf{g}\ell - M\mathbf{g}\left(\frac{\ell}{2}\right)$$

We could specify the length of the rod, but it is clear from the torque equation that the length can be eliminated. We can also immediately write

$$\mathbf{T} = \frac{m\mathbf{g}}{\sin \theta}$$

Suppose $M = 1$ kg, $m = 10$ kg, and $\theta = 30°$. Then $\mathbf{T} = 196$ N. Now we know that $\mathbf{R} \cos \phi = \mathbf{T} \cos \theta$ and that $\mathbf{R} \sin \theta = M\mathbf{g}$; thus we can write that $\tan \phi = M\mathbf{g}/(\mathbf{T} \cos \theta)$. Substituting in our values, we find that $\phi = 3.3$ degrees. Finally, we see that the magnitude of $\mathbf{R} = 170$ N.

It should be noted that the rod is considered to have mass and that the torque produced by the rod is its weight taken from the center of mass ($\ell/2$ in this case) and is always clockwise (negative).

Chapter Summary

- Torque is vector quantity equal to the product of (force × lever arm).
- The force and lever arm are perpendicular to each other.

- In equilibrium, the sum of the clockwise torques must be equal to the sum of the counterclockwise torques.
- Newton's laws apply in rotating systems just as they do in linear systems.
- The units for torque are $N \cdot m$, but they should *not* be confused with the scalar energy unit for joules, which can also be expressed as $N \cdot m$.

Problem-Solving Strategies for Torque

Solving torque problems is similar to solving static equilibrium problems with Newton's laws. In fact, Newton's laws provide the first condition for static equilibrium (the vector sum of all forces equals zero). This fact means that you should once again draw a free-body diagram of the situation.

In rotational static equilibrium, the vector sum of all torques must equal zero. Remember that torque is a vector even though its units are newton · meters. We take clockwise torques as negative and counterclockwise torques as positive. Any force going through the chosen pivot point (or fulcrum) does not contribute any torques. Therefore, it is wise to choose a point that eliminates the greatest number of forces. Also, remember that only the components of forces that are perpendicular to the direction of a radius displacement from the pivot (called the moment arm) are responsible, a situation that usually involves the sine of the force's orientation angle.

Practice Exercises

MULTIPLE-CHOICE

1. A 45-kg girl is sitting on a seesaw 0.6 m from the balance point, as shown below. How far, on the other side, should a 60-kg boy sit so that the seesaw will remain in balance?

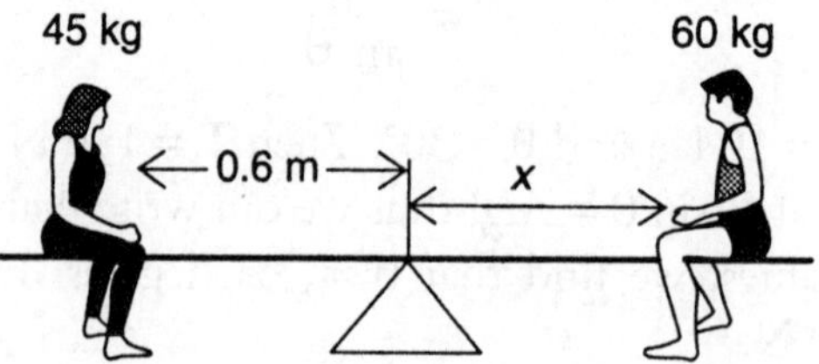

(A) 0.30 m
(B) 0.35 m
(C) 0.40 m
(D) 0.45 m
(E) 0.50 m

2. A balanced stick is shown below. The distance from the fulcrum is shown for each mass except the 10-g mass. What is the approximate position of the 10-g mass, based on the diagram?

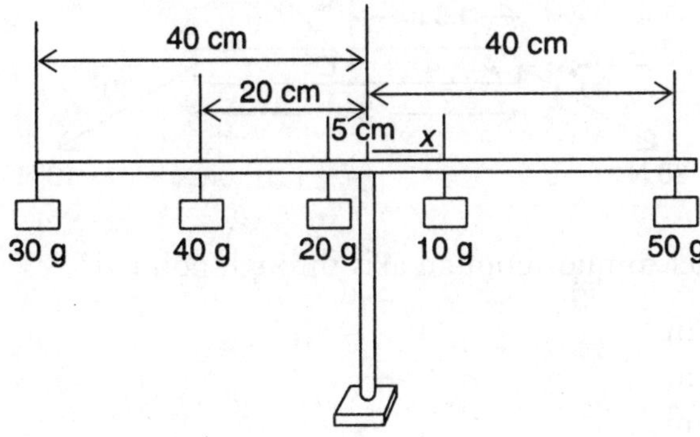

(A) 7 m
(B) 9 m
(C) 10 m
(D) 15 m
(E) 21 m

3. A solid cylinder consisting of an outer radius R_1 and an inner radius R_2 is pivoted on a frictionless axle as shown below. A string is wound around the outer radius and is pulled to the right with a force $\mathbf{F}_1 = 3$ N. A second string is wound around the inner radius and is pulled down with a force $\mathbf{F}_2 = 5$ N. If $R_1 = 0.75$ m and $R_2 = 0.35$ m, what is the net torque acting on the cylinder? **CHALLENGE**

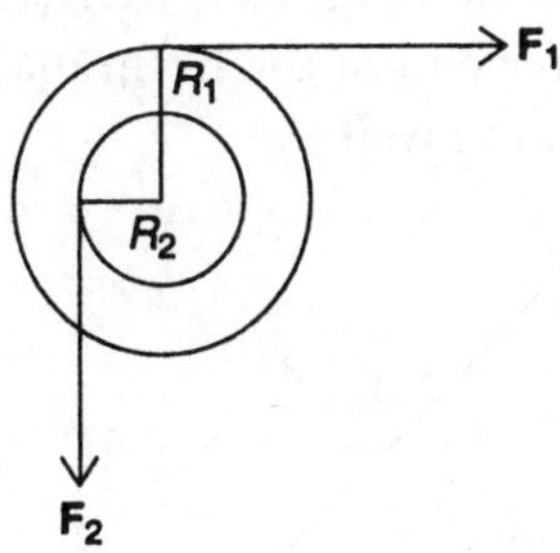

(A) 2.25 N · m
(B) −2.25 N · m
(C) 0.5 N · m
(D) −0.5 N · m
(E) 0 N · m

Answer questions 4 and 5 based on the following diagram. The rod is considered massless.

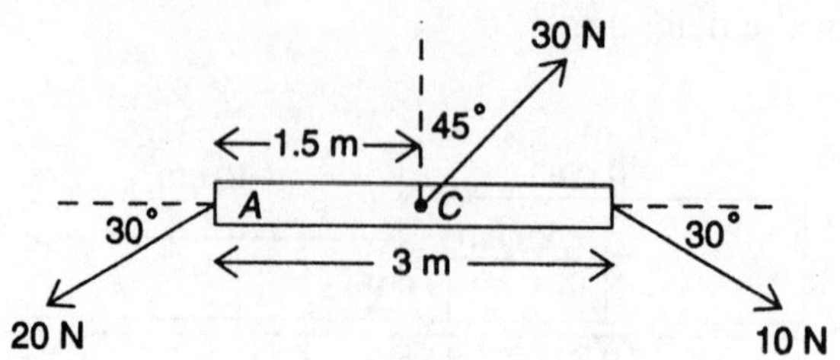

4. What is the net torque about an axis through point A?

   (A) 16.8 N · m
   (B) 15.2 N · m
   (C) −5.5 N · m
   (D) −7.8 N · m
   (E) 6 N · m

5. What is the net torque about an axis through point C?

   (A) 3.5 N · m
   (B) 7.5 N · m
   (C) −15.2 N · m
   (D) 5.9 N · m
   (E) 7 N · m

FREE-RESPONSE

1. A 500-N person stands 2.5 m from a wall against which a horizontal beam is attached. The beam is 6 m long and weighs 200 N (see diagram below). A cable attached to the free end of the beam makes an angle of 45° to the horizontal and is attached to the wall.

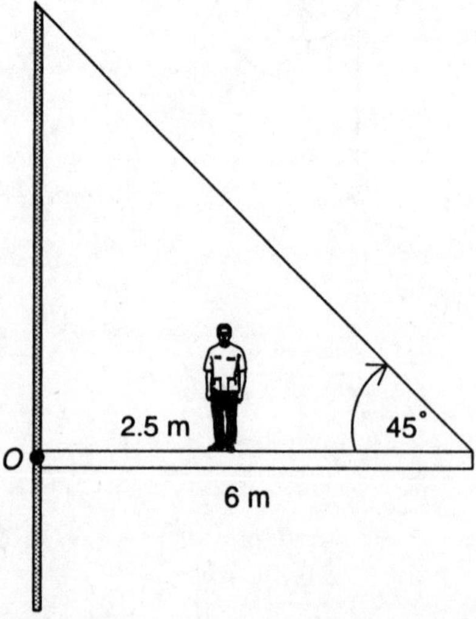

   (a) Draw a free-body diagram of the beam.
   (b) Determine the magnitude of the tension in the cable.
   (c) Determine the reaction force that the wall exerts on the beam.

2. A uniform ladder of length ℓ and weight 100 N rests against a smooth vertical wall. The coefficient of static friction between the bottom of the ladder and the floor is 0.5.
   (a) Draw a free-body diagram of this situation.
   (b) Find the minimum angle, θ, that the ladder can make with the floor so that the ladder will not slip.

ANSWERS EXPLAINED

MULTIPLE-CHOICE PROBLEMS

1. **(D)** To remain in balance, the two torques must be equal. The force on each side is given by the weight, $m\mathbf{g}$. The moment arm distances are 0.6 m and x. Since the factor $\mathbf{g}$ will appear on both sides of the torque balance equation, we can eliminate it and write

$$(45)(0.6) = (60)x \quad \text{implies} \quad x = 0.45 \text{ m}$$

2. **(C)** In this problem, again, the sum of the torques on the left must equal the sum of the torques on the right. We could convert all masses to kilograms and all distances to meters, but in the balance equation the same factors appear on both sides. Therefore, for simplicity and time efficiency, we can simply write

$$(30)(40) + (40)(20) + (20)(5) = (10)x + (50)(40)$$
$$x = 10 \text{ cm}$$

3. **(D)** The net torque is given by the vector sum of all torques. $\mathbf{F}_2$ provides a counterclockwise positive torque, while $\mathbf{F}_1$ provides a clockwise negative torque. Each radius is the necessary moment arm distance. Thus we have

$$\tau_{net} = (5)(0.35) - (3)(0.75) = -0.5 \text{ N} \cdot \text{m}$$

4. **(A)** In this problem, the net torque about point A implies that the force passing through point A does not contribute to the net torque. Also, we need the components of the remaining forces perpendicular to the beam. From the diagram, we see that the 30-N force acts counterclockwise (positive), while the 10-N force acts clockwise (negative). Thus:

$$\tau_{net} = (30)(\cos 45)(1.5) - (10)(\sin 30)(3) = 16.8 \text{ N} \cdot \text{m}$$

5. **(B)** In this problem, since the pivot is now set at point C, we can eliminate the 30-N force passing through point C as a contributor to the torque. Again, we see that the 20-N force will act in a counterclockwise direction, while the 10-N force will act clockwise. Each force is 1.5 m from the pivot. We also need the component of each force perpendicular to the beam. Thus:

$$\tau_{net} = (20)(\sin 30)(1.5) - (10)(\sin 30)(1.5) = 7.5 \text{ N} \cdot \text{m}$$

FREE-RESPONSE PROBLEMS

1. (a) The free-body diagram for this situation appears below:

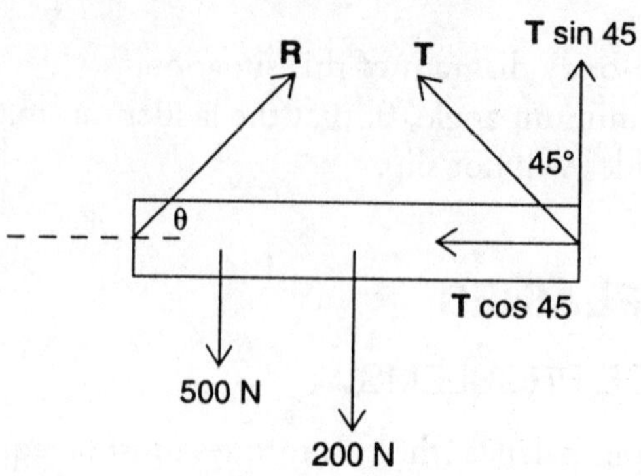

(b) The reaction force acts at some unknown angle, θ. We use our two conditions for equilibrium:

1. The sum of all forces in the *x*- and *y*-directions must equal zero:

$$0 = \mathbf{R} \cos \theta - \mathbf{T} \cos 45 \quad \text{and} \quad 0 = \mathbf{R} \sin \theta + \mathbf{T} \sin 45 - 500 \text{ N} - 200 \text{ N}$$

2. The sum of all torques through contact point *O* must be zero, thus eliminating the reaction force **R**:

$$0 = -(500)(2.5) - (200)(3) + \mathbf{T} \sin 45(6)$$

Both weights produce clockwise torques! Solving for **T** in the last equation, we get **T** = 436.05 N.

(c) Using the result from part (b), we can rewrite the first two equations as:

$$\mathbf{R} \cos \theta = (436.05) \cos 45 = 308.33 \text{ N} \quad \text{and} \quad \mathbf{R} \sin \theta = 391.67 \text{ N}$$

Taking the ratio of these two equations gives us tan θ = 1.27; therefore θ = 51.8°. Since we know the angle at which the reaction force acts, we can simply write

$$\mathbf{R} \cos 51.8 = 308.33 \quad \text{implies} \quad \mathbf{R} = 498.6 \text{ N}$$

2. (a) A sketch of the situation and its free-body diagram are given below:

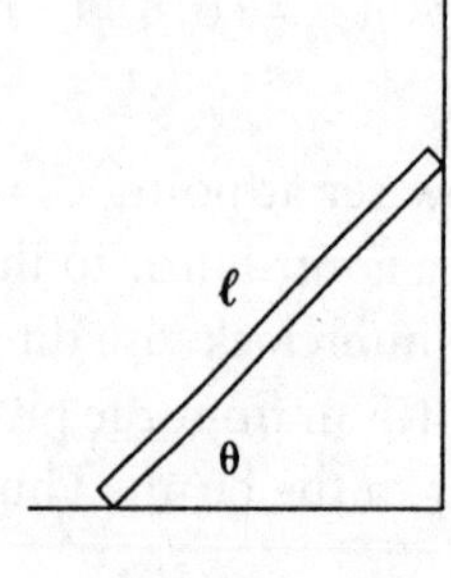

Situation

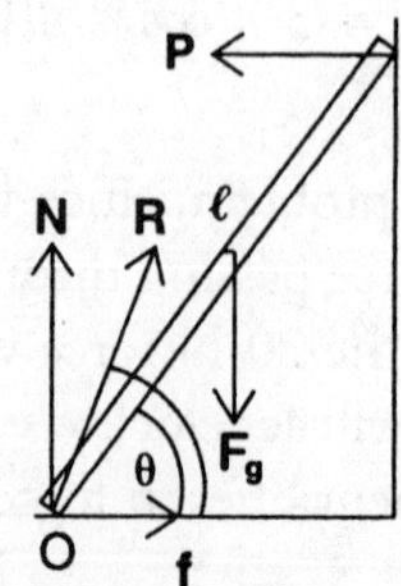

Free-body diagram

(b) The reaction force **R** is the vector resultant of the normal force **N** (from the floor) and the frictional force **f** that opposes slippage. The force **P** is the reaction force of the vertical wall; there is no friction on the vertical wall. The weight $\mathbf{F_g}$ acts from the center of mass, $\ell/2$, and initiates a clockwise torque (if allowed). Since we have equilibrium at this angle, we can write our equations for the conditions of static equilibrium.

1. The sum of all *x*- and *y*-forces must be equal to zero. Horizontally, friction is an opposing force, as well as the reaction force, **P**: $0 = \mathbf{f} - \mathbf{P}$. Vertically, we can write $0 = \mathbf{N} - \mathbf{F_g}$. Since $\mathbf{F_g} = 100$ N, then $\mathbf{N} = 100$ N. Now, for no slippage, $\mathbf{f} \leqq \mu\mathbf{N}$, thus $\mathbf{f} = (0.5)(100) = 50\text{ N} = \mathbf{P}$.

2. The sum of the torques must be zero. We need the component of **P** perpendicular to the ladder. From the geometry, we see that this is $\mathbf{P} \sin \theta$. Also, the component of weight perpendicular to the ladder is $\mathbf{F_g} \cos \theta$. Thus, for the sum of all torques equals zero, we write:

$$0 = \mathbf{P}\ell \sin \theta - \mathbf{F_g}\left(\frac{\ell}{2}\right)\cos \theta$$

The length of the ladder is therefore irrelevant and can be canceled out. Since we know that $\mathbf{P} = 50$ N and $\mathbf{F_g} = 100$ N, we find that $\tan \theta = \mathbf{F_g}/2\mathbf{P} = 1.0$. Therefore $\theta = 45°$.

CHAPTER 11

Oscillatory Motion

KEY CONCEPTS

- Simple Harmonic Motion: A Mass on a Spring
- Simple Harmonic Motion: A Simple Pendulum
- The Dynamics of Simple Harmonic Motion

11.1 SIMPLE HARMONIC MOTION: A MASS ON A SPRING

From Hooke's law in Chapter 7, we know that a spring will become elongated by an amount directly proportional to the force applied. The force constant, k, relates to the specific amount of force (in newtons) needed to stretch or compress the spring by 1 meter:

$$\mathbf{F} = -k\mathbf{x}$$

The negative sign indicates that the force is restorative. One could easily use Hooke's law without the negative sign in the proper context.

Suppose we have a spring with a mass attached to it horizontally in such a way that the mass rests on a flat frictionless surface as shown in Figure 11.1.

Remember

A spring will produce a restoring force if it is stretched or compressed.

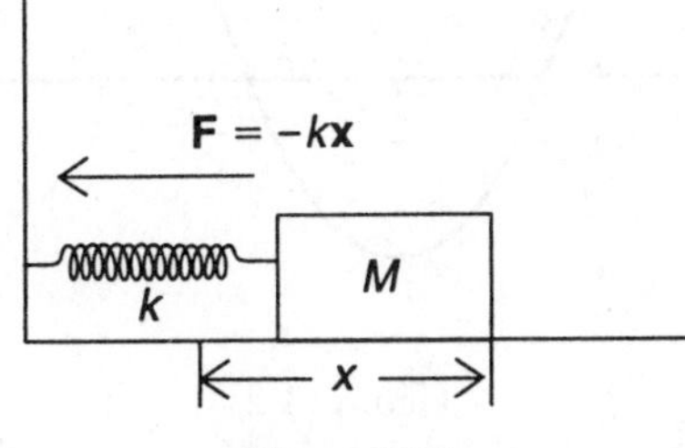

Figure 11.1

If the mass M is pulled a displacement $\mathbf{x}$, a restoring force of $\mathbf{F} = -k\mathbf{x}$ will act on it when released. However, as the mass accelerates past its equilibrium position, its momentum will cause it to keep going, thus compressing the spring. This action will slow the mass down until the same displacement, $\mathbf{x}$, is reached in compression. The same restoring force will then accelerate the mass back and forth, creating oscillatory motion with a certain period T (in seconds) and frequency f (in hertz or cycles per second). The SI unit for frequency is reciprocal seconds (s^{-1}).

Now, according to Newton's second law of motion, $\mathbf{F} = m\mathbf{a}$, so when the mass was originally extended, the restoring force $\mathbf{F} = -k\mathbf{x}$ would also produce an instantaneous acceleration, given by $\mathbf{F} = m\mathbf{a}$. In other words, $\mathbf{a} = -(k/m)\mathbf{x}$. The fact that the acceleration is directly proportional to the displacement (but in the opposite direction) is characteristic of a special kind of oscillatory motion called **simple harmonic motion**. From this expression, it can be shown that the acceleration is zero at the equilibrium point (where $x = 0$).

We can build up a qualitative picture of this type of motion by considering a displacement versus time graph for this mass-spring system. Suppose we have the system at rest so that the spring is unstretched. We now pull the mass to the right a distance A (called the **amplitude**) and release the mass. This action creates a restoring force that will pull the mass to the left toward the equilibrium point. The velocity will become greater and greater, reaching a maximum as the mass passes through the equilibrium point (at which $\mathbf{a} = 0$). The mass will then move toward the left, slowing down as it compresses the spring (since the acceleration is in the opposite direction). The mass stops momentarily when $x = -A$ (because of conservation of energy) and then accelerates again to maintain simple harmonic motion (in the absence of friction).

A graph of displacement versus time (see Chapter 5) would look like a graph of the **cosine** function since the mass is beginning at some distance from the origin. We could consider the motion in progress from the point of view of the origin, in which case the graph would be of the **sine** function (some textbooks use this format). Figure 11.2 shows the characteristics of acceleration and deceleration as the mass oscillates in a period T with a frequency f (for arbitrary units of displacement and time).

From Chapter 8, we know that the maximum energy of a compressed (or stretched) spring is given by $E = (1/2)kA^2$, for $x = A$ as in our example. Thus the constraining points $X = \pm A$ define the limits of oscillation for the mass.

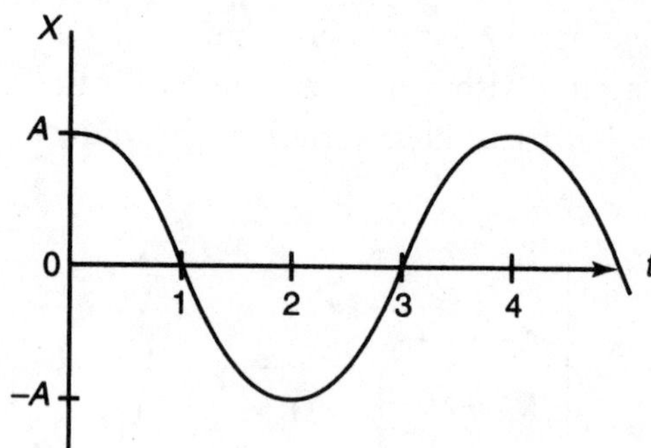

Figure 11.2

A graph of velocity versus time can be built up qualitatively in much the same way. Recall that the slope of the displacement versus time graph represents the instantaneous velocity. From Figure 11.2, you can see that, at $t = 0$, the graph is horizontal, indicating that $\boldsymbol{v} = 0$. The increasingly negative slope shows that the mass is accelerating "backward" until, at $t = 1$, the line is momentarily straight, indicating maximum velocity when $x = 0$. The slope now gradually approaches zero at $t = 2$, indicating that the mass is slowing down as it approaches $x = -A$. The cycle then repeats itself, producing a graph similar to Figure 11.3.

Finally, analyzing the velocity graph with slopes, we produce an acceleration versus time graph. When $t = 0$, the line is momentarily straight with a negative slope indicating a maximum negative acceleration. When the mass crosses the equilibrium point, velocity is maximum but acceleration is momentarily zero. The acceleration (proportional to displacement) reaches a maximum once again when $x = -A$. The cycle repeats itself, producing Figure 11.4. Notice that the acceleration graph is approximately shaped like the negative of the displacement graph, as expected!

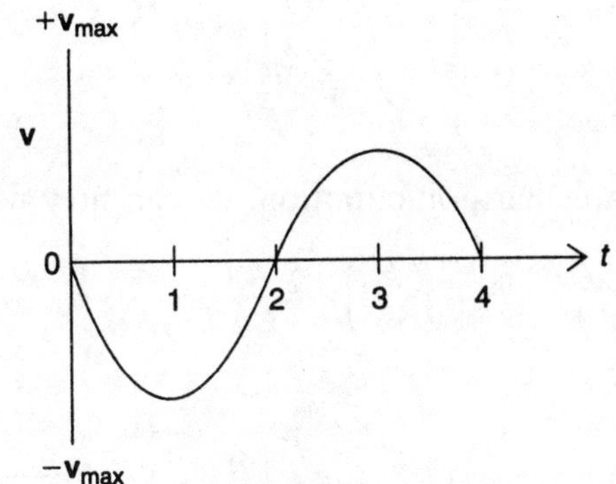

Figure 11.3

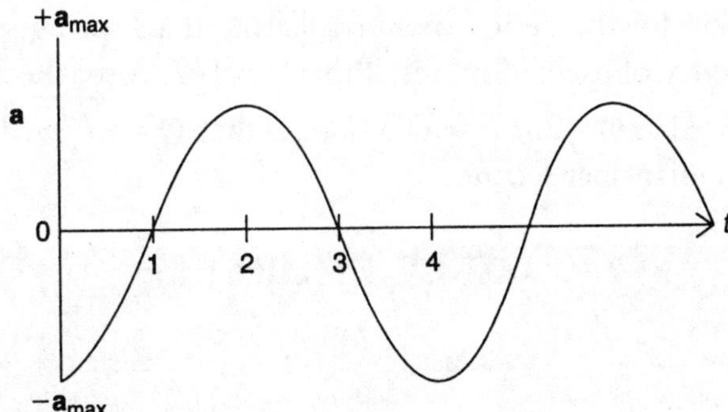

Figure 11.4

To derive the period of oscillation for the mass-spring system, we can consider another form of periodic motion already discussed: uniform circular motion. If a mass is attached to a rotating turntable and then turned onto its side, a projected shadow of the rotating mass simulates simple harmonic motion (Figure 11.5).

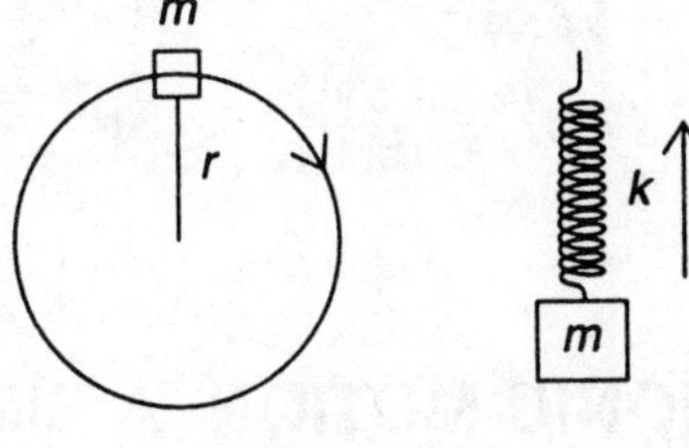

Figure 11.5

In this simulation, the radius r acts like the amplitude A, and the frequency and period of the rotation can be adjusted so that, when the shadow appears next to a real oscillating system, it is difficult to decide which one is actually rotating.

From our understanding of uniform circular motion, we know that the magnitude of the centripetal acceleration is given by

$$\mathbf{a}_c = \frac{4\pi^2 r}{T^2}$$

Since the projected sideways view of this motion appears to approximate simple harmonic motion, we can let the radius r be approximated by the linear displacement x and write

$$\mathbf{a}_x = \frac{4\pi^2 x}{T^2}$$

Since $\mathbf{a} = (k/m)\mathbf{x}$ in simple harmonic motion, we can now write

$$\left(\frac{k}{m}\right)x = \frac{4\pi^2 x}{T^2}$$

TIP

The period of a mass is independent of the acceleration due to gravity.

$$T = 2\pi\sqrt{\frac{m}{k}}$$

This is the equation for the period of an oscillating mass-spring system in seconds. To find the frequency of oscillation, recall that $f = 1/T$. Also, the angular frequency (velocity) is expressed as $\omega = 2\pi/T$, which implies that $\omega = \sqrt{k/m}$. The units of angular frequency are radians per second.

SAMPLE PROBLEM

A 0.5-kg mass is attached to a massless, elastic spring. The system is set into oscillation along a smooth horizontal surface. If the observed period is 0.5 s, what is the value of the force constant k?

Solution

The formula for the period is

$$T = 2\pi\sqrt{\frac{m}{k}}$$

Thus,

$$k = \frac{4\pi^2 m}{T^2}$$

$$k = \frac{(4\pi^2)(0.5\text{ kg})}{(0.5\text{ s})^2} = 79\text{ N/m}$$

11.2 SIMPLE HARMONIC MOTION: A SIMPLE PENDULUM

Imagine a pendulum consisting of a mass M (called a bob) and a string of length ℓ that is considered massless (Figure 11.6). The pendulum is displaced through an angle θ that is much less than 1 radian (about 57 degrees). Under these conditions, the pendulum approximates simple harmonic motion and the period of oscillation is independent of amplitude.

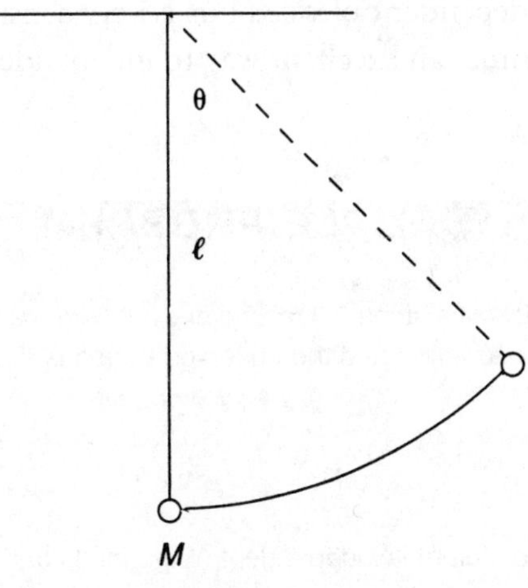

Figure 11.6

When the pendulum swings through an arc of length s, it appears to be following a straight path for a suitably chosen small period of time or small section of the arc. In this approximation, we can imagine the pendulum as being accelerated down an incline. The oscillations occur because gravity accelerates the pendulum back to its lowest position. Its momentum maintains the motion through that point, and then the constraining action of the string (providing a centripetal force) causes the pendulum to swing in an upward arc. Conservation of energy will bring the pendulum to the same vertical displacement (or cause it to swing through the same arc length) and then momentarily stop until gravity begins to pull it down again.

If we imagine that an incline of set angle θ is causing the acceleration, then, from our discussions of kinematics, we know that $\mathbf{a} = -\mathbf{g}\sin\theta$ (where the angle is measured in radians). Now, if θ is sufficiently less than 1 radian, we can write that $\sin\theta \approx \theta$ and therefore that

$$\mathbf{a} = -\mathbf{g}\sin\theta \approx -\mathbf{g}\theta \approx \frac{\mathbf{g}s}{\ell}$$

In the above equation we used the known relationship from trigonometry that, in radian measure, if s is the arc length and ℓ corresponds to the effective "radius" of swing, then $s = \ell\theta$.

It should be remembered that these are only approximations, but for angles of about 10 or 20 degrees the approximations are fairly accurate. Since the pendulum now approximates simple harmonic motion, we know that we can find a suitable rotational motion that, when viewed in projection (as well as the pendulum swing being viewed in projection), simulates our simple harmonic motion. Therefore, in a similar fashion to that in Section 11.1, we can write

$$\mathbf{a} = \frac{4\pi^2 s}{T^2} = \frac{\mathbf{g}s}{\ell}$$

where the arc length s is the amplitude of swing in this case.

If we solve for the period T, we finally get

$$T = 2\pi\sqrt{\frac{\ell}{\mathbf{g}}}$$

TIP 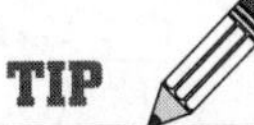

The period of a simple pendulum is independent of the mass.

Notice that period is independent of mass but is very sensitive to the local acceleration of gravity. This becomes an excellent way to independently measure the value of **g** in various locations.

Remember

The period of a system oscillating with simple harmonic motion is independent of the amplitude. Remember that for **simple harmonic motion**, the acceleration is proportional to the *negative* of the displacement!

SAMPLE PROBLEM

A simple pendulum consists of a string 0.4 m long and a 0.3 kg mass. On an unknown planet, the pendulum is set into oscillation and the observed period is 0.8 s. What is the value of **g** on this planet?

Solution

The period of a simple pendulum is independent of the mass and is given by

$$T = 2\pi\sqrt{\frac{L}{\mathbf{g}}}$$

Thus,

$$\mathbf{g} = \frac{4\pi^2 L}{T^2} = \frac{(4\pi^2)(0.4\text{ m})}{(0.8\text{ s})^2} = 24.67\text{ m/s}^2$$

11.3 THE DYNAMICS OF SIMPLE HARMONIC MOTION

From Chapter 8 we know that the work done to compress a spring is also equal to the potential energy stored in the spring. This value is expressed as $(1/2)kx^2$. If the system is set into oscillation by stretching the spring by an amount $x = A$, the maximum energy possessed by the oscillating system is given by $U = (1/2)kA^2$. As the mass oscillates, it reaches maximum velocity when it passes through $x = 0$. At that point, the potential energy of the spring is zero; and since we are treating the cases without friction, the loss of potential energy is balanced by this gain in kinetic energy:

$$\text{KE}_{\text{max}} = \frac{1}{2}m\mathbf{v}_{\text{max}}^2 = \frac{1}{2}kA^2$$

If we want to treat cases in which the mass is between the two extremes of $x = 0$ and $x = \pm A$, we note that the spring will still possess some elastic potential energy (no gravitational potential energy is involved since the mass is oscillating horizontally). This implies that

$$\frac{1}{2}kA^2 = \frac{1}{2}m\mathbf{v}^2 + \frac{1}{2}kx^2$$

If we solve for velocity, we get

$$\mathbf{v} = \pm\sqrt{\frac{k}{m}(A^2 - x^2)}$$

Chapter Summary

- Any periodic motion in which the acceleration is directly proportional to the negative of the displacement is called simple harmonic motion.
- The period of oscillation for a mass attached to a spring is independent of gravity and depends on the magnitude of the mass and force constant k.
- The period of oscillation for a simple pendulum is independent of the mass and depends on the length of the string and the value of the acceleration due to gravity **g**.
- The period of simple harmonic motion is independent of amplitude.
- Simple harmonic motion is related to uniform circular motion when viewed as a side component.
- The velocity of motion is always greatest at the middle part of the motion. The acceleration is always greatest at the ends of the motion.

Problem-Solving Strategies for Oscillatory Motion

Keep in mind the following facts:

1. Simple harmonic motion is related to circular motion and is characterized by the fact that the acceleration varies directly with the displacement (but in the opposite direction).
2. A simple pendulum approximates simple harmonic motion only if its displacement angle is $\leq 10°$. The period will be independent of amplitude and mass under these circumstances.
3. It is sometimes easier to use energy considerations since the equations involve scalars and no free-body diagrams have to be drawn.
4. For pendulum or pendulum-like problems, if $\theta \ll 1$ radian, then $\sin\theta \approx \tan\theta \approx \theta$.

Practice Exercises

MULTIPLE-CHOICE

1. What is the length of a pendulum whose period, at the Equator, is 1 s?

   (A) 0.15 m
   (B) 0.25 m
   (C) 0.30 m
   (D) 0.45 m
   (E) 1.0 m

2. On a planet, an astronaut determines the acceleration of gravity by means of a pendulum. She observes that the 1-m-long pendulum has a period of 1.5 s. The acceleration of gravity, in meters per second squared, on the planet is

   (A) 7.5
   (B) 15.2
   (C) 10.2
   (D) 26.3
   (E) 17.5

3. When a 0.05-kg mass is attached to a vertical spring, it is observed that the spring stretches 0.03 m. The system is then placed horizontally on a frictionless surface and set into simple harmonic motion. What is the period of the oscillations?

   (A) 0.75 s
   (B) 0.12 s
   (C) 0.35 s
   (D) 1.3 s
   (E) 2.3 s

CHALLENGE

4. A mass of 0.5 kg is connected to a massless spring with a force constant k of 50 N/m. The system is oscillating on a frictionless horizontal surface. If the amplitude of the oscillations is 2 cm, the total energy of the system is

   (A) 0.01 J
   (B) 0.1 J
   (C) 0.5 J
   (D) 0.3 J
   (E) 0.2 J

CHALLENGE

5. A mass of 0.3 kg is connected to a massless spring with a force constant k of 20 N/m. The system oscillates horizontally on a frictionless surface with an amplitude of 4 cm. What is the velocity of the mass when it is 2 cm from its equilibrium position?

   (A) 0.28 m/s
   (B) 0.08 m/s
   (C) 0.52 m/s
   (D) 0.15 m/s
   (E) 0.34 m/s

6. If the length of a simple pendulum is doubled, its period will

   (A) decrease by 2
   (B) increase by 2
   (C) decrease by $\sqrt{2}$
   (D) increase by $\sqrt{2}$
   (E) remain the same

7. The pendulums of two grandfather clocks have the same length. One clock (A) runs faster than the other clock (B). Which of the following statements is true?

   (A) Pendulum A is more massive.
   (B) Pendulum B is more massive.
   (C) Pendulum A swings through a smaller arc.
   (D) Pendulum B swings through a smaller arc.
   (E) None of the above statements is correct.

8. A 2-kg mass is oscillating horizontally on a frictionless surface when attached to a spring. The total energy of the system is observed to be 10 J. If the mass is replaced by a 4-kg mass, but the amplitude of oscillations and the spring remain the same, the total energy of the system will be **CHALLENGE**

   (A) 10 J
   (B) 5 J
   (C) 20 J
   (D) 3.3 J
   (E) 15.5 J

FREE-RESPONSE

1. A mass M is attached to two springs with force constants k_1 and k_2, respectively. The mass can slide horizontally over a frictionless surface. Two arrangements, (a) and (b), for the mass and springs are shown below. **CHALLENGE**

   (a) Show that the period of oscillation for situation (a) is given by

   $$T = 2\pi\sqrt{\frac{m(k_1 + k_2)}{k_1 k_2}}$$

   (b) Show that the period of oscillation for situation (b) is given by

   $$T = 2\pi\sqrt{\frac{m}{k_1 + k_2}}$$

   (c) Explain whether the effective spring constant for the system has increased or decreased.

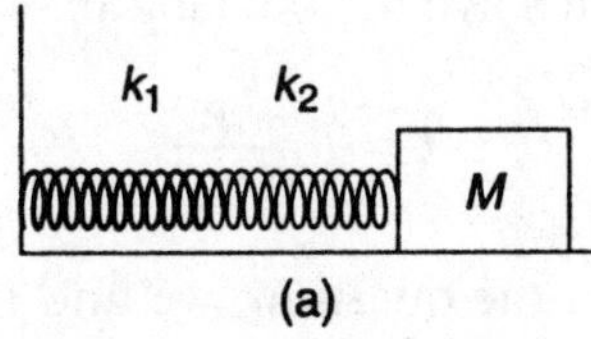

(a)

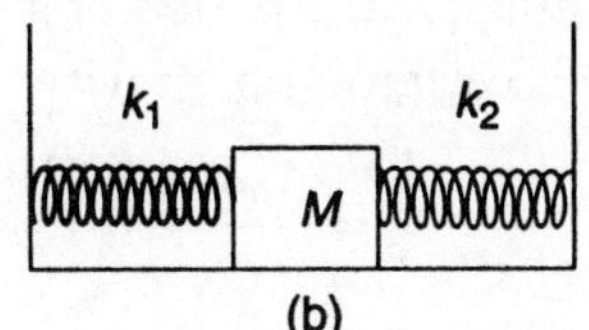

(b)

CHALLENGE

2. A mass M is attached to two light elastic strings both having length ℓ and both made of the same material. The mass is displaced vertically upward by a small displacement $\Delta\mathbf{y}$ such that equal tensions $\mathbf{T}$ exist in the two strings, as shown below. The mass is released and begins to oscillate up and down. Assume that the displacement is small enough so that the tensions do not change appreciably. (Ignore gravitational effects.)

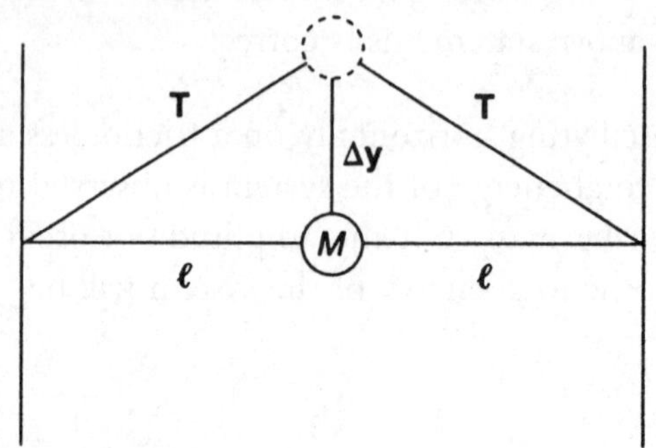

(a) Show that the restoring force on the mass can be given by (for small angles)

$$F = \frac{-2T\ \Delta y}{\ell}$$

(b) Derive an expression for the frequency of oscillation.

3. Explain why a simple pendulum undergoes only approximately simple harmonic motion.

4. Explain how the mass of an object can be determined in a free-fall orbit using the concept of simple harmonic motion.

5. Explain why soldiers are ordered to "break step" as they march over a bridge.

ANSWERS EXPLAINED

MULTIPLE-CHOICE PROBLEMS

1. **(B)** If we use the formula for the period of a pendulum, $T = 2\pi\sqrt{\ell/\mathbf{g}}$, and square both sides, we can solve for the length:

$$\ell = T^2 \frac{\mathbf{g}}{4\pi^2}$$

Using the values given in the question, we find that $\ell = 0.25$ m.

2. **(E)** Again using the formula for the period of a pendulum, this time we solve for the magnitude of the acceleration of gravity:

$$\mathbf{g} = \frac{4\pi\ell}{T^2}$$

Using the values given in the question, we find that $\mathbf{g} = 17.5\ \text{m/s}^2$.

3. **(C)** We first find the spring constant. From $\mathbf{F} = k\mathbf{x}$, we see that

$$\mathbf{F} = m\mathbf{g} = (0.05)(9.8) = 0.49 \text{ N}$$

Thus

$$k = \frac{\mathbf{F}}{x} = \frac{0.49}{0.03} = 16.3 \text{ N/m}$$

Now the period is given by the formula $T = 2\pi\sqrt{m/k}$. Using the values given in the question, we find that $T = 0.35$ s.

4. **(A)** The formula for total energy is $E = (1/2)kA^2$. Using the values given in the question, we get $E = 0.01$ J.

5. **(A)** The formula for any intermediary velocity is $\mathbf{v} = \sqrt{(k/m)(A^2 - x^2)}$. Before substituting the known values, we must change centimeters to meters: 4 cm = 0.04 m and 2 cm = 0.02 m. Then, using these values for amplitude and position, we find that $\mathbf{v} = 0.28$ m/s.

6. **(D)** Since the period of a pendulum varies directly as the square root of the length, if the length is doubled, the period increases by $\sqrt{2}$.

7. **(C)** If clock *A* runs faster, its period is quicker and therefore its pendulum swings through a smaller arc to maintain the faster period (especially since the lengths of pendulum *A* and pendulum *B* are the same).

8. **(A)** The maximum energy in the mass-spring system can be expressed independently of the mass; it depends on the force constant and the square of the amplitude. Thus, if the spring and the amplitude of oscillations remain the same, so will the total energy of the system (10 J).

FREE-RESPONSE PROBLEMS

1. (a) When mass M is stretched a distance x, spring k_1 is stretched distance x_1 and spring k_2 is stretched distance x_2. At the point of connection, Newton's third law states that the forces must be equal and opposite. Thus we can write

$$k_1x_1 = k_2x_2$$

Since $x = x_1 + x_2$, we have $x_2 = x - x_1$, and so

$$\begin{aligned} k_1x_1 &= k_2(x - x_1) \\ &= k_2x - k_2x_1 \\ k_1x_1 + k_2x_1 &= k_2x \\ k_2x &= (k_1 + k_2)x_1 \\ x_1 &= \left(\frac{k_2}{k_1 + k_2}\right)x \end{aligned}$$

is the expression for the displacement of k_1.

Now, for any spring, let's say spring 1:

$$\mathbf{F}_1 = k_1\mathbf{x}_1 = \left(\frac{k_1k_2}{k_1+k_2}\right)\mathbf{x}$$

Let's call $(k_1k_2/k_1 + k_2) = k'$, so

$$\mathbf{F}_1 = k'\mathbf{x}$$

This is in the form of the equation for simple harmonic motion and therefore is equal to $m\mathbf{a}$ by Newton's second law:

$$m\mathbf{a} = k'\mathbf{x}$$

Thus

$$T = 2\pi\sqrt{\frac{m}{k'}} = 2\pi\sqrt{\frac{m(k_1+k_2)}{k_1k_2}}$$

This is the period of oscillation for arrangement (a).

(b) In this case, each spring is displaced the same distance as the mass, since the springs are on either side of the mass. Thus we can write

$$\mathbf{F} = -(k_1 + k_2)\mathbf{x}_1$$

and set $\mathbf{k}' = k_1 + k_2$. We therefore have the equation for the period of oscillation

$$T = 2\pi\sqrt{\frac{m}{k_1+k_2}}$$

for arrangement (b).

(c) Situation (a) suggests that when the springs are connected in series, the effective spring constant decreases. Situation (b) suggests that when the springs are connected in parallel, the effective spring constant increases.

2. (a) In the diagram, we can see that, since $\Delta\mathbf{y}$ is small, $\theta = \Delta\mathbf{y}/\ell$. Now, since θ (in radians) is small, $\theta \approx \sin\theta \approx \Delta\mathbf{y}/\ell$. From the geometry, we see that the net restoring force is given by

$$\Sigma\mathbf{F} = -2\mathbf{T}\sin\theta = \frac{-2\mathbf{T}\,\Delta\mathbf{y}}{\ell}$$

(b) Since this situation is approximating simple harmonic motion, we can write

$$\mathbf{F} = -k\,\mathbf{y}$$

$$-2\mathbf{T}\frac{\Delta\mathbf{y}}{\ell} = -k\,\Delta\mathbf{y}$$

$$k = \frac{2\mathbf{T}}{\ell}$$

so we must then have

$$f = \frac{1}{2\pi}\sqrt{\frac{2T}{\ell m}}$$

3. A pendulum represents only approximately simple harmonic motion since only for small angles is the acceleration proportional to the displacement (angular displacement in this case).

4. Since the period of a horizontally oscillating mass on a spring is independent of the acceleration due to gravity, the object's mass can be determined using springs and horizontal oscillations.

5. The rhythmic marching of soldiers can set up resonance in a bridge where the frequency of the march matches the natural vibrating frequency of the bridge. If amplified, these resonant vibrations can cause damage to the bridge.

Gravitation

CHAPTER 12

KEY CONCEPTS

- Newton's Law of Universal Gravitation
- Gravitational Energy

12.1 NEWTON'S LAW OF UNIVERSAL GRAVITATION

We already know that all objects falling near the surface of Earth have the same acceleration, given by $\mathbf{g} = 9.8\ \text{m/s}^2$. The value of this constant can be determined from the independence of the period of a pendulum on the mass of the bob. This "empirical" verification is independent of any "theory" of gravity.

The weight of an object on Earth is given by $\mathbf{F_g} = m\mathbf{g}$, which represents the magnitude of the force of gravity due to Earth that is acting on the object. From our discussion in Chapter 6, we know that gravity causes a projectile to assume a parabolic path. Isaac Newton, in his book *Principia*, extended the idea of projectile motion to an imaginary situation in which the velocity of the projectile was so great that the object would fall and fall, but the curvature of the earth would bend away and leave the projectile in "orbit." Newton conjectured that this might be the reason why the Moon orbits Earth. This is shown in Figure 12.1.

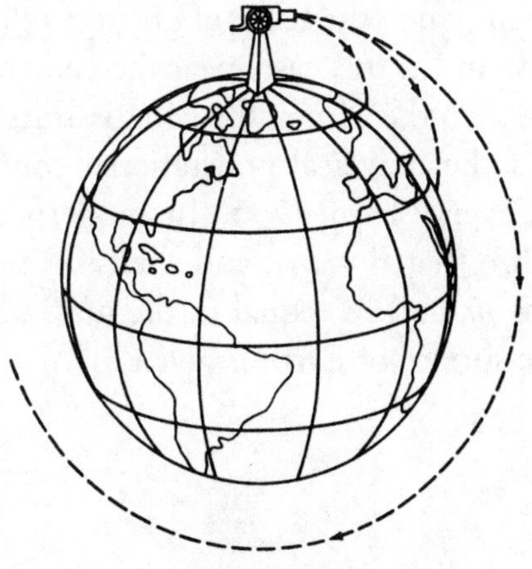

Figure 12.1

To answer this question, Newton first had to determine the centripetal acceleration of the Moon based on observations from astronomy. He knew the relationship between centripetal acceleration and period. We have written that relationship as

$$\mathbf{a}_c = \frac{4\pi^2 R}{T^2}$$

where R is the distance to the Moon (in meters) and T is the orbital period (in seconds). From astronomy we know that $R = 3.8 \times 10^8$ m, and since the Moon orbits Earth in 27.3 days, we have $T = 2.3 \times 10^6$ s. Using these values, we find that the magnitude of $\mathbf{a}_c = 2.8 \times 10^{-3}$ m/s^2.

Using the formula for centripetal acceleration, Newton was able to observe that, since $\mathbf{F} = m\mathbf{a}$ by his second law, the formula for centripetal force is given by (recall Chapter 7)

$$\mathbf{F}_c = \frac{M4\pi^2R}{T^2}$$

Using Kepler's third law of planetary motion, and setting M equal to the mass of the Moon, Newton could express the force of gravity in the form

$$\mathbf{F_g} = \frac{M_{\text{Moon}}4\pi^2K_{\text{Earth}}}{R^2}$$

Then Kepler's constant could be evaluated for the Earth-Moon system by using Kepler's third law.

TIP

Newton's law of universal gravitation is referred to as an inverse square law. For example, if the distance between the masses is doubled, the force between them is one-fourth.

Using his third law of motion, Newton realized that the force that Earth exerts on the Moon should be exactly equal, but opposite, to the force that the Moon exerts on Earth (observed as tides). This relationship implied that the constant K should be dependent on the mass of Earth, and this mutual interaction implied that the force of gravity should be proportional to the product of both masses. In other words, Newton's law of gravity could be expressed as

$$\mathbf{F_g} = \frac{GM_1M_2}{R^2}$$

This is a vector equation in which the force of gravity is directed inward toward the center of mass for the system (in this case, near the center of Earth). Since Earth is many times more massive than the Moon, the Moon orbits Earth, and not vice versa.

The value of G, called the universal gravitational constant, was experimentally determined by Henry Cavendish in 1795. In modern units $G = 6.67 \times 10^{-11}$ N · m^2/kg^2. If we recognize that $\mathbf{F} = m\mathbf{a}$, and then if we consider M_1 to equal the mass of an object of mass m, and M_2 equal to the mass of Earth, M_E, the acceleration of a mass m near the surface of Earth is given by

TIP

Notice that the **acceleration due to gravity** does not depend on the mass of the falling object. This is consistent with Galileo's observations of falling bodies and the period of a pendulum.

$$\mathbf{a} = \frac{GM_E}{R_E{}^2}$$

where R_E is the radius of Earth in meters. Using known values for these quantities, we discover that $\mathbf{a} = 9.8$ m/s^2 = $\mathbf{g}$!

Thus we have a theory that accounts for the value of the known acceleration due to gravity. In fact, if we replace the mass of Earth by the mass of any other planet, and the radius of Earth by the corresponding radius of the other planet, the above formula allows us to determine the value of $\mathbf{g}$ on any planet or astronomical object in the universe! For example, the value of **g** on the Moon is approximately 1.6 meters per second squared, or about one-sixth the value on Earth. Thus, objects on the Moon weigh one-sixth as much as they do on Earth.

Now, Newton's prediction for the acceleration of the Moon toward Earth is that the value of **g** decreases as the square of the distance from Earth. The ancient Greek astronomers discovered that the mean distance to the Moon is approximately equal to 60 times the radius of Earth. Thus, the acceleration of the Moon should be 1/3600 of the acceleration of an object near Earth's surface: $\mathbf{a} = 9.8/3600 = 0.0028\ \text{m/s}^2$! Newton's theory of gravity was confirmed in one simple, triumphant demonstration. Further proof came when his friend Edmund Halley used Newton's law to predict that a certain comet (now known as Halley's comet) would return every 76 years, appearing next in 1758.

SAMPLE PROBLEM

Calculate the gravitational force of attraction between a 2000-kg car and a 12,000-kg truck, separated by 0.5 m.

Solution

We use Newton's law of gravitation:

$$\mathbf{F_g} = \frac{GM_1M_2}{R^2}$$

$$\mathbf{F_g} = \frac{(6.67 \times 10^{-11}\ \text{N} \cdot \text{m}^2/\text{kg}^2)(2,000\ \text{kg})(12,000\ \text{kg})}{(0.5\ \text{m})^2} = 6.4 \times 10^{-3}\ \text{N}$$

12.2 GRAVITATIONAL ENERGY

In Chapter 8 we saw that the amount of work done (by or against gravity) when vertically displacing a mass is given by the change in the gravitational potential energy:

$$\Delta \text{PE} = \Delta m\mathbf{g}h$$

We now know that the value of **g** is not constant but varies inversely with the square of the distance from the center of Earth. Also, since we want the potential energy to become weaker the closer we get to Earth, we can use the results of Section 12.1 to rewrite the potential energy formula as

$$\text{PE} = \frac{-GM_0M_E}{R_E}$$

This equation is valid if an object is located near the surface of Earth. Actually, we should have used a distance variable $R = R_E + h$ (where h is the height above the surface). If h is very small compared with the radius of Earth, it can effectively be eliminated from the equation.

The "escape velocity" from the gravitational force of Earth can be determined by considering the situation where an object has just reached infinity, with zero final velocity, given some initial velocity at any direction away from the surface of Earth. We designate that escape velocity as $\mathbf{v}_{esc}$, and state that, when the final velocity is zero (at infinity), the total energy must be zero, which implies

$$\frac{1}{2}M_0\mathbf{v}_{esc}{}^2 = \frac{GM_0M_E}{R_E}$$

The mass of the object can be eliminated from the relationship, leaving

$$\mathbf{v}_{esc} = \sqrt{\frac{2GM_E}{R_E}}$$

Think About It

Notice that both the **escape velocity** and the **orbital velocity** depend on the mass of Earth and not on the mass of the object.

An object can leave the surface of Earth with any velocity. There is, however, one minimum velocity at which, if the spacecraft coasted, it would not fall back to Earth because of gravity.

The orbital velocity can be determined by assuming that we have an approximately circular orbit (see Chapter 6). In this case, we can set the centripetal force equal to the gravitational force:

$$\frac{GM_0M_E}{R_E^2} = \frac{M_0\mathbf{v}_0{}^2}{R_E}$$

Eliminating the mass of the object from the equation leaves

$$\mathbf{v}_{orbit} = \sqrt{\frac{GM_E}{R_E}}$$

SAMPLE PROBLEM

(a) What is the magnitude of the acceleration due to gravity at an altitude of 400 km above the surface of Earth?

(b) What percentage loss in the weight of an object results?

Solution

(a) The formula for the acceleration due to gravity above Earth's surface is

$$\mathbf{g} = \frac{GM_E}{(R_E + h)^2}$$

We have $M_E = 5.98 \times 10^{24}$ kg, $R_E = 6.38 \times 10^6$ m, $h = 400$ km $= 0.4 \times 10^6$ m. Substituting these values, as well as the known value for G, given in Section 12.1, we get $\mathbf{g} = 8.67$ m/s^2.

(b) The fractional change in weight is found by comparing the value of **g** at 400 km to its value at Earth's surface: 8.67/9.8 = 0.885. Thus, there is an 11.5% loss of weight at that height.

SAMPLE PROBLEM

Three uniform spheres of masses 1 kg, 2 kg, and 4 kg are placed at the corners of a right triangle as shown below. The positions relative to the coordinate system indicated are also shown. What is the magnitude of the resultant gravitational force on the 4-kg mass if we consider that mass to be fixed?

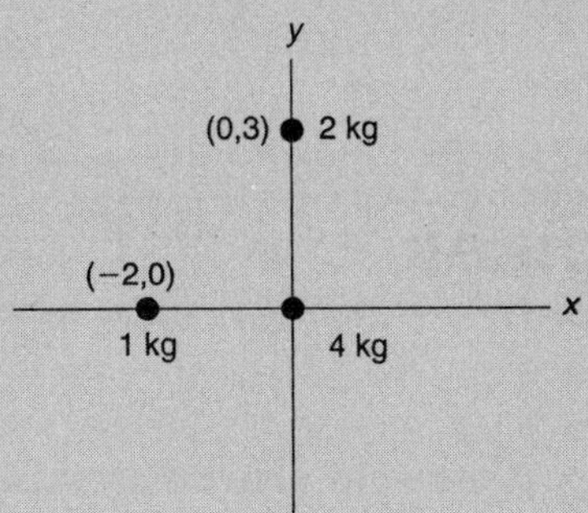

Solution

In this problem, we have to determine separately the force of gravitational attraction between each of the smaller masses and the 4-kg mass. The 1-kg mass will be attracted to the right along the *x*-axis, and the 2-kg mass will be attracted downward along the *y*-axis.

We begin by calculating the force between the 1-kg mass and the 4-kg mass:

$$\mathbf{F}_{1\text{-}4} = \frac{(6.67 \times 10^{-11})(1)(4)}{(-3)^2} = 6.67 \times 10^{-11}\text{ N}$$

The direction is to the right and is therefore considered positive.

For the 2-kg mass and the 4-kg mass, the downward direction is taken as negative:

$$\mathbf{F}_{2\text{-}4} = \frac{(6.67 \times 10^{-11})(2)(4)}{(-3)^2} = 5.93 \times 10^{-11}\text{ N}$$

The magnitude of the resultant force is given by the Pythagorean theorem:

$$\mathbf{F} = \sqrt{(6.67 \times 10^{-11})^2 + (5.93 \times 10^{-11})^2} = 8.92 \times 10^{-11}\text{ N}$$

Chapter Summary

- Newton's law of gravitation states that there is a force of attraction between any two masses and the magnitude of this force is directly proportional to the product of the masses and inversely proportional to the square of the distance between them.
- Newton's law of gravitation is sometimes referred to as an inverse square law similar to Coulomb's law for electrostatics (see Chapter 16).

Problem-Solving Strategies for Gravitation

Keep in mind that gravity is a force and therefore a vector quantity. Distances are measured in meters from the center of masses of each object; therefore if an object is above the surface of Earth, you must add the height to the radius of Earth. If you remember that the force of gravity is an inverse-square-law relationship, you may be able to deduce an answer from logic rather than algebraic calculations, which take time.

Practice Exercises

MULTIPLE-CHOICE

1. What is the value of **g** at a height above Earth's surface that is equal to the radius of Earth?

   (A) 9.8 N/kg
   (B) 4.9 N/kg
   (C) 6.93 N/kg
   (D) 2.45 N/kg
   (E) 1.6 N/kg

2. A planet has half the mass of Earth and half the radius. Compared to the acceleration due to gravity near the surface of Earth, the acceleration of gravity near the surface of this other planet is

   (A) twice as much
   (B) one-fourth as much
   (C) half as much
   (D) the same
   (E) zero

CHALLENGE

3. Which of the following is an expression for the acceleration of gravity with uniform density ρ and radius R?

   (A) $G(4\pi\rho/3R^2)$
   (B) $G(4\pi\rho R^2/3)$
   (C) $G(4\pi\rho/3R)$
   (D) $G(4\pi R\rho/3)$
   (E) None of these is correct.

CHALLENGE

4. What is the orbital velocity of a satellite at a height of 300 km above the surface of Earth? (The mass of Earth is approximately 6×10^{24} kg, and its radius is 6.4×10^6 m.)

   (A) 5.42×10^1 m/s
   (B) 1.15×10^6 m/s
   (C) 7.7×10^3 m/s
   (D) 6×10^6 m/s
   (E) 3×10^8 m/s

5. What is the escape velocity from the Moon, given that the mass of the Moon is 7.2×10^{22} kg and its radius is 1.778×10^{6} m? **CHALLENGE**

   (A) 1.64×10^3 m/s
   (B) 2.32×10^3 m/s
   (C) 2.69×10^6 m/s
   (D) 5.38×10^6 m/s
   (E) 3×10^8 m/s

6. A "black hole" theoretically has an escape velocity that is greater than or equal to the velocity of light (3×10^8 m/s). If the effective mass of the black hole is equal to the mass of the Sun (2×10^{30} kg), what is the effective "radius" (called the "Schwarzchild radius") of the black hole? **CHALLENGE**

   (A) 3×10^3 m
   (B) 1.5×10^3 m
   (C) 8.9×10^6 m
   (D) 4.45×10^6 m
   (E) 0 m

7. What is the gravitational force of attraction between two trucks, each of mass 20,000 kg, separated by a distance of 2 m?

   (A) 0.057 N
   (B) 0.013 N
   (C) 0.0067 N
   (D) 1.20 N
   (E) 0 N

8. The gravitational force between two masses is 36 N. If the distance between masses is tripled, the force of gravity will be

   (A) the same
   (B) 18 N
   (C) 9 N
   (D) 4 N
   (E) 27 N

FREE-RESPONSE

1. Find the magnitude of the gravitational field strength **g** at a point P along the perpendicular bisector between two equal masses, M and M, that are separated by a distance $2b$ as shown below:

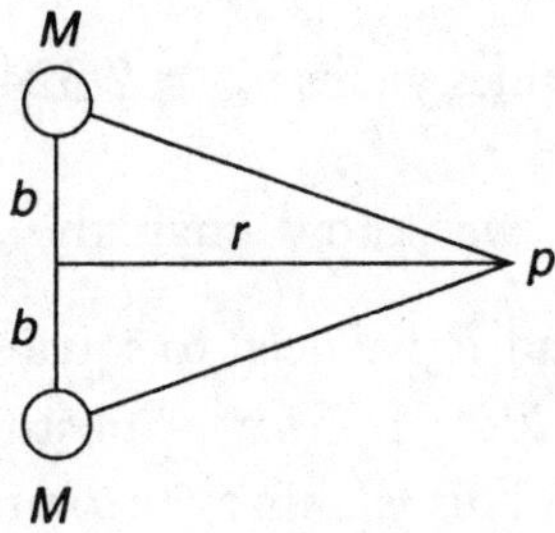

2. Explain why a heavier object near the surface of Earth does not fall faster than a lighter object (neglect air resistance).
3. Show that the units for **g** in N/kg are equivalent to m/s^2. Why is **g** referred to as the "gravitational field strength"?
4. Explain why objects in orbit appear to be "weightless."

ANSWERS EXPLAINED

MULTIPLE-CHOICE PROBLEMS

1. **(D)** The value of **g** varies inversely with the square of the distance from the center of Earth; therefore, if we double the distance from the center (as in this case), the value of **g** decreases by one-fourth: 1/4(9.8) = 2.45. Since $\mathbf{g} = \mathbf{F}/m$, alternative units are newtons per kilogram.

2. **(A)** If, using the formula for **g**, we take half the mass, the value decreases by one-half. If we decrease the radius by half, the value will increase by four times. Combining both effects results in an overall increase of two times.

3. **(D)** The formula for **g** is $\mathbf{g} = GM/R^2$. The planet is essentially a sphere of mass M, radius R, and density ρ (with $M = V_\rho$; where V is the volume). The volume of the planet is given by $V = 4/3\pi R^3$. Making the substitutions yields:

$$\mathbf{g} = G\frac{4\pi R_\rho}{3}$$

4. **(C)** The formula for orbital velocity is

$$\mathbf{v}_{\text{orbit}} = \sqrt{\frac{GM}{R}}$$

where R is the distance from the center of Earth. In this case we must add 300 km = 300,000 m to the radius of Earth. Thus, $R = 6.7 \times 10^6$ m. Substituting the given values yields $\mathbf{v}_{\text{orbit}} = 7728$ or 7.7×10^3 m/s.

5. **(B)** The formula for escape velocity is

$$\mathbf{v}_{\text{esc}} = \sqrt{\frac{2GM_{\text{Moon}}}{R_{\text{Moon}}}}$$

Substituting the given values yields $\mathbf{v}_{\text{esc}} = 2{,}324$ or 2.32×10^3 m/s.

6. **(A)** From question 6, we know that the escape velocity is given by $\mathbf{v}_{\text{esc}} = \sqrt{2GM/R}$. To find R, we need to square both sides, and then solve for the radius. This yields $R = 2GM/\mathbf{v}^2$. Substituting the given values yields $R = 3000$ or 3×10^3 m. (This is only a "theoretical" size for the black hole. As an interesting exercise, try calculating the value of **g** on such an object!)

7. **(C)** We use the formula for gravitational force:

$$\mathbf{F} = \frac{GM_1M_2}{R^2}$$

Substituting the given values (don't forget to square the distance!) yields $\mathbf{F} = 0.0067$ N.

8. **(D)** The force of gravity is an inverse-square-law relationship. This means that, as the distance is tripled, the force is decreased by one-ninth. One-ninth of 36 N is 4 N.

FREE-RESPONSE PROBLEMS

1. Let's designate as $\mathbf{g}_1$ the field strength at *P* caused by the top mass and designate as $\mathbf{g}_2$ the field strength at point *P* due to the bottom mass. Both of these field strengths are accelerations and therefore vectors. The distance from point *P* to the line connecting the masses is *r*, and the midpoint distance connecting the masses is *b*. Therefore, the distance from point *P* to each mass is given by the Pythagorean theorem and is equal to $\sqrt{r^2 + b^2}$.

   The direction of each acceleration **g** is directed toward each mass from point *P*. Since each mass is identical and the distance to each mass is the same, the angles formed by the vectors to the *x*-axis are the same. Let's call each angle θ such that tan $\theta = b/r$ in magnitude.

   From the above analysis we can conclude that

$$\mathbf{g}_1 = \mathbf{g}_2 = \frac{GM}{r^2 + b^2}$$

   The vector components of each field strength result in a symmetrical cancellation of the *y*-components. This is true since the direction of $\mathbf{g}_1$ is toward the upper left, and thus its *x*-component is directed to the left and its *y*-component is directed upward. Field strength $\mathbf{g}_2$ has an *x*-component that is also directed to the left and equal in magnitude to the *x*-component of $\mathbf{g}_1$. The *y*-component of $\mathbf{g}_2$ is directed downward and is also equal in magnitude to the *y*-component of $\mathbf{g}_1$. Since these two vectors are equal and opposite, they will sum to zero and will not contribute to the net resultant field (which is just directed horizontally to the left).

   What remains to be done is to determine the expression for the *x*-component of $\mathbf{g}_1$ or $\mathbf{g}_2$ and then multiply by 2. Since $\mathbf{g}_{1x} = \mathbf{g}_1 \cos\theta$, we can see from the geometry that

$$\cos\theta = \frac{r}{\sqrt{r^2 + b^2}}$$

   Combining results gives

$$\mathbf{g}_{\text{net}} = \frac{2GMr}{(r^2 + b^2)^{3/2}}$$

2. The acceleration due to gravity near the surface of Earth is given by

$$\mathbf{g} = \frac{GM_E}{R^2}$$

   Hence, this acceleration is independent of the mass of an object.

3. We know that $\mathbf{F} = m\mathbf{a}$, and for gravity, $\mathbf{F} = m\mathbf{g}$. Hence,

$$\mathbf{g} = \frac{\mathbf{F}}{m} = \text{N/kg}$$

   But we also know that $\text{N} = \text{kg} \cdot \text{m/s}^2$. Thus, in units,

$$\text{N/kg} = \text{m/s}^2$$

   A gravitational field measures the amount of force per unit mass. Thus, $\mathbf{g}$ is referred to as the gravitational field strength since $\mathbf{g} = \mathbf{F}/m$.

4. Objects in orbit are in free fall. Thus, all objects fall together and appear to be weightless.

CHAPTER 13

Fluids

KEY CONCEPTS

- Static Fluids
- Pascal's Principle
- Static Pressure and Depth
- Buoyancy and Archimedes' Principle
- Fluids in Motion
- Bernoulli's Equation

13.1 STATIC FLUIDS

Fluids represent states of matter that take the shape of their containers. Liquids are referred to as *incompressible fluids*, while gases are referred to as *compressible fluids*. In a Newtonian sense, liquids do work by being displaced, while gases do work by compressing or expanding. As we shall see later, the compressibility of gases leads to other effects described by the subject of thermodynamics.

Remember

Liquids are *incompressible* fluids, while **gases** are *compressible* fluids.

Fluids can exert pressure by virtue of their weight or force of motion. We have already defined the unit of pressure to be the pascal, which is equivalent to 1 N of force per square meter of surface area. An additional unit used in physics is the **bar**, where 1 bar = 100,000 Pa. Atmospheric pressure is sometimes measured in millibars.

13.2 PASCAL'S PRINCIPLE

In a fluid, static pressure is exerted on the walls of the container. Within the fluid these forces act perpendicular to the walls. If an external pressure is applied to the fluid, this pressure will be transmitted uniformly to all parts of the fluid. The last sentence is also known as **Pascal's principle** since it was developed by the French physicist Blaise Pascal.

Pascal's principle refers only to an external pressure. Within the fluid, the pressure at the bottom of the fluid is greater than at the top. We can also state that the pressure exerted on a small object in the fluid is the same regardless of the orientation of the object.

As an example of Pascal's principle consider the hydraulic press shown in Figure 13.1. The small-area piston A_1 has an external force $\mathbf{F}_1$ applied to it. At the other end, the large-area piston A_2 has some unknown force $\mathbf{F}_2$ acting on it. How do these forces compare? According to Pascal's principle, the force per unit area represents an external pressure which will be transmitted uniformly through the fluid. Thus, we can write

$$\frac{\mathbf{F}_1}{A_1} = \frac{\mathbf{F}_2}{A_2}$$

SAMPLE PROBLEM

Referring to Figure 13.1, suppose a force of 10 N is applied to the small piston of area 0.05 m^2. If the large piston has an area of 0.15 m^2, what is the maximum weight the large piston can lift?

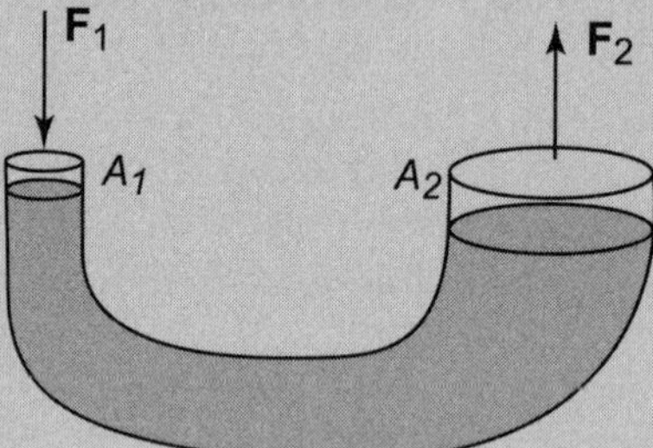

Figure 13.1

Solution

Since the secondary force is proportional to the ratio of the areas, $\mathbf{F}_2 = 30$ N.

13.3 STATIC PRESSURE AND DEPTH

Figure 13.2 shows a tall column of liquid in a sealed container. What is the pressure exerted on the bottom of the container? To answer this question, we first consider the weight of the column of liquid of height h. Since $\mathbf{F_g} = m\mathbf{g}$ and $m = \rho V$, the weight is $\rho\mathbf{g}V$. This is the force applied to the bottom of the container.

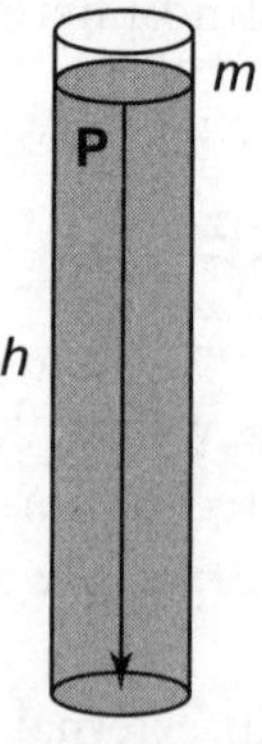

Figure 13.2

Now, in a container with a regular shape, $V = AH$, where A is the cross-sectional area (in this case, we have a cylinder whose cross sections are uniform circles). Thus, $\mathbf{F} = \rho\mathbf{g}V = \rho\mathbf{g}Ah$. Using the definition of pressure, we obtain

$$p = \frac{\mathbf{F}}{A} = \rho\mathbf{g}h$$

If the container is open at the top, then air pressure adds to the pressure of the column of liquid. The total pressure can therefore be written as $\mathbf{p} = \mathbf{p}_{ext} + \rho \mathbf{g} h$.

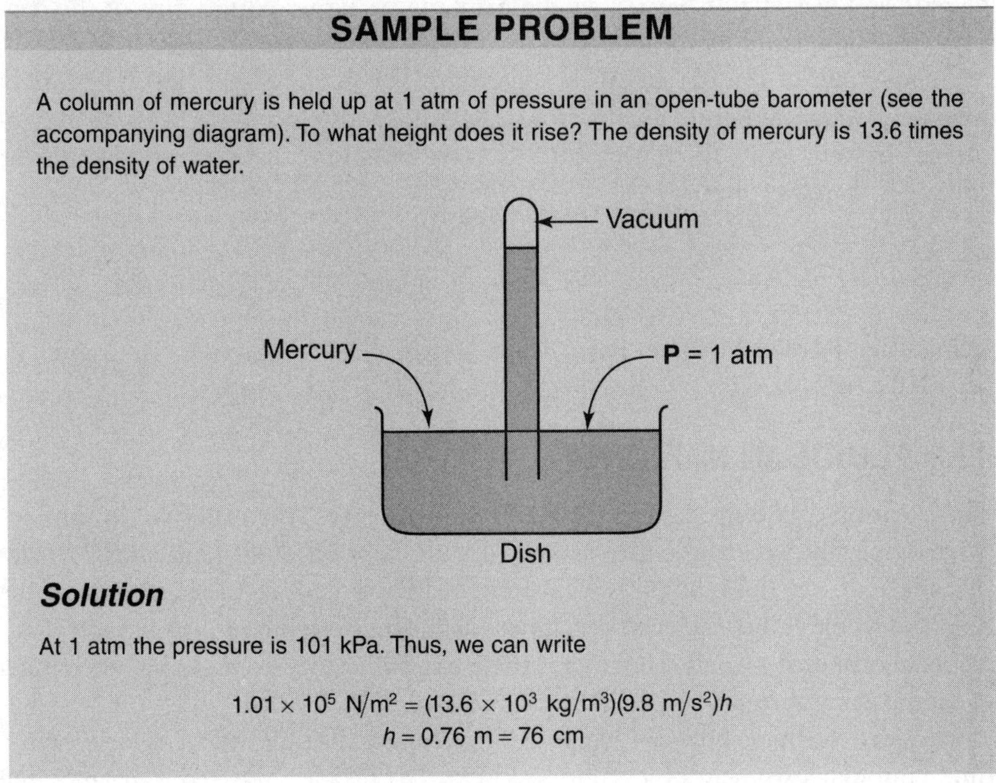

SAMPLE PROBLEM

A column of mercury is held up at 1 atm of pressure in an open-tube barometer (see the accompanying diagram). To what height does it rise? The density of mercury is 13.6 times the density of water.

Solution

At 1 atm the pressure is 101 kPa. Thus, we can write

$$1.01 \times 10^5 \text{ N/m}^2 = (13.6 \times 10^3 \text{ kg/m}^3)(9.8 \text{ m/s}^2)h$$
$$h = 0.76 \text{ m} = 76 \text{ cm}$$

13.4 BUOYANCY AND ARCHIMEDES' PRINCIPLE

When an object is immersed in water, it feels lighter. In a cylinder filled with water, the action of inserting a mass in the liquid causes it to displace upward. The volume of the water displaced is equal to the volume of the object (even if it is irregularly shaped), as illustrated in Figure 13.3.

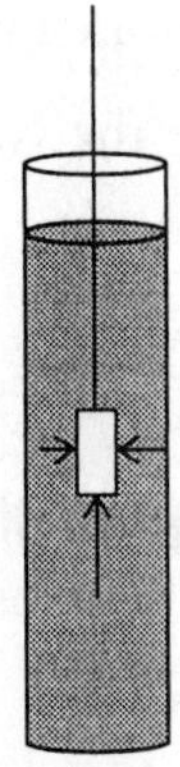

Figure 13.3 Measuring Volume by Water Displacement

Archimedes' principle states that the upward force of the water (called the **buoyant force**) is equal to the weight of the water displaced. Normally, one might think that an object floats if its density is less than water. This statement is only

TIP

A submerged object displaces a volume of water equal to its own volume.

partially correct. A steel needle floats because of surface tension, and a steel ship floats because it displaces a volume of water equal to its weight.

The weight of the water displaced can be found mathematically. The fluid displaced has a weight $\mathbf{F_g} = m\mathbf{g}$. Now, the mass can be expressed in terms of the density of the liquid and its volume, $m = \rho V$. Hence, $\mathbf{F_g} = \rho V\mathbf{g}$.

The volume of the object can be determined in terms of the apparent loss of weight in water. Suppose an object weighs 5 N in the air and 4.5 N when submerged in water. The difference of 0.5 N is the weight of the water displaced. The volume is therefore given by

$$V_f = \frac{\Delta m}{\rho_{\text{fluid}}} = \frac{\Delta \mathbf{F_g}}{\mathbf{g}\rho}$$

Using this relationship, we have $V_f = (0.5\text{ N})/(9.8\text{ N/kg})(1 \times 10^3\text{ kg/m}^3) = 5.1 \times 10^{-5}\text{ m}^3$.

13.5 FLUIDS IN MOTION

The situation regarding static pressures in fluids changes when they are in motion. Microscopically, we could try to account for the motion of all molecular particles that make up the fluid, but this would not be very practical. Instead, we treat the fluid as a whole and consider what happens as the fluid passes through a given cross-sectional area each second. Sometimes the word *flux* is used to describe the volume of fluid passing through a given area each second.

Consider the fluid shown in Figure 13.4 moving uniformly with a velocity **v** in a time t through a segment of a cylindrical pipe. The distance traveled is given by the product **v***t*. Since the motion is ideally smooth, there is no resistance offered by the fluid as different layers move relative to one another. This resistance is known as **viscosity**, and the type of fluid motion we are considering here is called **laminar flow**.

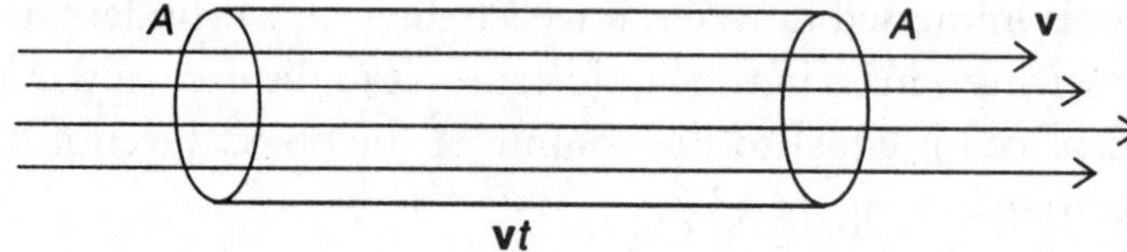

Figure 13.4

The rate of flow R is defined to be the volume of fluid flowing out of the pipe each second (in m^3/s):

$$R = \frac{\mathbf{v}tA}{t} = \mathbf{v}A$$

If the flow is laminar, then the **equation of continuity** states that the rate of flow R will remain constant. Therefore, as the cross-sectional area decreases, the velocity must increase:

$$\mathbf{v}_1 A_1 = \mathbf{v}_2 A_2$$

13.6 BERNOULLI'S EQUATION

Consider a fluid moving through an irregularly shaped tube at two different levels given by h_1 and h_2 as shown in Figure 13.5. At the lower level, the fluid exerts a pressure $\mathbf{P}_1$ while moving through an area A_1 with a velocity $\mathbf{v}_1$. At the top, the fluid exerts a pressure $\mathbf{P}_2$ while moving through an area A_2 with a velocity $\mathbf{v}_2$. Bernoulli's equation is related to changes in pressure as a function of velocity.

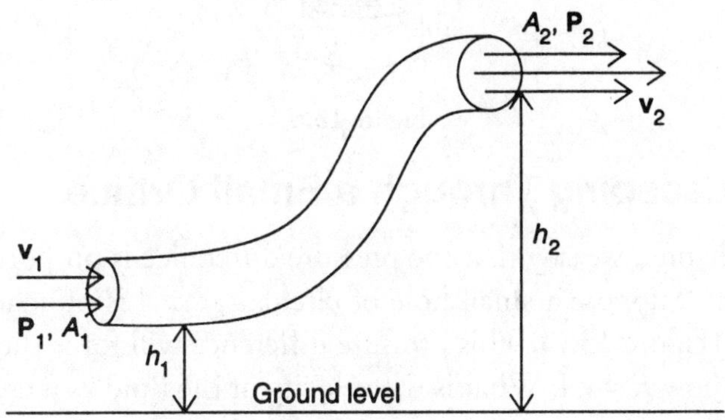

Figure 13.5

Let us begin by considering the work done in moving the fluid from position 1 to position 2. The power generated is equal to the product of the pressure and the rate of flow (**P**R), and so the work, which is equal to the product of power and time, can be written as

$$\Delta W = P_1A_1\mathbf{v}_1t - P_2A_2\mathbf{v}_2t = \frac{P_1m}{\rho} - \frac{P_2m}{\rho}$$

The change in the potential energy is given by

$$\Delta \text{PE} = m\mathbf{g}h_2 - m\mathbf{g}h_1$$

The change in kinetic energy is given by

$$\Delta \text{KE} = \frac{1}{2}m\mathbf{v}_2^2 - \frac{1}{2}m\mathbf{v}_1^2$$

Adding up both changes in energy and equating it with the work done, we obtain

$$\frac{\mathbf{P}_1m}{\rho} - \frac{\mathbf{P}_2m}{\rho} = m\mathbf{g}h_2 - m\mathbf{g}h_1 - \frac{1}{2}m\mathbf{v}_2^2 - \frac{1}{2}m\mathbf{v}_1^2$$

$$\mathbf{P}_1 + \rho\mathbf{g}h_1 + \frac{1}{2}\rho\mathbf{v}_1^2 = \mathbf{P}_2 + \rho\mathbf{g}h_2 + \frac{1}{2}\rho\mathbf{v}_2^2$$

TIP

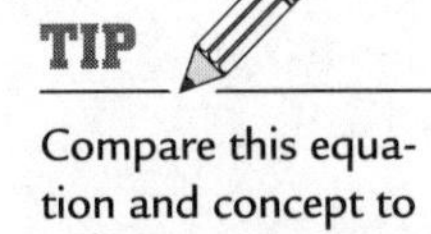

Compare this equation and concept to the equation for the conservation of mechanical energy.

The last equation is known as **Bernoulli's equation**. Now let us consider some applications of this equation.

A. A Fluid at Rest

In Figure 13.6, we see a static fluid. The two layers at heights h_1 and h_2 have static pressures $\mathbf{P}_1$ and $\mathbf{P}_2$. Since the fluid is at rest, Bernoulli's equation reduces to

$$\Delta\mathbf{P} = \mathbf{P}_2 - \mathbf{P}_1 = \rho\mathbf{g}\ \Delta h = \rho\mathbf{g}(h_2 - h_1)$$

The difference in pressure is just proportional to the difference in levels.

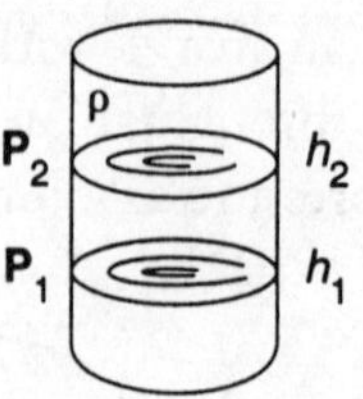

Figure 13.6

B. A Fluid Escaping Through a Small Orifice

Earlier in this chapter, we saw that the pressure difference is proportional to the difference in height. Suppose a small hole of circular area A is punched into the container below h_1 (Figure 13.7). This pressure difference will force the fluid out of the hole at a rate of flow $R = \mathbf{v}A$. What is the velocity of the fluid as it escapes? And what is the rate of flow?

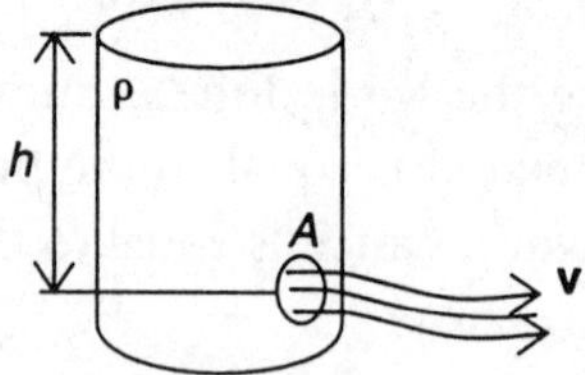

Figure 13.7

To answer these questions, we consider Bernoulli's equation as an analog of the conservation of energy equation. Let's choose the potential energy to be zero at h_1. The top fluid at h_2 is essentially at rest. Thus, using Bernoulli's equation, we can solve for the flow rate:

$$\frac{1}{2}\rho\mathbf{v}^2 = \rho\mathbf{g}\,\Delta h$$

$$\mathbf{v} = \sqrt{2\mathbf{g}\,\Delta h}$$

$$R = \mathbf{v}A = A\sqrt{2\mathbf{g}\,\Delta h}$$

C. A Fluid Moving Horizontally

Consider a fluid moving horizontally through a tube that narrows in area. Bernoulli's equation states that as the velocity of a moving fluid increases, its static pressure decreases. We can analyze this in Figure 13.8.

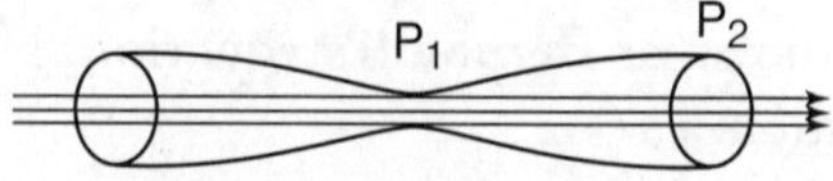

Figure 13.8 Venturi Tube

Since the level is horizontal, there is no change in pressure resulting from a change in level. We can therefore write

$$\mathbf{P}_1 + \frac{1}{2}\rho\mathbf{v}_1{}^2 = \mathbf{P}_2 + \frac{1}{2}\rho\mathbf{v}_2{}^2$$

$$\Delta\mathbf{P} = \mathbf{P}_1 - \mathbf{P}_2 = \frac{1}{2}\rho(\mathbf{v}_2^2 - \mathbf{v}_1^2)$$

SAMPLE PROBLEM

Water (ρ = 1000 kg/m^3) is flowing smoothly through a horizontal pipe that tapers from 1.5×10^{-3} m^2 to 0.8×10^{-3} m^2 in cross-sectional area. The pressure difference between the two sections is equal to 5,000 Pa. What is the volume flow rate of the water?

Solution

Since the water flows smoothly, we know that the volume flow rate is constant:

$$R = \mathbf{v}_1 A_1 = \mathbf{v}_2 A_2$$

We also know that from Bernoulli's equation

$$\mathbf{P}_1 + \frac{1}{2}\rho\mathbf{v}_1^2 + \rho\mathbf{g}y_1 = \mathbf{P}_2 + \frac{1}{2}\rho\mathbf{v}_2^2 + \rho\mathbf{g}y^2$$

But, since the pipe is horizontal, $y_1 = y_2$, and so the equation simplifies to

$$\mathbf{P}_1 + \frac{1}{2}\rho\mathbf{v}_1^2 = \mathbf{P}_2 + \frac{1}{2}\rho\mathbf{v}_2^2$$

We can simplify this by eliminating the velocities in each expression since

$$\mathbf{v}_1 = \frac{R}{A_1} \quad \text{and} \quad \mathbf{v}_2 = \frac{R}{A_2}$$

Thus,

$$\mathbf{P}_1 + \frac{1}{2}\rho\frac{R^2}{A_1^2} = \mathbf{P}_2 + \frac{1}{2}\rho\frac{R^2}{A_2^2}$$

We also know that $\mathbf{P}_1 > \mathbf{P}_2$ since pressure decreases with increasing velocity, and velocity increases with decreasing area. Thus we can write:

$$(\mathbf{P}_1 - \mathbf{P}_2) = \frac{1}{2}\rho R^2 \left(\frac{1}{A_2^2} - \frac{1}{A_1^2}\right)$$

Since we know all values (including the pressure difference $\mathbf{P}_1 - \mathbf{P}_2$), we can substitute and solve for R, and we obtain $R = 2.99 \times 10^{-3}$ m^3/s .

In aerodynamics, a wing moving in level flight has a lifting force acting on it exactly equal to its load. This force is caused by the pressure difference between the upper and lower surfaces of the wing. The wing is curved or *cambered* on top (see Figure 13.9), which has the effect of lowering the pressure on the top of wing (caused by faster-moving air molecules). It is this shape that causes the pressure difference and lift.

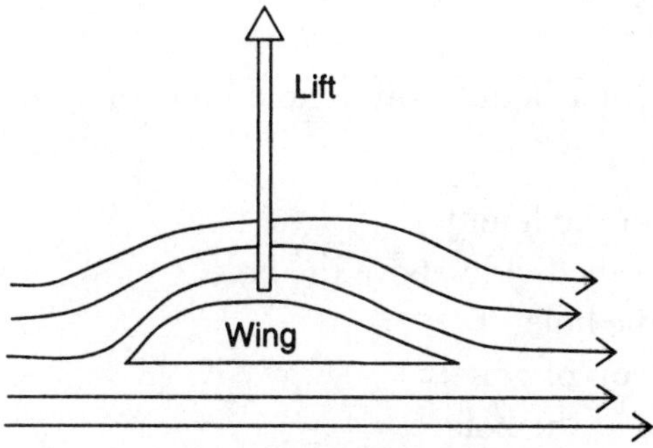

Figure 13.9 Airflow Pattern Over a Wing Section

Chapter Summary

- Liquids are called incompressible fluids whereas gases are called compressible fluids.
- Pascal's principle states that in a confined fluid at rest, any change in pressure is transmitted undiminished throughout the fluid.
- The pressure in a confined fluid is proportional to the density of the fluid and its depth ($\mathbf{P} = \rho \mathbf{g} h$).
- Archimedes' principle states that a submerged object will displace a volume of water equal to its own volume. A submerged object also experiences an upward force called the buoyant force, which is equal to the weight of the water displaced.
- An object will float if it displaces a volume of water equal in weight to its own weight in the air.
- Bernoulli's principle states that for a fluid in motion, the static pressure acting at right angles to the direction of the flow will decrease with an increase in velocity. This principle helps explain the lifting force of an airplane wing.

Problem-Solving Strategies for Fluids

Solving fluid problems is similar to solving particle problems. We treat the fluid as a whole unit (i.e., macroscopically) as opposed to microscopically. The units of pressure must be either pascals or N/m^2 in order to use the formulas derived.

The concepts of Bernoulli's principle and Archimedes' principle should be thoroughly understood, as well as their applications and implications. Buoyancy is an important physical phenomenon and an important part of your overall physics education.

As always, drawing a sketch helps. Often, conceptual knowledge will be enhanced if you understand how the variables in a formula are related. Consider questions that involve changing one variable and observing the effect on others.

Practice Exercises

MULTIPLE-CHOICE

1. The rate of flow of a liquid from a hole in a container depends on all of the following except

   (A) the density of the liquid
   (B) the height of the liquid above the hole
   (C) the area of the hole
   (D) the acceleration of gravity
   (E) the diameter of the hole

2. A person is standing on a railroad station platform when a high-speed train passes by. The person will tend to be

   (A) pushed away from the train
   (B) pulled in toward the train
   (C) pushed upward into the air
   (D) pulled down into the ground
   (E) unaffected by the passage of the train

3. Bernoulli's equation is based on which law of physics?

   (A) conservation of linear momentum
   (B) conservation of angular momentum
   (C) Newton's first law of motion
   (D) Newton's law of gravity
   (E) conservation of energy

4. Which of the following expressions represents the power generated by a liquid flowing out of a hole of area A with a velocity $\mathbf{v}$?

   (A) (pressure) × (rate of flow)
   (B) (pressure)/(rate of flow)
   (C) (rate of flow)/(pressure)
   (D) (pressure) × (velocity)/(area)
   (E) (rate of flow)/(area)

5. Which of the following statements is an expression of the equation of continuity?

   (A) Rate of flow equals the product of velocity and cross-sectional area.
   (B) Rate of flow depends on the height of the fluid above the hole.
   (C) Pressure in a static fluid is transmitted uniformly throughout.
   (D) Fluid flows faster through a narrower pipe.
   (E) None of these statements.

6. A moving fluid has an average pressure of 600 Pa as it exits a circular hole with a radius of 2 cm at a velocity of 60 m/s. What is the approximate power generated by the fluid?

   (A) 32 W
   (B) 62 W
   (C) 1200 W
   (D) 45 W
   (E) 12 W

7. Alcohol has a specific gravity of 0.79. If a barometer consisting of an open-ended tube placed in a dish of alcohol is used at sea level, to what height in the tube will the alcohol rise?

   (A) 8.1 m
   (B) 7.9 m
   (C) 15.2 m
   (D) 13.1 m
   (E) 0.79 m

8. An ice cube is dropped into a mixed drink containing alcohol and water. The ice cube sinks to the bottom. From this, you can conclude

(A) that the drink is mostly alcohol
(B) that the drink is mostly water
(C) that the drink is equally mixed with water and alcohol
(D) nothing unless you know how much liquid is present
(E) nothing unless you know the temperature of the drink

9. A 2-N force is used to push a small piston 10 cm downward in a simple hydraulic machine. If the opposite large piston rises by 0.5 cm, what is the maximum weight the large piston can lift?

(A) 2 N
(B) 40 N
(C) 20 N
(D) 4 N
(E) 1 N

CHALLENGE

10. Balsa wood with an average density of 130 kg/m_3 is floating in pure water. What percentage of the wood is submerged?

(A) 87%
(B) 13%
(C) 50%
(D) 25%
(E) 5%

FREE-RESPONSE

CHALLENGE

1. A U-tube open at both ends is partially filled with water. Benzene ($\rho = 0.897 \times 10^3$ kg/m^3) is poured into one arm, forming a column 4 cm high. What is the difference in height between the two surfaces?

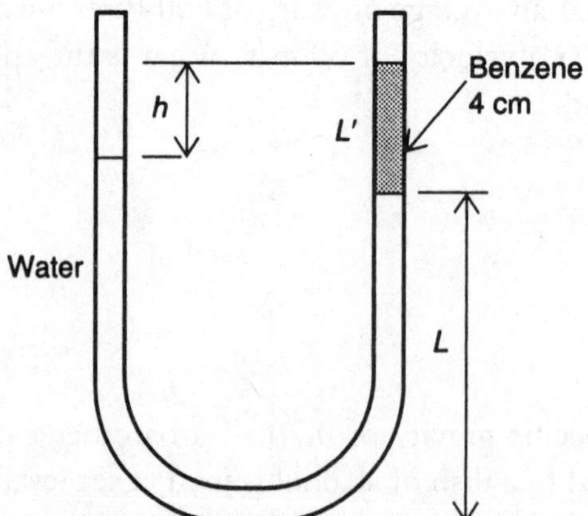

2. A Venturi tube has a pressure difference of 15,000 Pa. The entrance radius is 3 cm, while the exit radius is 1 cm. What are the entrance velocity, exit velocity, and flow rate if the fluid is gasoline ($\rho = 700$ kg/m^3)?

3. A cylindrical tank of water (height H) is punctured at a height h above the bottom. How far from the base of the tank will the water stream land (in terms of h and H)? What must the value of h be such that the distance at which the stream lands will be equal to H?

CHALLENGE

4. Which has more pressure on the bottom? A large tank of water 30 cm deep or a cup of water 35 cm deep? Explain your answer.

5. Two paper cups are suspended by strings and hung near each other. They are separated by about 10 cm. When you blow air between them, the cups are attracted to one another. Explain why this occurs.

6. A cup of water is filled to the brim and held above the ground. Two holes on either side of the cup and horizontally level with one another are punched into the cup, causing water to run out. When the cup is released and allowed to fall, it is observed that no water escapes during the fall. Explain why this occurs.

ANSWERS EXPLAINED

MULTIPLE-CHOICE PROBLEMS

1. **(A)** The flow rate of a liquid from a hole does not depend on the density of the liquid.

2. **(B)** Because of the Bernoulli effect, the speeding train reduces the air pressure between the person and the train. This pressure difference creates a force tending to pull the person into the train. Be very careful when standing on a train platform!

3. **(E)** Bernoulli's principle was developed as an application of conservation of energy.

4. **(A)** Power is expressed in J/s. The product of pressure (force divided by area) and flow rate (m^3/s) leads to newtons times meters over seconds (J/s).

5. **(D)** The equation of continuity states that the product of velocity and area is a constant for a given fluid. A consequence of this is that a fluid moves faster through a narrower pipe.

6. **(D)** The power is equal to the product of the pressure and the flow rate. The flow rate is equal to the product of the velocity and the cross-sectional area (which is a circle of radius 0.02 m). The area is given by $\pi r^2 = 1.256 \times 10^{-3}$ m^2. When we multiply this area by the velocity and the pressure, we get 45.2 W as a measure of the generated power.

7. **(D)** The pressure exerted by the air is balanced by the column of liquid alcohol in equilibrium:

$$h = \frac{\mathbf{P}}{\rho \mathbf{g}} = \frac{1.01 \times 10^5 \text{ N/m}^2}{(790 \text{ kg/m}^3)(9.8 \text{ m/s}^2)} = 13.1 \text{ m}$$

8. **(A)** Since the density of alcohol is less than that of water, ice floats "lower" in alcohol than in water. From the given information, the drink appears to be mostly alcohol.

9. **(B)** We need to use the conservation of work-energy in this problem. The work done on the small piston must equal the work done by the large piston. Since the ratio of displacements is 20:1, the large piston will be able to support a maximum load of 40 N (since 2 N × 20 = 40 N).

10. **(B)** The percentage submerged is given by the ratio of its density to that of pure water (1000 kg/m^3). Thus, 130/1000 = 0.13 = 13%.

FREE-RESPONSE PROBLEMS

1. Let L' be the level of benzene that will float on top of the water, let L be the level of water, and let h be the difference in levels. Since the tube is open, the pressures are equalized at both ends. Thus, we can write

$$(L + L' - h)\mathbf{g}\rho_w = L'\mathbf{g}\rho_b + L\mathbf{g}\rho_w$$
$$h = L'\left(1 - \frac{\rho_b}{\rho_w}\right)$$

Since the ratio of the densities is 0.879 and $L' = 4$ cm, $h = 0.484$ cm.

2. Using Bernoulli's equation and the equation of continuity, we can write

$$\Delta \mathbf{P} = \frac{\rho}{2}(\mathbf{v}_2^2 - \mathbf{v}_1^2)$$
$$\mathbf{v}_1 A_1 = \mathbf{v}_2 A_2$$
$$\mathbf{v}_2 = \mathbf{v}_1 \frac{A_1}{A_2}$$
$$\Delta \mathbf{P} = \frac{\rho}{2}\mathbf{v}_1^2\left(1 - \frac{A_1^2}{A_2^2}\right)$$

From the given information, we know that the ratio of areas $A_1/A_2 = 9$. Thus, substituting all given values into the last equation yields $\mathbf{v}_1 = 0.732$ m/s, $R = 0.0021$ m^3/s, and $\mathbf{v}_2 = 6.56$ m/s.

3. The change in potential energy must be equal to the change in horizontal kinetic energy:

$$m\mathbf{g}(H-h) = \left(\frac{1}{2}\right)m\mathbf{v}_x^{\,2}$$
$$\mathbf{v}_x = \sqrt{2\mathbf{g}(H-h)}$$

Now if we assume, as in projectile motion, that the horizontal velocity remains the same and the only acceleration of the stream is vertically downward because of gravity, we can write

$$\mathbf{x} = \mathbf{v}_x t$$
$$y = h = \frac{1}{2}\mathbf{g}t^2$$
$$t = \sqrt{\frac{2h}{\mathbf{g}}}$$
$$\mathbf{x} = 2\sqrt{h(H-h)}$$

For $\mathbf{x} = H$, we must have $h = H/2$, as can be easily verified using the preceding range formula.

4. The pressure at the bottom of a container of water depends on its depth and not on its volume. Thus, the pressure at the bottom of the 35-cm cup is greater.

5. Bernoulli's principle states that as the velocity of a moving fluid increases, the pressure it exerts decreases. Thus, blowing between the cups reduces the air pressure between them, causing a net force which pushes them together.

6. The water streams out of the holes because of the difference in pressure between the top and bottom of the cup. However, during free fall, both water and cup fall together and this prevents the water from coming out of the holes.

CHAPTER 14

Thermodynamics

KEY CONCEPTS

- Temperature and Its Measurement
- Pressure
- Molar Quantities
- The Ideal Gas Law Equation of State
- The First Law of Thermodynamics
- The Second Law of Thermodynamics and Heat Engines
- The Kinetic Theory of Gases
- Heat Transfer

14.1 TEMPERATURE AND ITS MEASUREMENT

Temperature can be defined as the relative measure of heat flow. In other words, temperature is a measure of the extent to which two bodies differ in heat content. We can state even more simply that temperature is a measure of how "hot" or "cold" a body is (relative to your own body or another standard or arbitrary reference). Human touch is unreliable in measuring temperature since what feels hot to one person may feel cool to another.

During the seventeenth century, Galileo invented a method for measuring temperature using water and its ability to react to changes in heat content by expanding or contracting. Figure 14.1 shows a simple "thermometer" that consists of a beaker of water in which is inverted a tube with a circular bulb at the top. When heat is added to the beaker, the level of water will rise.

TIP 

Heat and temperature are not the same thing.

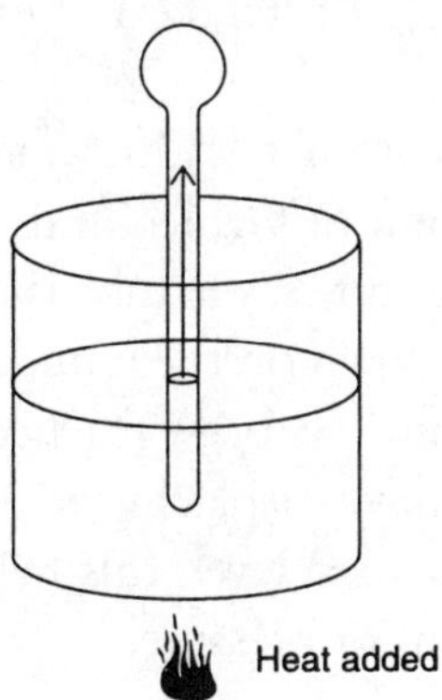

Figure 14.1

For two reasons, however, water is not the best substance to use to measure temperature. First, water has a relatively high freezing and low boiling point; second, it has a peculiar variation in density as a function of temperature because of its chemical structure. Even so, the phenomenon of "thermal expansion" is an important topic not only in this context but also for engineering buildings, roads, and bridges. We will discuss this aspect in more detail later in the chapter.

Most modern thermometers (see Figure 14.2) use mercury or alcohol in a small capillary tube. In the metric system, the standard sea-level points of reference are the freezing point (0 degree) and boiling point (100 degrees) of water. This temperature scale, formerly known as the centigrade scale, is now called the Celsius scale after Anders Celsius. (The English Fahrenheit system is not used in physics.)

Figure 14.2

One difficulty with the Celsius scale is that the zero mark is not a true zero since some substances freeze at temperatures below zero. Some gases, such as oxygen and hydrogen, do not even condense until their temperatures reach far below 0 degree Celsius. The discovery of a theoretical "absolute zero" temperature resulted in the so-called absolute or Kelvin temperature scale. A temperature of −273 degrees Celsius is redefined as "absolute zero" on the Kelvin scale. An easy way to convert from Celsius to Kelvin is to add the number 273 to the Celsius temperature. This just adds a scale factor that preserves the metric scale of the Celsius reading. In other words, there is still a 100-degree difference in temperature between melting ice and boiling water—273 K and 373 K, respectively. (Note that neither the word "degrees" nor the degree sign is used in expressing Kelvin temperatures.)

14.2 PRESSURE

Pressure is defined to be the measure of the force per unit area applied to matter. Mathematically, we can write this as

$$P = \frac{\mathbf{F}}{A}$$

and the units are newtons per square meter, or **pascals**. The choice of pascals (Pa) as the unit of pressure reflects the use of standard SI units in this review book and most textbooks.

Pressure is measured using a barometer. One of the first barometers was developed by Evangelista Torricelli (a student of Galileo's); it consists of a dish of mercury and an inverted tube. As Figure 14.3 shows, the tube is inverted into the dish of mercury. The outside air pressure pushes down on the liquid in the dish, forcing some of it up into the tube. When equilibrium has been reached (a pressure difference creates a force and does work on a liquid by displacing it), the level of mercury can be calibrated to read the air pressure. At sea level, this height corresponds to 76 cm and is referred to as 1 atmosphere (atm) of pressure.

We can calculate the amount of air pressure in such a tube at zero degree Celsius and sea level (a condition referred to by chemists as STP, an acronym for standard temperature and pressure) by recognizing that the air pressure forcing the level up is balanced by the weight of the fluid pushing the liquid down. Thus, in equilibrium, if $P = \mathbf{F}/A$, the force is equal to the weight of the column of liquid (given by the product $m\mathbf{g}$). In the tube, we have a liquid with density ρ and volume V at a height h. In the tube, $V = Ah$, and so $A = V/h$. The mass of liquid in the tube is related to the density and volume ($\rho = m/V$) so that $m = \rho V$.

Combining all these ideas, we can rewrite the pressure formula (for a column of liquid exposed to the air) as

$$P = \rho \mathbf{g} h$$

If we substitute the density of mercury, which is 13.6×10^3 kilograms per cubic meter, we find that 1 atmosphere equals 1.01×10^5 newtons per square meter, or 101 kilopascals (kPa) of pressure.

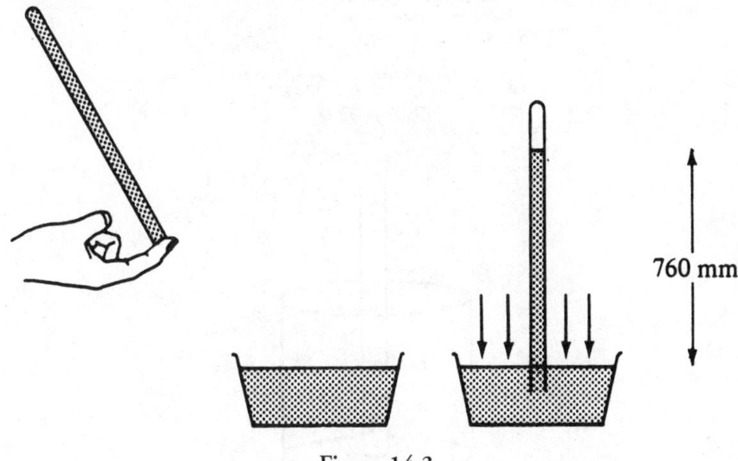

Figure 14.3

14.3 MOLAR QUANTITIES

Under the conditions of STP, most gases behave according to a simple relationship among their pressures, volumes, and absolute temperatures. This **equation of state** applies to what is referred to as an **ideal gas**. A real gas, however, is very complex, and a large number of variables are required to understand all of the intricate movements taking place within it.

To help us understand ideal gases, we introduce a quantity called the **molar mass**. In thermodynamics (and chemistry), the molar mass or **gram-molecular mass** is defined to be the mass of a substance that contains 6.02×10^{23} molecules. This number is called **Avogadro's number** (N_A), and 1 mole of an ideal gas occupies a volume of 22.4 liters at zero degree Celsius. For example, the gram-molecular mass of carbon is 12 grams, while the gram-molecular mass of oxygen is 32 grams. If m represents the actual mass, the number of moles of gas is given by the relationship $n = m/M$, where M equals the gram-molecular weight. For example, 64 grams of oxygen equals 2 moles, and 36 grams of carbon equals 3 moles.

14.4 THE IDEAL GAS LAW EQUATION OF STATE

Imagine that we have an ideal gas in a sealed, insulated chamber such that no heat can escape or enter. Figure 14.4 shows an external force placed on the piston at the top of the chamber so that the pressure change causes the piston to move down. The chamber is a cylinder with cross-sectional area A and height h. If a constant force **F** is applied, work is done to move the piston (frictionless) a distance h. Thus, $W = \mathbf{F}h$, and since $h = V/A$, the work done is

$$W = \frac{\mathbf{F}V}{A} = PV$$

In other words, the work done (in this case) is equal to the product of the pressure and the volume.

If the temperature of the gas remains constant, this relationship is known as **Boyle's law** and can be expressed as

$$PV = \text{constant}$$

or

$$P_1V_1 = P_2V_2$$

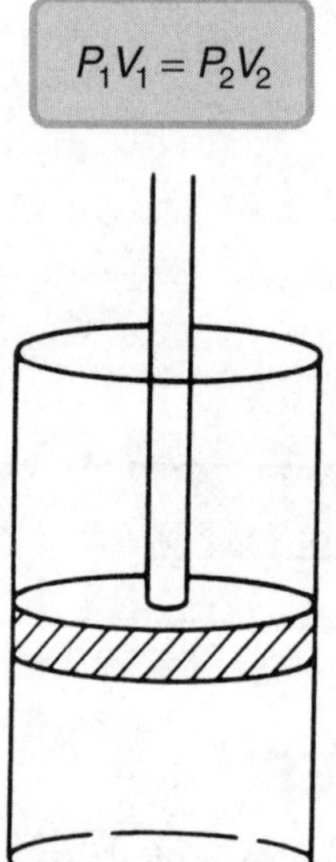

Figure 14.4

Experiments demonstrate that, if a gas is enclosed in a chamber of constant volume, there is a direct relationship between the pressure in the chamber and the absolute Kelvin temperature of the gas: the ratio of pressure to absolute temperature in an ideal gas always remains constant (indicated by a diagonal, straight-line graph at constant volume). We can write this relationship algebraically as:

$$\frac{P}{T} = \text{constant}$$

or

$$\frac{P_1}{T_1} = \frac{P_2}{T_2}$$

where the temperatures must be in kelvins.

French scientist Jacques Charles (and, independently, Joseph Gay-Lussac) developed a relationship, commonly known as **Charles's law**, between the volume and the absolute temperature of an ideal gas at constant pressure. This law states that the ratio of volume and absolute temperature in an ideal gas remains constant; that is,

$$\frac{V_1}{T_1} = \frac{V_2}{T_2}$$

The three gas laws—pressure-temperature law, Boyle's, and Charles's—can be summarized by the one **ideal gas law**, which has four forms:

The first form is a direct consequence of all three gas laws:

$$\frac{P_1 V_1}{T_1} = \frac{P_2 V_2}{T_2}$$

where the temperatures must be in kelvins.

The second form states that the product of pressure and volume, divided by the absolute temperature, remains a constant:

$$\frac{PV}{T} = \text{constant}$$

The third and more general form of the ideal gas law (and the one found in most advanced textbooks) includes the number of moles of gas involved and a universal gas constant R:

$$PV = nRT$$

where $R = 8.31$ joules per mole · kelvin (J/mol · K).

Finally, if $n = N/N_A$, where N is the actual number of molecules and N_A is Avogadro's number, a constant $k = R/N_A$, called **Boltzmann's constant**, can be introduced, and the ideal gas law takes the form

$$PV = NkT$$

Again, the temperatures must always be in kelvins. Additionally, if we express the volume in liters, the pressure in atmospheres, and the temperature in kelvins, the number of moles, n, is given by

$$n = (12.2)\frac{PV}{T}$$

SAMPLE PROBLEM

A confined gas at constant temperature has a volume of 50 m^3 and a pressure of 500 Pa. If it is compressed to a volume of 20 m^3, what is the new pressure?

Solution

We use Boyle's law:

$$\mathbf{P}_1 V_1 = \mathbf{P}_2 V_2$$
$$(500 \text{ Pa})(50 \text{ m}^3) = \mathbf{P}_2(20 \text{ m}^3)$$
$$\mathbf{P}_2 = 1{,}250 \text{ Pa}$$

SAMPLE PROBLEM

A confined gas is at a temperature of 27°C when it has a pressure and volume of 1,000 Pa and 30 m^3, respectively. If the volume is decreased to 20 m^3 and the temperature is raised to 30°C, what is the new pressure?

Solution

We will use the ideal gas law, but we must remember that the temperatures must be in Kelvin!

$$27^\circ\text{C} = 300 \text{ K}$$
$$30^\circ\text{C} = 323 \text{ K}$$

Now,

$$\frac{(\mathbf{P}_1 V_1)}{T_1} = \frac{(\mathbf{P}_2 V_2)}{T_2}$$
$$\frac{(1{,}000 \text{ Pa})(30 \text{ m}^3)}{300 \text{ K}} = \frac{\mathbf{P}_2(20 \text{ m}^3)}{323 \text{ K}}$$

Solving, we obtain

$$\mathbf{P}_2 = 1{,}615 \text{ Pa}$$

14.5 THE FIRST LAW OF THERMODYNAMICS

If work is being done to an ideal gas at constant temperature, Boyle's law states that the work done is equal to the product of the pressure and the volume (this product remaining constant). A graph of Boyle's law, seen in Figure 14.5, is a hyperbola. Pressure and volume changes that include temperature changes and are adiabatic obey a different kind of inverse relationship, in which the volume is raised to a higher exponential power (making the hyperbola steeper).

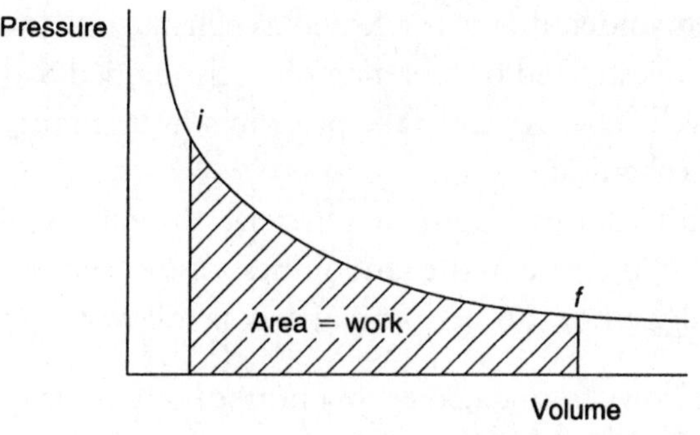

Figure 14.5

The pressure versus volume diagram for Boyle's law shows that the area enclosed by the curve from some initial state to a final state is equal to the work done.

The **first law of thermodynamics** is basically the law of conservation of energy. Let ΔQ represent the change in heat energy supplied to a system, ΔU represent the change in internal molecular energy, and ΔW represent the amount of net work done on the system. Then the first law of thermodynamics states that

$$\Delta U = \Delta Q + \Delta W$$

In thermodynamics, certain terms are often used and are worth defining:

Isothermal means "constant temperature."
Isobaric means "constant pressure."
Isochoric means "constant volume."
Adiabatic means "no gain or loss of heat to the system."

An example of the application of each term is shown in Figure 14.6 in a pressure versus volume diagram.

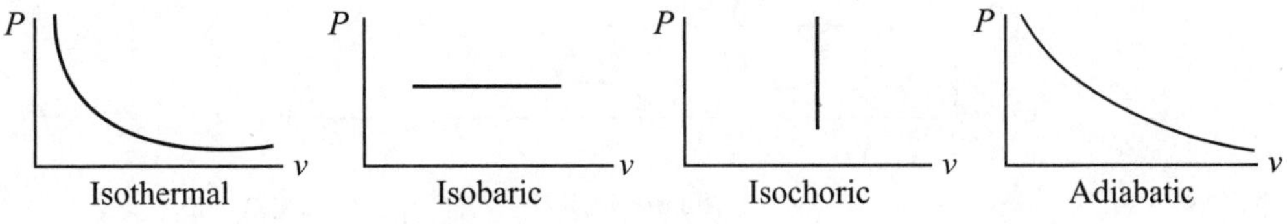

Figure 14.6

14.6 THE SECOND LAW OF THERMODYNAMICS AND HEAT ENGINES

No system is ideal. When a machine is manufactured and operated, heat is always lost because of friction. This friction creates heat that can be accounted for by the law of conservation of energy. Recall that heat flows from a hotter body to a cooler body. Thus, if there is a reservoir of heat at high temperature, the system will draw energy from that reservoir (hence its name). Only a certain amount of energy is needed to do work, so the remainder, if any, is released as exhaust at a lower temperature. This phenomenon was investigated by a French physicist named Sadi Carnot. Carnot was interested in the cyclic changes that take place in a heat engine, which converts thermal energy into mechanical energy.

In the nineteenth century, German physicist Rudolf Clausius coined the term *entropy* to describe an increase in the randomness or disorder of a system. These studies by Clausius and Carnot can be summarized as follows:

1. Heat never flows from a cooler to a hotter body of its own accord.
2. Heat can never be taken from a reservoir without something else happening in the system.
3. The entropy of the universe is always increasing.

These three statements embody the **second law of thermodynamics**: it is impossible to construct a cyclical machine that produces no other effect than to transfer heat continuously from one body to another at higher temperature.

Connected with the second law of thermodynamics is the concept of a reversible process. Heat flows spontaneously from hot to cold, and this action is irreversible unless acted upon by an outside agent. A system may be reversible if it passes from the initial to the final state by way of some intermediary equilibrium states. This reversible process, which can be represented on a pressure versus volume diagram, was studied extensively by Carnot.

Consider the simple Carnot cycle shown in Figure 14.7.

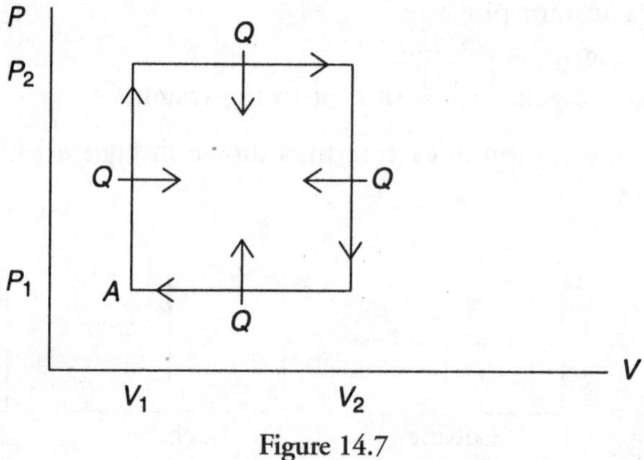

Figure 14.7

To change the system from P_1 to P_2 isochorically, heat must be supplied and the temperature will rise. Now, because of the pressure change, we allow the gas to expand isobarically from (P_2, V_1) to (P_2, V_2). The gas is doing work to expand so we supply energy to replace whatever is lost. Now, suppose we allow the system to cool by placing it in contact with a reservoir at lower temperature. If cooling is to occur isochorically from P_2 back to P_1 at the same volume, additional energy must again be added to the system from the outside. Finally, we allow the gas to contract from V_2 to V_1 isobarically. The temperature will decrease according to Charles's law, so additional energy must be supplied.

In each isochoric change, $\Delta W = 0$ since there were no changes in volume. During isobaric changes,

$$\Delta W_2 = P_2(V_2 - V_1) \quad \text{and} \quad \Delta W_1 = P_1(V_1 - V_2)$$

The net work is equal to $(P_2 - P_1)(V_2 - V_1)$ and would be equal to the area of the rectangle in Figure 14.7.

A Carnot cycle in which adiabatic temperature changes occur is shown in Figure 14.8.

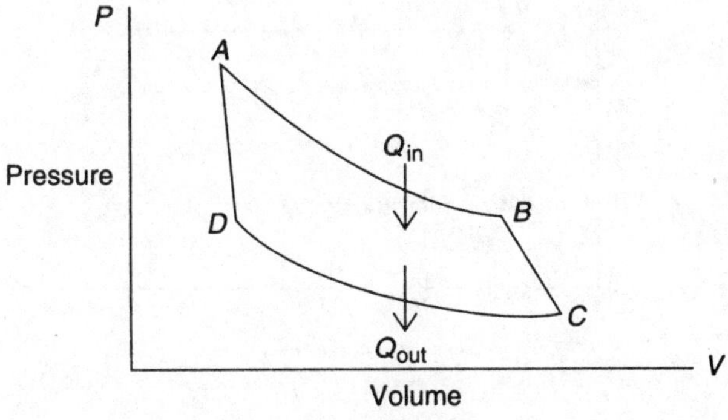

Figure 14.8

The isothermal changes occur from *A* to *B* and from *C* to *D*. The adiabatic changes are represented by segments *BC* and *AD*, in which no heat is gained or lost. Therefore, heat must be added in segment *AB* and released in *CD*. In other words, there is an isothermal expansion in *AB* at a temperature T_1. Heat is absorbed from the reservoir, and work is done. *BC* represents an adiabatic expansion in which an insulator has been placed in the container of the gas. The temperature falls, and work is done. In *CD*, we have an isothermal compression when the container is connected to a reservoir of temperature T_2 that is lower than T_1. Heat is released in this process, and work is done. Finally, in *DA*, the gas is compressed adiabatically and work is done. The total work done would be equal to the area of the figure enclosed by the graph.

The efficiency of this "heat engine" is given by the fractional ratio between the total work done and the original amount of heat added:

$$\text{Efficiency} = \frac{W}{Q_1} = 1 - \frac{Q_2}{Q_1}$$

The reason is that the net work done in one cycle is equal to the net heat transferred $(Q_1 - Q_2)$, since the change in internal energy is zero. If we use Kelvin temperatures, we can write

$$\text{Efficiency} = 1 - \frac{T_2}{T_1}$$

where $T_1 > T_2$ and both temperatures are in kelvins.

SAMPLE PROBLEM

A confined gas undergoes changes during a thermodynamic process as shown in the P-V diagram. Calculate the net energy added as heat during one cycle.

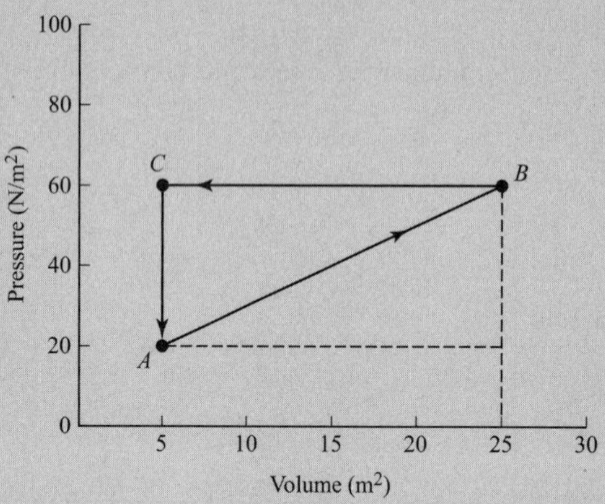

Solution

We can calculate the net energy by calculating the area *under* each segment of the graph. From $A \rightarrow B$, the work done *below* the diagonal segment is equal to 800 J. From $B \rightarrow C$, the area below the straight line segment is equal to 1,200 J. No work is done from $C \rightarrow A$ since we do not have a change in volume. Thus, the net energy is 800 J − 1,200 J = −400 J.

14.7 THE KINETIC THEORY OF GASES

Six assumptions are made for the kinetic theory of gases:

1. All gases consist of very large numbers of molecules, and the separations between molecules are large compared with the molecular sizes.
2. Collectively the molecules obey Newton's laws of motion, but individually the molecules move randomly.
3. The molecules undergo elastic collisions with each other.
4. The forces between molecules are negligible except during a collision.
5. The gas involved at any time is a pure gas.
6. The gas is in thermal equilibrium with the walls of the container.

Accepting these assumptions, we imagine that such a gas is contained in a cubical container of length ℓ (see Figure 14.9). We suppose that N molecules are moving randomly about the container. Under the assumption of large numbers, we can assume also that approximately one-third of the molecules will be moving along one of the principal coordinate axes, x, y, or z, at any given time.

We imagine one molecule of mass M moving from one side of the container horizontally to the other with a constant horizontal velocity $\mathbf{v}$. The molecule has a momentum $M\mathbf{v}$ in the x-direction. If an elastic collision takes place with one of the walls, the magnitude of the change in momentum will be $2M\mathbf{v}$.

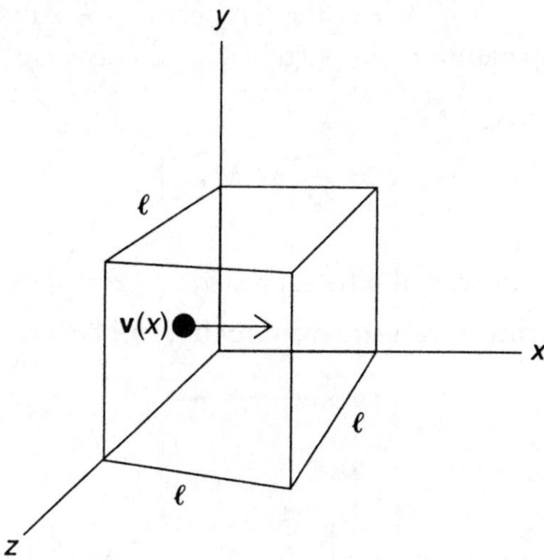

Figure 14.9

In a time t, the distance traveled by the molecule will be equal to $\mathbf{v}t$. Since the round-trip distance is 2ℓ, the total number of round trips is given by $\mathbf{v}t/2\ell$. Thus, in time t, the total change in momentum for this molecule is given by $M\mathbf{v}^2t/\ell$. The change in momentum is equal to the impulse applied to the molecule. Hence, if we divide the change in momentum by the time, the result is equal to the average force applied and we can state that

$$F = M\frac{\mathbf{v}^2}{\ell}$$

The total force on the wall is the sum of all such terms, and the total pressure will be the total force divided by the area of the wall. Since we have $N/3$ molecules in the x-direction moving with some average velocity $\overline{\mathbf{v}_x}$, what we need is the average of the squares of the velocities. If we divide the force by the area, we end up with the cube of the length of a side of the container, ℓ^3, which equals the volume V.

The final expression will be the pressure exerted on the wall in the x-direction. Now, we know that, for all three directions, we have equally probable average velocities. Thus the average of the squares of the velocities in the three directions is just one-third of the average of the square of the actual velocity. Therefore, the final expression for the pressure will be

$$P = \left(\frac{N}{3}\right)\left(\frac{M}{V}\right)\overline{\mathbf{v}^2}$$

where $\mathbf{v}$ is the actual velocity. Rearranging this expression, we see that the pressure is proportional to twice the average kinetic energy of the molecules (at constant volume):

$$PV = \frac{2}{3}N(\mathrm{KE}_{\mathrm{avg}})$$

where $\mathrm{KE}_{\mathrm{avg}} = (1/2)M\overline{\mathbf{v}^2}$. This is the kinetic theory equivalent of the ideal gas law derived in Section 14.4.

If we recall that $PV = NkT$, where the temperature T is in kelvins, we can write that the absolute temperature is proportional to the average kinetic energy of the molecules:

$$T = \frac{2}{3k}\left(\frac{1}{2}M\overline{\mathbf{v}^2}\right)$$

The average kinetic energy can also be expressed as $KE_{avg} = (3/2)kT$, and the square root of $\overline{\mathbf{v}^2}$, called the **root mean square velocity**, can be written as

$$\mathbf{v}_{rms} = \sqrt{\frac{3kT}{M}}$$

If we let m represent the molecular mass, usually expressed in units of grams per mole, the magnitude of the root mean square velocity can be written as:

$$\mathbf{v}_{rms} = \sqrt{\frac{3RT}{m}}$$

Finally, the change in internal energy, ΔU, is given by the expression

$$\Delta U = \frac{3}{2}nRT$$

where $R = 8.31$ joules per mole.

SAMPLE PROBLEM

Twenty cubic meters of an ideal gas is at 50°C and at a pressure of 50 kPa. If the temperature is raised to 200°C, and the volume is compressed by half, what is the new pressure of the gas?

Solution

We must first change all temperatures to kelvins:

$$50°\text{C} = 323\ \text{K} \quad \text{and} \quad 200°\text{C} = 473\ \text{K}$$

We now use the ideal gas law for two sets of conditions:

$$\frac{P_1V_1}{T_1} = \frac{P_2V_2}{T_2}$$

Substituting the appropriate values, we get

$$\frac{(50)(20)}{323} = \frac{(P_2)(10)}{473}$$

Solving for the pressure gives $P_2 = 146.4$ kPa.

14.8 HEAT TRANSFER

Heat can be transferred by conduction, convection, or radiation.

Because metals contain a large number of free electrons, heat **conduction** is the main mode of heat transfer. If you have ever left a spoon in a cup of hot water for a long time, you have experienced heat conduction.

The rate of heat transfer ($\Delta Q/t$) through a solid slab of material (see Figure 14.10) depends on four quantities:

1. The temperature difference on either side of the slab (ΔT).
2. The thickness of the slab (L).
3. The frontal area of the slab (A).
4. The thermal conductivity (k) of the material. This is a number in units of watts per meter per degree Celsius (W/m · °C).

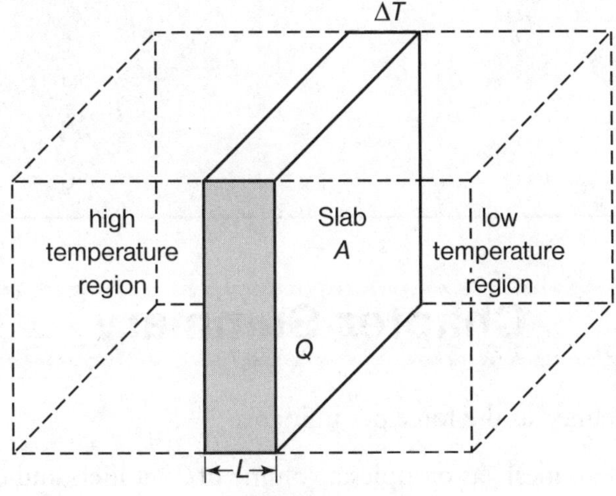

Figure 14.10

$$H = \frac{\Delta Q}{t} = \frac{kA\Delta T}{L}$$

Table 14.1 lists some typical thermal conductivities.

Heat can be transferred by **convection** in a gas or liquid as matter circulates throughout the system. For example, hot air is less dense and rises while cold air is more dense and sinks.

If the air above the ground is heated by the Sun, that air will rise and cool. This colder air will sink, and circulation will occur. A room can be heated by convection as hot air displaces cold air and circulates throughout the room. Convection currents can also be seen in water being heated. If a dye is placed into water that is being heated from below, the patterns formed will show the flow of convection.

If heat is transferred by means of electromagnetic waves (see Chapter 20), the process is known as **radiation**. Infrared waves from the Sun are a typical mechanism for heating the atmosphere. Radiation, like all waves, transfers energy only from place to place. Radiation does not require a material medium for propagation.

Table 14.1

Thermal Conductivities

Material	k (W/m · °C)
Air	0.024
Aluminum	2.50
Brick	0.6
Brass	109
Concrete	0.8
Cork	0.04
Copper	385
Glass	0.8
Ice	1.6
Iron, steel	50
Silver	406

Chapter Summary

- Pressure is defined as the force per unit area.
- One mole of an ideal gas occupies a volume of 22.4 liters and contains 6.02×10^{23} molecules (Avogadro's number) at STP.
- Boyle's law states that if a confined ideal gas is kept at constant temperature, then the product of the pressure and the volume remains constant ($\mathbf{P}V=$ constant).
- Charles's law states that if a confined ideal gas is kept at constant pressure, then there is a direct relationship between the volume and absolute temperature ($V/T=$ constant).
- For an ideal gas, the relationship $\mathbf{P}V/T$ is equal to a constant.
- The first law of thermodynamics is essentially a restatement of the law of conservation of energy: $\Delta U = \Delta Q + \Delta W$.
- The second law of thermodynamics relates to an increase in entropy (randomness) in the universe and the fact that heat is always generated when work is done. It is therefore impossible to create an ideal machine without any losses due to heat.
- Ideal gases can be described using a statistical theory known as the kinetic theory of gases in which the molecules of a gas undergo elastic collisions and the temperature is a measure of the average kinetic energy of the molecules.
- Heat can be transferred by conduction, convection, or radiation.

Problem-Solving Strategies for Thermodynamics and the Kinetic Theory of Gases

It is important to remember that all temperatures must be converted to kelvins before doing any calculations. Also, since the units for the gas constant R and Boltzmann's constant k are in SI units, all masses or molecular masses must be in kilograms. For example, the molecular mass of nitrogen is usually given as 28 grams per mole. However, in an actual calculation, this must be converted to 0.028 kilogram per mole.

You also should remember that, unless a comparison relationship is used (such as Boyle's law or the preceding sample problem), all pressures must be in units of pascals (not kilopascals, atmospheres, or centimeters of mercury) and all volumes must be in cubic meters. However, in a comparison relationship, the chosen units are irrelevant (except for temperature, which is always in kelvins), and you can use atmospheres for pressure, or liters (1 liter = 1000 cubic centimeters) for volume. The molar specific heat capacities depend on the number of moles, not the number of grams (the specific heat capacity).

Practice Exercises

MULTIPLE-CHOICE

1. At constant temperature, an ideal gas is at a pressure of 30 cm of mercury and a volume of 5 L. If the pressure is increased to 65 cm of mercury, the new volume will be

   (A) 10.8 L
   (B) 2.3 L
   (C) 0.43 L
   (D) 1.7 L
   (E) 5.6 L

2. At constant volume, an ideal gas is heated from 75°C to 150°C. The original pressure was 1.5 atm. After heating, the pressure will be

   (A) doubled
   (B) halved
   (C) the same
   (D) less than doubled
   (E) quadrupled

3. At constant pressure, 6 m^3 of an ideal gas at 75°C is cooled until its volume is halved. The new temperature of the gas will be

   (A) 174°C
   (B) 447°C
   (C) −99°C
   (D) 37.5°C
   (E) 100°C

4. Water is used in an open-tube barometer. If the density of water is 1000 kg/m^3, what will be the level of the column of water at sea level?

(A) 10.3 m
(B) 12.5 m
(C) 13.6 m
(D) 11.2 m
(E) 9.7 m

5. As the temperature of an ideal gas increases, the average kinetic energy of its molecules

(A) increases, then decreases
(B) decreases
(C) remains the same
(D) decreases, then increases
(E) increases

6. The product of pressure and volume is expressed in units of

(A) pascals
(B) kilograms per newton
(C) watts
(D) joules
(E) newtons

7. Which of the following is equivalent to 1 Pa of gas pressure?

(A) 1 kg/s^2
(B) 1 kg · m/s
(C) 1 kg · m^2/s^2
(D) 1 kg/m · s^2
(E) 1 kg/m

8. What is the efficiency of a heat engine that performs 700 J of useful work from a reservoir of 2,700 J?

(A) 74%
(B) 26%
(C) 35%
(D) 65%
(E) 50%

9. The first law of thermodynamics is a restatement of which law?

(A) Conservation of charge
(B) Conservation of energy
(C) Conservation of entropy
(D) Conservation of momentum
(E) None of the above

10. Which of the following graphs represents the relationship between pressure and volume for an ideal confined gas at constant temperature?

(A)

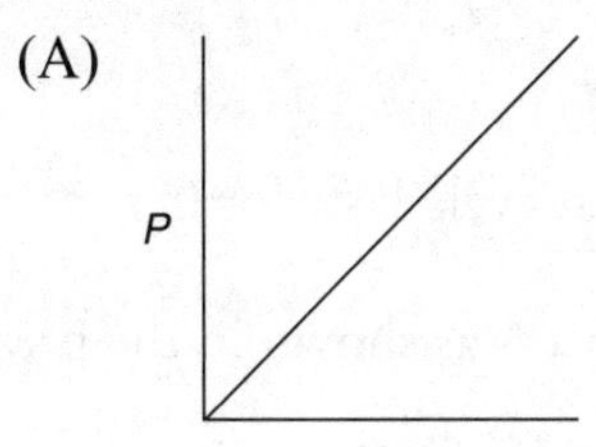

(B)

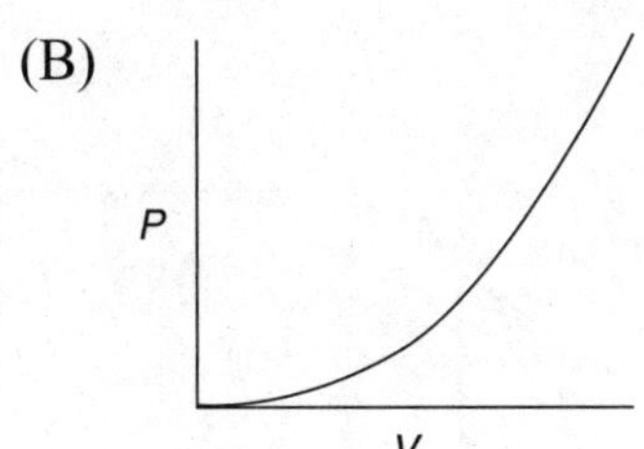

(C)

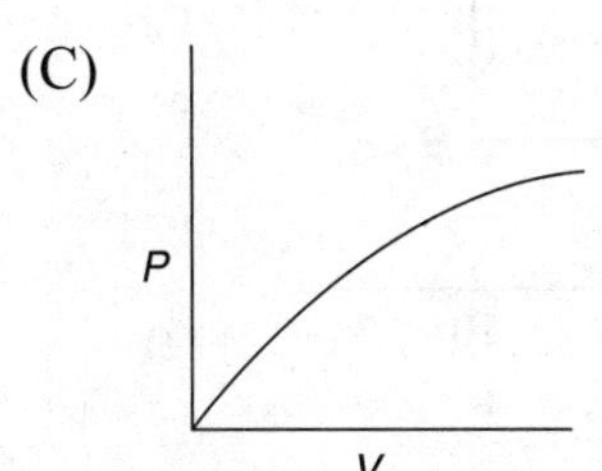

(D)

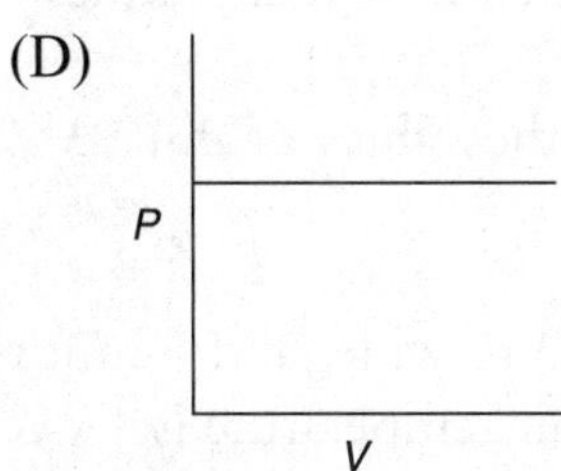

(E)

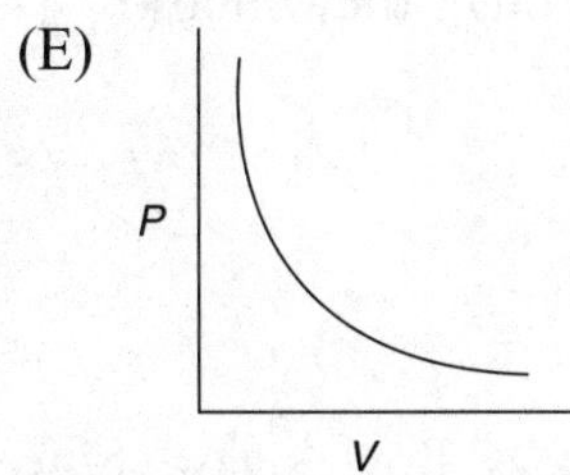

FREE-RESPONSE

1. (a) How many molecules of helium are required to fill a balloon with a diameter of 50 cm at a temperature of 27°C? **CHALLENGE**
   (b) What is the average kinetic energy of each molecule of helium?
   (c) What is the average velocity of each molecule of helium?

2. An engine absorbs 2000 J of heat from a hot reservoir and expels 750 J to a cold reservoir during each operating cycle.

   (a) What is the efficiency of the engine?
   (b) How much work is done during each cycle?
   (c) What is the power output of the engine if each cycle lasts for 0.5 s?

CHALLENGE

3. A monatomic ideal gas undergoes a reversible process as shown in the pressure versus volume diagram below:

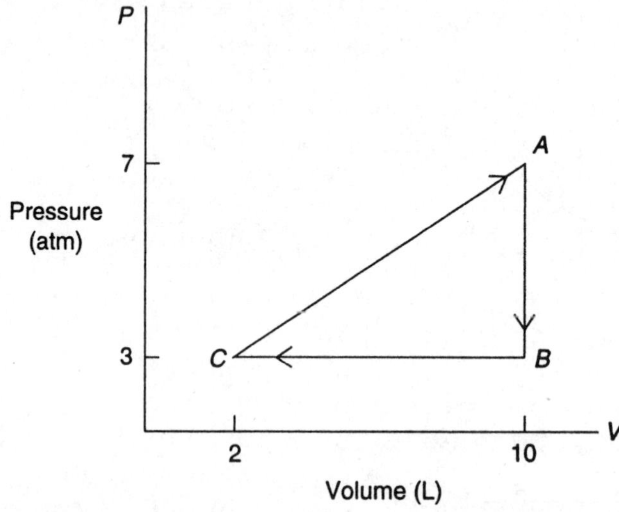

   (a) If the temperature of the gas is 500 K at point *A*, how many moles of gas are present?
   (b) For each cycle (*AB*, *BC*, and *CA*), determine the values of ΔW, ΔU, and ΔQ.

4. A glass window has dimensions of 1.2 m high, 1.0 m wide, and 5 mm thick. The outside temperature is 5°C, while the inside air temperature is 15°C.

   (a) What is the conductive rate of heat transfer through the window?
   (b) How much energy is transferred in one hour?

ANSWERS EXPLAINED

MULTIPLE-CHOICE PROBLEMS

1. **(B)** At constant temperature, an ideal gas obeys Boyle's law. Since we are making a two-position comparison, we can use the given units for pressure and volume. Substituting the given values into Boyle's law, we obtain $(30)(5) = (65)(V_2)$, which implies that $V_2 = 2.3$ L.

2. **(D)** At constant volume, the changes in pressure are directly proportional to the changes in absolute Kelvin temperature. Therefore, even though the Celsius temperature is being doubled, the Kelvin temperature is not! Thus, the new pressure must be less than doubled.

   Alternatively, you can calcluate the new pressure since 75°C = 348 K and 150°C = 423 K. Thus $1.5/348 = P_2/423$, and $P_2 = 1.8$ atm (less than double).

Again, since a comparison relationship is involved, we can keep the pressure in atmospheres for convenience.

3. **(C)** This is a Charles's law problem, and the temperatures must be in kelvins. Since 75°C = 348 K, we can write that $6/348 = 3/T_2$. However, the volume change is directly proportional to the Kelvin temperature change. Thus $T_2 = 174$ K = −99°C.

4. **(A)** The formula for pressure is $P = \rho\mathbf{g}h$. The density of water is 1000 kg/m^3, and $\mathbf{g}$ = 9.8 m/s^2. We also know that at sea level P = 101 kPa, which must be converted to pascals or newtons per square meter. Thus $h = P/\rho\mathbf{g}$, and making the necessary substitutions gives h = 10.3 m (a very large barometer over 35 ft tall!).

5. **(E)** The absolute temperature of an ideal gas is directly proportional to the average kinetic energy of its molecules. Hence, the average kinetic energy increases with temperature.

6. **(D)** The units of the product PV are joules, the units of energy.

7. **(D)** One pascal is equivalent to 1 N/m^2, but 1 N is equal to 1 kg · m/s^2. Thus the final equivalency is $\dfrac{1 \text{ kg}}{\text{m} \cdot \text{s}^2}$.

8. **(B)** Efficiency = (700 J/2700 J) × 100% = 26%.

9. **(B)** The first law of thermodynamics expresses the conservation of energy.

10. **(E)** Boyle's law relates pressure and volume in an ideal gas. This is an inverse relationship that looks like a hyperbola.

FREE-RESPONSE PROBLEMS

1. (a) From the ideal gas formula, we know that $PV = NkT$, where N equals the number of molecules and k is Boltzmann's constant. For a 50-cm-diameter balloon, the radius will be 0.25 m. One atmosphere is equal to 101,000 Pa of pressure. The volume of a sphere is given by $V = (4/3)\pi R^3$, which equals 0.065 m^3 upon substitution.

   We know also that $N = PV/kT$, where T is equal to 300 K in this problem. Substituting all known values gives

$$N = \frac{(101{,}000)(0.065)}{(1.38 \times 10^{-23})(300)} = 1.59 \times 10^{24} \text{ molecules}$$

   (b) The average kinetic energy of each molecule is equal to

$$\overline{\text{KE}} = \frac{3}{2}kT = (1.5)(1.38 \times 10^{-23})(300) = 6.21 \times 10^{-21} \text{ J}$$

(c) The average velocity of each molecule is given by

$$v_{\text{rms}} = \sqrt{\frac{3RT}{m}} = \sqrt{\frac{(3)(8.31)(300)}{0.004}} = 1367 \text{ m/s}$$

2. (a) The efficiency of the heat engine is given by the formula

$$e = \frac{\Delta Q}{Q_{\text{hot}}} = \frac{2000-750}{2000} = 0.625 \text{ or } 62.5\%$$

(b) The work done during each cycle is equal to

$$\Delta Q = 2000 - 750 = 1250 \text{ J}$$

(c) The power output each cycle equals the work done divided by the time:

$$P = \frac{1250}{0.5} = 2500 \text{ W}$$

3. (a) Since we know the pressure, volume, and absolute temperature of the ideal (monatomic) gas at point *A*, we can use the ideal gas law, $PV = nRT$, to find the number of moles, *n*.

First, however, we must be in standard units: 7 atm = 707,000 Pa, and since 1 L = 0.001 m^3, then 10 L = 0.01 m^3. Thus:

$$n = \frac{(707{,}000)(0.01)}{(8.31)(500)} = 1.7 \text{ mol of gas}$$

(b) i. The cycle *AB* is isochoric, and so

$$\Delta W = 0$$

Thus, $\Delta U = \Delta Q$. Now, as the pressure decreases from 7 atm to 3 atm, the absolute temperature decreases proportionally. Thus we can write

$$\frac{7}{500} = \frac{3}{T_B}$$

which implies that the absolute temperature at point *B* is equal to 214.29 K. To find the change in the internal energy, we recall that

$$\Delta U = \frac{3}{2} nR\, \Delta T = (1.5)(1.7)(8.31)(-285.7)$$
$$= -6054 \text{ J} = \Delta Q \text{ also!}$$

ii. The cycle *BC* is isobaric, and the decrease in volume from 10 L to 2 L is accompanied by a corresponding decrease in absolute temperature. We know from step (i) that the temperature at *B* is 214.29 K, so we can use Charles's law to find the temperature at point *C*. Also, the work done is negative since we are expelling energy to reduce pressure and volume. We can therefore write

$$\frac{10}{214.29} = \frac{2}{T_C}$$

which implies that the absolute temperature at point *C* is equal to 42.86 K. Therefore $\Delta T = -171.43$ K.

Now, in this process, work is done to compress the gas, and $W = P(V_C - V_B)$. The pressure is constant at $P = 3$ atm = 303,000 Pa, and $\Delta V = -8$ L = -0.008 m^3.

Thus

$$\Delta W = (303{,}000)(-0.008) = -2424 \text{ J}$$

To find ΔU, we note that the temperature change of the 1.7 moles was -171.43 K, and so

$$\Delta U = (1.5)(1.7)(8.31)(-171.43) = -363.69 \text{ J}$$

To find ΔQ, we use the first law of thermodynamics:

$$\Delta Q = \Delta U + \Delta W = -3632.69 - 2424 = -6057 \text{ J}$$

iii. The last part of the reversible cycle puts back all the work and energy, a total of 12,111 J of heat (the sum of ΔQ in steps i and ii), that was removed in the first two steps. Thus, for *CA*,

$$\Delta Q = +12{,}111 \text{ J}$$

In a similar way, the total change in internal energy equaled -9686.69 J. Thus, in cycle *CA*,

$$\Delta U = +9686.69 \text{ J}$$

Finally,

$$\Delta W = \Delta Q - \Delta U = 12{,}111 - 9686.69 = 2424.31 \text{ J}$$

Notice that, since no work was done in step i, this is approximately equal to the work done in step ii!

4. (a) The formula for conductive heat transfer is $H = \dfrac{kA\Delta T}{L}$. First we need to change the glass thickness to meters

$$5 \text{ mm} = 0.005 \text{ m}$$

From Table 14.1, we see that the thermal conductivity of glass is

$$k = 0.8 \ W/\text{m} \cdot {}^\circ\text{C}$$

The area of the glass window is simply

$$A = (1.2 \text{ m})(1.0 \text{ m}) = 1.2 \text{ m}^2$$

By substituting in all the values and noting that $\Delta T = 10^\circ$C, we get

$$H = \frac{(0.8 \ W/\text{m} \cdot {}^\circ\text{C})(1.2 \text{ m}^2)(10^\circ\text{C})}{(0.005 \text{ m})}$$

$$= 1920 \ W$$

b) Since energy = power × time, the energy transferred in 1 hour (3600 s) is

$$E = (1920 \text{ J/s})(3600 \text{ s}) = 6.912 \times 10^6 \text{ J}.$$

Electrostatics

CHAPTER 15

KEY CONCEPTS

- The Nature of Electric Charges
- The Detection and Measurement of Electric Charges
- Coulomb's Law
- The Electric Field
- Electric Potential
- Capacitance

15.1 The Nature of Electric Charges

The ancient Greeks used to rub pieces of amber on wool or fur. The amber was then able to pick up small objects that were not made of metal. The Greek word for amber was *elektron* (ελεκτρον)—hence the term *electric.* The pieces of amber would retain their attractive property for some time, so the effect appears to have been **static**. The amber acted differently from magnetic ores (lodestones), naturally occurring rocks that attract only metallic objects.

In modern times, hard rubber, such as ebonite, is used with cloth or fur to dramatically demonstrate the properties of electrostatic force. If you rub an ebonite rod with cloth and then bring it near a small cork sphere painted silver (called a "pith ball") and suspended on a thin thread, the pith ball will be attracted to the rod. When the ball and rod touch each other, the pith ball will be repelled. If a glass rod that has been rubbed with silk is then brought near the pith ball, the ball will be attracted to the rod. If you touch the pith ball with your hand, the pith ball will return to its normal state. This is illustrated in Figure 15.1.

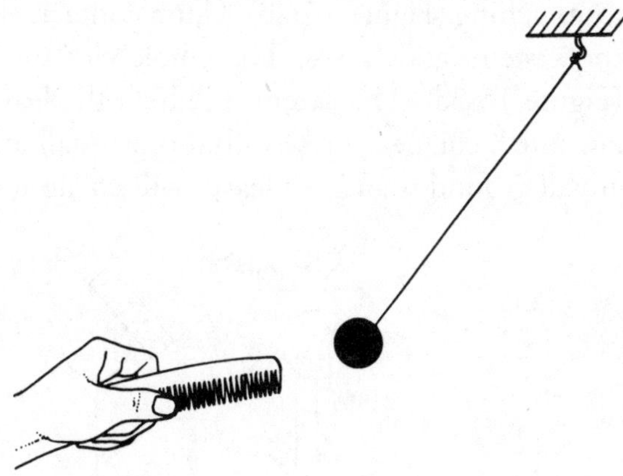

Figure 15.1

In the nineteenth century, chemical experiments to explain these effects showed the presence of molecules called **ions** in solution. These ions possessed similar affinities for certain objects, such as carbon or metals, placed in the solution. These objects are called **electrodes**. The experiments confirmed the existence of two types of ions, **positive** and **negative**. The effects they produce are similar to the two types of effects produced when ebonite and glass are rubbed. Even though both substances attract small objects, these objects become **charged** oppositely when rubbed, as indicated by the behavior of the pith ball. Further, chemical experiments coupled with an atomic theory demonstrated that in solids it is the negative charges that are transferred. Additional experiments by Michael Faraday in England during the first half of the nineteenth century suggested the existence of a single, fundamental carrier of **electric charge**, which was later named the **electron**. The corresponding carrier of positive charge was termed the **proton**.

15.2 The Detection and Measurement of Electric Charges

When ebonite is rubbed with cloth, only the part of the rod in contact with the cloth becomes charged. The charge remains localized for some time (hence the name *static*). For this reason, among others, rubber, along with plastic and glass, is called an **insulator**. A metal rod held in your hand cannot be charged statically for two reasons. First, metals are **conductors**; that is, they allow electric charges to flow through them. Second, your body is a conductor, and any charges placed in the metal rod are conducted out through you (and into the earth). This effect is called **grounding**. The silver-coated pith balls mentioned in Section 15.1 can become statically charged because they are suspended by thread, which is an insulator. They can be used to detect the presence and sign of an electric charge, but they are not very helpful in obtaining a qualitative measurement of the magnitude of charge they possess.

An instrument that is often used for qualitative measurement is the **electroscope**. One form of electroscope consists of two "leaves" made of gold foil (Figure 15.2a). The leaves are vertical when the electroscope is uncharged. As a negatively charged rod is brought near, the leaves diverge. If we recall the hypothesis that only negative charges move in solids, we can understand that the electrons in the knob of the electroscope are repelled down to the leaves through the conducting stem. The knob becomes positively charged, as can be verified with a charged pith ball, as long as the rod is near but not touching (Figure 15.2b). Upon contact, electrons are directly transferred to the knob, stem, and leaves. The whole electroscope then becomes negatively charged (Figure 15.2c). The extent to which the leaves are spread apart is an indication of how much charge is present (but only qualitatively). If you touch the electroscope, you will ground it and the leaves will collapse together.

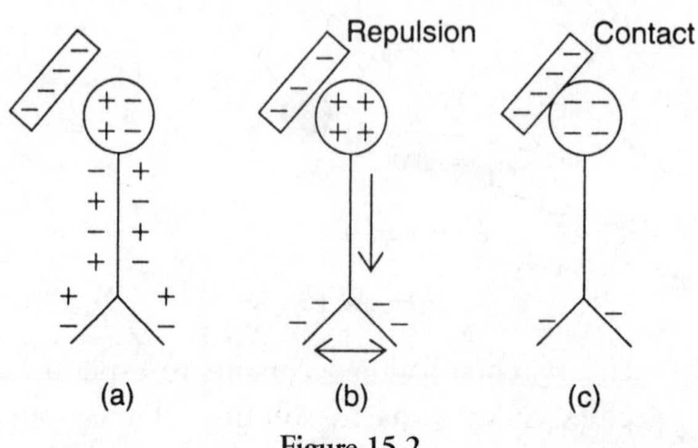

Figure 15.2

The electroscope can also be charged by **induction**. If you touch the electroscope shown in Figure 15.2b with your finger when the electroscope is brought near (see Figure 15.3a), the repelled electrons will be forced out into your body. If you remove your finger, keeping the rod near, the electroscope will be left with an overall positive charge by induction (Figure 15.3b).

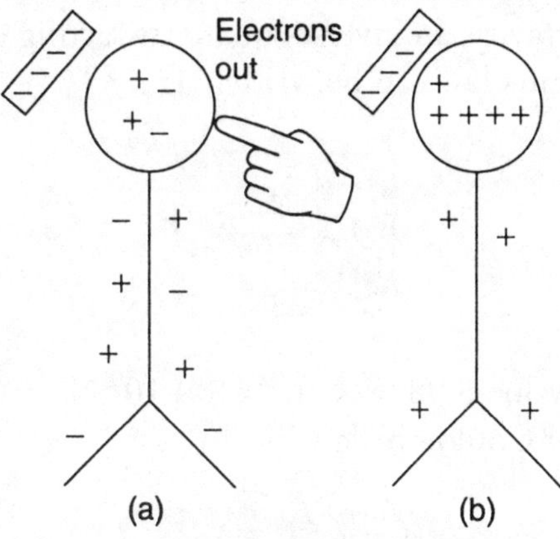

Figure 15.3

Finally, we can state that electric charges, in any distribution, obey a conservation law. When we transfer charge, we always maintain a balanced accounting. Suppose we have two charged metal spheres. Sphere *A* has +5 elementary charges, and sphere *B* has a +1 elementary charge (thus both are positively charged). The two spheres are brought into contact. Which way will charges flow? Negative charges always flow from a higher concentration to a lower one, and sphere *B* has an overall greater negative tendency (since it is less positive). Upon contact, it will give up negative charges until equilibrium is achieved. Notice that it is only the excess charges that flow out of the countless numbers of total charges in the sphere! A little arithmetic shows that sphere *B* will give up two negative charges, leaving it with an overall charge of +3. Sphere *A* will gain the two negative charges, reducing its charge to +3. If the two spheres are separated, each will have +3 elementary charges.

15.3 Coulomb's Law

From Sections 15.1 and 15.2, we can conclude that like charges repel each other while unlike charges attract (see Figure 15.4). The electrostatic force between two charged objects can act through space and even a vacuum. This property makes electrostatic force similar to the force of gravity.

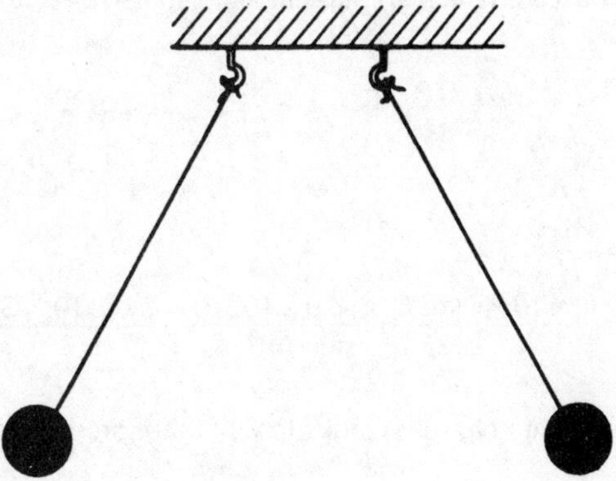

Figure 15.4

Think About It

Coulomb's law is an *inverse square* law that is very similar to Newton's law of gravitation. However, the electrostatic force can be neutralized while gravity is always present. Additionally, with the electrostatic force, there can be repulsion as well as attraction.

In the SI system of units, charge is measured in **coulombs** (C). In the late eighteenth century, the nature of the electrostatic force was studied by French scientist Charles Coulomb. He discovered that the force between two point charges (designated as q_1 and q_2), separated by a distance r, experienced a mutual force along a line connecting the charges that varied directly as the product of the charges and inversely as the square of the distance between them. This law, known as **Coulomb's law**, is, like the law of gravity, an inverse square law acting on matter at a distance. Mathematically, Coulomb's law can be written as

$$\mathbf{F} = \frac{kq_1q_2}{r^2}$$

The constant k has the value $8.9875 \times 10^9 \approx 9 \times 10^9 \text{ N} \cdot \text{m}^2/\text{C}^2$.

An alternative form of Coulomb's law is

$$\mathbf{F} = \left(\frac{1}{4\pi\epsilon_0}\right)\frac{q_1q_2}{r^2}$$

The new constant, ϵ_0, is called the **permittivity of free space** or **permittivity of the vacuum** and has a value of $8.8542 \times 10^{-12} \text{ C}^2/\text{N} \cdot \text{m}^2$. In the SI system, one elementary charge is designated as e and has a magnitude of 1.6×10^{-19} C.

Coulomb's law, like Newton's law of universal gravitation, is a vector equation. The direction of the force is along a radial vector connecting the two point sources. It should be noted that Coulomb's law applies only to point sources (or to sources that can be treated as point sources, such as charged spheres). If we have a distribution of point charges, the net force on one charge is the vector sum of all the other electrostatic forces. This aspect of force addition is sometimes termed **superposition.**

SAMPLE PROBLEM

Calculate the static electric force between a $+6.0 \times 10^{-6}$ C charge and a -3.0×10^{-6} C charge separated by 0.1 m. Is this an attractive or repulsive force?

Solution

We use Coulomb's law:

$$Fe = \frac{kq_1q_2}{R^2}$$

$$Fe = \frac{(9 \times 10^9 \text{ N} \cdot \text{m}^2/\text{C}^2)(+6.0 \times 10^{-6} \text{ C})(-3.0 \times 10^{-6} \text{ C})}{(0.1 \text{ m})^2} = -16.2 \text{ N}$$

The negative sign (opposite charges) indicates that the force is attractive.

15.4 The Electric Field

Another way to consider the force between two point charges is to recall that the force can act through free space. If a charged sphere has charge $+Q$, and a small test charge $+q$ is brought near it, the test charge will be repelled according to Coulomb's law. Everywhere, the test charge will be repelled along a radial vector out from charge $+Q$. We can state that, even if charge $+Q$ is too small to be visible, the influence of the electrostatic force can be observed and measured (since charges have mass and $\mathbf{F} = m\mathbf{a}$ as usual).

In this way, charge $+Q$ is said to set up an electric field, which pervades the space surrounding the charge and produces a force on any other charge that enters the field (just as a gravitational field does). The strength of the electric field, **E**, is defined to be the measure of the force per unit charge experienced at a particular location. In other words,

$$\mathbf{E} = \frac{\mathbf{F}}{q}$$

TIP

Compare this formula for the electric field to the formula for the gravitational field strength $\mathbf{g} = \mathbf{F}/m$.

and the units are newtons per coulomb (N/C). The electric field strength **E** is a vector quantity since it is derived from the force (a vector) and the charge (a scalar).

An illustration of this situation is given in Figure 15.5. Charge $+Q$ is represented by a charged sphere drawn as a circle. A small positive test charge, $+q$, is brought near and repelled along a radial vector **r** drawn outward from charge $+Q$. In fact, anywhere in the vicinity of charge $+Q$, the test charge will be repelled along an outward radial vector. We therefore draw these **force field lines** as radial vectors coming out from charge $+Q$.

By convention, we always consider the test charge to be positive. Since the force varies according to Coulomb's law, the electric field strength **E** will also vary, depending on the location of the test charge (we assume that it and the main charge remain constant in magnitude). Thus, we can write

$$\mathbf{E} = \frac{\mathbf{F}}{q} = \frac{kQ}{\mathbf{r}^2} = \left(\frac{1}{4\pi\epsilon_0}\right)\frac{Q}{\mathbf{r}^2}$$

Remember

The electric field lines show which way an imaginary positive test charge would move.

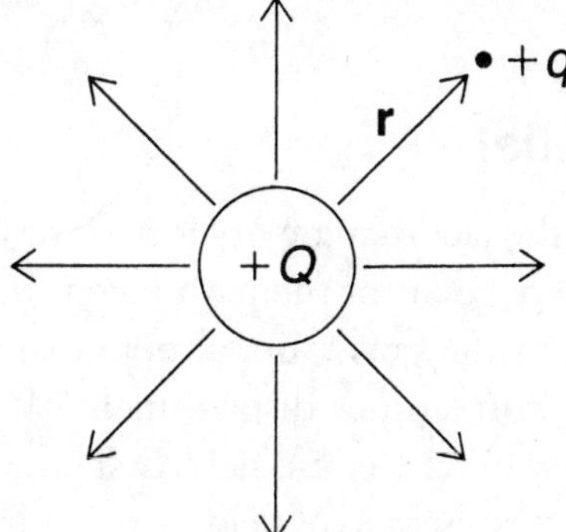

Figure 15.5

We can also interpret field strength by observing the density of field lines per square meter. With point charges, the radial nature of their construction causes the field lines to converge near the surface of the charge, Q, indicating a relative increase in field strength. In Section 15.5, we will encounter a configuration in which the field strength remains constant throughout.

There are several other configurations of electric field lines that we can consider. In Figure 15.6, the field between two point charges, with different arrangements of signs, is illustrated. Notice that the field lines are always perpendicular to the "surface" of the source and that they never cross each other. Additionally, between two like charges the electric field is zero. This fact can be extended to the idea that the electric field inside a hollow conducting sphere is zero, and all of the charges reside on the surface of the sphere.

The arrows on the field lines always indicate the direction in which a positive test charge would move.

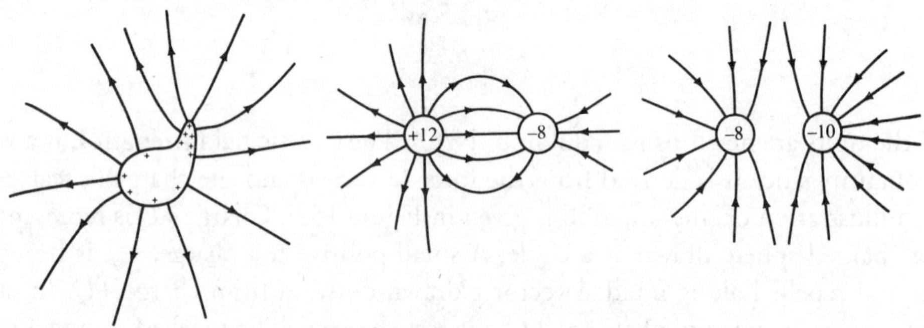

Figure 15.6

SAMPLE PROBLEM

What is the force acting on an electron placed in an external electric field $\mathbf{E} = 100$ N/C ?

Solution

We know that $\mathbf{E} = \mathbf{F}/q$

Thus,

$$\mathbf{F} = \mathbf{E}q = (100 \text{ N/C})(-1.6 \times 10^{-19} \text{ C}) = -1.6 \times 10^{-17} \text{ N}$$

15.5 Electric Potential

In Chapter 7, we reviewed the fact that gravity is a conservative force. Any work done by or against gravity is independent of the path taken. In Chapter 8, the work done was measured as a change in the gravitational potential energy. Recall that work is equal to the magnitude of force times displacement, $W = \mathbf{F}\,\Delta x$. Suppose that the force is acting on a charge q in an electric field $\mathbf{E}$. The work done by or against the electric field, which is also a conservative field, is given by

$$W = \mathbf{E}q\,\Delta r \text{ (in one dimension, radially)}$$

Since the electrostatic force is conservative, the work done should be equal to the change in the electrical potential energy between two points, *A* and *B*.

These two points cannot be just anywhere in the field. Consider the original electric field diagram in Figure 15.5. If test charge $+q$ is a distance r from source charge $+Q$, it will experience a certain radial force whose magnitude is given by Coulomb's law. At any position around the source, at a distance r, we can observe that the test charge will experience the same force. The localized potential energy will be the same as well. The set of all such positions defines a circle of radius r (actually a sphere in three dimensions), called an **equipotential surface**.

Figure 15.7 shows a series of equipotential "curves" or varying radii. Since the electric field is conservative, work is done by or against the field only when a charge is moved from one equipotential surface to another. To better understand this effect, we define a quantity called the **electric potential**, which is a measure of the magnitude of electrical potential energy per unit charge at a particular location in the field. Thus, the work done per unit charge, in moving from equipotential surface *A* to equipotential surface *B*, is a measure of what is called the **potential difference** or **voltage**.

Thus we define V to be the potential difference ($V_A - V_B$) and to be equal to W/q, and so the units of potential difference are joules per coulomb (J/C) or **volts**. Since $W = \mathbf{E}_q \, \Delta r$, calculating the work can be difficult if the electric field can vary (as in the case of a point source). However, if we can find a configuration in which the electric field is uniform, **E** becomes a constant. Since potential difference is the work done per charge, we see that $V = \mathbf{E} \, \Delta r$. The units of electric field strength can now be defined in terms of the potential difference between two points. Usually, one of those points is infinity, and the potential is defined to be zero there. The localized electric potential is then given by $V = kQ/r$, using Coulomb's law. In a uniform electric field, we can now define the electric field strength, $\mathbf{E} = V/d$, and the units are volts/meter (V/m).

TIP

Try not to confuse electric potential with electric potential energy. Additionally, the ratio of volts per meter (V/m) is often referred to as a gradient.

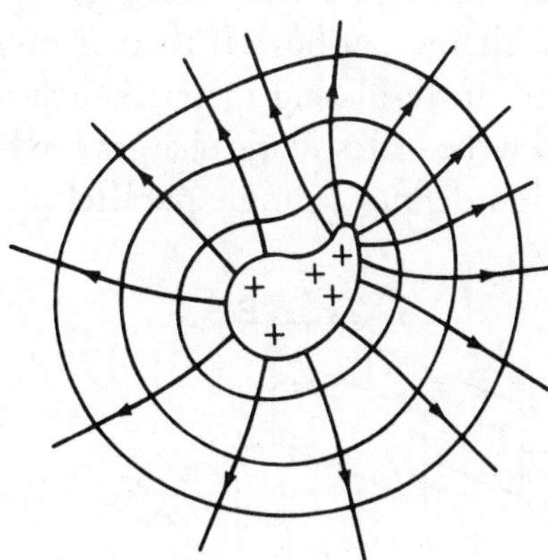

Figure 15.7

The desired configuration is one in which two charged parallel plates are separated by a distance *d*. If each side is oppositely charged, then (see Figure 15.8) the electric field (and hence the electric force on a test charge within the field) is uniform. If a charge q and mass m enters the electric field with a velocity **v**, the charge will be subject to an electrical force **F** such that $\mathbf{F} = m\mathbf{a}$. In this case, $\mathbf{F} = \mathbf{E}q$, and the kinematics of charged particles in uniform electric fields can be developed in an analogous manner to the kinematics of particles in a uniform gravitational field. The applications of this kinematics extends into the fields of electronics and atomic theory.

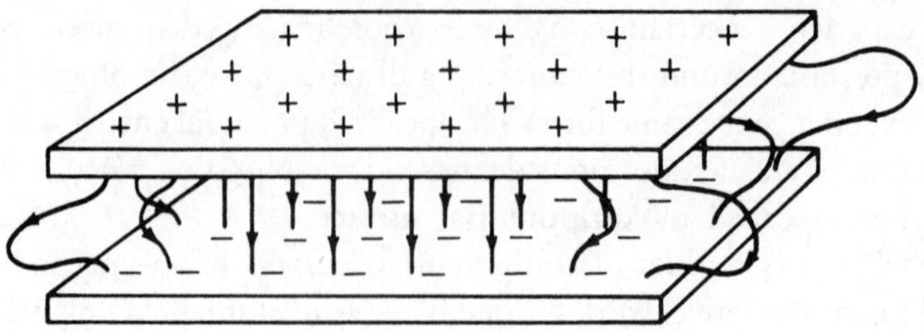

Figure 15.8

Some sign conventions have been developed for dealing with electric potential. By agreement, we state that a positive test charge, in a uniform electric field, will lose potential energy as it moves in the direction of the field. In other words, if the change in electrical potential energy is given by ΔU, then $\Delta U = -q\mathbf{E}d$. Consequently, the charge, because of the conservation of energy, will gain kinetic energy as it moves in the direction of the field.

Another convenient unit of electrical energy is the **electron volt** (eV). By definition, if one elementary charge experiences a potential difference of 1 volt, its energy (or the work done to transfer the charge) is equal to 1 electron volt. Consequently, we can state that 1 electron volt = 1.6×10^{-19} joule.

Charges flow when a difference in potential exists, just as water flows when there is a difference in pressure. To maintain a constant flow of charge (current), a continuous potential difference must be maintained. Chemically, this is achieved through the use of a **battery** and attached conducting wires, which can create a potential difference between two points. In the case of two parallel plates (see Figure 15.9), the potential difference sets up an electric field between the plates. If a conductor is placed between the plates, charge can flow. If there is no conductor, charge is simply stored on the plates. If an insulating material such as wax or paper (called a **dielectric**) is placed between the plates, static charges can be stored to a much higher concentration. This situation defines a simple **parallel plate capacitor**.

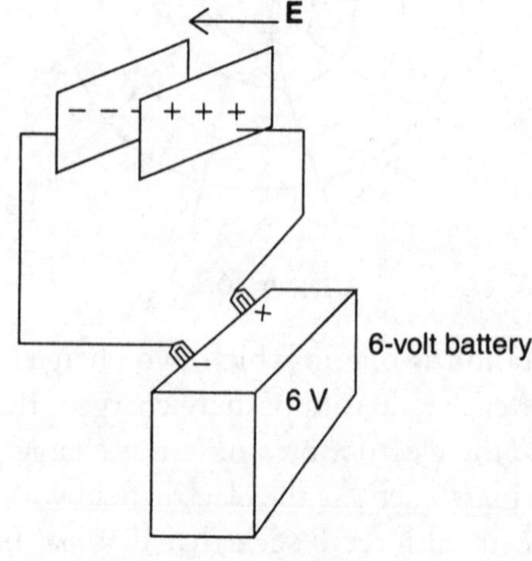

Figure 15.9

SAMPLE PROBLEM

Given that the charge on a proton is $+1.6 \times 10^{-19}$ C,

(a) Calculate the electric potential at a point 2.12×10^{-10} m from the proton.
(b) If an electron (and no other charges) is placed at that point, what will be its electric potential energy. Assume that the potential energy at infinity is equal to zero.

Solution

(a) For the proton, we use

$$V = \frac{kQ}{r}$$

$$V = \frac{(9 \times 10^9 \text{ N}\cdot\text{m}^2/\text{C}^2)(1.6 \times 10^{-19}\text{ C})}{2.12 \times 10^{-10}\text{ m}} = 6.79 \text{ V}$$

(b) For the electron, the potential energy

$$U = qV = (-1.6 \times 10^{-19}\text{ C})(6.79\text{ V}) = -1.09 \times 10^{-18}\text{ J}$$

15.6 Capacitance

When there are two charged parallel metal plates in a configuration called a **capacitor**, there is a proportional relationship between the potential difference and the total charge. The capacitance, C, of a capacitor is defined to be the ratio of the charge on either conductor and the potential difference between the conductors:

$$C = \frac{Q}{V}$$

This is a positive scalar quantity that is essentially a measure of the capacitor's ability to store charge. The reason is that, for a given capacitor, the ratio Q/V is a constant since potential difference increases as the charge on either capacitor increases.

The SI unit of capacitance is the **farad** (F). The types of capacitors used in electronic devices have capacitances that range from microfarads to picofarads.

In this section we deal only with simple parallel-plate capacitors. However, in general, if there is a spherical distribution of charge, the electric field, at some point r, is given by:

$$\mathbf{E} = \left(\frac{1}{4\pi \epsilon_0}\right)\left(\frac{Q}{r^2}\right) = \left(\frac{Q}{\epsilon_0}\right)\left(\frac{1}{4\pi r^2}\right) = \frac{Q}{\epsilon_0 A}$$

Now, with two parallel plates, the electric field is uniform and is given by $\mathbf{E} = Vd$, and A is the area of the rectangular plates. The potential difference is then given by:

$$V = \mathbf{E}d = \frac{Qd}{\epsilon_0 A}$$

Finally, given the definition of capacitance, we can write

$$C = \frac{Q}{V} = \frac{Q}{Qd/\epsilon_0 A} = \frac{\epsilon_0 A}{d}$$

The implication of the last formula is that the capacitance of the parallel-plate capacitor is directly proportional to the area of the plates and inversely proportional to their separation.

The energy stored in a capacitor can be studied by recognizing that the definition of capacitance can be written as $Q = CV$. This direct relationship between plate charge and voltage will produce a diagonal straight line if we plot a graph of charge versus voltage (see Figure 15.10).

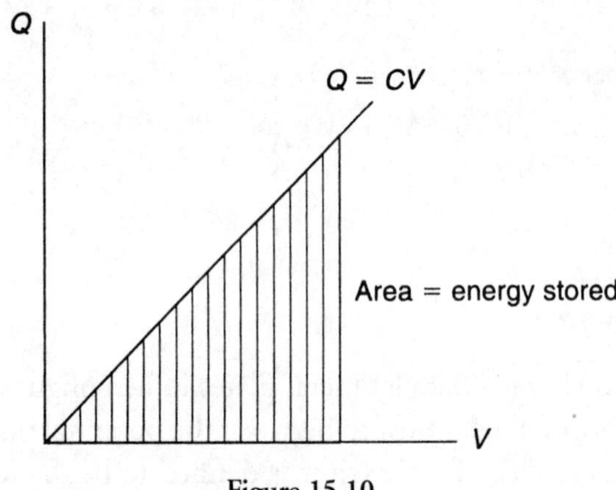

Figure 15.10

Notice that the units of QV are joules (the units of work). The area under the diagonal line is equal to the amount of energy stored in the capacitor. This energy is given by $E = (\frac{1}{2})CV^2$.

SAMPLE PROBLEM

A point charge $+Q$ is fixed at the origin of a coordinate system. A second charge $-q$ is uniformly distributed over a spherical mass m and orbits around the origin at a radius r and with a period T. Derive an expression for the radius.

Solution

If we neglect gravity and any other external forces, the orbit is due to the electrostatic force between the charges. This sets up a centripetal force for the mass m.

Thus, we can write

$$\frac{1}{4\pi \epsilon_0}\left(\frac{Qq}{r^2}\right) = \frac{m4\pi^2 r}{T^2}$$

and

$$r = \left(\frac{T^2 Qq}{m \epsilon_0 16\pi^3}\right)^{1/3}$$

Chapter Summary

- There are two kinds of electric charges: positive and negative.
- Electrons are the fundamental carriers of negative charge.
- Protons are the fundamental carriers of positive charge.
- Like charges repel, while unlike charges attract.
- An electroscope can be used to detect the presence of static charges.
- Objects become charged through the transfer of electrons.
- Coulomb's law describes the nature of the force between two static charges. It states that the force of either attraction (−) or repulsion (+) is proportional to the product of the charges and inversely proportional to the square of the distance between them. This is similar to Newton's law of gravitation (see Chapter 12).
- The electrical potential difference is defined to be equal to the work done per unit charge (in units of joules per coulomb or volts).
- A capacitor is a device that can store charge. The capacitance of a capacitor is defined as the amount of charge stored per unit volt ($C = Q/V$).

Problem-Solving Strategies for Electrostatics

The preceding sample problem reminds us that we are dealing with vector quantities when the electrostatic force and field are involved. Drawing a sketch of the situation and using our techniques from vector constructions and algebra were most effective in solving the problem.

Be sure to keep track of units. If a problem involves potential or capacitance, remember to maintain the standard SI system of units, which are summarized in the appendices.

Remember that electric field lines are drawn as though they followed a "positive" test charge and that Coulomb's law applies only to point charges. The law of superposition allows you to combine the effects of many forces (or fields) using vector addition. Additionally, symmetry may allow you to work with a reduced situation that can be simply doubled or quadrupled. In this aspect, a careful look at the geometry of the situation is called for.

If the electric field is uniform, the charges in motion will experience a uniform acceleration that will allow you to use Newtonian kinematics to analyze the motion. Electrons in oscilloscopes are an example of this kind of application.

Potential and potential difference (voltage) are scalar quantities usually measured relative to the point at infinity in which the localized potential is taken to be zero. In a practical sense, it is the change in potential that is electrically significant, not the potential itself.

Practice Exercises

MULTIPLE-CHOICE

1. An insulated metal sphere, *A*, is charged to a value of $+Q$ elementary charges. It is then touched and separated from an identical but neutral insulated metal sphere, *B*. This second sphere is then touched to a third identical and neutral insulated metal sphere *C*. Finally, spheres *A* and *C* are touched and then separated. Which of the following represents the distribution of charge on each sphere, respectively, after the above process has been completed?

   (A) $Q/3$, $Q/3$, $Q/3$
   (B) $Q/4$, $Q/2$, $Q/4$
   (C) $3Q/8$, $Q/4$, $3Q/8$
   (D) $3Q/8$, $Q/2$, $Q/4$
   (E) None of these is correct.

2. Electrical appliances are usually grounded in order to

   (A) maintain a balanced charge distribution
   (B) prevent the buildup of heat on them
   (C) run properly using household electricity
   (D) prevent the buildup of an overloaded circuit
   (E) prevent the buildup of static charges on their surfaces

CHALLENGE

3. Two point charges experience a force of repulsion equal to 3×10^{-4} N when separated by 0.4 m. If one of the charges is 5×10^{-4} C, what is the magnitude of the other charge?

   (A) 1.07×10^{-11} C
   (B) 2.7×10^{-11} C
   (C) 3.2×10^{-9} C
   (D) 5×10^{-4} C
   (E) 0 C

4. A parallel-plate capacitor has a capacitance of *C*. If the area of the plates is doubled, while the separation between the plates is halved, the new capacitance will be

   (A) $2C$ F
   (B) $4C$ F
   (C) $C/2$ F
   (D) C F
   (E) $C/4$ F

CHALLENGE

5. What is the capacitance of a parallel-plate capacitor made of two aluminum plates, 4 cm in length on a side and separated by 5 mm?

   (A) 2.832×10^{-11} F
   (B) 2.832×10^{-10} F
   (C) 2.832×10^{-12} F
   (D) 2.832×10^{-9} F
   (E) 1 F

6. If 10 J of work is required to move 2 C of charge in a uniform electric field, the potential difference present is equal to

(A) 20 V
(B) 12 V
(C) 8 V
(D) 5 V
(E) 10 V

7. Which of the following diagrams represents the equipotential curves in the region between a positive point charge and a negatively charged parallel plate?

(A) 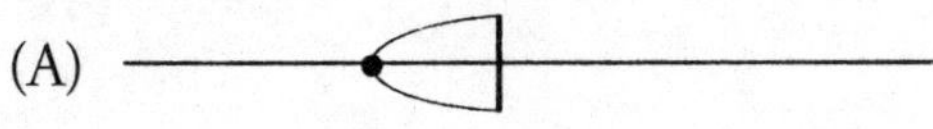

(B) 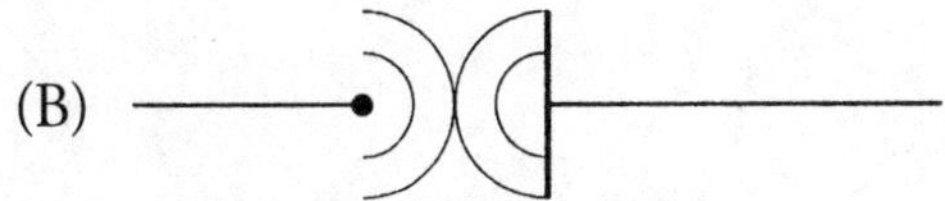

(C) 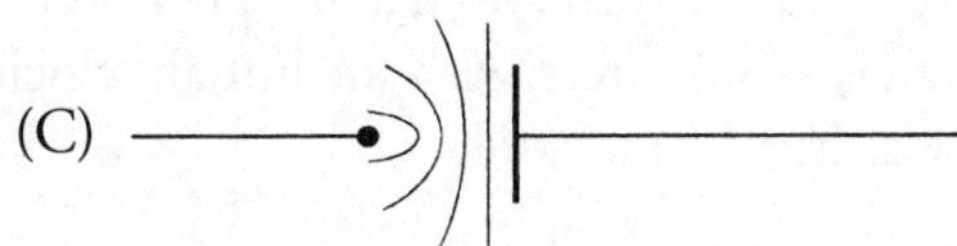

(D) 

(E) None of these is correct.

8. An electron is placed between two charged parallel plates as shown below. Which of the following statements is true?

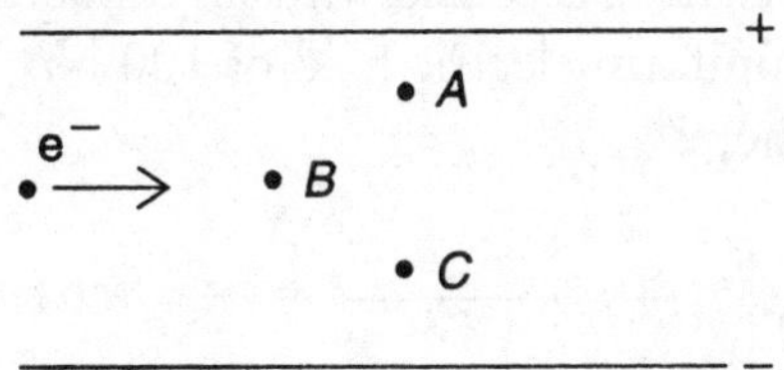

I. The electrostatic force is greater at *A* than at *B*.
II. The work done from *A* to *B* to *C* is the same as the work done from *A* to *C*.
III. The electrostatic force is the same at points *A* and *C*.
IV. The electric field strength decreases as the electron is repelled upward.

(A) I and II
(B) I and III
(C) II and III
(D) II and IV
(E) I and IV

9. How much kinetic energy is given to a doubly ionized helium atom due to a potential difference of 1000 V? **CHALLENGE**

(A) 3.2×10^{-16} eV
(B) 1000 eV
(C) 2000 eV
(D) 3.2×10^{-19} eV
(E) Not enough information is given.

10. Which of the following is equivalent to 1 F of capacitance?

(A) $\dfrac{C^2 \cdot s^2}{kg \cdot m^2}$

(B) $\dfrac{kg \cdot m^2}{C^2 \cdot s^2}$

(C) $\dfrac{C}{kg \cdot s}$

(D) $\dfrac{C \cdot m}{kg^2 \cdot s}$

(E) $C \cdot m/s$

FREE-RESPONSE

CHALLENGE

1. An electron enters a uniform electric field between two parallel plates vertically separated with $\mathbf{E}$ = 400 N/C. The electron enters with an initial velocity of 4×10^7 m/s. The length of each plate is 0.15 m.
   (a) What is the acceleration of the electron?
   (b) How long does the electron travel through the electric field?
   (c) If it is assumed that the electron enters the field level with the top plate, which is negatively charged, what is the vertical displacement of the electron during that time?
   (d) What is the magnitude of the velocity of the electron as it exits the field?

2. A 1-g cork sphere on a string is coated with silver paint. It is in equilibrium, making a 10° angle in a uniform electric field of 100 N/C as shown below. What is the charge on the sphere?

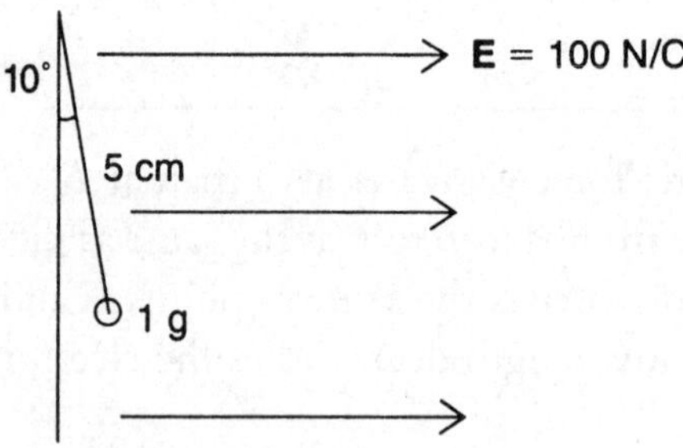

3. Four identical point charges ($Q = +3\ \mu C$) are arranged in the form of a rectangle as shown below. What are the magnitude and the direction of the net electric force on the charge in the lower left corner due to the other three charges?

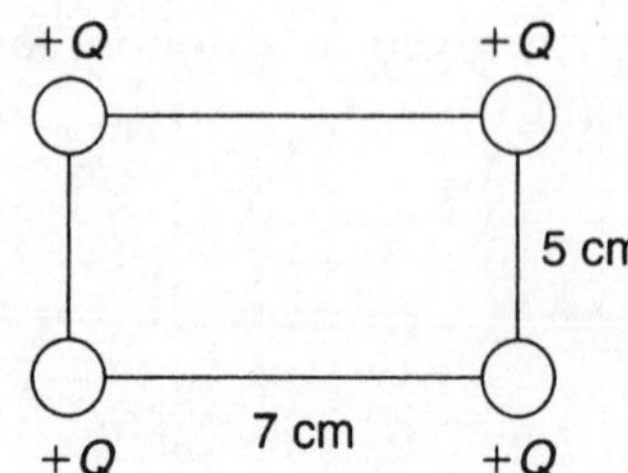

4. An electrostatically charged rod attracts a small, suspended sphere. What conclusions can you make about the charge on the suspended sphere?

5. Why is it safe to be inside a car during a lightning storm?

6. Why are most electrical appliances grounded?

ANSWERS EXPLAINED

MULTIPLE-CHOICE PROBLEMS

1. **(C)** We use the conservation of charge to determine the answer. Sphere A has $+Q$ charge, and B has zero. When these spheres are touched and separated, each will have one-half of the total charge, which was $+Q$. Thus, each now has $+Q/2$.

   When B is touched with C, which also has zero charge, we again distribute the charge evenly by averaging. Thus, B and C will now each have $+Q/4$, while A still has $+Q/2$.

   Finally, when A and C are touched, we take the average of $+Q/4$ and $+Q/2$, which is $+3Q/8$.

   The final distribution is therefore $3Q/8$, $Q/4$, $3Q/8$.

2. **(E)** Electrical appliances are grounded to prevent the buildup of static charges on their surfaces.

3. **(A)** We use Coulomb's law to calculate the answer. Everything is given except one charge. The working equation therefore looks like this:

$$3 \times 10^{-4} = \frac{(9 \times 10^{9})(5 \times 10^{-4})q}{(0.4)^2}$$
$$q = 1.07 \times 10^{-11} \text{ C}$$

4. **(B)** The formula for the capacitance of a parallel-plate capacitor is $C = \epsilon_0 A/d$. Therefore, if the area is doubled and the separation distance halved, the capacitance increases by four times and is equal to $4C$ F.

5. **(C)** We use the formula from question 4, but first all measurements must be in meters or square meters. Converting the given lengths yields 4 cm = 0.04 m and 5 mm = 0.005 m. Then $A = (0.04)(0.04) = 0.0016 \text{ m}^2$ and $d = 0.005$ m. Using the formula and the value for the permittivity of free space given in the chapter, we get $C = 2.832 \times 10^{-12}$ F.

6. **(D)** Potential difference is the work done per unit charge. Thus

$$V = \frac{W}{q} = \frac{10}{2} = 5 \text{ V}$$

7. **(C)** The equipotential curves around a point charge are concentric circles. The equipotential curves between two parallel plates are parallel lines (since the electric field is uniform). The combination of these two produces the diagram shown in choice C.

8. **(C)** The electric field is the same at all points between two parallel plates since it is uniform. Thus the force on an electron is the same everywhere. Also, we stated in the chapter that the electric field, like gravity, is a conservative field. This means that work done to or by the field is independent of the path taken. Thus both statements II and III are true.

9. **(C)** One electron volt is defined to be the energy given to one elementary charge through 1 V of potential difference. A doubly ionized helium atom has +2 elementary charges. Since the potential difference is 1000 V, the kinetic energy is 2000 eV. When using electron-volt units, we eliminate the need for the small numbers associated with actual charges of particles.

10. **(A)** One farad is defined to be 1 C/V. One volt is 1 J/C. One joule is 1 kg · m^2/s^2. The combination results in

$$\frac{C^2 \cdot s^2}{kg \cdot m^2}$$

FREE-RESPONSE PROBLEMS

1. (a) Since the electron is in a uniform electric field, the kinematics of the electron follows the standard kinematics of mechanics. Since $\mathbf{F} = m\mathbf{a}$ and $\mathbf{F} = -e\mathbf{E}$, $\mathbf{a} = -e\mathbf{E}/m$ (the negative sign is needed since an electron is negatively charged). Using the information given in the problem and the values for the charge and mass of the electron from the Appendix, we get

$$\mathbf{a} = \frac{-(1.6 \times 10^{-19})(400)}{9.1 \times 10^{-31}} = -7.03 \times 10^{13} \text{ m/s}^2$$

(b) The electron is subject to a uniform acceleration in the downward direction. The path will therefore be a parabola, and the electron will be a projectile in the field. The initial horizontal velocity remains constant, and therefore the time is given by the expression

$$t = \frac{\ell}{\mathbf{v}_0} = \frac{0.15}{4 \times 10^7} = 3.75 \times 10^{-9} \text{ s}$$

(c) If we take the initial entry level as zero in the vertical direction, the electron has initially zero velocity in that direction. Thus, the vertical displacement is given by $\mathbf{y} = (1/2)\mathbf{a}t^2$. Using the information from parts (a) and (b), we get

$$\mathbf{y} = -(1/2)(7.03 \times 10^{13})(3.75 \times 10^{-9})^2 = -4.94 \times 10^{-4} \text{ m}$$

(d) The magnitude of the velocity as the electron exits the field, after time t has elapsed, is given by the vector nature of two-dimensional motion. The horizontal velocity is the same throughout the encounter and is equal to 4×10^7 m/s. The vertical velocity is given by $\mathbf{v}_y = -\mathbf{a}t$ since there is no initial vertical velocity. Using the information from the previous parts, we find that

$$\mathbf{v}_y = -(7.03 \times 10^{13})(3.75 \times 10^{-9}) = -2.6 \times 10^5 \text{ m/s}$$

Now the magnitude of the velocity is given by the Pythagorean theorem since the velocity is the vector sum of these two component velocities:

$$\mathbf{v} = \sqrt{(4 \times 10^7)^2 + (2.6 \times 10^5)^2} = 40,000,845 \text{ m/s} \approx 4 \times 10^7 \text{ m/s}$$

The result is not a significant change in the magnitude of the initial velocity. There is of course a significant change in direction. The angle of this change can be determined by the tangent function if desired.

2. If the sphere is in equilibrium, the vector sum of all forces acting on it must equal zero. There are several forces involved. Mechanically, gravity acts to create a tension in the string. This tension has an upward component that directly balances the force of gravity ($m\mathbf{g}$) acting on the sphere. A second horizontal component acts to the left and counters the horizontal electrical force established by the field and the charge on the sphere. A free-body diagram looks like this:

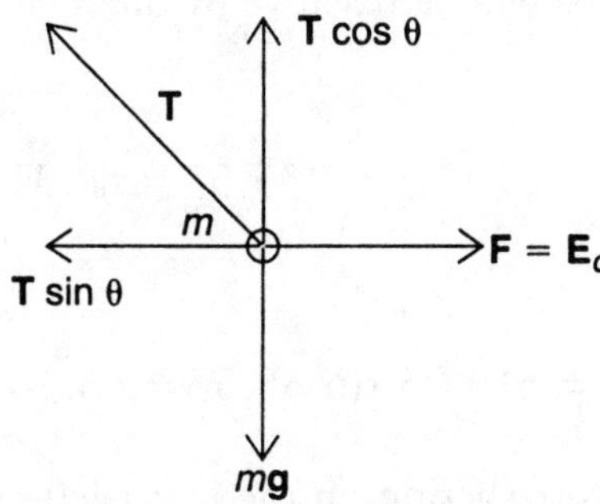

From the diagram, we see that $\mathbf{T} \cos \theta = m\mathbf{g}$, where $\theta = 10°$ and $m = 1$ g = 0.001 kg. Thus, $\mathbf{T} = 0.00995$ N, using our known value for the acceleration due to gravity, $\mathbf{g}$. Now, in the horizontal direction, $\mathbf{T} \sin \theta = \mathbf{E}_q$ and $\mathbf{E} = 100$ N/C. Using all known values given and derived, we arrive at

$$q = 1.75 \times 10^{-5} \text{ C}$$

3. Since all the charges are positive, the lower-left-corner charge will experience mutual repulsions from the other three. A diagram showing these forces appears below:

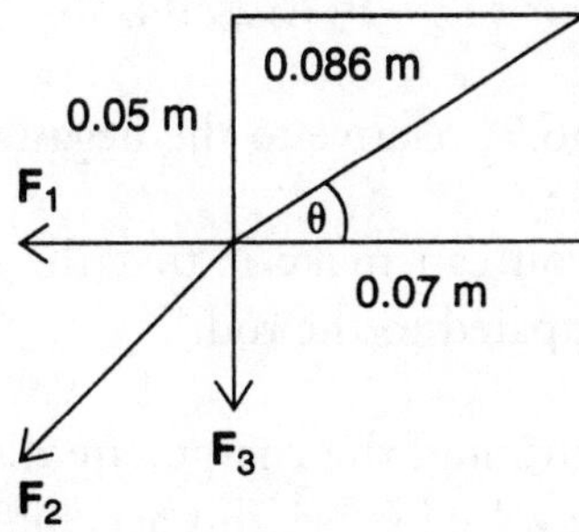

The magnitude of each force is given by Coulomb's law. The diagonal distance was determined using the Pythagorean theorem. The three force magnitudes are therefore calculated to be:

$$F_1 = \frac{(9\times 10^9)(3\times 10^{-6})(3\times 10^{-6})}{(0.07)^2} = 16.53\ \text{N}$$

$$F_2 = \frac{(9\times 10^9)(3\times 10^{-6})(3\times 10^{-6})}{(0.086)^2} = 10.95\ \text{N}$$

$$F_3 = \frac{(9\times 10^9)(3\times 10^{-6})(3\times 10^{-6})}{(0.05)^2} = 32.4\ \text{N}$$

Now, $\mathbf{F}_1$ acts to the left and $\mathbf{F}_3$ acts downward, so no further vector reductions are necessary. However, $\mathbf{F}_2$ acts at some angle toward the lower left, given by θ. To find the angle, we recall that $\tan\theta = 5/7$, which implies that $\theta = 35.53°$. Both components of $\mathbf{F}_2$ will be negative by our sign convention of Chapter 4.

We can now write

$$\mathbf{F}_{2x} = \pm 10.95\cos 35.53 = \pm 8.91\ \text{N}$$

and

$$\mathbf{F}_{2y} = \pm 10.95\sin 35.53 = \pm 6.36\ \text{N}$$

The resultant or net force acting on the lower-left-corner charge is the vector sum of these charges. We can find the components of this force by simply adding up the forces in the same directions:

$$\mathbf{F}_x = -16.53 - 8.91 = -25.44\ \text{N}$$

$$\mathbf{F}_y = -32.40 - 6.36 = -38.76\ \text{N}$$

The magnitude of the net force $\mathbf{F}$ is given by the Pythagorean theorem and is equal to 46.3 N. The direction of this force is given by

$$\tan\phi = \frac{\mathbf{F}_y}{\mathbf{F}_x} = \frac{38.76}{25.44} = 1.5$$

This implies that $\phi = 56.7°$ relative to the negative *x*-axis.

4. The only conclusion you can make is that the sphere is either "neutral" or oppositely charged compared to the rod.

5. If your car is hit by lightning, the charges are distributed around the outside and then dissipated away. In a sense, this acts like an electrostatic shield.

6. Appliances are grounded to prevent the buildup of static charges on their outside surfaces.

CHAPTER 16

Electric Circuits

KEY CONCEPTS

- Current and Electricity
- Electric Resistance
- Electric Power and Energy
- Series and Parallel Circuits
- Networks and Combination Circuits
- Combination Circuits
- Capacitors in Circuits

16.1 CURRENT AND ELECTRICITY

In Chapter 15 we observed that, if two points have a potential difference between them, and they are connected with a conductor, negative charges will flow from a higher concentration to a lower one. This aspect of charge flow is very similar to the flow of water in a pipe due to a pressure difference.

Moving electric charges are referred to as electric **current**, which measures the amount of charge passing a given point every second. The units of measurement are coulombs per second (C/s). Algebraically, we designate current by the capital letter I and state that

$$I = \frac{\Delta Q}{\Delta t}$$

In electricity, it is the battery that supplies the potential difference needed to maintain a continuous flow of charge. In the nineteenth century, physicists thought that this potential difference was an electric force, called the **electromotive force** (emf), that pushed an electric fluid through a conductor. Today, we know that an emf is not a force, but rather a potential difference measured in volts. Do not be confused by the designation emf in the course of reviewing or of solving problems!

From chemistry, recall that a battery uses the action of acids and bases on different metals to free electrons and maintain a potential difference. In the process, two terminals, designated positive and negative, are created. When a conducting wire is attached and looped around to the other end, a complete circle of wire (a **circuit**) is produced, allowing for the continuous flow of charge. The battery acts like a pump, forcing electrons off the positive side onto the negative side using chemical reactions (see Figure 16.1). The moving electrons can then do work since they carry energy. This work is the electricity with which we have become so familiar in our modern world.

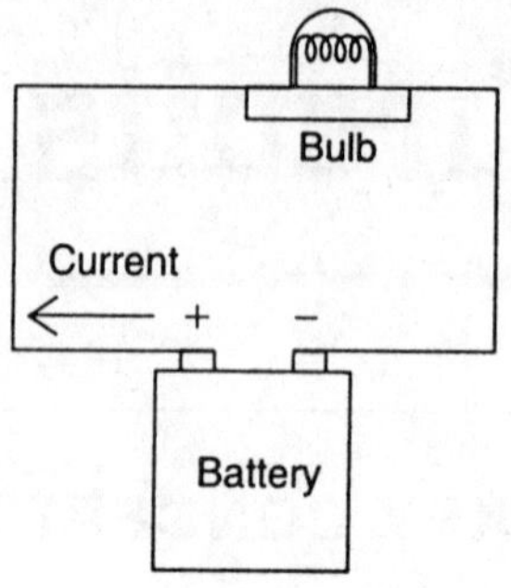

Figure 16.1

The diagram in Figure 16.1 shows a simple electric circuit. The direction of the conventional current, in most textbooks, is from the positive terminal. To maintain a universal acceptance of concepts and ideas (recall our earlier discussions of concepts and labels), schematic representations for electrical devices were developed and accepted by physicists and electricians worldwide. These schematics are used when drawing or diagramming an electric circuit, and it is important that you be able to interpret and draw them to fully understand this topic. In Figure 16.2, schematics for some of the most frequently encountered electrical devices are presented.

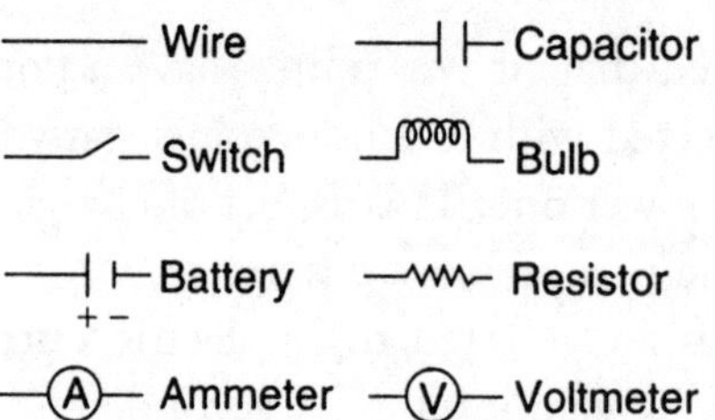

Figure 16.2 Electrical Schematic Diagrams

The simple circuit shown in Figure 16.1 can now be diagrammed schematically as in Figure 16.3.

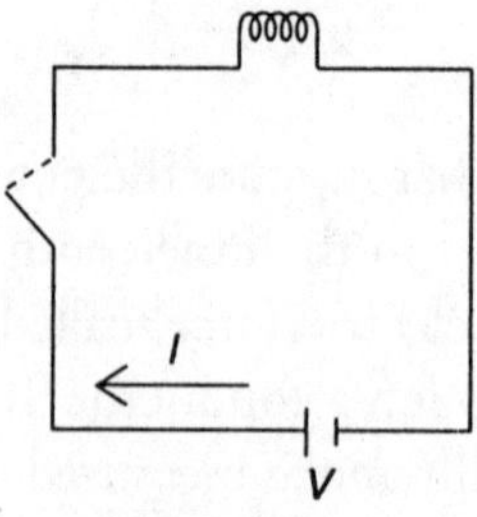

Figure 16.3

A switch has been added to the schematic; of course the charge would not flow unless the switch was closed.

In Figure 16.2, three schematics appear that we have not yet discussed. A **resistor** is a device whose function is to oppose the current. We will investigate resistors in more detail in the next section.

An **ammeter** is a device that measures the current. You can locate the water meter in your house or apartment building and notice that it is placed within the flow line. The reason is that the meter must measure the flow of water per second through a given point. An ammeter is placed within an electric circuit in much the same way. This is referred to as a "series connection," and it maintains the singular nature of the circuit. In practical terms, you can imagine cutting a wire in Figure 16.3, and hooking up the bare leads to the two terminals of the ammeter.

A **voltmeter** is a device that measures the potential difference, or voltage, between two points. Unlike an ammeter, the voltmeter cannot be placed within the circuit since it will effectively be connected to only one point. A voltmeter is therefore attached in a "parallel connection," creating a second circuit through which only a small amount of current flows to operate the voltmeter. In Figure 16.4, the simple circuit is redrawn with the ammeter and voltmeter placed.

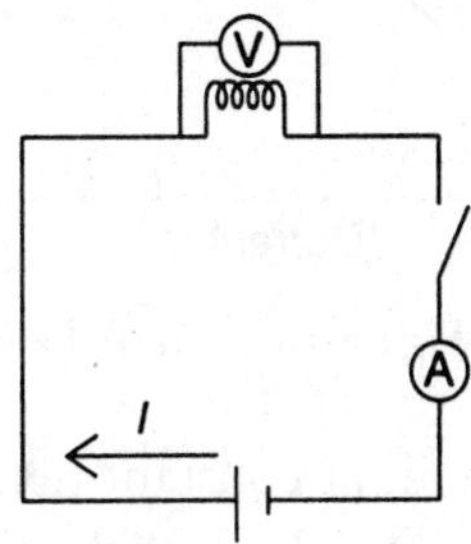

Figure 16.4

For simple circuits, there will be no observable difference in readings if the ammeter and voltmeter are moved to different locations. However, there is a slight difference in the emf across the terminals of the battery and the emf across the bulb. The first emf reflects the work done by the battery not only to move charge through the circuit but also to move charge across the terminals of the battery. This is sometimes referred to as the **terminal emf** or **internal emf**.

16.2 ELECTRIC RESISTANCE

In Figure 16.4, a simple electric circuit is illustrated with measuring devices for voltage and current. If a light bulb is left on for a long time, two observations can be made. First, the bulb gets hot because of the action of the electricity and the filament. The light produced by the bulb is caused by the heat of the filament. Second, the current in the ammeter will begin to decrease.

These two observations are linked to the idea of electrical resistance. Friction generates heat. The interaction of flowing electrons and the molecules of a wire (or bulb filament) creates an electrical friction called **resistance**. This resistance is temperature dependent since, from the kinetic theory, an increase in temperature will increase the molecular activity and therefore interfere to a greater extent with the flow of current.

In the case of a light bulb, it is this resistance that is desired in order for the bulb to do its job. Resistance along a wire or in a battery, however, is unwanted and must be minimized. In more complicated circuits, a change in current flow is required to protect devices, and so special resistors are manufactured that are small enough to easily fit into a circuit. While resistance is the opposite of conductance, we do not want to use insulators as resistors since insulators will stop the flow altogether. Therefore, a range of materials, indexed by "resistivity," is catalogued in electrical handbooks to assist scientists and electricians in choosing the proper resistor for a given situation.

If the temperature can be maintained at a constant level, a simple relationship between the voltage and the current in a circuit is revealed: as the voltage is increased, a greater flow of current is observed. This direct relationship was investigated theoretically by a German physicist named Georg Ohm and is called **Ohm's law** (see Figure 16.5).

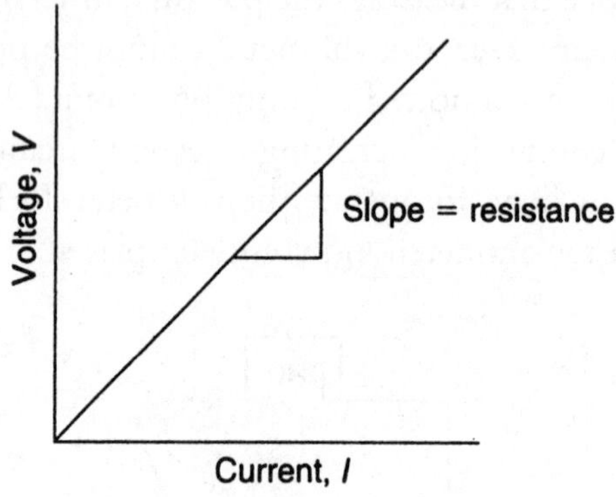

Figure 16.5 Ohm's Law

Ohm's law states that in a circuit at constant temperature the ratio of the voltage to the current remains constant. The slope of the line in Figure 16.5 represents the resistance of the circuit, which is measured in units of ohms (Ω). Algebraically, we can write Ohm's law as

$$R = \frac{V}{I} \quad \text{or} \quad V = IR$$

Not all conductors obey Ohm's law. Semiconductors and liquid conductors, for example, do not.

Other factors also affect the resistance of a conductor. We have already discussed the effects of temperature and material type (resistivity). Resistivity is designated by the Greek letter ρ (rho). In a wire, electrons try to move through, interacting with the molecules that make up the wire. If the wire has a small cross-sectional area, the chances of interacting with a bound molecule increase and thus the resistance of the wire increases. If the length of the wire is increased, the greater duration of interaction time will also increase the resistance of the wire. In summary, these resistance factors all contribute to the overall resistance of the circuit.

Algebraically, we may write these relationships (at constant temperature) in the following form:

$$R = \frac{\rho L}{A}$$

where L is the length in meters and A is the cross-sectional area in units of m^2. The resistivity (ρ) is measured in units of ohm · meters (Ω · m) and is usually rated at 20 degrees Celsius.

In Table 16.1, the resistivities of various materials are presented. Since the ohm is a standard (SI) unit, be sure that all lengths are in meters and areas are in square meters.

Table 16.1

Resistivities of Selected Materials at 20°C

Substance	Resistivity, ρ ($\Omega \cdot$ m)
Aluminum	2.83×10^{-8}
Carbon	3.5×10^{-5}
Copper	1.69×10^{-8}
Gold	2.44×10^{-8}
Nichrome	100×10^{-8}
Platinum	10.4×10^{-8}
Silver	1.59×10^{-8}
Tungsten	5.33×10^{-8}

SAMPLE PROBLEM

Copper wire is being used in a circuit. The wire is 1.2 m long and has a cross-sectional area of 1.2×10^{-8} m^2 at a constant temperature of 20°C.

(a) Calculate the resistance of the wire.
(b) If the wire is connected to a 10-V battery, what current will flow through it?

Solution

(a) We use

$R = \rho L/A$ where in this case, ρ is the resistivity at 20°C

$$R = \frac{(1.69 \times 10^{-8} \Omega \cdot \text{m})(1.2 \text{ m})}{1.2 \times 10^{-8} \text{ m}^2} = 169\ \Omega$$

(b) Now, we use Ohm's law

$$I = \frac{V}{R} = \frac{10 \text{ V}}{1.69\ \Omega} = 5.92 \text{ A}$$

16.3 ELECTRIC POWER AND ENERGY

Electricity produces energy that, in turn, can be used to produce light and heat. Electricity can do work to turn a motor. Having measured the voltage and current in a circuit, we can determine the amount of power and energy being produced in the following way. The units of voltage (potential difference) are joules per coulomb (J/C) and are a measure of the energy supplied to each coulomb of charge flowing in the circuit. The current, *I*, measures the total number of coulombs per second flowing at any given time. The product of the voltage and the current (VI) is therefore a measure of the total power produced since the units will be joules per second (watts):

$$P = VI$$

The electrical energy expended in a given time t (in seconds) is simply the product of the power and the time:

$$\text{Energy} = Pt = VIt = I^2Rt$$

In the second equation we used Ohm's law ($V = IR$) to express the energy in another form. This energy could be related to heat if we considered, for example, a direct transfer of energy into some water. The rise in temperature could then be computed. Note that the electrical energy is expressed in joules, while specific heats involve kilojoules and kilograms. To be consistent, you should convert either to joules or to kilojoules, but maintain the use of kilograms for mass!

SAMPLE PROBLEM

How much energy is used by a 10-Ω resistor connected to a 24-V battery for 30 minutes?

Solution

We know that the energy is given by $E = VIt = (V^2/R)t$, where t must be in seconds.

Thus,

$$E = \frac{(10\ \text{V})^2(1{,}800\ \text{s})}{10\ \Omega} = 18{,}000\ \text{J}$$

16.4 SERIES AND PARALLEL CIRCUITS

A. Series Circuits

A series circuit consists of two or more resistors sequentially placed within one circuit. An example is seen in Figure 16.6.

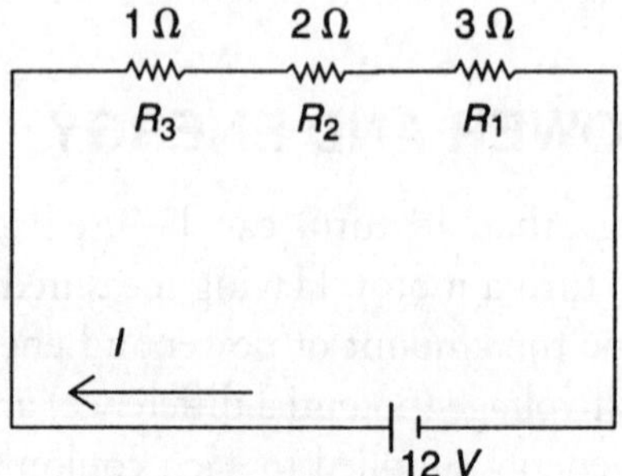

Figure 16.6 Series Circuit

We need to ask two questions. First, what is the effect of adding more resistors in series to the overall resistance in the circuit? Second, what effect does adding the resistors in series have on the current flowing in the circuit?

A look at the series circuit shows that it is still one continuous loop. Therefore, if there is a break in one place (let's say one light bulb goes out), the entire circuit will shut down. Experiments reveal that the current is the same through the entire circuit, because actually there is only one circuit. However, these experiments show also that the terminal emf across the battery is essentially equal to the sum of the emf's across all of the resistors. There exists a "potential drop" across each resistor.

One way to think about this circuit is to imagine a series of doors, one after the other. As people exit through one door, they must wait to open another. The result is to drain energy out of the system and decrease the number of people per second leaving the room. In an electric circuit, adding more resistors in series decreases the current (with the same voltage) by increasing the resistance of the circuit.

Algebraically, all of these observations can be summarized as follows. In resistors R_1, R_2, and R_3, we have currents I_1, I_2, and I_3. Each of these three currents is equal to each other current and to the circuit current I:

$$I = I_1 = I_2 = I_3$$

However, the voltage across each of these resistors is less than the source voltage V. If all three resistors were equal, each voltage would be equal to one-third of the total source voltage. In any case, we have

$$V = V_1 + V_2 + V_3$$

Using Ohm's law ($V = IR$), we can rewrite this expression as

$$IR = I_1R_1 + I_2R_2 + I_3R_3$$

However, the three currents are equal, so they can be canceled out of the expression. This leaves us with

$$R = R_1 + R_2 + R_3$$

In other words, when resistors are added in series, the total resistance of the circuit increases as the sum of all the resistances. This fact explains why the current decreases as more resistors are added. In our example, the total resistance of the circuit is

$$1\ \Omega + 2\ \Omega + 3\ \Omega = 6\ \Omega$$

Since the source voltage is 12 V, Ohm's law provides for a circuit current of 12/6 = 2 A. The voltage across each resistor can now be determined from Ohm's law since each gets the same current of 2 A. Thus,

$$V_1 = (1)(2) = 2\ \text{V}; \quad V_2 = (2)(2) = 4\ \text{V}; \quad V_3 = (3)(2) = 6\ \text{V}$$

The voltages in a series circuit add up to the total, and not unexpectedly we have

$$2\text{ V} + 4\text{ V} + 6\text{ V} = 12\text{ V}$$

It should be noted that, when batteries are connected in series (positive to negative), the effective voltage increases additively as well.

B. Parallel Circuits

A parallel circuit consists of two or more resistors connected across each other to a common point of potential difference. An example of a parallel circuit is seen in Figure 16.7.

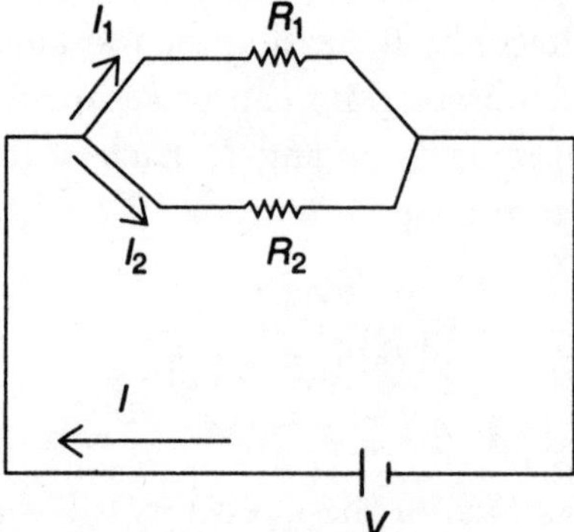

Figure 16.7 Parallel Circuit

In this circuit, a branch point is reached in which the current I is split into I_1 and I_2. Since each resistor is connected to a common potential point, experiments verify that the voltage across each resistor equals the source voltage V. Thus, in this circuit it is the current that is shared; the voltage is the same. Another feature of the parallel circuit is the availability of alternative paths. If one part of the circuit is broken, current can flow through the other path. While each branch current is less than the total circuit current I, the effect of adding resistors in parallel is to increase the effective circuit current by decreasing the circuit resistance R.

To understand this effect further, imagine a set of doors placed next to each other along a wall in a room. Unlike the series connection (in which the doors are placed one after the other), the parallel-circuit analogy involves placing the doors next to each other. Even though each door will have fewer people per second going through it at any given time, the overall effect is to allow, in total, more people to exit from the room. This is analogous to reducing the circuit resistance and increasing the circuit current (at the same voltage).

Algebraically, we can express these observations as follows. We have two resistors, R_1 and R_2, with currents I_1 and I_2. Voltmeters placed across the two resistors would indicate voltages V_1 and V_2, which would be essentially equal to the source voltage V; that is, in this example,

$$V = V_1 = V_2$$

Ammeters placed in the circuit would reveal that the circuit current I is equal to the sum of the branch currents I_1 and I_2; that is,

$$I = I_1 + I_2$$

Using Ohm's law, we see that

$$\frac{V}{R} = \frac{V_1}{R_1} + \frac{V_2}{R_2}$$

Since all voltages are equal, they can be canceled from the expression, leaving us with the following expression for resistors in parallel:

$$\frac{1}{R} = \frac{1}{R_1} + \frac{1}{R_2}$$

This expression indicates that the total resistance is determined "reciprocally," which in effect reduces the total resistance of the circuit as more resistors are added in parallel. If, for example, $R_1 = 10\ \Omega$ and $R_2 = 10\ \Omega$, the total resistance is $R = 5\ \Omega$ (in parallel)! It should be noted that connecting batteries in parallel (positive to positive and negative to negative) has no effect on the overall voltage of the combination.

SAMPLE PROBLEM

A 20-Ω resistor and a 5-Ω resistor are connected in parallel. If a 16-V battery is used, calculate the equivalent resistance of the circuit, the circuit current, and the amount of current flowing through each resistor.

Solution

We know that the equivalent resistance is given by

$$\frac{1}{R_{eq}} = \frac{1}{R_1} + \frac{1}{R_2} = \frac{1}{20\ \Omega} + \frac{1}{5\ \Omega}$$

In this case, it is easily seen that $R_{eq} = 4\ \Omega$.

Thus, using Ohm's law,

$$I = \frac{V}{R_{eq}} = \frac{16\ \text{V}}{4\ \Omega} = 4\ \text{A}$$

To find the current in each branch, we recall that the voltage drop across each resistor is the same as the source voltage. In this case:

$$I_{20} = \frac{16\ \text{V}}{20\ \Omega} = 0.8\ \text{A}$$

$$I_5 = \frac{16\ \text{V}}{5\ \Omega} = 3.2\ \text{A}$$

Notice that the total current is equal to 4 A as expected.

Gustav Kirchoff stated that the sum of the currents flowing into a branch must equal the sum of the currents leaving the branch. He also stated that the sum of the emfs around a loop is equal to the sum of the *IR* potential drops around the loop.

16.5 COMBINATION CIRCUITS

SAMPLE PROBLEM

In Figure 16.8, a circuit that consists of resistors in series and parallel is presented. The key to reducing such a circuit is to decide whether it is, overall, a series or parallel circuit. In Figure 16.8, an 8-Ω resistor is placed in series with a parallel branch containing two 4-Ω resistors. The problem is to reduce the circuit to only one resistor and then to determine the circuit current, the voltage across the branch, and the current in each branch.

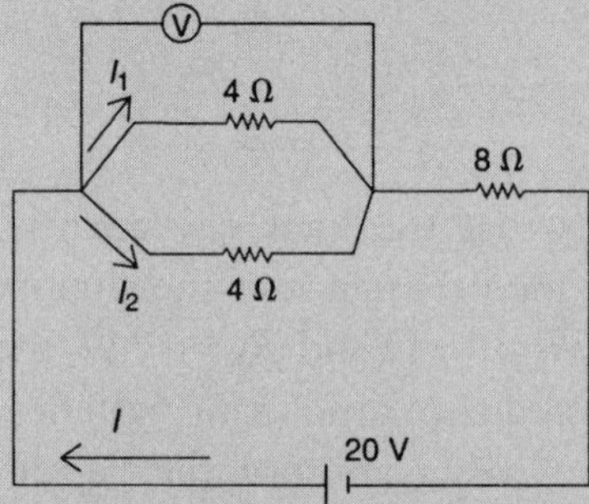

Figure 16.8

Solution

To reduce the circuit and find the circuit resistance R, we must first reduce the parallel branch. Thus, if R_e is the equivalent resistance in the branch, then

$$\frac{1}{R_e} = \frac{1}{4} + \frac{1}{4}$$

This implies that $R_e = 2\ \Omega$ of resistance. The circuit can now be thought of as a series circuit between a 2-Ω resistor and an 8-Ω resistor. Thus,

$$R = 2\ \Omega + 8\ \Omega = 10\ \Omega$$

Since the circuit voltage is 20 V, Ohm's law states that the circuit current is

$$I = \frac{V}{R} = \frac{20}{10} = 2\ \text{A}$$

In a series circuit, the voltage drop across each resistor is shared proportionally since the same current flows through each resistor. Thus, for the branch,

$$V = IR_e = (2)(2) = 4\ \text{V}$$

To find the current in each part of the parallel branch, we recall that in a parallel circuit the voltage is the same across each resistor. Thus, since the two resistors are equal, each will get half of the circuit current; thus each current is 1 A. Note how we used Kirchoff's rule.

16.6 CAPACITORS IN CIRCUITS

A. Capacitors in Series

We know from Chapter 15 that the capacitance of a parallel-plate capacitor is given by $C = Q/V$. Suppose we have a series of capacitors connected to a potential difference V as shown in Figure 16.9.

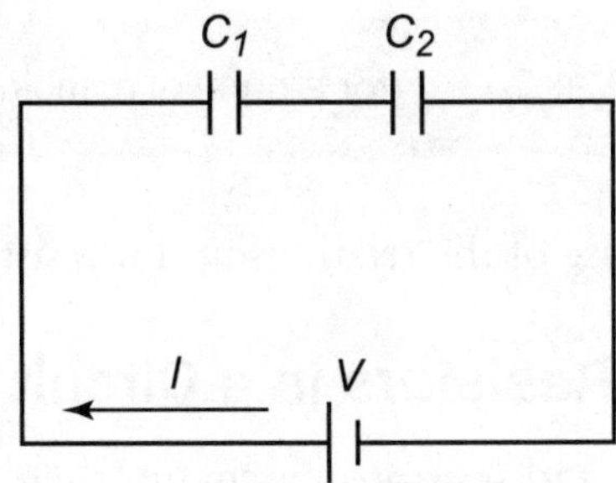

Figure 16.9

In this circuit, the magnitude of the charge must be the same on all plates because the left plate of C_1 becomes $+Q$ and the right plate of C_2 becomes $-Q$. By induction, the right plate of C_1 becomes $-Q$ and the left plate of C_2 becomes $+Q$. Additionally, the voltage across each capacitor is shared proportionally so that $V_1 + V_2 = V$ (just like the situation with resistors in series). Thus,

$$\frac{Q}{C_1} + \frac{Q}{C_2} = \frac{Q}{C}$$

and

$$\frac{1}{C_1} + \frac{1}{C_2} = \frac{1}{C} \quad \text{(for a series combination)}$$

Note that this is the opposite of the relationship for resistors in series. Be sure not to confuse them!

B. Capacitors in Parallel

A parallel combination of capacitors is shown in Figure 16.10.

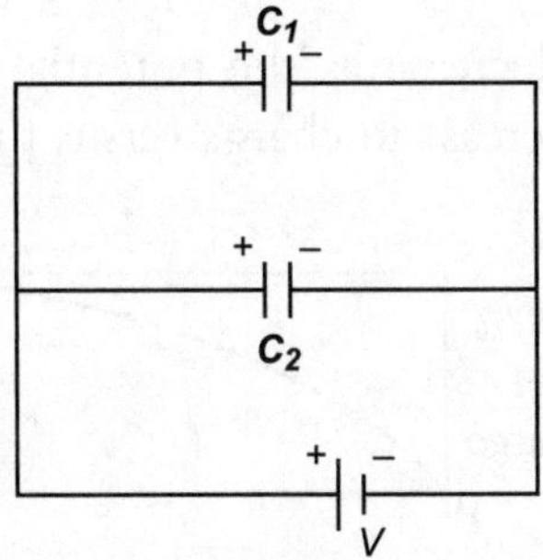

Figure 16.10

In this circuit, the current branches off, and so each capacitor is charged to a different total charge, Q_1 and Q_2. The equivalent total charge is $Q = Q_1 + Q_2$, and the voltage across each capacitor is the same as the source emf voltage. Thus,

$$CV = C_1V + C_2V$$

and

$$C = C_1 + C_2 \quad \text{(for a parallel combination)}$$

Note that this is the opposite of the relationship for resistors in parallel.

C. Capacitors and Resistors in a Circuit

If there is both a capacitor and resistor in a circuit, then there is a delay in the way in which the circuit responds to an applied emf. A typical RC circuit consists of a resistor and capacitor connected in series together with a switch and a battery (see Figure 16.11).

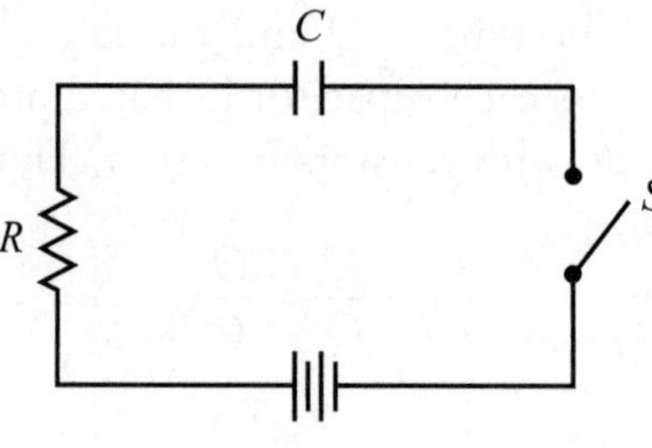

Figure 16.11

When the switch is closed, current flows through the resistor given by Ohm's law:

$$I = \frac{\text{emf}}{R}$$

As the capacitor (with capacitance C) becomes charged, a potential difference appears across it given by

$$V = \frac{Q}{C}$$

The current begins to decrease as the potential difference across the capacitor increases. A graph of the increase in charge versus time for the capacitor is sketched in Figure 16.12:

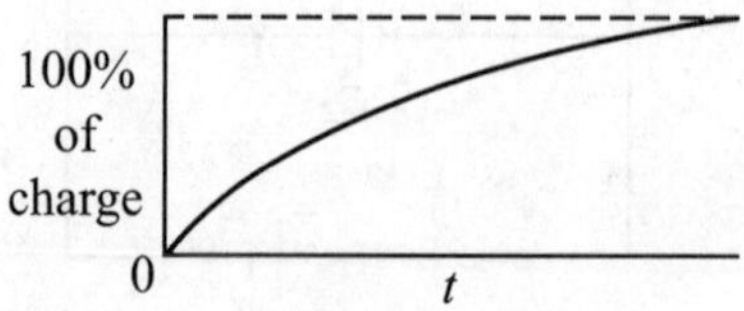

Figure 16.12

SAMPLE PROBLEM

What is the equivalent capacitance of a 5-F capacitor and a 15-F capacitor connected in parallel?

Solution

We recall that in parallel, the total capacitance is equal to the sum of the individual capacitances:

$$C_T = C_1 + C_2 = 5\text{ F} + 15\text{ F} = 20\text{ F}$$

Chapter Summary

- Electric current is a measure of the flow of charge in units of amperes (1A = 1 C/s).
- The conventional current is based on the direction of positive charge flow.
- Ohm's law states that at constant temperature the ratio of voltage and current is a constant in a conductor ($V/I = R$).
- Electrical resistance is based on the material used ("resistivity"), the length, and the cross-sectional area at constant temperature ($R = \rho L/A$).
- Ammeters measure current and are placed "in series" within the circuit.
- Voltmeters measure potential difference (voltage) and are placed in parallel (across segments of the circuit).
- A closed path and a source of potential difference are needed for a simple circuit.
- Resistors connected in series have an equivalent resistance equal to their numerical sum ($R_{eq} = R_1 + R_2$).
- Resistors connected in parallel have an equivalent resistance equal to their reciprocal sum[$1/R_{eq} = (1/R_1) + (1/R_2)$].
- If capacitors are connected in series, then their equivalent capacitance is equal to their reciprocal sum (the opposite of resistors).
- If capacitors are connected in parallel, then their equivalent resistance is equal to their numerical sum (the opposite of resistors).
- Kirchoff's rules describe the flow of current in circuit branches.

Problem-Solving Strategies for Electric Circuits

Several techniques for solving electric circuit problems have been discussed in this chapter. For resistances with one source of emf, we use the techniques of series and parallel circuits. Ohm's law plays a big role in measuring the potential drops across resistors and the determination of currents within the circuit.

When working with a combination circuit, try to reduce all subbranches first, as well as any series combinations. The goal is to be left with only one equivalent resistor and then use Ohm's law to identify the various missing quantities. Drawing a sketch of the circuit (if one is not already provided) is essential. Also, remember that in most college textbooks, the direction of current is taken from the positive terminal of a battery. This is opposite to the way current is presented in most high school physics textbooks.

Practice Exercises

MULTIPLE-CHOICE

1. How many electrons are moving through a current of 2 A for 2 s?

   (A) 3.2×10^{-19}
   (B) 6.28×10^{18}
   (C) 4
   (D) 2.5×10^{19}
   (E) 1

2. How much electrical energy is generated by a 100-W light bulb turned on for 5 min?

   (A) 20 J
   (B) 500 J
   (C) 3000 J
   (D) 15,000 J
   (E) 30,000 J

3. What is the current in a 5-Ω resistor due to a potential difference of 20 V?

   (A) 25 A
   (B) 4 A
   (C) 0.20 A
   (D) 5 A
   (E) 20 A

4. A 20-Ω resistor has 10 A of current in it. The power generated is

   (A) 2000 W
   (B) 2 W
   (C) 200 W
   (D) 4000 W
   (E) 20 W

5. What is the value of resistor *R* in the circuit shown below?

   (A) 20 Ω
   (B) 2 Ω
   (C) 4 Ω
   (D) 12 Ω
   (E) 10 Ω

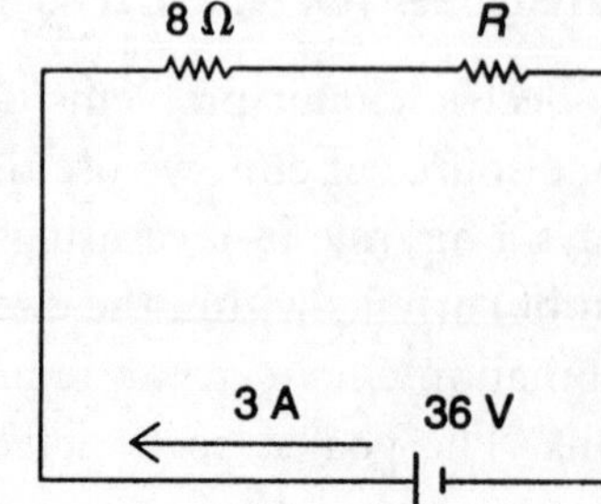

6. What is the equivalent capacitance of the circuit shown below?

(A) 2 F
(B) 8 F
(C) 5 F
(D) 10 F
(E) 15 F

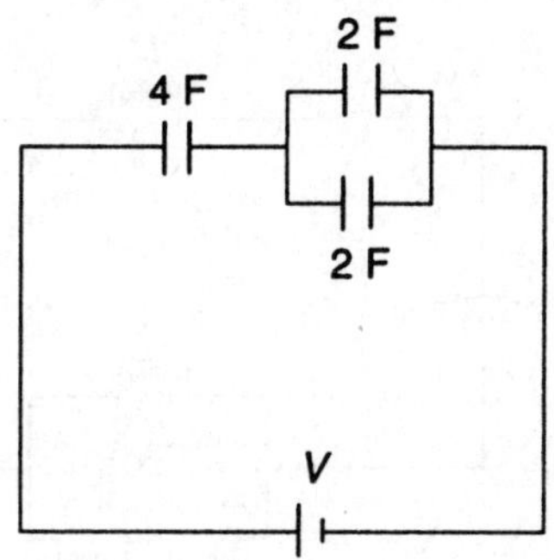

Answer questions 7 and 8 based on the circuit below:

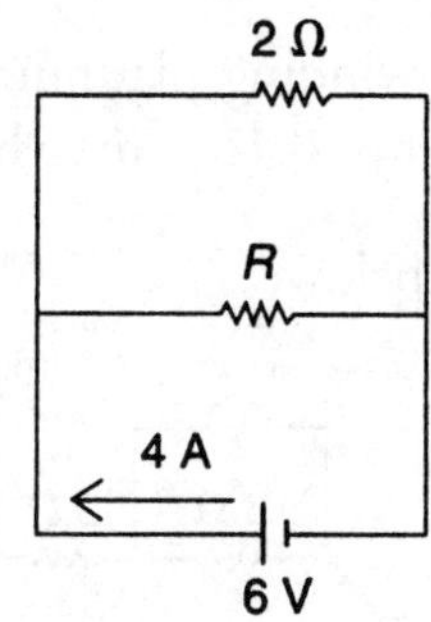

7. If the current in the circuit above is 4 A, then the power generated by resistor *R* is

(A) 6 W
(B) 18 W
(C) 24 W
(D) 7 W
(E) 16 W

8. What is the value of resistor *R* in the circuit above?

(A) 4 Ω
(B) 3 Ω
(C) 5 Ω
(D) 6 Ω
(E) 8 Ω

9. A 5-Ω and a 10-Ω resistor are connected in series with one source of emf of negligible internal resistance. If the energy produced in the 5-Ω resistor is *X*, then the energy produced in the 10-Ω resistor is

(A) *X*
(B) 2*X*
(C) *X*/2
(D) *X*/4
(E) 4*X*

CHALLENGE

10. Four equivalent light bulbs are arranged in a circuit as shown below. Which bulb(s) will be the brightest?

(A) *A*
(B) *B*
(C) *C*
(D) *D*
(E) Both *C* and *D*

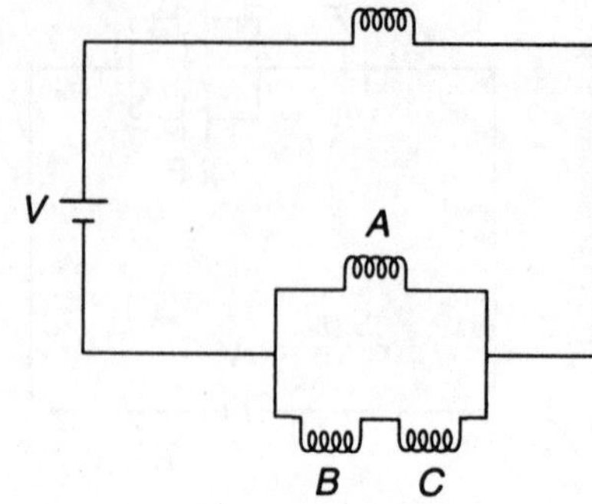

FREE-RESPONSE

1. A 60-Ω resistor is made by winding platinum wire into a coil at 20°C. If the platinum wire has a diameter of 0.10 mm, what length of wire is needed?

2. Based on the circuit shown below:

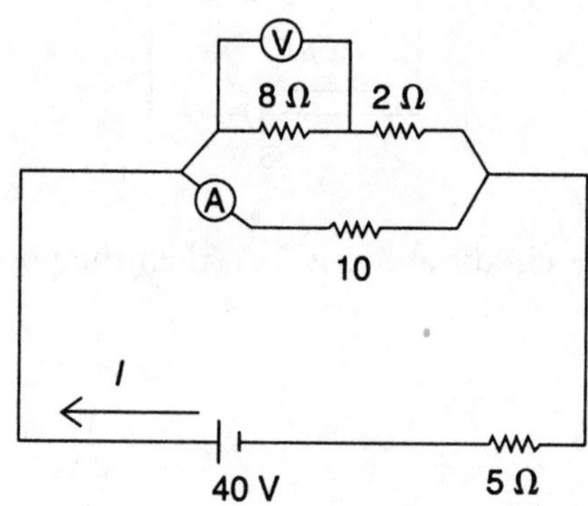

(a) Determine the equivalent resistance of the circuit.
(b) Determine the value of the circuit current *I*.
(c) Determine the value of the reading of ammeter A.
(d) Determine the value of the reading of voltmeter V.

3. Show that for two resistors in parallel, the equivalent resistance can be written as

$$R_{eq} = \frac{R_1 R_2}{R_1 + R_2}$$

4. Explain why connecting two batteries in parallel does not increase the effective emf.

5. You have five 2-Ω resistors. Explain how you can combine them in a circuit to have equivalent resistances of 1, 3, and 5 Ω.

6. A laboratory experiment is performed using lightbulbs as resistors. After a while, the current readings begin to decline. What can account for this?

ANSWERS EXPLAINED

MULTIPLE-CHOICE PROBLEMS

1. **(D)** By definition, 1 A of current measures 1 C of charge passing through a circuit each second. One coulomb is the charge equivalent of 6.28×10^{18} electrons. In this question, we have 2 A for 2 s. This is equivalent to 4 C of charge or 2.5×10^{19} electrons.

2. **(E)** Energy is equal to the product of power and time. The time must be in seconds; 5 min = 300 s. Thus

$$\text{Energy} = (100)(300) = 30{,}000 \text{ J}$$

3. **(B)** Using Ohm's law, we have

$$I = \frac{V}{R} = \frac{20}{5} = 4 \text{ A}$$

4. **(A)** We know that $P = VI$, and using Ohm's law ($V = IR$), we get

$$P = I^2R = (100)(20) = 2000 \text{ W}$$

5. **(C)** Using Ohm's law, we can determine the equivalent resistance of the circuit: 36/3 = 12 Ω. The two resistors are in series, so 12 = 8 + R. Thus, $R = 4\ \Omega$.

6. **(A)** We reduce the parallel branch first. In parallel, capacitors add up directly. Thus, C_1 = 4 F. Now, this capacitor would be in series with the other 4-F capacitor. In series, capacitors add up reciprocally; thus the final capacitance is

$$\frac{1}{C} = \frac{1}{4} + \frac{1}{4}$$

This implies that $C = 2$ F.

7. **(A)** In a parallel circuit, all resistors have the same potential difference across them. The branch currents must add up to the source current (4 A, in this case). Thus, using Ohm's law, we can see that the current in the 2-Ω resistor is $I = 6/2 = 3$ A. Thus, the current in resistor R must be equal to 1 A. Since $P = VI$, the power in resistor R is

$$P = (6)(1) = 6 \text{ W}$$

8. **(D)** Using Ohm's law, we have,

$$R = \frac{V}{I} = \frac{6}{1} = 6\ \Omega$$

9. **(B)** In a series circuit, the two resistors have the same current but proportionally shared potential differences. Thus, since the 10-Ω resistor has twice the resistance of the 5-Ω resistor in series, it has twice the potential difference. Therefore, with the currents equal, the 10-Ω resistor will generate twice as much energy, that is, $2X$, when the circuit is on.

10. **(D)** The fact that all the resistors (bulbs) are equal means that the parallel branch will reduce to an equivalent resistance less than that of resistor *D*. The remaining series circuit will produce a greater potential difference across *D* than across that equivalent resistor. With the currents equal at that point, resistor *D* will generate more power than the equivalent resistance. Thus, with the current split for the actual parallel part, even less energy will be available for resistors *A*, *B*, and *C*. Bulb *D* will be the brightest.

FREE-RESPONSE PROBLEMS

1. The formula we want to use is

$$R = \frac{\rho L}{A}$$

From Table 16.1, the resistivity of platinum at 20°C is equal to

$$\rho = 10.4 \times 10^{-8}\ \Omega \cdot \text{m}$$

The resistance needed is $R = 60\ \Omega$. We need the cross-sectional area. The diameter of the wire is 0.10 mm, so the radius of the wire must be equal to

$$r = \frac{\text{diameter}}{2} = 0.05\ \text{mm} = 0.00005\ \text{m}$$

The cross-sectional area is

$$A = \pi r^2 = \pi(0.00005\ \text{m})^2 = 7.85 \times 10^{-9}\ \text{m}^2$$

The length of the wire is therefore

$$60\ \Omega = \frac{(10.4 \times 10^{-8} \cdot \Omega \cdot \text{m})L}{(7.85 \times 10^{-9}\ \text{m}^2)}$$

$$L = 4.53\ \text{m}$$

2. (a) The first thing we need to do is to reduce the series portion of the parallel branch. The equivalent resistance there is $8\ \Omega + 2\ \Omega = 10\ \Omega$. Now, this resistance is in parallel with the other 10-Ω resistor. The equivalent resistance is found to be 5 Ω for the entire parallel branch. This equivalent resistance is in series with the other 5-Ω resistor, making a total equivalent resistance of 10 Ω.

(b) We can determine the circuit current using Ohm's law:

$$I = \frac{V}{R} = \frac{40}{10} = 4\ \text{A}$$

(c) To determine the reading of ammeter A, we must first understand that the potential drop across the 5-Ω resistor on the bottom is equal to 20 V since

$$V = IR = (4)(5) = 20$$

Now, the equivalent resistance of the parallel branch is also 5 Ω, and therefore across the entire branch there is also a potential difference of 20 V (in a series circuit, the voltages must add up—in this case, to 40 V). In a parallel circuit, the potential difference is the same across each portion. Thus, the 10-Ω resistor has 20 V across it. Using Ohm's law, we find that this implies a current reading of 2 A for ammeter A.

(d) If, from part (c), ammeter A reads 2 A, the top branch must be getting 2 A of current as well (since the source current is 4 A). Thus, using Ohm's law, we can find the voltage across the 8-Ω resistor:

$$V = IR = (2)(8) = 16\ \text{V}$$

3. We start with the regular formula for two resistors in parallel and solve:

$$\frac{1}{R_{eq}} = \frac{1}{R_1} + \frac{1}{R_2}$$
$$\frac{1}{R_{eq}} = \frac{R_2}{R_2R_1} + \frac{R_1}{R_1R_2}$$
$$\frac{1}{R_{eq}} = \frac{R_1 + R_2}{R_1R_2}$$
$$R_{eq} = \frac{R_1R_2}{R_1 + R_2}$$

4. When two batteries are connected in parallel, the emf of the system is not changed. The storage capacity of the battery system, however, is increased since the parallel connection of the two batteries is similar to the parallel connection of two capacitors.

5. An equivalent resistance of 1 Ω can be achieved by connecting two 2-Ω resistors in parallel. An equivalent resistance of 3 Ω can be achieved by connecting two 2-Ω resistors in parallel and then connecting this system to one 2-Ω resistor in series. The addition of one more 2-Ω resistor in series to the combination above achieves an equivalent resistance of 5 Ω.

6. Lightbulbs produce their luminous energy by heating up the filament inside them. Thus, after a while, this increase in temperature reduces the current flowing through them since the resistance has increased.

CHAPTER 17

Magnetism

KEY CONCEPTS

- Magnetic Fields and Forces
- Magnetic Force on a Moving Charge
- Magnetic Fields Due to Currents in Wires
- Magnetic Force on a Wire
- Magnetic Force Between Two Wires

17.1 MAGNETIC FIELDS AND FORCES

When two statically charged point objects approach each other, each exerts a force on the other that varies inversely with the square of their separation distance. We can also state that one of the charges creates an electrostatic field around itself, and this field transmits the force through space.

When we have a wire carrying current, we no longer consider the effects of static fields. A simple demonstration reveals that, when we arrange two current-carrying wires parallel to each other so that the currents are in the same direction, the wires will be attracted to each other with a force that varies inversely with their separation distance and directly with the product of the currents in the two wires. If the currents are in opposite directions, the wires are repelled. Since this is not an electrostatic event, we must conclude that a different force is responsible.

The nature of this force can be further understood when we consider the fact that a statically charged object, such as an amber stone, cannot pick up small bits of metal. However, certain naturally occurring rocks called **lodestones** can attract metal objects. The lodestones are called **magnets** and can be used to **magnetize** metal objects (especially those made from iron or steel).

If a steel pin is stroked in one direction with a magnetic lodestone, the pin will become magnetized as well. If an iron rail is placed in the earth for many years, it will also become magnetized. If cooling steel is hammered while lying in a north-south line, it too will become magnetized (heating will eliminate the magnetic effect). If the magnetized pin is placed on a floating cork, the pin will align itself along a general north-south line. If the cork's orientation is shifted, the pin will oscillate around its original equilibrium direction and eventually settle down along this direction. Finally, if another magnet is brought near, the "compass" will realign itself toward the new magnet. This is illustrated in Figure 17.1.

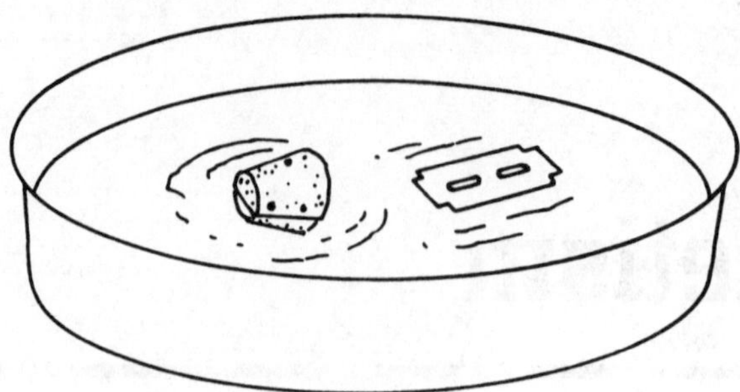

Figure 17.1

Each of these experiences suggests the presence of a magnetic force, and we can therefore consider the actions of this force through the description of a magnetic field. If a magnet is made in the shape of a rectangular bar (see Figure 17.2), the orientation of a compass, moved around the magnet, demonstrates that the compass aligns itself tangentially to some imaginary field lines that were first discussed by English physicist Michael Faraday. In general, we label the north-seeking pole of the magnet as N and the south-seeking pole as S. We also observe that these two opposite poles behave in a similar way to opposite electric charges: like poles repel, while unlike poles attract.

A fundamental difference between these two effects is that magnet poles never appear in isolation. If a magnet is broken in two pieces, each new magnet has a pair of poles. This phenomenon continues even to the atomic level. There appear, at present, to be no magnetic monopoles naturally occurring in nature. (New atomic theories have postulated the existence of magnetic monopoles, but none has as yet been discovered.)

Remember

A compass needle will align itself tangent to the external magnetic field.

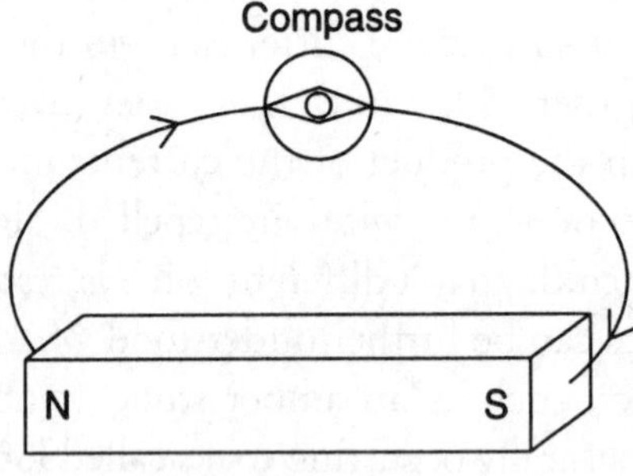

Figure 17.2

Small pieces of iron (called "filings") can be used as miniature compasses to illustrate the characteristics of the magnetic field around magnets made of iron or similar alloys. The ability of a metal to be magnetized is called its **permeability**; substances such as iron, cobalt, and nickel have among the highest permeabilities. These substances are sometimes called **ferromagnets**.

In Figure 17.3, several different magnetic field configurations are shown. By convention, the direction of the magnetic field is taken to be from the north pole of the magnet to the south pole.

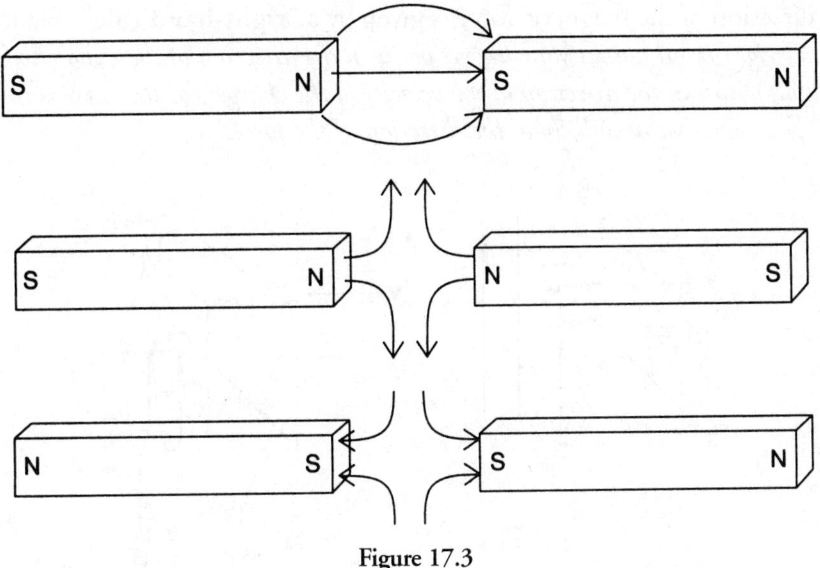

Figure 17.3

17.2 MAGNETIC FORCE ON A MOVING CHARGE

If a moving electric charge enters a magnetic field, it will experience a force that depends on its velocity, charge, and orientation with respect to the field. The force will also depend on the magnitude of the magnetic field strength (designated as **B** algebraically). This field strength is a vector quantity, just like the gravitational and electrical field strengths.

Experiments with moving charges in a magnetic field reveal that the resultant force is a maximum when the velocity, **v**, is perpendicular to the magnetic field and zero when the velocity is parallel to the field. With this varying angle of orientation θ and a given charge *Q*, we find that the magnitude of the magnetic force is given by:

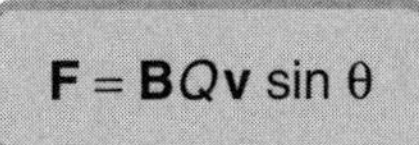

$$\mathbf{F} = \mathbf{B}Q\mathbf{v} \sin \theta$$

Since the units of force must be newtons, the units for the magnetic field strength **B** must be newton · seconds per coulomb · meter (N · s/C · m). Recall that the newton · second is the unit for an impulse. We can think of the magnetic field as a measure of the impulse given to 1 coulomb of charge moving a distance of 1 meter in a given direction. Additionally, if we have a beam of charges, we effectively have an electric current. The units (seconds per coulomb) can be interpreted as the reciprocal of amperes, and so the strength of the magnetic field in a current-carrying wire is given in units of newtons per ampere · meter (N/A · m).

In the SI system, this combination is called a **tesla**; 1 tesla is equal to 1 newton per ampere · meter. Thus, an electric current carries a magnetic field. This is consistent with the earliest experiments by Hans Oersted, who first showed that an electric current can influence a compass. Oersted's discovery, in 1819, established the new science of **electromagnetism**.

The direction of the magnetic force is given by a "**right-hand rule**" (Figure 17.4): *Open your right hand so that your fingers point in the direction of the magnetic field and your thumb points in the direction of the velocity of the charges (or the current if you have a wire). Your open palm will show the direction of the force.*

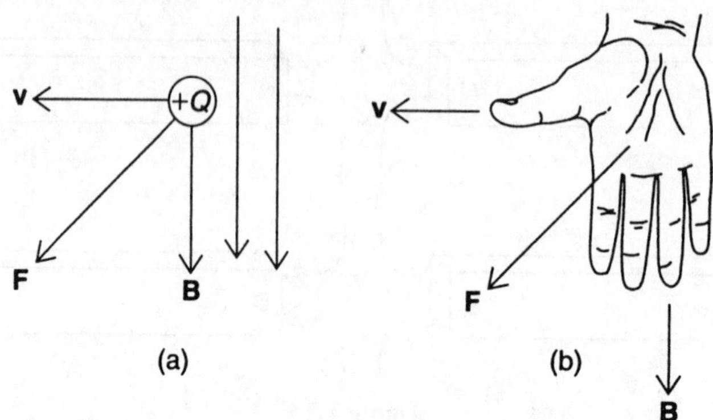

Figure 17.4

Figure 17.5 illustrates that, if the velocity and magnetic field are perpendicular to each other, the path of the charge in the field is a circle of radius *R*. The direction, provided by the right-hand rule, is the direction that a positive test charge would take. Thus, the direction for a negative charge would be opposite to what the right-hand rule predicted.

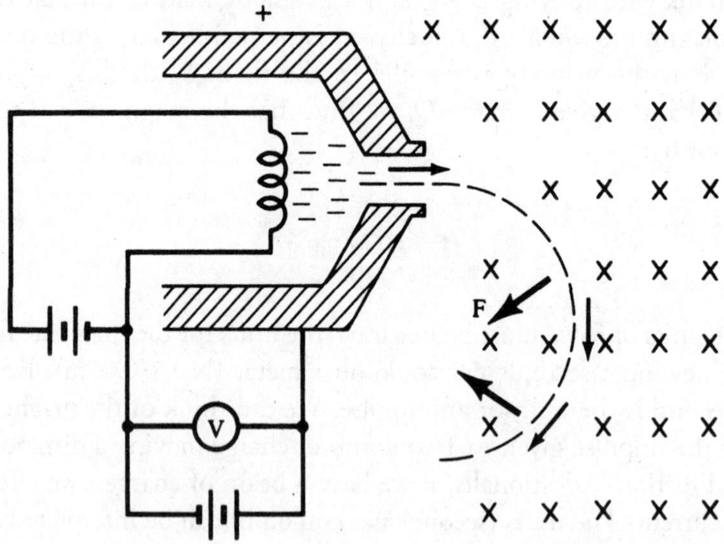

Figure 17.5

The circular path arises because the induced force is always at right angles to the deflected path when the velocity is initially perpendicular to the magnetic field. This path is similar to the circular path of a stone being swung in an overhead horizontal circle while attached to a string.

In the frame of reference of the moving charge, the inward magnetic force is apparently balanced by an outward centrifugal force. This relationship is given by:

$$m\frac{\mathbf{v}^2}{R} = \mathbf{B}\,Qv$$

where m is the mass of the charge.

This relationship can be used to determine the radius of the path:

$$R = m\frac{\mathbf{v}}{Q\mathbf{B}}$$

This expression assumes that the velocity and mass of the charge are known independently. Even if we wish to use this expression to find the mass of the charge (as used in a mass spectrograph):

$$m = \frac{RQ\mathbf{B}}{\mathbf{v}}$$

we still must know what the velocity is independent of the mass of the charge.

One way to determine the velocity of the charge, independently from the mass, is to pass the charge through both an electric and a magnetic field, as in a cathode ray tube. As shown in Figure 17.6, if the electric field is horizontal, the charges will be attracted or repelled along a horizontal line. If the magnetic field is vertical, the right-hand rule will force the charges along the same horizontal line. Careful adjustment of both fields (keeping the velocity perpendicular to the magnetic field) can lead to a situation where the effect of one field balances out the effect from the other and the charges remain undeflected. In that case, we can write that the magnitudes of the electric and magnetic forces are equal:

$\mathbf{E}q = \mathbf{B}q\mathbf{v}$ which implies that the velocity $\mathbf{v} = \mathbf{E}/\mathbf{B}$ (independent of mass)!

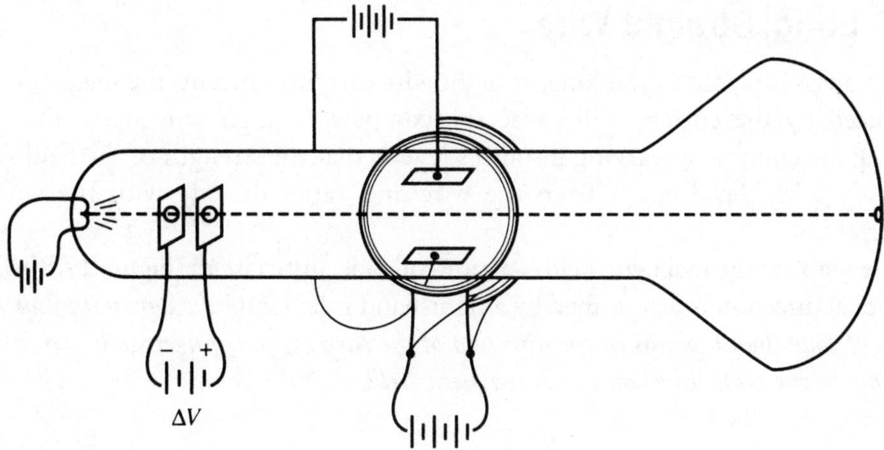

Figure 17.6

If the velocity is not perpendicular to the magnetic field when it first enters, the path will be a spiral because the velocity vector will have two components. One component will be perpendicular to the field (and create a magnetic force that will try to deflect the path into a circle), while the other component will be parallel to the field (producing no magnetic force and maintaining the original direction of motion). This situation is illustrated in Figure 17.7.

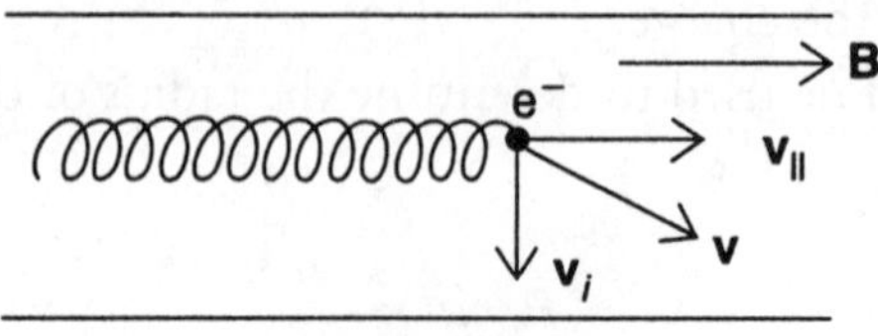

Figure 17.7

SAMPLE PROBLEM

An electron ($m = 9.1 \times 10^{-31}$ kg) enters a uniform magnetic field $\mathbf{B} = 0.4$ T at right angles and with a velocity of 6×10^{7} m/s.

(a) Calculate the magnitude of the magnetic force on the electron.
(b) Calculate the radius of the path followed.

Solution

(a) We use

$$\mathbf{F}_m = \mathbf{B}q\mathbf{v}$$
$$\mathbf{F}_m = (0.4\text{ T})(-1.6 \times 10^{-19}\text{ C})(6 \times 10^{7}\text{ m/s}) = 3.84 \times 10^{-12}\text{ N}$$

(b) We use

$$r = \frac{(m\mathbf{v})}{(\mathbf{B}q)} = \frac{(9.1\times10^{-31}\text{ kg})(6\times10^{7}\text{ m/s})}{(0.4\text{ T})(-1.6\times10^{-19}\text{C})} = 8.53\times10^{-4}\text{ m}$$

17.3 MAGNETIC FIELDS DUE TO CURRENTS IN WIRES

A. A Long, Straight Wire

If a compass is placed near a long, straight wire carrying current, the magnetic field produced by the current will cause the compass to align tangent to the field. Placing the compass at varying distances reveals that the strength of the field varies inversely with the distance from the wire and varies directly with the amount of current.

The shape of the magnetic field is a series of concentric circles (Figure 17.8) whose rotational direction is determined by a right-hand rule: *Grasp the wire with your right hand. If your thumb points in the direction of the current, your fingers will curl around the wire in the same direction as the magnetic field.*

A schematic for the three-dimensional nature of the field is sometimes used. For an emerging field, we use a series of dots; for an inward field, we use a series of X's.

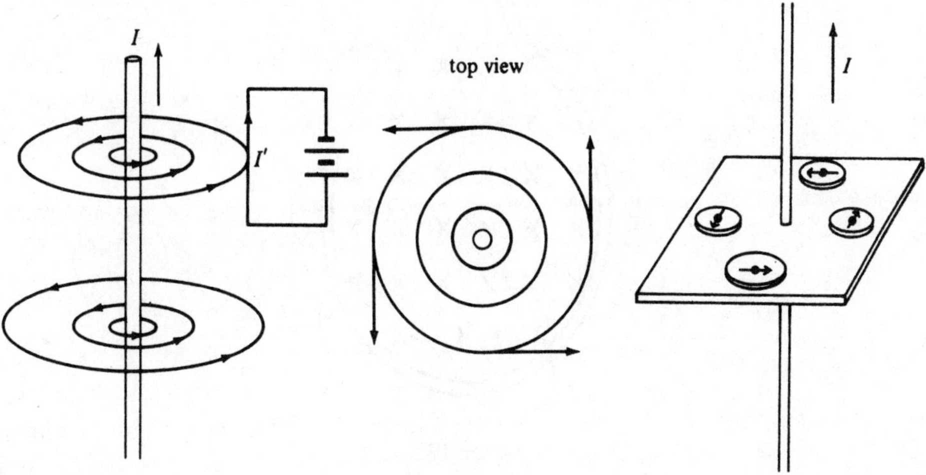

Figure 17.8

The magnitude of the magnetic field strength at some distance *r* from the wire is given by

$$\mathbf{B} = \frac{\mu I}{2\pi r}$$

where μ is the permeability of the material around the wire in units of newtons per square ampere (N/A^2). In air, the value of μ is approximately equal to the permeability of a vacuum:

$$\mu_0 = 4\pi \times 10^{-7} \frac{\text{N}}{\text{A}^2}$$

This equation is therefore usually written in the form

$$\mathbf{B} = \frac{\mu_0 I}{2\pi r}$$

B. A Loop of Wire

Imagine a straight wire with current that has been bent into the shape of a loop with radius *r*. Each of the circular magnetic fields now interacts in the center of the loop. In Figure 17.9, we see that the effect of all of these subfields is to produce one concentrated field, pointing inward or outward, at the center.

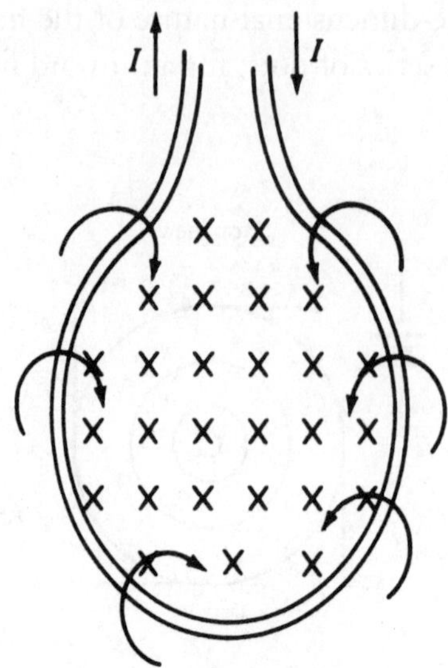

Figure 17.9

The general direction of the emergent magnetic field is given by a right-hand rule: *Grasp the loop in your right hand with your fingers curling around the loop in the same direction as the current. Your thumb will now show the direction of the magnetic field.* Figure 17.10 illustrates this rule.

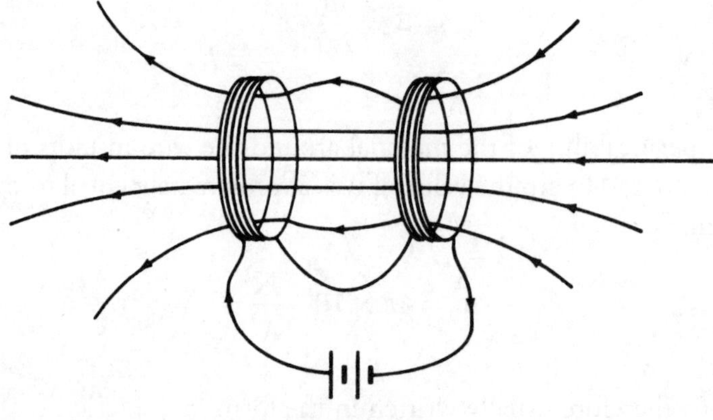

Figure 17.10

If the loop consists of N turns of wire, the magnetic field strength will be increased N times. The magnitude of the field strength, at the center of the loop (of radius r), is given by

$$\mathbf{B} = \frac{\mu N I}{2r}$$

The strength of the field can also be increased by changing the permeability of the core. If the wire is looped around iron, the field is stronger than if the wire were looped around cardboard. Also, if the loop's radius is smaller, the field is more concentrated at the center and thus stronger.

If the loop described in Section B is stretched out to a length L, in the form of a spiral, we have what is called a **solenoid** (Figure 17.11). This is the basic form for an electromagnet, in which wire is wrapped around an iron nail and then connected to a battery. The strength of the field will increase with the number of turns, the permeability of the core, and the amount of current in the wire:

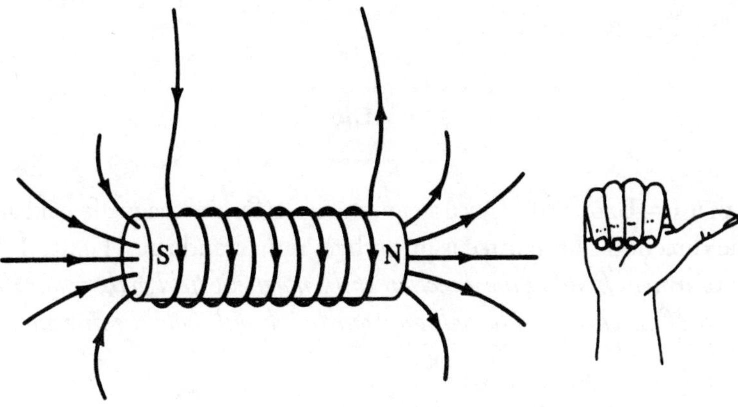

Figure 17.11

The direction of the magnetic field is given by another right-hand rule: *Grasp the solenoid with your right hand so that your fingers curl in the same direction as the current. Your thumb will point in the direction of the emergent magnetic field.*

SAMPLE PROBLEM

What is the strength of the magnetic field at a point 0.05 m away from a straight wire carrying a current of 10 A?

Solution

We use

$$\mathbf{B} = \frac{\mu_0 I}{2\pi r} = \frac{(4\pi \times 10^{-7}\ \text{N/A}^2)(10\ \text{A})}{(2\pi)(0.05\ \text{m})} = 4 \times 10^{-5}\ \text{T}$$

17.4 MAGNETIC FORCE ON A WIRE

The magnetic force on a charge Q, moving in a magnetic field **B** with a velocity v, is given by

$$\mathbf{F} = \mathbf{B}qv \sin \theta$$

If we have a collection of charges, then effectively we have an electric current given by $I = Q/t$, where Q is now the total charge present. If the current passes through a length L in time t, we can state that the induced magnetic force is given by:

$$\mathbf{F} = \mathbf{B}IL \sin \theta$$

If the charges are contained within a wire and the wire is perpendicular to the magnetic field, the force is given simply by

$$\mathbf{F} = \mathbf{B}IL$$

This means that $\mathbf{F} = \mathbf{B}IL$ at 90°. Once again, the units for the magnetic field are N/A • m.

Again, the direction of the force is given by a right-hand rule (Figure 17.12): *Point the fingers of your right hand in the direction of the magnetic field so that your thumb points in the direction of the current. The palm of your hand will show the direction of the force.*

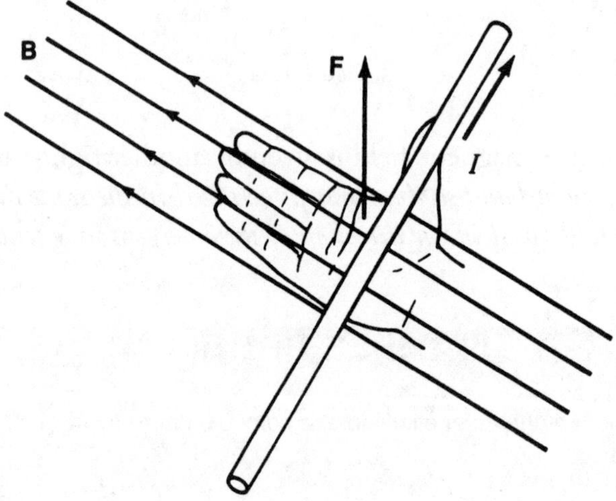

Figure 17.12

SAMPLE PROBLEM

A wire 1.2 m long and carrying a current of 60 A is lying at right angles to a magnetic field $\mathbf{B} = 5 \times 10^{-4}$ T. Calculate the magnetic force on the wire.

Solution

We use

$$\mathbf{F}_m = \mathbf{B}IL = (5 \times 10^{-4}\ \text{T})(60\ \text{A})(1.2\ \text{m}) = 0.036\ \text{N}$$

17.5 MAGNETIC FORCE BETWEEN TWO WIRES

Imagine that you have two straight wires with currents going in the same direction (Figure 17.13). Each wire creates a circular magnetic field around itself so that, in the region between the wires, the net effect of each field is to weaken the other field relative to the other side. This difference produces a net inward attractive force.

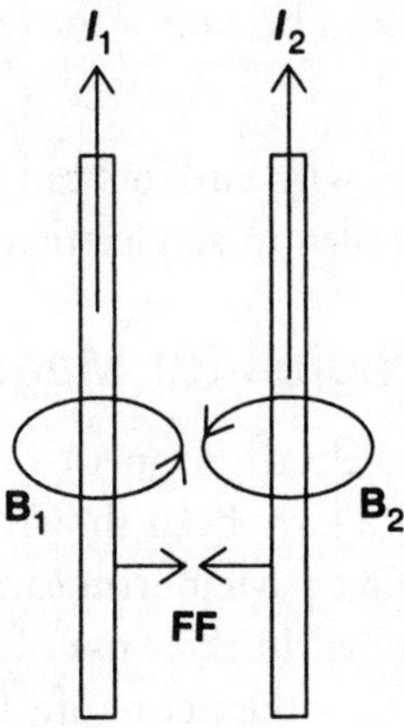

Figure 17.13

If the current in one wire is reversed, both forces will change directions and the wires will repel each other.

The magnitude of the force between the wires, at a distance r, can be determined if we consider the fact that one wire sets up the external field that acts on the other. If we want the magnetic force on wire 2 due to wire 1, we can write

$$\mathbf{B}_1 \frac{\mu_0 I_1}{2\pi r}$$

Since $\mathbf{F} = \mathbf{B}IL$, we can write

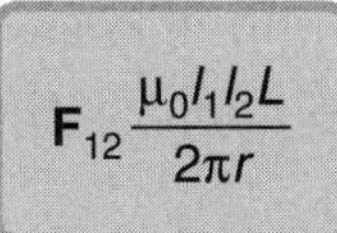

$$\mathbf{F}_{12} \frac{\mu_0 I_1 I_2 L}{2\pi r}$$

Chapter Summary

- Magnets consist of both north and south poles.
- Isolated magnetic poles do not exist.
- The ability of a metal to become magnetized is called permeability.
- Iron, cobalt, and nickel are metals with the highest permeabilities. Magnets made from these materials are called ferromagnets.
- Magnetic field lines go from north to south and follow the direction of a compass.
- An electric charge moving in an external magnetic field has a force induced on it. The direction of that force is determined by a right-hand rule.
- The path of a charged particle, traveling at right angles to an external magnetic field, is a circle.
- Metal wires carrying current generate magnetic fields. The magnetic field directions around wires can be found using a right-hand rule.

- Two wires carrying current can become attracted or repelled, depending on the directions of the currents.
- The forces induced on wires with currents can be used to make loops of wires spin in a motor or be controlled in an electric meter.

Problem-Solving Strategies for Magnetism

The key to solving magnetic and electromagnetic field problems is remembering the right-hand rules. These heuristics have been shown to be useful in determining the directions of field interaction forces. Memorize each one, and become comfortable with its use. Some rules are familiar in their use of fingers, thumb, and open palm. Make sure, however, that you know the net result. Drawing a sketch always helps.

Keeping track of units is also useful. The tesla is an SI unit and as such requires lengths to be in meters and velocities to be in meters per second (forces are in newtons). Most force interactions are angle dependent and have maximum values when two quantities (usually velocity and field or current and field) are perpendicular. Be sure to read each problem carefully and to remember that magnetic field **B** is a vector quantity!

Practice Exercises

MULTIPLE-CHOICE

1. A charge moves in a circular orbit of radius R due to a uniform magnetic field. If the velocity of the charge is doubled, the orbital radius will become

   (A) $2R$
   (B) R
   (C) $R/2$
   (D) $4R$
   (E) $R/4$

2. Inside a solenoid, the magnetic field

   (A) is zero
   (B) decreases along the axis
   (C) increases along the axis
   (D) is uniform
   (E) increases, then decreases along the axis

3. A proton enters a magnetic field of 10 T and an electric field of 2000 N/C in such a way that it passes through both fields undeflected. The velocity of the proton will become

   (A) 2000 m/s
   (B) 200 m/s
   (C) 200,000 m/s
   (D) 20,000 m/s
   (E) 2,000,000 m/s

4. Which of the following measures will decrease the strength of the magnetic field of a solenoid?

   (A) Increase the permeability of the core.
   (B) Increase the temperature of the solenoid.
   (C) Increase the current.
   (D) Increase the number of turns of wire.
   (E) All of these measures will decrease the strength.

5. An electron crosses a perpendicular magnetic field as shown below. The direction of the induced magnetic force is

   (A) to the right
   (B) to the left
   (C) up the page
   (D) out of the page
   (E) into the page

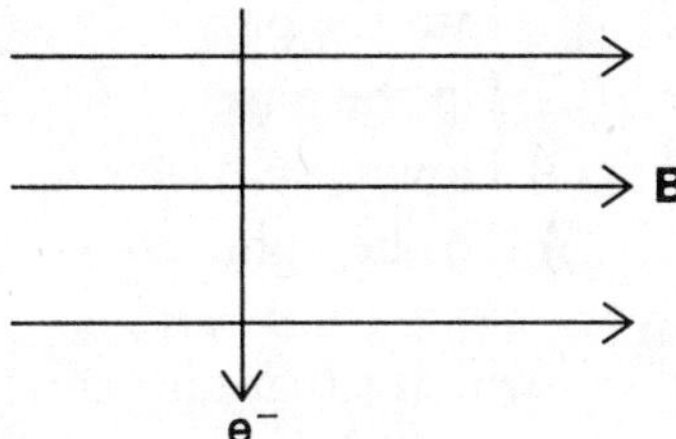

6. Three centimeters from a long, straight wire, the magnetic field produced by the current is determined to be equal to 3×10^{-5} T. The current in the wire must be

   (A) 2.0 A
   (B) 4.5 A
   (C) 1.5 A
   (D) 3 A
   (E) 5.0 A

7. In which direction does the compass needle point when it is placed near the straight wire carrying current in the direction shown below?

   (A) *A*
   (B) *B*
   (C) *C*
   (D) *D*
   (E) The direction will vary.

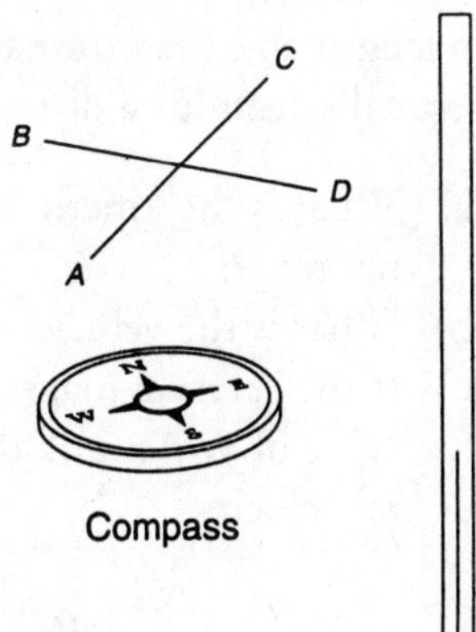

8. Magnetic field lines determine

   (A) only the direction of the field
   (B) the relative strength of the field
   (C) both the relative strength and the direction of the field
   (D) only the shape of the field
   (E) All of the preceding are correct.

9. What is the direction of the magnetic field at point A above the wire carrying current?

   (A) Out of the page
   (B) Into the page
   (C) Up the page
   (D) Down the page
   (E) To the right

10. In which situation below did an electron enter the magnetic field at a non-right angle?

    (A) A
    (B) B
    (C) C
    (D) D
    (E) None of the preceding is correct.

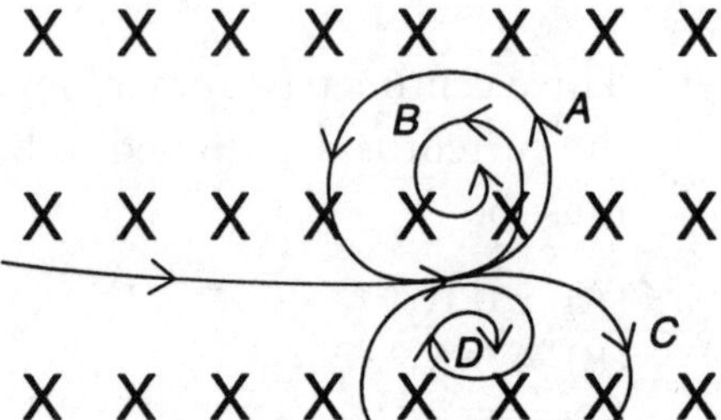

FREE-RESPONSE

CHALLENGE

1. An electron is accelerated by a potential difference of 12,000 V as shown in the following diagram. The electron enters a cathode ray tube that is 20 cm in length and the external magnetic field causes the path to deflect along a vertical screen for a distance y.

   (a) What is the kinetic energy of the electron as it enters the cathode ray tube?
   (b) What is the velocity of the electron as it enters the cathode ray tube?
   (c) If the external magnetic field has a strength of 4×10^{-5} T, what is the value of y? Assume that the field has a negligible effect on the horizontal velocity.

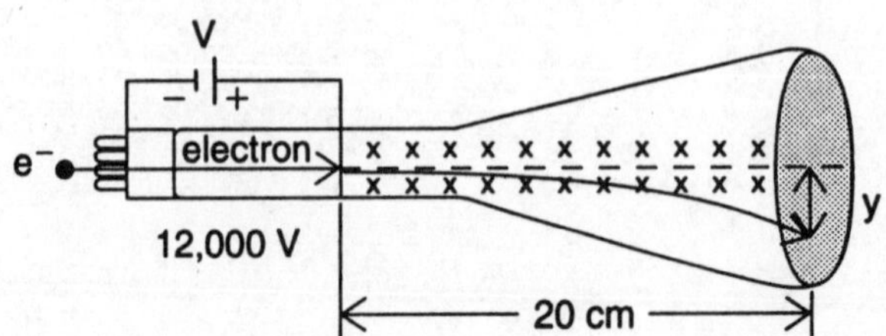

2. A straight conductor has a mass of 15 g and is 4 cm long. It is suspended from two parallel and identical springs as shown below. In this arrangement, the springs stretch a distance of 0.3 cm. The system is attached to a rigid source of potential difference equal to 15 V, and the overall resistance of the circuit is 5 Ω. When current flows through the conductor, an external magnetic field is turned on and it is observed that the springs stretch an additional 0.1 cm. What is the strength of the magnetic field?

CHALLENGE

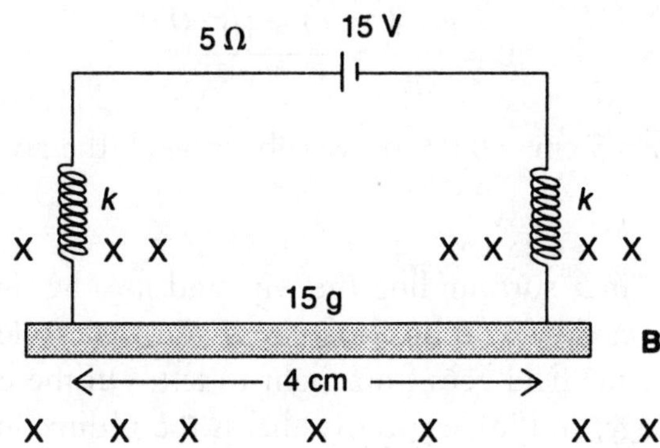

3. Show that the units of the magnetic induction **B** are equivalent to the units of (impulse per coulomb) per meter.

4. What happens to the maximum voltage read on a voltmeter as the resistance placed in series with its coil is changed?

5. Explain why a bar magnet loses its magnetic strength if it is struck too many times.

ANSWERS EXPLAINED

MULTIPLE-CHOICE PROBLEMS

1. **(A)** The formula is

$$R = \frac{mv}{q\mathbf{B}}$$

If the velocity is doubled, so is the radius.

2. **(D)** Inside a solenoid, the effect of all the coils is to produce a long, uniform magnetic field.

3. **(B)** Using the formula and substituting the appropriate numbers gives

$$v = \frac{E}{\mathbf{B}} = \frac{2000}{10} = 200 \text{ m/s}$$

4. **(B)** Increasing the temperature of the solenoid will increase the resistance of the coils. This, in turn, will decrease the current and hence decrease the strength of the magnetic field.

5. **(E)** Using the right-hand rule, we place the fingers of the right hand along the line of the magnetic field and point the thumb in the direction of the velocity. The palm points outward. However, since the particle is an electron, and the right-hand rule is designed for a positive charge, we must reverse the direction and say that the force is directed inward (into the page).

6. **(B)** The formula is

$$\mathbf{B} = \frac{\mu_0 I}{2\pi r} = \frac{(2 \times 10^{-7})I}{r}$$

Recalling that $r = 3$ cm $= 0.03$ m, we substitute all the given values to obtain $I = 4.5$ A.

7. **(C)** Imagine a circle surrounding the wire and use the right-hand rule. The direction of the circle would be clockwise if you were to look down on it. In this orientation, the field goes from right to left. On the left side, the needle will enter the page, and in perspective that is the *C* direction.

8. **(C)** Magnetic field lines were introduced by Michael Faraday to determine both the direction of the field and its relative strength (a stronger field is indicated by a greater line density).

9. **(A)** Using the right-hand rule, we find that the circle surrounding the wire will have its field coming out of the page at point *A*.

10. **(D)** When a charged particle (here, an electron) enters a magnetic field at an angle other than 90°, it takes a spiral path. Using the right-hand rule and reversing for the negative charge gives us *D* for the desired situation (the path is seen in perspective since the field is inward).

FREE-RESPONSE PROBLEMS

1. (a) The kinetic energy of the electron is the product of its charge and potential difference. Thus,

$$\text{KE} = (1.6 \times 10^{-19})(12{,}000) = 1.92 \times 10^{-15}\text{J}$$

(b) The formula for the velocity, when the mass and kinetic energy are known, is

$$\mathbf{v} = \sqrt{\frac{2\text{KE}}{m}} = \sqrt{\frac{(2)(1.92 \times 10^{-15})}{9.1 \times 10^{-31}}} = 6.5 \times 10^{7} \text{ m/s}$$

(c) The magnetic force will cause a downward uniform acceleration, which can be found from Newton's second law, $\mathbf{F} = m\mathbf{a}$. The force will be due to the magnetic force, $\mathbf{F} = qv\mathbf{B}$. Thus, $\mathbf{a} = qv\mathbf{B}/m$. Using our known values, we get

$$\mathbf{a} = 4.57 \times 10^{14} \text{ m/s}^2$$

Now, for linear motion downward (recall projectile motion), $\mathbf{y} = (1/2)\mathbf{a}t^2$. The time to drop y meters is the same time required to go 20 cm (0.20 m), assuming no change in horizontal velocity. Thus,

$$t = \frac{x}{\mathbf{v}} = \frac{0.20}{6.5 \times 10^7} = 3 \times 10^{-9}\ \text{s}$$

This implies that, upon substitution, $y = 0.00205$ m = 2.05 mm.

2. The effect of the conductor is to stretch both springs. Since we have two parallel springs of equal force constant k, we know from our work on oscillatory motion (Chapter 11) that the effective spring constant is equal to $2k$. Thus:

$$\mathbf{F} = 2k\,\Delta\text{x}$$

where $\Delta x = 0.003$ m. Thus, the weight of the conductor is

$$\mathbf{W} = mg = (0.015)(9.8) = 0.147\ \mathbf{N}$$

and k = 24.5 N/m. Now, the magnetic force, $\mathbf{F} = \mathbf{B}IL$, is responsible for another elongation, x. Since $\mathbf{F} = k\mathbf{x}$, we have (with both springs attached)

$$\mathbf{B}IL = (24.5)(24.5) = 0.0490\ \text{N}$$

The length of the conductor is 4 cm = 0.04 m; and using Ohm's law, we know that the current is $I = 15/5 = 3$ A. Thus, we find that

$$\mathbf{B}\,(3)(0.04) = 0.0490 \text{ implies } \mathbf{B} = 0.4\ \text{T}$$

3. The units of **B** are (newtons per ampere) per meter. An ampere is equivalent to coulombs per second. Hence, we see that the units for magnetic induction are equivalent to (newtons times seconds per coulomb) per meter. Recall that the units of newtons times seconds correspond to impulse.

4. The maximum voltage measured by a voltmeter varies directly with the resistance placed in series with its coil. To measure a high voltage, we use high resistance. To measure a low voltage, we use low resistance.

5. Striking a bar magnet or even heating it disrupts the magnetic domains that have aligned to magnetize it, causing the magnet to lose its strength.

Electromagnetic Induction

CHAPTER 18

KEY CONCEPTS

- Induction and EMF in a Wire
- Faraday's Law of Induction

18.1 INDUCED MOTIONAL EMF IN A WIRE

We know from Oersted's experiments, mentioned in Chapter 17, that a current in a wire generates a magnetic field. We also know that the field of a solenoid approximates that of an ordinary bar magnet. A third fact is that the magnetic property of a material (called its **permeability**) influences the strength of its magnetic field. Additionally, the interactions between fields lead to magnetic forces that can be used to operate an electric meter or motor.

A question can now be raised: Can a magnetic field be used to induce a current in a wire? The answer, investigated in the early nineteenth century by French scientist Andre Ampere, is yes. However, the procedure is not as simple as placing a wire in a magnetic field and having a current "magically" arise!

A simple experimental setup is shown in Figure 18.1. A horseshoe magnet is arranged with a conducting wire attached to a galvanometer. When the wire rests in the magnetic field, the galvanometer registers zero current. If the wire is then moved through the field in such a way that its motion "cuts across" the imaginary field lines, the galvanometer will register the presence of a small current. If, however, the wire is moved through the field in such a way that its motion remains parallel to the field, again no current is present. This experiment also reveals that a maximum current, for a given velocity, occurs when the wire moves perpendicularly through the field (which suggests a dependency on the sine of the orientation angle).

The phenomenon described above is known as **electromagnetic induction**. The source of the induced current is the establishment of a **motional emf** due to the change in something called the **magnetic flux**, which we will define shortly.

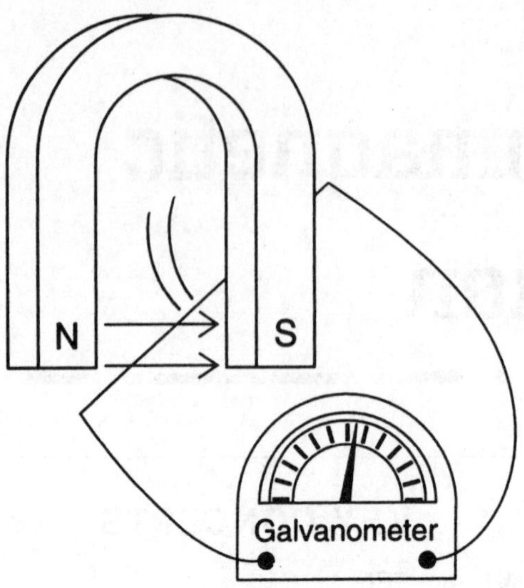

Figure 18.1

If the wire is moved up and down through the field in Figure 18.1, the galvanometer shows that the induced current alternates back and forth according to a right-hand rule. The fingers of the right hand point in the direction of the magnetic field; the thumb points in the direction of the velocity of the wire. Finally, the open palm indicates the direction of the induced current.

Now, since a parallel motion through the field implies no induced current, if the wire is moved at any other angle, only the perpendicular component contributes to the induction process. Additionally, it can be demonstrated that the magnitude of the induced current varies directly with the velocity of the wire, the length of wire in the field, and the strength of the external field. Since current is related to potential difference (emf) through Ohm's law ($V = IR$), we can write

$$\text{emf} = \mathbf{B}\ell\mathbf{v} \tag{1}$$

where $\mathbf{v}$ is the perpendicular velocity component.

The derivation of this expression involves a reconsideration of the induced magnetic force on a charge in an external magnetic field. Recall that, if $\mathbf{v}$ is a perpendicular velocity component, $\mathbf{F} = \mathbf{B}q\mathbf{v}$ is the expression for the magnitude of the induced magnetic force on a charge q. Now, the wire in our situation contains charges that will experience a force $\mathbf{F}$ if we can get them to move through an external field. Physically moving the entire wire accomplishes this task. The direction of the induced force, determined by the right-hand rule, is along the length of the wire. If we let Q represent the total charge per second, we can write

$$\mathbf{F} = \mathbf{B}Q\mathbf{v} \tag{2}$$

The charges experience a force as long as they are in the magnetic field. This situation occurs for a length ℓ, and the work done by the force is given by $W = \mathbf{F}\ell$. Thus

$$W = \mathbf{B}Q\mathbf{v}\ell \tag{3}$$

The induced potential difference (emf) is a measure of the work done per unit charge (W/Q), which gives us emf = $\mathbf{b}\ell\mathbf{v}$. This emf is sometimes called a "motional emf" since it is due to the motion of a wire through a field.

An interesting fact about electromagnetic induction is that the velocity in the above expression is a relative velocity, that is, the induced emf can be produced whether the wire moves through the field or the field changes over the wire! Additionally, an emf can be induced even if there is no relative velocity as long as the magnetic field is changing.

Figure 18.2 is a simple illustration of this effect.

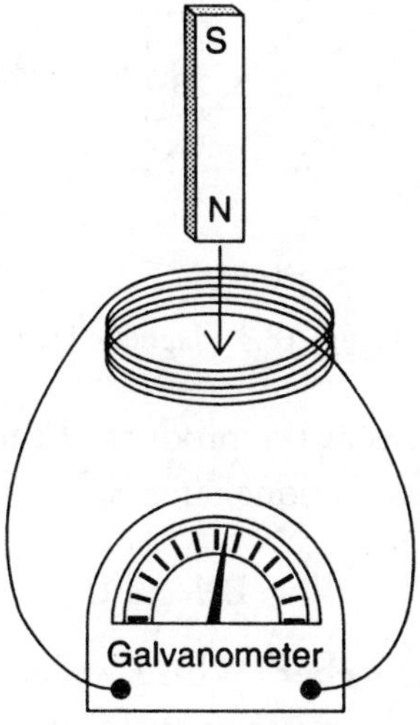

Figure 18.2

In Figure 18.2, a bar magnet is thrust into and out of a coil in which there are N turns of wire. The number of coils is related to the overall length ℓ, in equation 1. The galvanometer registers the alternating current in the coil as the magnet is inserted and withdrawn. Varying the velocity affects the amount of current as predicted. Moving the magnet around the outside of the coils generates only a weak current. The concept of relative velocity is illustrated if the magnet is held constant and the coil is moved up and down over the magnet.

SAMPLE PROBLEM

What is the induced emf in a wire 0.5 m long, moving at right angles to a 0.04-T magnetic field with a velocity of 5 m/s?

Solution

We use

$$\text{emf} = \mathbf{B}\ell\mathbf{v}$$

$$\text{emf} = (0.04\text{ T})(0.5\text{ m})(5\text{ m/s}) = 0.1\text{ V}$$

18.2 MAGNETIC FLUX AND FARADAY'S LAW OF INDUCTION

In Figure 18.3, there is a circular region of cross-sectional area A. An external magnetic field **B** passes through the region at an angle θ to the region. The perpendicular component of the field is given by

$$\mathbf{B}_{\perp} = \mathbf{B} \cos \theta$$

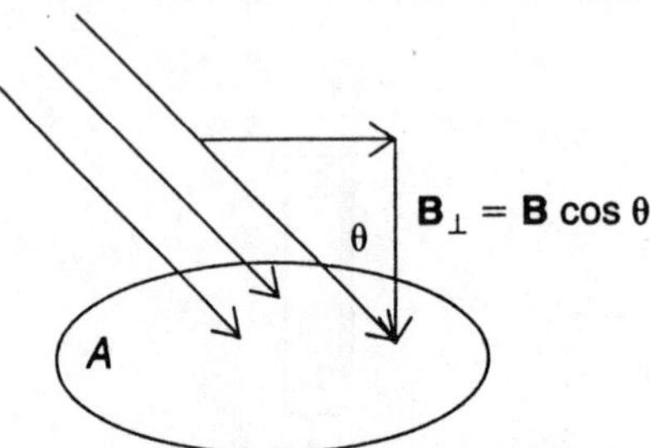

Figure 18.3 Magnetic Flux

The magnetic flux, Φ, is defined as the product of the perpendicular component of the magnetic field and the cross-sectional area A:

$$\Phi = \mathbf{B}A \cos \theta$$

The unit of magnetic flux is the **weber** (Wb); 1 weber equals 1 tesla per square meter.

On the basis of these ideas, the magnetic field strength is sometimes referred to as the **magnetic flux density** and can be expressed in units of webers per square meter (Wb/m^2).

English scientist Michael Faraday demonstrated that the induced motional emf was due to the rate of change of the magnetic flux. We now call this relationship **Faraday's law of electromagnetic induction** and express it in the following way:

$$\text{emf} = -\left(\frac{\Delta\Phi}{\Delta t}\right)$$

The negative sign is used because another relationship, known as **Lenz's law**, states that an induced current will always flow in a direction such that its magnetic field opposes the magnetic field that induced it. Lenz's law is just another way of expressing the law of conservation of energy. Consider the experiment in Figure 18.2. When the magnet is inserted through the coil, an induced current flows. This current, in turn, produces a magnetic field that is directed either into or out of the coil, along the same axis as the bar magnet.

If the current had a direction such that the "pole" of the coil attracted the pole of the bar magnet, more energy would be derived from the effect than is possible in nature. Thus, according to Lenz's law, the current in the coil will have a direction so that the new magnetic field will oppose the magnetic field of the bar magnet. The more one tries to overcome this effect, the greater it becomes. The opposition of fields in this way prevents violation of conservation of energy and can generate a large amount of heat in the process.

If there are N turns of wire in the coil, Faraday's law becomes:

$$\text{emf} = -N\left(\frac{\Delta\Phi}{\Delta t}\right)$$

SAMPLE PROBLEM

A coil is made of 10 turns of wire and has a diameter of 5 cm. The coil is passing through the field in such a way that the axis of the coil is parallel to the field. The strength of the field is 0.5 T. What is the change in the magnetic flux? Also, if an average emf of 2 V is observed, for how long was the flux changing?

Solution

The change in the magnetic flux is given by

$$\Delta\Phi = -\mathbf{B}A = -\mathbf{B}\pi r^2 = -(0.5)(3.14)(0.025)^2 = -9.8 \times 10^{-4} \text{ Wb}$$

The time is therefore given by

$$\Delta t = N\left(\frac{\Delta\Phi}{\text{emf}}\right) = 10\left(\frac{9.8 \times 10^{-4}}{2}\right) = 4.9 \times 10^{-3} \text{ s}$$

Chapter Summary

- A wire moving across an external magnetic field will have an emf induced in it.
- The induced emf will be a maximum if the wire cuts across the magnetic field at right angles to it.
- A wire moving parallel through an external magnetic field will not have an emf induced in it.
- Faraday's law states that the induced emf is equal to the rate of change of magnetic flux.
- Lenz's law states that an induced current will always flow in a direction such that its magnetic field opposes the magnetic field that induced it.

Problem-Solving Strategies for Electromagnetic Induction

When solving electromagnetic problems, keep in mind that vector quantities are involved. The directions of these vectors are usually determined by the right-hand rules. Additionally, remember Lenz's law: **An induced current will always flow in a direction such that its magnetic field opposes the magnetic field that induced it**, which plays a role in many applications.

When solving problems in electromagnetic induction, it is important to remember the right-hand rules. Also, keep in mind that the induced emf is proportional to the change in the magnetic flux, not the magnetic field.

Practice Exercises

MULTIPLE-CHOICE

1. The back emf of a motor is at a maximum when

   (A) the speed of the motor is constant
   (B) the speed of the motor is at its maximum value
   (C) the motor is first turned on
   (D) the speed of the motor is increasing
   (E) the motor is turned off

2. A bar magnet is pushed through a flat coil of wire. The induced emf is greatest when

   (A) the north pole is pushed through first
   (B) the magnet is pushed through quickly
   (C) the magnet is pushed through slowly
   (D) the south pole is pushed through first
   (E) the magnet rests in the coil

3. The magnetic flux through a wire loop is independent of

   (A) the shape of the loop
   (B) the area of the loop
   (C) the strength of the magnetic flux
   (D) the orientation of the magnetic field and the loop
   (E) none of the preceding

4. A flat, 300-turn coil has a resistance of 3 Ω. The coil covers an area of 15 cm^2 in such a way that its axis is parallel to an external magnetic field. At what rate must the magnetic field change in order to induce a current of 0.75 A in the coil?

   (A) 0.0075 T/s
   (B) 2.5 T/s
   (C) 0.0005 T/s
   (D) 5 T/s
   (E) 15 T/s

5. When a loop of wire is turned in a magnetic field, the direction of the induced emf changes every

   (A) one-quarter revolution
   (B) two revolutions
   (C) one revolution
   (D) one-half revolution
   (E) The direction never changes.

6. A wire of length 0.15 m is passed through a magnetic field with a strength of 0.2 T. What must be the velocity of the wire if an emf of 0.25 V is to be induced?

   (A) 8.3 m/s
   (B) 6.7 m/s
   (C) 0.0075 m/s
   (D) 0.12 m/s
   (E) 7.6 m/s

FREE-RESPONSE

1. A horizontal conducting bar is free to slide along a pair of vertical bars, as shown below. The conductor has length ℓ and mass M. As it slides vertically downward under the influence of gravity, it passes through an outward-directed, uniform magnetic field. The resistance of the entire circuit is R. Find an expression for the terminal velocity of the conductor (neglect any frictional effects due to sliding). **CHALLENGE**

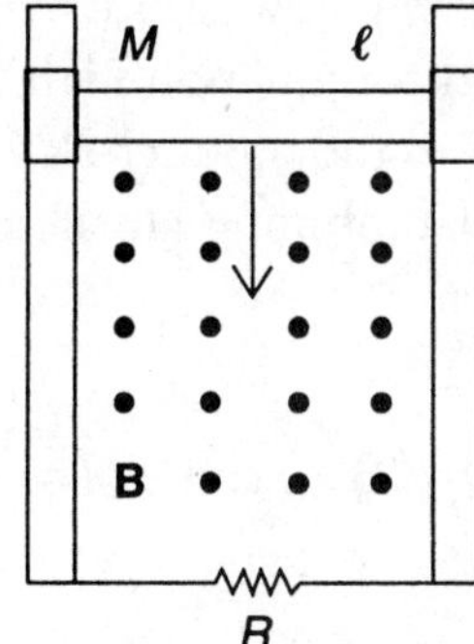

2. A small cylindrical magnet is dropped into a long copper tube. The magnet takes longer to emerge than the predicted free-fall time. Give an explanation for this effect.

ANSWERS EXPLAINED

MULTIPLE-CHOICE PROBLEMS

1. **(B)** The back emf is proportional to the rate of change of magnetic flux and opposes the ability of the motor to turn. It is at its greatest value when the speed of the motor is at maximum.

2. **(B)** The motional emf is equal to $\mathbf{B}\ell\mathbf{v}$, where $\mathbf{v}$ is the velocity of the bar magnet. Thus, the induced emf is greatest when the magnet is pushed through quickly.

3. **(A)** The magnetic flux is independent of the shape of the wire loop.

4. **(D)** The formula for the rate of change of magnetic flux is emf $= -N(\Delta\Phi/\Delta t)$. Using the given values and Ohm's law, we get for the magnitude of the flux change

$$\left(\frac{\Delta\Phi}{\Delta t}\right) = \frac{(0.75)(3)}{300} = 0.0075 \text{ Wb/s}$$

Our question concerns the rate of change of the magnetic field. In a situation in which the axis of the coil is parallel to the magnetic field, $\Phi = \mathbf{B}A$, where $A = 0.0015 \text{ m}^2$. Thus $\Delta\mathbf{B}/\Delta t = 5$ T/s.

5. **(D)** In an ac generator, the emf reverses direction every one-half revolution.

6. **(A)** The motional emf is given by emf $= \mathbf{B}\ell\mathbf{v}$. Thus, using the given values, we find that $\mathbf{v} = 8.3$ m/s.

FREE-RESPONSE PROBLEMS

1. Initially, the bar is accelerated downward by the force of gravity, given by $\mathbf{F} = M\mathbf{g}$. The resistance R, in conjunction with the induced current I, produces an emf equal to IR and also to the product $\boldsymbol{B}\ell\mathbf{v}$. The magnetic force due to the current I is given by $\mathbf{B}I\ell$. Combining all these ideas, we get that at terminal velocity

$$M\mathbf{g} = I\ell\mathbf{B} \quad \text{and} \quad I = \frac{\mathbf{B}\ell\mathbf{v}}{R}$$

Thus

$$\mathbf{v}_t = \frac{M\mathbf{g}R}{\ell^2\mathbf{B}^2}$$

2. Copper is not highly magnetic. However, because of Lenz's law, the currents set up in the tube produce an upward magnetic field opposing the original field. The result is to slow down the magnet's terminal velocity, as compared to the normal uniformly accelerated motion.

CHAPTER 19

Waves and Sound

KEY CONCEPTS

- Pulses
- Wave Motion
- Types of Waves
- Standing Waves and Resonance
- Sound
- The Doppler Effect

19.1 PULSES

A pulse is a single vibratory disturbance in a medium. An example of a pulse is seen in Figure 19.1. If a string fixed at both ends and made taut, under a tension T (in newtons), is given a quick up and down snap, an upwardly pointing pulse is directed from the left to the right. The pulse appears to travel with a velocity **v** down the string. In actuality, the energy transferred to the string causes segments to vibrate up and down successively. This effect produces the illusion of a continuous pulse. Only energy is transferred by the pulse, and because the vibration is perpendicular to the apparent direction of motion, the pulse is said to be **transverse**. The **amplitude** of the pulse is the displacement of the disturbance above the level of the string.

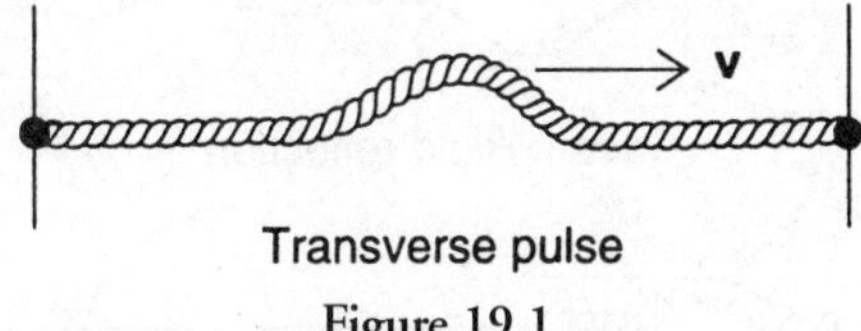

Transverse pulse

Figure 19.1

If the tension in the string is increased, the velocity appears to increase. If a heavier string is used (greater mass per unit length), the pulse appears to move slower. Careful measurements of these observations leads to the following equation for the velocity of the pulse:

$$\mathbf{v} = \sqrt{\frac{T}{M/\ell}}$$

where T is the tension in newtons, M is the mass in kilograms, and ℓ is the length of the string in meters.

When the pulse reaches one boundary, the energy transferred is directed upward against the wall, creating an upward force. Since the wall is rigid, the reaction to this action is a downward force. The string responds by being displaced downward and is reflected back. The result of this boundary interaction, illustrated in Figure 19.2, is observed as an inversion of the pulse upon reflection. Some energy is lost in the interaction, but most is reflected back with the pulse.

TIP

Waves transfer only energy.

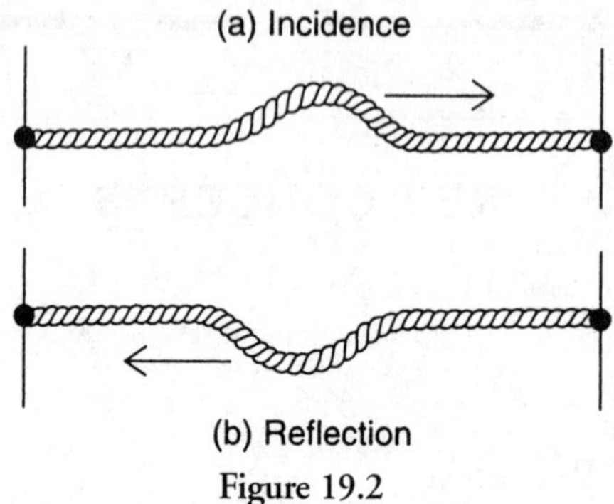

Figure 19.2

At a nonrigid boundary, say between two strings of different masses (but equal tensions), a different effect occurs. When the pulse reaches the nonrigid boundary, the upward displacement causes the second string to also be displaced upward. The magnitude of the second displacement depends on the mass density and tension in the second string. Thus, the transmitted pulse has an upward orientation and may travel with a larger or smaller velocity, depending, in this case, on the mass density.

The reflected pulse will be upward in orientation since no inversion takes place with a nonrigid boundary unless the difference between the two media (as measured by the second string's inertia) is large enough to behave as a semirigid boundary. As an example, consider a light string attached to a heavy rope, with the pulse traveling from the string to the rope. Figure 19.3 illustrates both cases.

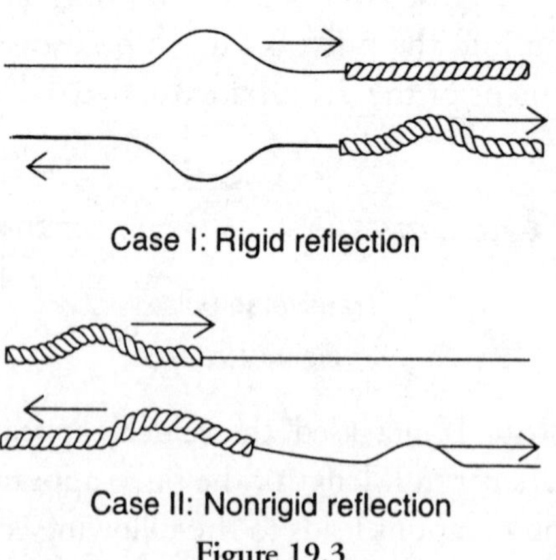

Figure 19.3

Since pulses transmit only energy, when two pulses interact, they obey a different set of physical laws than when two pieces of matter interact. Pulses are governed by the **principle of superposition**, which states that, when two pulses interact at the same point and at the same time, the interaction produces a single pulse whose amplitude displacement is equal to the sum of the displacements of the original two pulses. After the interaction, however, both pulses continue in their original directions of motion, unaffected by the interaction. Sometimes, this interaction is called **interference.**

When a pulse with an upward displacement interacts with another pulse with an upward displacement, the resulting pulse has a displacement larger than that of either original pulse (equal to the numerical sum of these pulses). This interaction is called **constructive interference.**

When a pulse with an upward displacement interacts with a pulse that has a downward displacement, the interaction is called **destructive interference**, and the resulting pulse has a smaller displacement than either original pulse (equal to the numerical difference of these pulses). Figure 19.4 illustrates these two types of interferences; *a* and *b* designate the two original amplitudes and *p* is a point on the string.

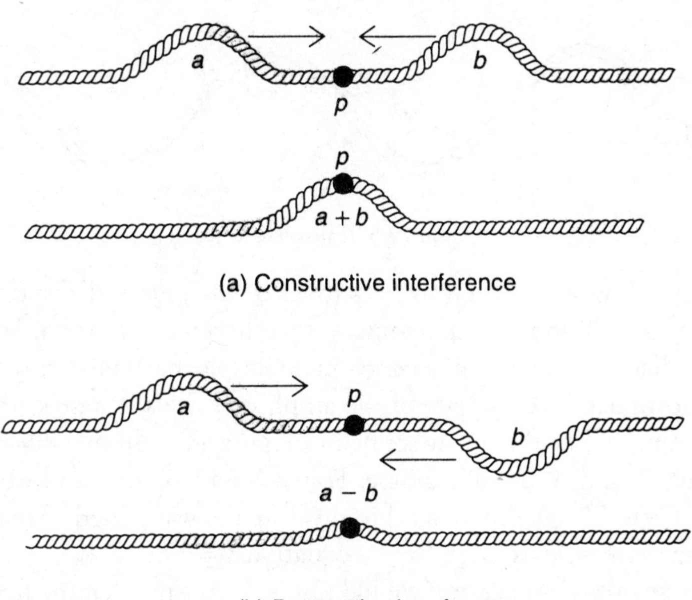

(a) Constructive interference

(b) Destructive interference

Figure 19.4

In the destructive interference case, it is possible, if the amplitudes are equal, that the interaction will momentarily cancel out both pulses. In any case, the orientation of the resulting pulse depends on the magnitude of each amplitude.

19.2 WAVE MOTION

If a continuous up and down vibration is given to the string in Figure 19.1, the resulting set of transverse pulses is called a **wave train** or just a **series of waves.** Whereas the pulse was just an upward- or downward-oriented, transverse disturbance, a wave consists of a complete up and down segment. Waves are thus periodic disturbances in a medium. The number of waves per second is called the **frequency**, designated by the letter f and expressed in units of reciprocal seconds (s^{-1}) or **Hertz** (Hz). The time required to complete one wave cycle is called the **period**, designated by the capital letter T and expressed in seconds. The frequency and period of a wave are reciprocals of each other, so that $f = 1/T$.

Like pulses, waves transmit only energy. The particles in the medium vibrate only up and down as the transverse waves approach and pass a given point. Since there is a contiuous series of waves, many points in the medium will be moving up or down in unison. The **phase** of a wave is defined to be the relative position of a point on a wave with respect to another point on the same wave.

The distance between any two successive points in phase is a measure of the **wavelength**, designated by the Greek letter λ (lambda) and measured in meters. The peaks of the wave are called **crests**; the valleys, **troughs**. Thus, one wavelength can also be measured as the distance between any two successive crests or troughs. The **amplitude**, *A*, of the wave is the distance, measured in meters, of the maximum displacement, up or down, above the normal rest position. Figure 19.5 illustrates a typical transverse wave.

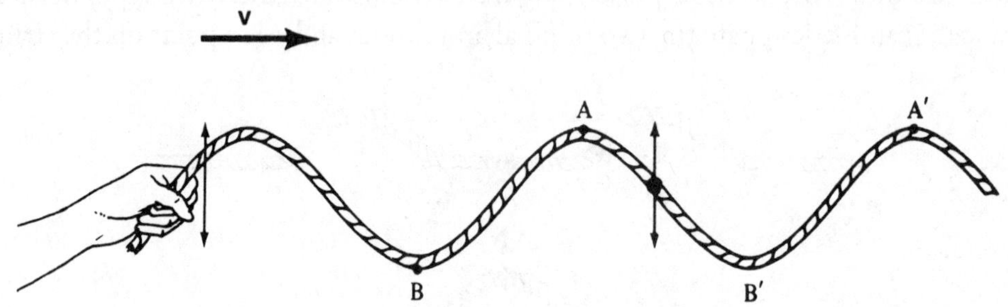

Figure 19.5 Transverse Wave

The sinusoidal nature of the wave is similar to the graph of displacement versus time for a mass on a spring undergoing simple harmonic motion. We recall from Chapter 11 that the maximum energy in a simple harmonic motion system is directly proportional to the square of the amplitude. This is indeed the case with a simple wave motion like the one described in Figure 19.5. In this illustration we can identify points A and A′ as all in phase. Points B and B′ are also in phase, and the distance from B to B′ can also be used to measure the wavelength. Finally, the velocity of the transverse wave is given by the equation $\mathbf{v} = f\lambda$.

Since each wave carries energy, it should not be surprising that the frequency is also proportional to the energy of a wave. The amplitude is related to what we might consider the wave's **intensity**. In sound, this property might be called **volume**. The frequency, however, measures the **pitch** of the wave. Waves of higher frequency transmit, at the same amplitude, more energy per second than waves of lower frequency.

There are many different kinds of frequencies that interact with human senses. It is worthwhile, therefore, to recall that the prefix *kilo-* represents 10^3, the prefix *mega-* represents 10^6, and the prefix *giga-* represents 10^9.

19.3 TYPES OF WAVES

In physics we usually deal with two types of waves, mechanical and electromagnetic. Waves that result from the vibration of a physical medium (a drum, a string, water, etc.) are called **mechanical waves**. Waves that result from electromagnetic interactions (light, X rays, radio waves, etc.) are called **electromagnetic waves**. These electromagnetic waves are special since they do not require a physical medium to carry them.

Also, all electromagnetic waves are transverse in nature. Physicists know this because of an analogous behavior with mechanical waves. Imagine a long string as shown in Figure 19.5. We can set up transverse vibrations in the spring by shaking it up and down. The resulting transverse waves can be analyzed as before.

Another way to set up periodic disturbances in the spring is to pull the spring back and forth along its longitudinal axis. The resulting disturbances consist of regions of compressions and expansions that appear to travel parallel to the disturbances themselves. These waves are called **longitudinal** (Figure 19.6) or **compressional** waves. Sound is an example of a longitudinal wave. The vibrations of an object in the air create pressure differences that alternately expand and contract the air and that, when impacted on our ears, create that phenomenon known as **sound**.

Reminder

Make sure you know the differences between **transverse** and **longitudinal waves.** You should also know the differences between **mechanical** and **electromagnetic waves.**

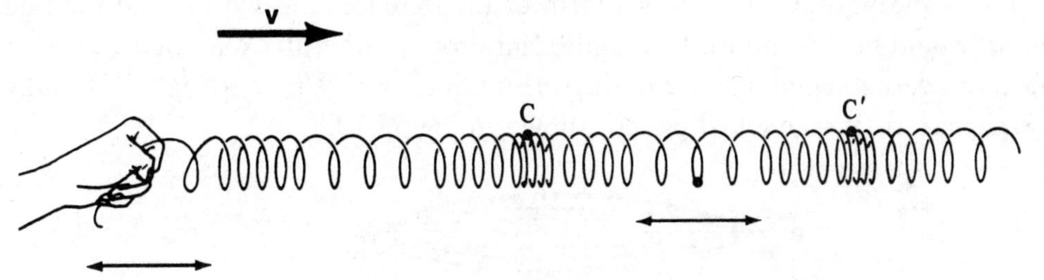

Figure 19.6 Longitudinal Wave

In a transverse wave, the vibrations must be perpendicular to the direction of apparent motion. However, if we cut a plane containing the directions of vibration, we find that there are many possible three-dimensional orientations in which the vibrations can still be considered perpendicular to the direction of travel. Thus, we can select one or more planes containing our chosen vibrational orientation. For example, a sideways vibration is just as "transverse" as an up and down or diagonal vibration. This selection process, known as **polarization** (see Figure 19.7), applies only to transverse waves.

In a longitudinal wave, there is only one way to make the vibrations parallel to the direction of travel. Special devices have been developed to test for polarization in mechanical and electromagnetic waves. Since all electromagnetic waves can be polarized, physicists conclude that they must be transverse waves. Since sound cannot be polarized, physicists conclude that sound waves are longitudinal.

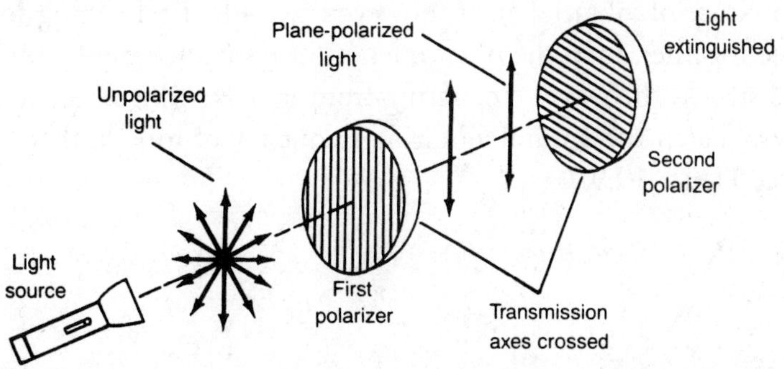

Figure 19.7 Used with permission of Prentice Hall, Needham, MA.

19.4 STANDING WAVES AND RESONANCE

Consider Figure 19.1 again. There are several ways in which we can make the taut string vibrate transversely. One way is make the entire string move up and down as a single unit. This is the easiest frequency mode of vibration possible and is called the **fundamental mode**. If we increase the frequency, something interesting happens. The waves reaching a boundary reflect off it inverted and match perfectly, in amplitude and frequency, the remaining incoming waves. In this situation, the apparent horizontal motion of the wave stops and we have a **standing wave**, which appears to be segmented by **nodal points**. These nodal points correspond to points where no appreciable displacement takes place.

Figure 19.8 shows a standing wave in various vibrational modes. Figure 19.8a illustrates the fundamental mode. In Figure 19.8b we see the second mode, in which one nodal point appears. This nodal point occurs at the one-half wavelength mark, so that, for a string of length ℓ, $\lambda = \ell$ (at this frequency). In the fundamental mode of Figure 19.8a, $\lambda = 2\ell$. In Figure 19.8c, there are two nodal points and $\lambda = 2\ell/3$.

This standing-wave pattern is a form of interference, and with sound the nodal points would be determined by a significant drop in intensity. With light, we would observe a darker region relative to the surrounding area. With a string, we would see the characteristic segmented pattern shown in Figure 19.8.

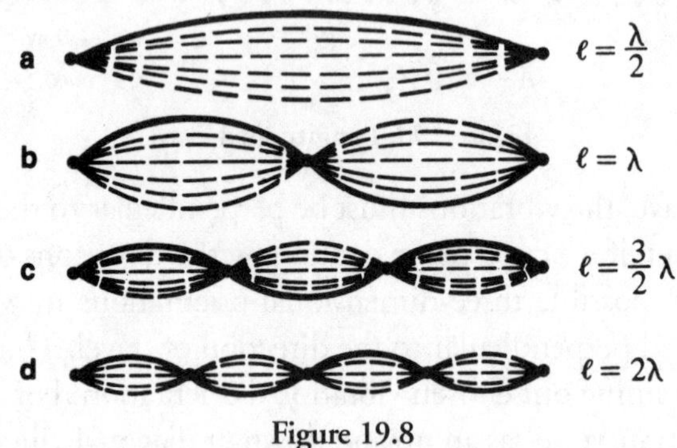

Figure 19.8

Standing waves can be manipulated so that the maximum points continuously reinforce themselves. This buildup of wave energy due to the constructive interference of standing waves is called **resonance**. Resonance can also be induced by an external agent.

All objects have a natural vibrating frequency. When a glass or bell is struck, only one fundamental characteristic frequency of vibration is heard (actually, the sound is a complex mixture of vibrations). If, however, two identical tuning forks, *A* and *B*, are held close together, then the phenomenon of resonance can be observed. If tuning fork *A* is struck, the waves emanating from it strike fork *B*. Since the frequency of these waves match the natural vibrating frequency of fork *B*, that fork will begin to vibrate (see Figure 19.9).

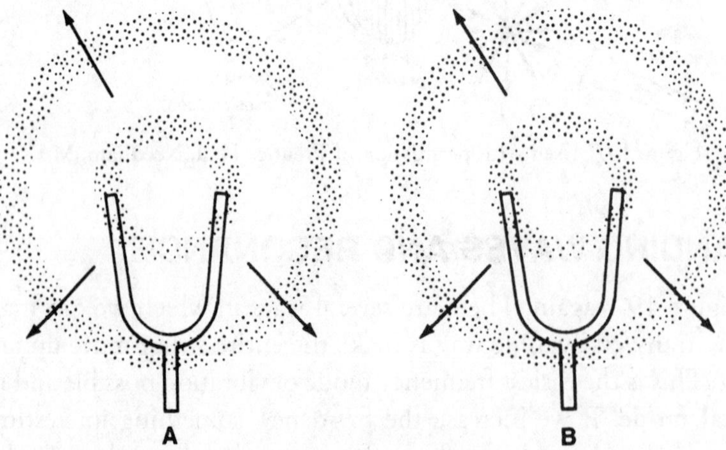

Figure 19.9 Resonance in Two Identical Tuning Forks

Resonance also explains how an opera singer can shatter a crystal wine glass. More important, it explains why soldiers are ordered to "break march" when crossing a bridge. The rhythmic marching can set up a resonance vibration and perhaps collapse the bridge. In November 1940, a resonance vibration caused by a light gale wind generated catastrophic torsional vibrations in the Tacoma Narrows Bridge in Washington State. The violent convulsions were photographed and are shown as a classic illustration of the principle of resonance. Nicknamed "Galloping Gertie," the bridge has become a testament to the need for engineers to be very careful when considering the possible effects of resonance.

19.5 SOUND

Sound is a longitudinal mechanical wave. Air molecules are alternately expanded and contracted as pressure differences move through the air. When lightning occurs, the expansion of air due to the very high temperature of the lightning bolt creates the violent sound known as thunder.

Since sound is "carried" by air molecules, it is temperature dependent. At 0 degree Celsius, the velocity of sound in air is 331 meters per second. For each 1-degree rise in temperature, the velocity of sound increases by 0.6 meter per second. The study of sound is called **acoustics**, and longitudinal waves in matter are sometimes termed **acoustical waves**.

The ability of sound waves to pass through matter depends on the structure of the molecules involved. Thus, sound travels slower in a gas, in which the molecules have more random motion, than in a liquid or a solid, which has a more "rigid" structure. In Table 19.1 the velocities of sound in selected substances are listed.

Think About It

Since sound is a longitudinal wave, it cannot be polarized. Polarization can be used to test whether a wave has a transverse nature. For example, since light exhibits the properties of interference and polarization, it contains transverse electromagnetic waves.

Table 19.1

Velocities of Sound in Selected Substances

Substance	Velocity (m/s)
Gases (0°C)	
Carbon dioxide	259
Air	331
Helium	965
Liquids (25°C)	
Ethyl alcohol	1207
Water, pure	1498
Water, sea	1531
Solids	
Lead	1200
Wood	~4300
Iron and steel	~5000
Aluminum	5100
Glass (Pyrex)	5170

Human hearing can detect sound in a range from 20 to 20,000 hertz. Sound waves exceeding a frequency of 20,000 hertz are called **ultrasonic** waves. The amplitude of the sound wave is the intensity of **loudness** of the wave, while the frequency relates to the **pitch**. When two sound waves interfere, the regions of constructive and destructive interference produce the phenomena known as **beats**. The number of beats per second is equal to the frequency difference.

Interference, for any kind of wave, is dependent on what is called the **path-length difference**. Consider two point sources of sound, *A* and *B*, and a receiver some distance away. If the distance from source *A* to the receiver is ℓ_A and the distance from source *B* to the receiver is ℓ_B, the receiver will be at a point of constructive interference if the path-length difference, $\ell_A - \ell_B$, is equal to a whole multiple of the wavelength, $n\lambda$. Destructive interference will occur if the path-length difference is equal to an odd multiple of the half-wavelengths, $(n + 1/2)\lambda$.

Another phenomenon associated with interference is **diffraction**. When a wave encounters a boundary, the wave appears to bend around the corners of the boundary. This effect, known as **diffraction**, occurs because at the corners the wave behaves like a point source and creates circular waves. These waves, because of their shape, reach behind the corners, and this continuous effect gives the illusion of wave bending. In Figure 19.10a, we observe a series of straight waves, made in a water tank, and the resulting diffraction at a corner. Figure 19.10b shows the diffraction of waves as they pass through a narrow opening.

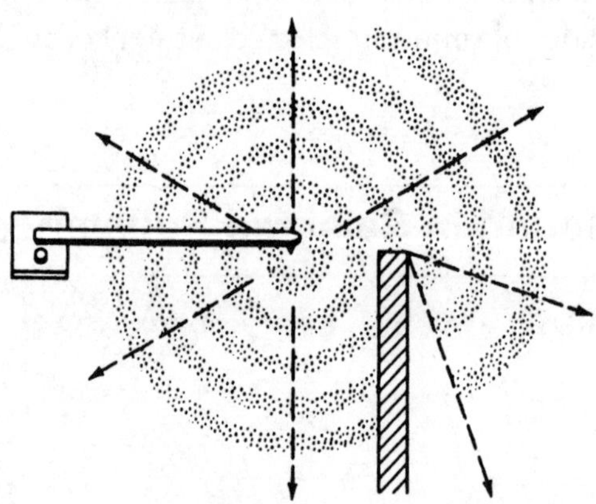

(a) Diffraction around a corner

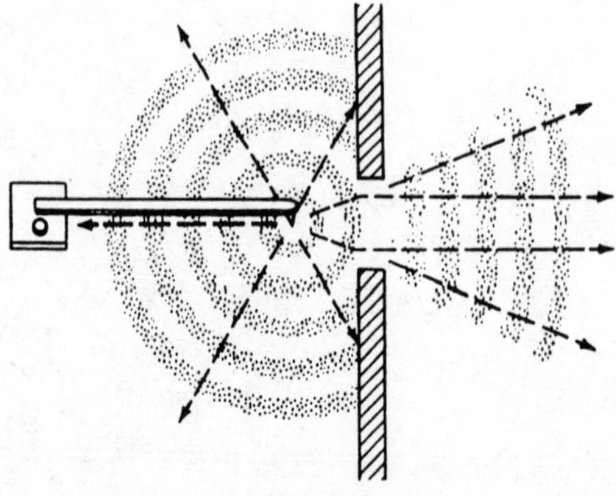

(b) Diffraction through a narrow opening

Figure 19.10

19.6 THE DOPPLER EFFECT

Almost everyone has had the experience of hearing a siren pass by. Even though the siren is emitting a sound at a single frequency, the changing position of the siren, relative to the hearer, produces an apparent change in the frequency. As the siren approaches, the pitch is increased; as the siren passes, the pitch is decreased. This phenomenon is called the **Doppler effect**.

In Figure 19.11 the source is moving; let's say toward point *A*. Then, for each period of time *T* between waves, the circular waves will not be concentric. In other words, the spacing between each two successive waves will be reduced by an amount equal to the distance traveled by the source in time *T*. Thus, an apparent increase in frequency is experienced at point *A*, while at point *B* there is an apparent decrease in frequency.

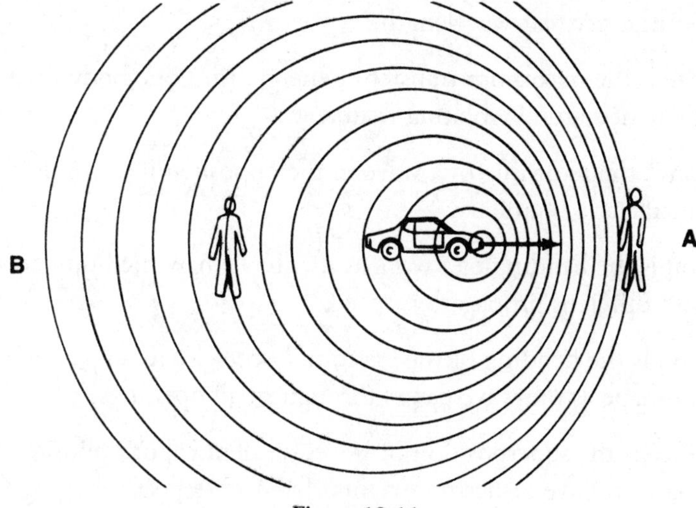

Figure 19.11

Chapter Summary

- Mechanical waves are periodic vibratory disturbances in a medium.
- A pulse is a single vibratory disturbance.
- A pulse or wave is transverse if its vibrations are perpendicular to the direction of propagation.
- A pulse or wave is longitudinal if its vibrations are parallel to the direction of motion.
- Transverse waves can be polarized. This means that we can select the preferred orientation of vibrations relative to the direction of propagation. Longitudinal waves cannot be polarized.
- Waves transmit only energy. The principle of superposition states that waves can be added together.
- The amplitude of a wave is its maximum displacement from its equilibrium position.
- The frequency of a wave is the number of cycles per second.

- The period of a wave is the time that it takes to complete one cycle and is equal to the reciprocal of the frequency.
- Particles in a medium that display the same state of motion, simultaneously, are said to be in phase.
- The "wavelength" of a wave is the distance between any two successive points in phase.
- The wave velocity is equal to the product of the frequency times the wavelength.
- A standing wave is a wave in which incident and reflected waves combine to produce a wave that appears to be "standing" in one place.
- Standing wave nodes are points where there is no appreciable displacement. Nodes occur every half-wavelength.
- Resonance is the maximum transfer of energy from one body to another that shares the same natural vibrating frequency.
- Reflection is the returning of a wave in the opposite direction due to a difference in media at a boundary.
- Refraction is the bending of a wave as it enters a new medium, at an oblique angle, with different propagation speed.
- Diffraction is the bending of waves around obstacles in a given medium or the change in shape as the wave passes through small openings.
- Interference is the superposition of waves of identical or opposite phase to produce constructive or destructive interference, respectively.
- Sound is a longitudinal mechanical wave.
- The Doppler effect states that when a source is moving relatively toward an observer, there is an apparent increase in frequency. If the relative motion is away from source or observer, there is an apparent decrease in frequency.

Problem-Solving Strategies for Waves and Sound

Solving wave motion problems involves remembering the basic physical concepts involved. For transverse waves, the only particle motion in the medium will be up and down. If a series of waves is presented, you must resist the temptation of viewing labeled points as beads on a wire. As a wave approaches a point, it will first go up and then go down. Points that are in phase are either going up at the same time or going down at the same time.

Transverse waves can be polarized, meaning that you can select a preferred axis to contain the vibrations and have them remain perpendicular to the direction of apparent wave motion. Longitudinal waves cannot be polarized, and the vibrations remain parallel to the apparent direction of wave motion.

Sound waves travel through a medium, and their speed is determined by the molecular structure of the matter involved. In air, the speed of sound increases with increasing temperature. Sound wave interference is manifested by the beat pulsations heard when two frequencies are experienced simultaneously.

Practice Exercises

MULTIPLE-CHOICE

1. A stretched string has a length of 1.5 m and a mass of 0.25 kg. What must be the tension in the string in order for pulses in the string to have a velocity of 5 m/s? CHALLENGE

   (A) 2.45 N
   (B) 12.5 N
   (C) 4.2 N
   (D) 150 N
   (E) 250 N

2. A stretched string is vibrated in such a way that a standing wave with two nodes appears. The distance between the nodes is 0.2 m. What is the wavelength of the standing wave?

   (A) 0.4 m
   (B) 0.3 m
   (C) 0.2 m
   (D) 0.6 m
   (E) 0.8 m

3. What is the wavelength of sound produced at a frequency of 300 Hz when the air temperature is 20°C? CHALLENGE

   (A) 1.10 m
   (B) 1.30 m
   (C) 0.80 m
   (D) 1.14 m
   (E) 1.2 m

4. Two tuning forks are vibrating simultaneously. One fork has a frequency of 256 Hz; the other, a frequency of 280 Hz. The number of pulsational beats per second heard will be

   (A) 256
   (B) 536
   (C) 1.1
   (D) 0.91
   (E) 24

5. In a stretched string with a constant tension T, as the frequency of the waves increases, the wavelength

   (A) increases
   (B) decreases
   (C) remains the same
   (D) increases, then decreases
   (E) decreases, then increases

6. What is the period of a wave that has a frequency of 12,000 Hz?
   (A) 36.25 s
   (B) 0.000083 s
   (C) 0.0275 s
   (D) 12,000 s
   (E) 0.012 s
7. Sound waves travel fastest in
   (A) a vacuum
   (B) air
   (C) water
   (D) wood
   (E) Not enough information is provided.
8. The amplitude of a sound wave is related to its
   (A) pitch
   (B) loudness
   (C) frequency
   (D) resonance
   (E) wavelength

CHALLENGE

9. At 25°C, a sound wave takes 3 s to reach a receiver. How far away is the receiver from the source?
   (A) 1038 m
   (B) 993 m
   (C) 1215 m
   (D) 1068 m
   (E) 887 m

FREE-RESPONSE

1. What must be the tension in a 0.25-m string with a mass of 0.30 kg so that its fundamental frequency mode is 400 Hz?
2. Why should people not march in unison when crossing over a bridge?
3. Apartment dwellers are used to hearing the "boom-boom" sound of bass tones when neighbors play their stereos too loudly. What can account for this phenomenon?
4. People who live near airports claim that they hear planes that are on the ground more during the summer. Explain why this might occur.

ANSWERS EXPLAINED

MULTIPLE-CHOICE PROBLEMS

1. **(C)** The formula for wave velocity in a string is

$$\mathbf{v} = \sqrt{\frac{\mathbf{T}}{M/\ell}}$$

Using the given information and solving for tension leads to $\mathbf{T} = 4.2$ N.

2. **(A)** Standing-wave nodes occur at half-wavelength intervals. Thus,

$$\lambda = 2(0.2) = 0.4 \text{ m}$$

3. **(D)** At 20°C, the speed of sound is $330 + (0.6)(20) = 343$ m/s. Since $\mathbf{v} = f\lambda$, we get a wavelength of $\lambda = 1.14$ m.

4. **(E)** The number of beats per second is equal to the frequency difference: $280 - 256 = 24$.

5. **(B)** Since $\mathbf{v} = f\lambda$ and since, for constant tension, the velocity is constant, as the frequency increases, the wavelength decreases.

6. **(B)** Frequency and period are reciprocals of each other. Thus

$$T = \frac{1}{12{,}000} = 0.000083 \text{ s}$$

7. **(D)** Because of molecular arrangement, sound travels faster in solids than in liquids or gases.

8. **(B)** Amplitude is related to the intensity of a wave. In sound, wave intensity is perceived as loudness.

9. **(A)** At 25°C, the speed of sound is $331 + (0.6)(25) = 346$ m/s. Since $\mathbf{x} = \mathbf{v}t$,

$$\mathbf{x} = (346)(3) = 1038 \text{ m}$$

FREE-RESPONSE PROBLEMS

1. At fundamental frequency, we have

$$\lambda = 2\ell = (2)(0.25) = 0.5 \text{ m}$$

Now, $\mathbf{v} = f\lambda$, which means that

$$\mathbf{v} = (400)(0.5) = 200 \text{ m/s}$$

We can now use the formula for the velocity in a fixed string:

$$\mathbf{v} = \sqrt{\frac{\mathbf{T}}{M/\ell}}$$

Solving for the tension gives **T** = 48,000 N.

2. People should not march in unison when crossing over a bridge because the uniform vibrations may induce resonance, which can be "destructive."

3. The resonant frequency of apartment building walls is quite low, and so low-frequency bass tones resonate through the walls as the "boom-boom" sound so familiar to apartment building dwellers.

4. Since the velocity of sound increases with temperature, the hot summer air carries the sound waves farther.

CHAPTER 20

Light

KEY CONCEPTS

- Electromagnetic Waves
- Reflection
- Refraction
- Applications of Light Refraction
- Interference and Diffraction of Light

20.1 ELECTROMAGNETIC WAVES

In Chapter 18, we reviewed aspects of electromagnetic induction. In one example, we observed how the changing magnetic flux in a solenoid can influence a second solenoid not physically attached to the primary. This **mutual induction** involves the transfer of energy through space by means of an oscillating magnetic field. Experiments in the late nineteenth century by German physicist Heinrich Hertz confirmed what Scottish physicist James Clerk Maxwell had asserted theoretically in 1864: that oscillating electromagnetic fields travel through space as transverse waves and that they travel with the same speed as does light. See Figure 20.1.

TIP 

Electromagnetic waves do not need a medium through which to propagate. All electromagnetic waves travel with the velocity of light in a vacuum.

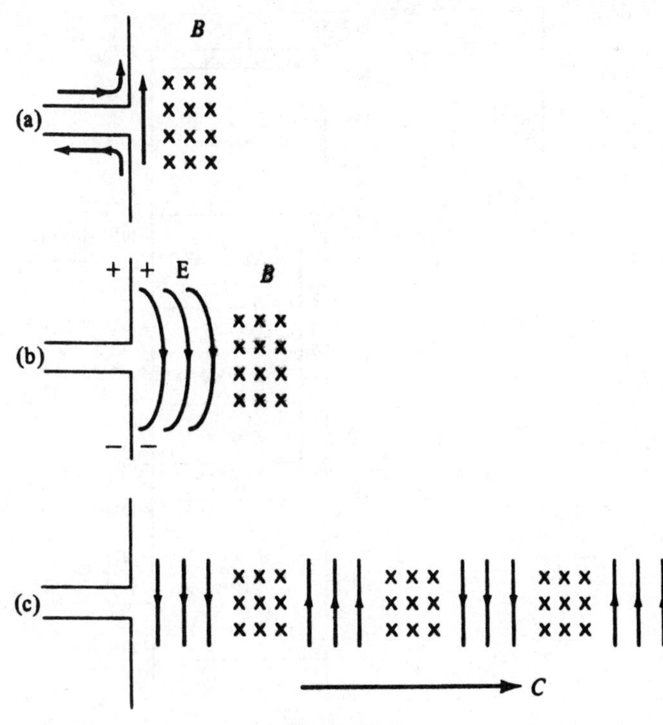

Figure 20.1

Light is just one form of electromagnetic radiation that travels in the form of transverse waves. We know that these waves are transverse because they can be polarized, as discussed in Chapter 19. The speed of light is approximately equal to 3×10^8 meters per second and is designated by the letter *c*.

Light waves can be coherent if they are produced in a way that maintains a constant phase-amplitude relationship. **Lasers** produce coherent light of one wavelength. (The word *laser* is an acronym for *l*ight *a*mplification by the *s*timulated *e*mission of *r*adiation.)

Experiments by Sir Isaac Newton in the seventeenth century showed that "white light," when passed through a prism, contains the colors red, orange, yellow, green, blue, and violet (abbreviated as ROYGBV). Each color of light is characterized by a different wavelength and frequency. The wavelengths range from about 3.5×10^{-7} meter for violet light to about 7.0×10^{-7} meter for red. Small wavelengths are sometimes measured in **angstroms**, (Å); 1 angstrom is equal to 1×10^{-10} meter. In more modern textbooks, the units **nanometers** (nm) are used; 1 nanometer is equal to 1×10^{-9} meter.

TIP 

Make sure you know the correct order of the **electromagnetic spectrum**.

Together with other electromagnetic waves, such as radio waves, X rays, and infrared waves, light occupies a special place in the so-called **electromagnetic spectrum** because we can "see" it. A sample electromagnetic spectrum is presented in Figure 20.2.

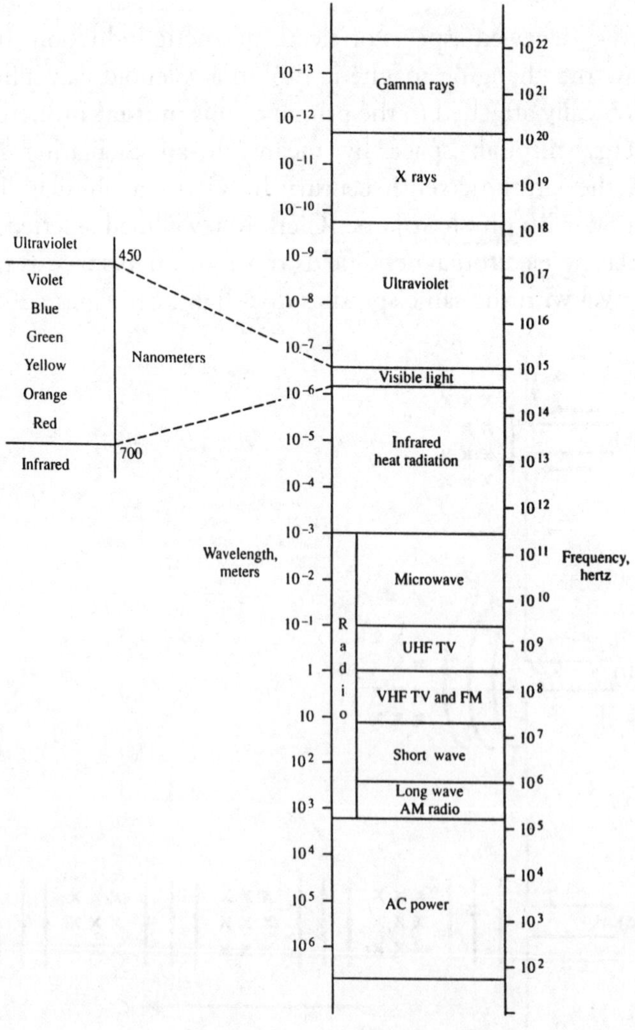

Figure 20.2 Electromagnetic Spectrum

Since electromagnetic waves are "waves," they all obey the relationship $c = f\lambda$, where c is the speed of light, discussed above. With this in mind, gamma rays have frequencies in the range of 10^{25} hertz and wavelengths in the range of 10^{-13} meter. These properties make gamma rays very small but very energetic. They are sometimes referred to as "cosmic rays" since they often originate in the cores of exploding stars and galaxies.

20.2 REFLECTION

Reflection is the ability of light to seemingly bounce off a surface. This phenomenon does not reveal the wave nature of light, and the notion of wavelength or frequency rarely enters into a discussion of reflection. If light is incident on a flat mirror, the angle of incidence is measured with respect to a line perpendicular to the surface of the mirror and called the **normal**.

In Figure 20.3 the **law of reflection**, which states that the angle of incidence equals the angle of reflection, is illustrated. Note that the angles of incidence and reflection and the normal are all coplanar.

Remember

The **law of reflection** states that the angle of incidence is equal to the angle of reflection. Remember that the angles are measured relative to the normal line.

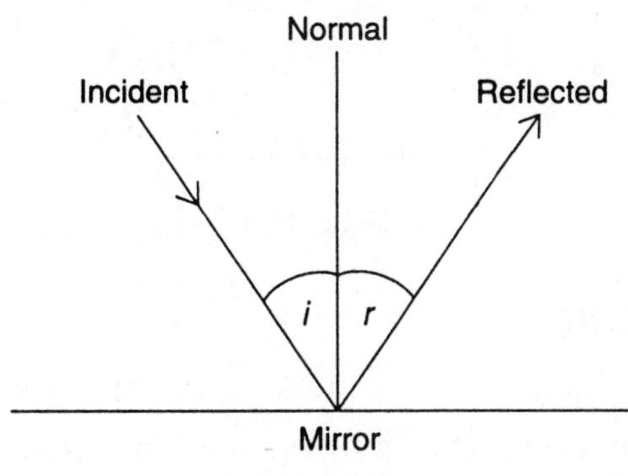

Figure 20.3

Reflection helps explain the colors of opaque objects. Ordinary light contains many different colors all blended together (so-called white light). A "blue" object looks blue because of the selective reflection of blue light due to the chemical dyes in the painted material. White paper assumes the color of the light incident on it because "white" reflects all colors. Black paper absorbs all colors (and of course nothing can be painted a "pure" color). If light of a single wavelength can be isolated (using a laser, for example), the light is said to be **monochromatic**. If all of the waves of light are moving in phase, the light is said to be **coherent**. The fact that laser light is both monochromatic and coherent contributes to its strength and energy.

If a concave or convex mirror is used, the law of reflection still holds, but the curved shapes affect the direction of the reflected rays. Figure 20.4a shows that a concave mirror converges parallel rays of light to a **focal point** that is described as **real** since the light rays really cross. In Figure 20.4b, we see that a convex mirror causes the parallel rays to diverge away from the mirror. If we extend the rays backward in imagination, they appear to originate from a point on the other side of the mirror. This point is called the **virtual focal point** since it is not real. The human eye will always trace a ray of light back to its source in a line. This deception of the eye is responsible for images in some mirrors appearing to be on the "other side" of the mirror (**virtual images**).

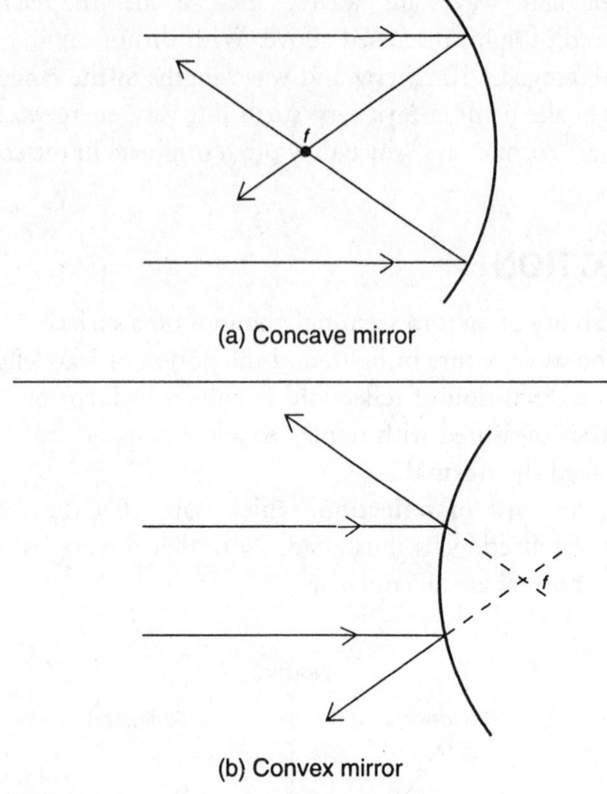

(a) Concave mirror

(b) Convex mirror

Figure 20.4

20.3 REFRACTION

Place a pencil in a glass of water and look at the pencil from the side. The apparent bending of the pencil is due to refraction (Figure 20.5a). When light passes obliquely from one transparent medium to another, there is a uniform change in the speed of light. When this occurs, the light appears to bend toward or away from the normal. If there is no change in speed, as when light passes from benzene to Lucite, for example, there will be no refraction at any angle (Figure 20.5b). If the light is incident at an angle of zero degrees to the normal, there will again be no refraction, whether or not there is a change in speed (Figure 20.5c).

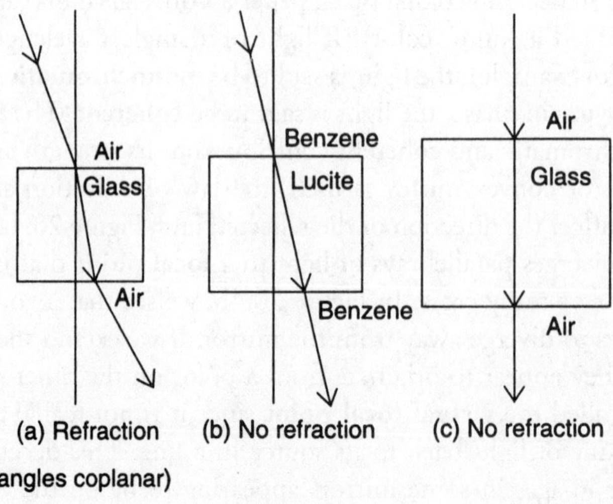

(a) Refraction (all angles coplanar) (b) No refraction (c) No refraction

Figure 20.5

When light goes from one medium to another and slows down, at an oblique angle, the angle of refraction will be less than the angle of incidence, and we say that the light has been refracted **toward the normal**. Notice, in Figure 20.5a, that, when the light reemerges into the air, it will be parallel to its original direction, but slightly offset. This is due to the fact that the light is speeding up when it reenters the air. In that case, the angle of refraction will be larger than the angle of incidence, and we say that the light has been refracted **away from the normal**. Remember that, if the optical properties of the two media through which light passes are the same, no refraction occurs since there is no change in the speed of the light. In that case, the angle of incidence will be equal to the angle of refraction (no deviation).

TIP

When light refracts, its frequency does not change.

The extent to which a medium is a good refracting medium is measured by how much change there is in the speed of light passing through it. This "physical" characteristic is manifested by a "geometric" characteristic, namely, the angle of refraction. The relationship between these quantities is expressed by **Snell's law**. With the angle of incidence designated as *i* and the angle of refraction as *r*, and with $\mathbf{v}_1$ the velocity of light in medium 1 (equal to **c** if medium 1 or 2 is air) and $\mathbf{v}_2$ the velocity of light in medium 2, Snell's law states that

$$\frac{\sin \theta\, i}{\sin \theta\, r} = \frac{\mathbf{v}_1}{\mathbf{v}_2} = \frac{n_2}{n_1}$$

The ratio n_2/n_1 is called the **relative index of refraction**. If medium 1 is air, then, by definition, the **absolute index of refraction**, *n*, of air (and a vacuum) is taken to be equal to 1.00, although it is actually slightly larger (1.00029; see Table 20.1). Snell's law can now be written in the form

$$\frac{\sin \theta\, i}{\sin \theta\, r} = \frac{\mathbf{c}}{\mathbf{v}_2} = n_2$$

Note that, when light is refracted, there is no change in the frequency of the light, only a change in wavelength.

Table 20.1

Absolute Indices of Refraction of Selected Media

Substance	Index of Refraction
Air (vacuum)	1.00
Water	1.33
Alcohol	1.36
Quartz	1.46
Lucite	1.50
Benzene	1.50
Glass	
Crown	1.52
Flint	1.61
Diamond	2.42

The particular colors of visible light all have specific frequencies. When these colors are used in a refraction experiment, they produce angles of refraction since they travel at different speeds in media other than air (or a vacuum). Substances that allow the frequencies of light to travel at different speeds are called **dispersive media**. This aspect of refraction explains why a prism allows one to see a colored, "continuous" spectrum and, additionally, why red light (at the low-frequency end) emerges on top (see Figure 20.6).

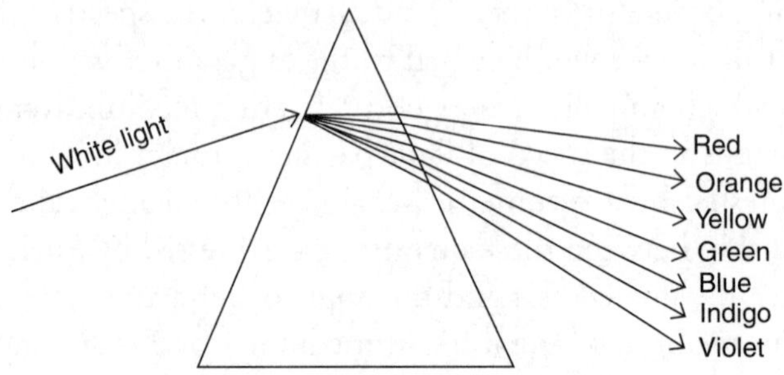

Figure 20.6 Prismatic Dispersion of Light

SAMPLE PROBLEM

A ray of light is incident from the air onto the surface of a diamond ($n = 2.42$) at a 30° angle to the normal.

(a) Calculate the angle of refraction in the diamond.
(b) Calculate the speed of light in the diamond.

Solution

(a) We use Snell's law:

$$n_1 \sin \theta_1 = n_2 \sin \theta_2$$
$$(1.00) \sin (30°) = (2.42) \sin \theta_2$$
$$\theta_2 = 12°$$

b) We use

$$\mathbf{v} = \frac{\mathbf{c}}{n}$$
$$\mathbf{v} = \frac{3 \times 10^8 \text{ m/s}}{2.42} = 1.24 \times 10^8 \text{ m/s}$$

20.4 APPLICATIONS OF LIGHT REFRACTION

When light is refracted from a medium with a relatively large index of refraction to one with a low index of refraction, the angle of refraction can be quite large. Note, however, that, although the relative index of refraction can be less than 1.00, the absolute index of refraction cannot.

Figure 20.7 illustrates a situation in which the angle of incidence is at some critical value θ_c such that the angle of refraction equals 90 degrees. This can occur only when the relative index of refraction is less than 1.00, as stated above.

If the angle of incidence exceeds this critical value, the angle of refraction will exceed 90 degrees and the light will appear to be internally reflected, a phenomenon aptly called **total internal reflection**. The ability of diamonds to sparkle in sunlight is due to total internal reflection and a relatively small critical angle of incidence (due to a diamond's high index of refraction). Fiber optics communication, in which information is processed along hair-thin glass fibers, works because of total internal reflection.

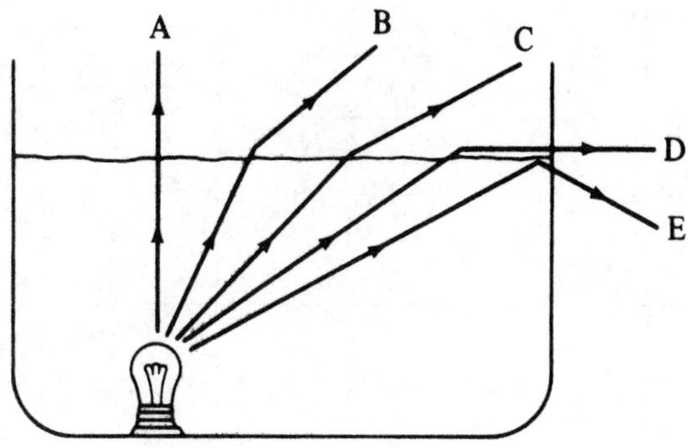

Figure 20.7

The critical angle can be determined from the relationship

$$\sin \theta_c = \frac{n_2}{n_1} \quad (n_2 < n_1)$$

SAMPLE PROBLEM

(a) Find the critical angle of incidence for a ray of light going from a diamond to air.
(b) Find the critical angle of incidence for a ray of light going from a diamond to water.

Solution

(a) We use

$$\sin \theta_c = \frac{n_2}{n_1}$$

$$\sin \theta_c = \frac{1.00}{2.42} = 0.4132$$

$$\theta_c = 24.4°$$

(b) We again use

$$\sin \theta_c = \frac{n_2}{n_1}$$

$$\sin \theta_c = \frac{1.33}{2.42} = 0.5496$$

$$\theta_c = 33.33°$$

As a further example of light refraction, consider the refraction due to a prism. If monochromatic light is used, Figure 20.8a demonstrates what happens when two prisms are arranged base to base and two parallel rays of light are incident on them. The ability of the prism to disperse white light does not apply in this example where monochromatic light is used. However, we can observe that the light rays converge to a real focal point some distance away. This situation simulates the effect of refraction by a double convex lens, as shown in Figure 20.8b.

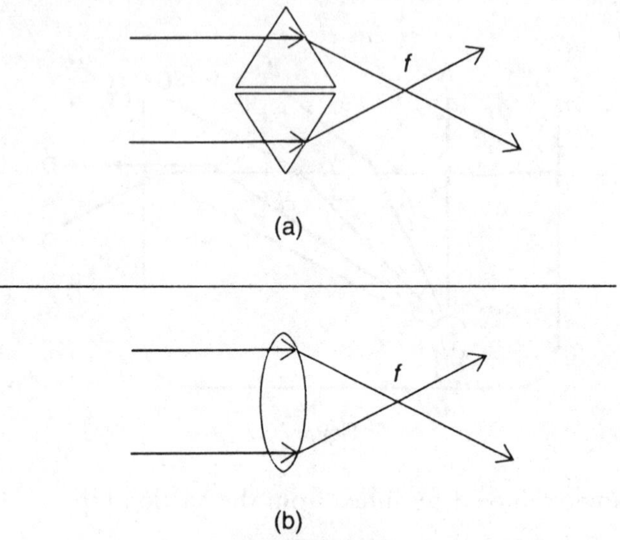

Figure 20.8

The focal length is dependent on the frequency of light used; red light will produce a greater focal length than violet light in a convex lens.

If the prisms are placed vertex to vertex, as in Figure 20.9a, the parallel rays of light will be diverged away from an apparent virtual focal point. This simulates, as shown in Figure 20.9b, the effect of a double concave lens.

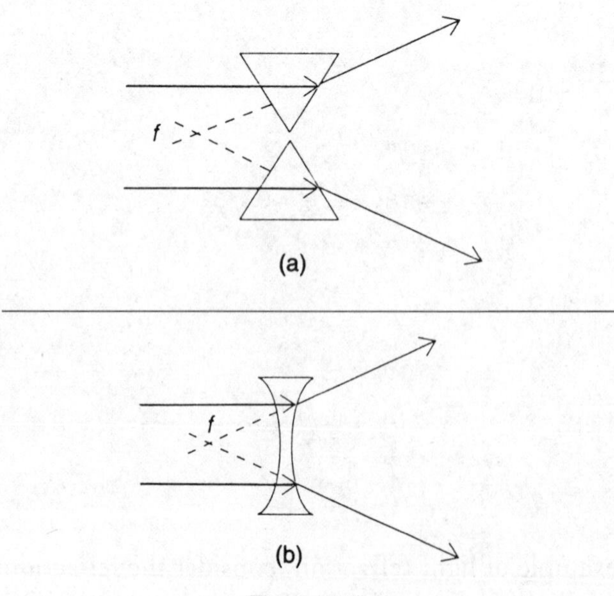

Figure 20.9

20.5 INTERFERENCE AND DIFFRACTION OF LIGHT

The ability of light to diffract and exhibit an interference pattern is evidence of the wave nature of light. Light interference can be observed by using two or more narrow slits. Multiple-slit diffraction is achieved with a plastic "grating" onto which over 5000 lines per centimeter are scratched.

Figure 20.10 shows that, if white light is passed through the grating, a series of continuous spectra appears. Interestingly, this continuous spectrum is reversed from the way it appears in dispersion. The reason for this difference is that diffraction is wavelength dependent. When a narrow opening is used, the short-wavelength violet rays are diffracted least and the longer red rays most.

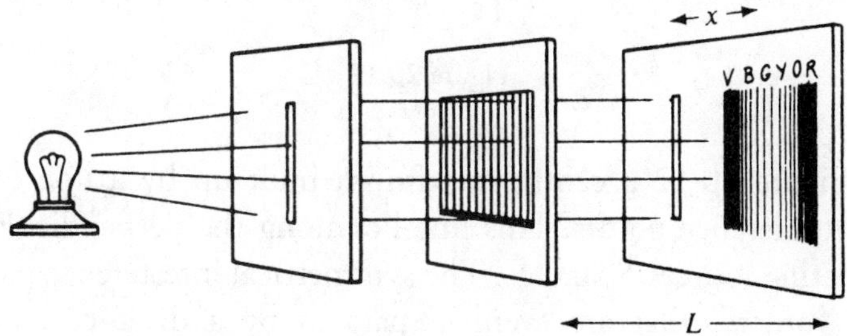

Figure 20.10 Multiple-Slit Diffraction

With monochromatic light, an alternating pattern of bright and dark regions appears. If a laser is used, the pattern appears as a series of dots representing regions of constructive and destructive interference (see Figure 20.11). The dots are evenly spaced throughout.

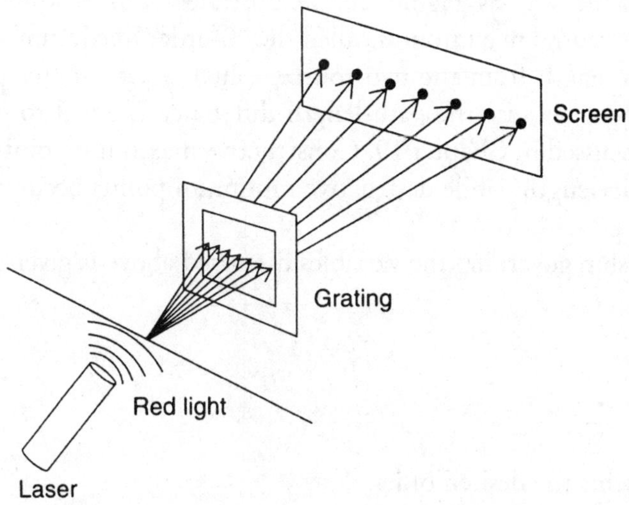

Figure 20.11 Monochromatic Diffraction Pattern

The origin of this pattern can be understood if we consider a water-tank analogy. Suppose two point sources are vibrating in phase in a water tank. Each source produces circular waves that overlap in the region in front of the sources. Where crests meet crests, there is constructive interference; where crests meet troughs, destructive interference. This situation is illustrated in Figure 20.12.

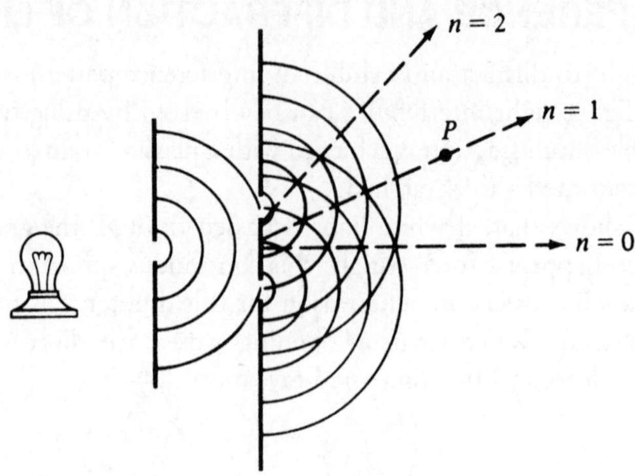

Figure 20.12

Notice in Figure 20.12 a central maximum built up by a line of intersecting constructive interference points. This line lies along the perpendicular bisector of the line connecting sources S_1 and S_2. The symmetrical interference pattern consists of numbered "orders" that are evenly separated by a distance *x*. The distance from the sources to the screen along the perpendicular bisector is labeled *L*. The wavelength of each wave, measured by the spacing between each two concentric circles, is of course represented by λ, and the separation between the sources is designated as *d*.

In Young's double-slit experiment, the two explicit sources in the water tank are replaced by two narrow slits in front of a monochromatic ray of light. Each slit acts as a new point source of light that interferes with the waves from the other slit in much the same way as Figure 20.12 illustrates (this is known as **Huygen's principle**). The central maximum is called the "0-order maximum" because the difference in path length from the two sources (slits) is zero at this point. The next order is first order because the path-length difference is equal to one wavelength, and so on as discussed in Chapter 19. Constructive maximum points occur at whole multiples of wavelength, while destructive minimum points occur at odd multiples of half-wavelengths.

The relationship governing the variables discussed above is given by the equation

$$m\frac{\lambda}{d} = \frac{x}{L}$$

where *m* represents the desired order.

If θ is an angle measured from the midpoint between the slits and a particular order, the ratio *x*/*L* is approximately equal to sin θ, and the diffraction formula becomes:

$$m\lambda = d \sin \theta$$

SAMPLE PROBLEM

A ray of monochromatic light is incident on a pair of double slits separated by 8×10^{-5} m. On a screen 1.2 m away, a set of dark and bright lines appear separated by 0.009 m. What is the wavelength of the light used?

Solution

We use

$$\lambda = \frac{dx}{L}$$

$$\lambda = \frac{(8 \times 10^{-5}\text{ m})(0.009\text{ m})}{1.2\text{ m}} = 6 \times 10^{-7}\text{ m}$$

Chapter Summary

- Electromagnetic waves are produced by oscillating electromagnetic fields.
- Electromagnetic waves can travel through a vacuum and do not need a material medium for propagation.
- In a vacuum, all electromagnetic waves travel with the speed of light; $c = 3 \times 10^8$ m/s.
- Electromagnetic waves may be represented on a chart called the electromagnetic spectrum.
- Light waves are transverse since they can be polarized.
- Light rays travel in straight lines and produce shadows when incident on opaque objects.
- Luminous objects emit their own light. Opaque objects are seen by illumination; that is, they reflect light.
- The color of an opaque object is due to the selective reflection of certain colors.
- Light rays exhibit all wave characteristics such as reflection, refraction, diffraction, and interference.
- The law of reflection states that the angle of incidence is equal to the angle of reflection (as measured relative to a line normal to the surface).
- In optics, all angles are measured relative to the normal to a surface.
- Snell's law governs the refraction of light through transparent media.
- The ratio of the speed of light in air to the speed of light in a transparent medium is called the absolute index of refraction.
- The speed of light in a transparent medium can be obtained using Snell's law and is inversely proportional to the absolute index of refraction for the medium.

Problem-Solving Strategies for Light

Remember that light is an electromagnetic wave. Therefore it can travel through a vacuum. Reflection and refraction do not, by themselves, verify the wave nature of light. Diffraction and interference are evidence of the wave nature of light, and the ability of light to be polarized, using special Polaroid filters, is evidence that light is a transverse wave.

Keep in mind that, when light refracts, the frequency of the light is not affected. Since the velocity changes, so does the wavelength in the new medium. If the velocity in the medium is frequency dependent, the medium is said to be "dispersive." Light waves can be coherent if they are produced in a fashion that maintains constant phase-amplitude relationships. Lasers produce coherent light of one wavelength.

When doing refraction problems, remember that the light refracts toward the normal if it enters a medium of higher index of refraction and refracts away from the normal if it enters a medium of lower index of refraction. The absolute index of refraction of a substance can never be less than 1.00.

Practice Exercises

MULTIPLE-CHOICE

1. What is the frequency of a radio wave with a wavelength of 2.2 m?

(A) 3×10^8 Hz
(B) 1.36×10^8 Hz
(C) 7.3×10^{-9} Hz
(D) 2.2×10^8 Hz
(E) 2.2 Hz

CHALLENGE

2. A ray of light is incident from a layer of crown glass ($n = 1.52$) upon a layer of water ($n = 1.33$). The critical angle of incidence for this situation is equal to

(A) 32°
(B) 41°
(C) 49°
(D) 61°
(E) 75°

3. The relative index of refraction between two media is 1.20. Compared to the velocity of light in medium 1, the velocity of light in medium 2 will be

(A) greater by 1.2 times
(B) reduced by 1.2 times
(C) the same
(D) The velocity will depend on the two media.
(E) The velocity will depend on the angle of incidence.

4. What is the approximate angle of refraction for a ray of light incident from air on a piece of quartz at a 37° angle? **CHALLENGE**

   (A) 24°
   (B) 37°
   (C) 42°
   (D) 66°
   (E) 75°

5. What is the velocity of light in alcohol ($n = 1.36$)?

   (A) 2.2×10^8 m/s
   (B) 3×10^8 m/s
   (C) 4.08×10^8 m/s
   (D) 1.36×10^8 m/s
   (E) None of the preceding values is correct.

6. If the velocity of light in a medium depends on its frequency, the medium is said to be

   (A) coherent
   (B) refractive
   (C) resonant
   (D) diffractive
   (E) dispersive

7. If the intensity of a monochromatic ray of light is increased while the ray is incident on a pair of narrow slits, the spacing between maxima in the diffraction pattern will be

   (A) increased
   (B) decreased
   (C) the same
   (D) increased or decreased, depending on the frequency
   (E) increased or decreased, depending on the slit separation

8. In the diagram below, a source of light (S) sends a ray toward the boundary between two media in which the relative index of refraction is less than 1. The angle of incidence is indicated by i. Which ray best represents the path of the refracted light?

   (A) A
   (B) B
   (C) C
   (D) D
   (E) Either A or C

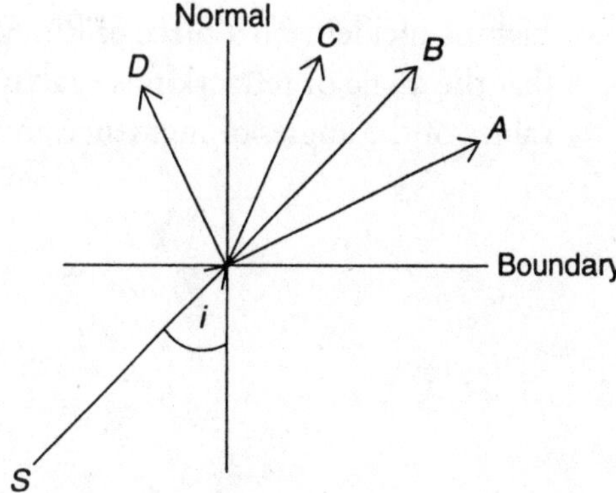

9. A coin is placed at the bottom of a clear trough filled with water ($n = 1.33$) as shown below. Which point best represents the approximate location of the coin as seen by someone looking into the water?

(A) *A*
(B) *B*
(C) *C*
(D) *D*
(E) The perceived location depends on the depth of the water.

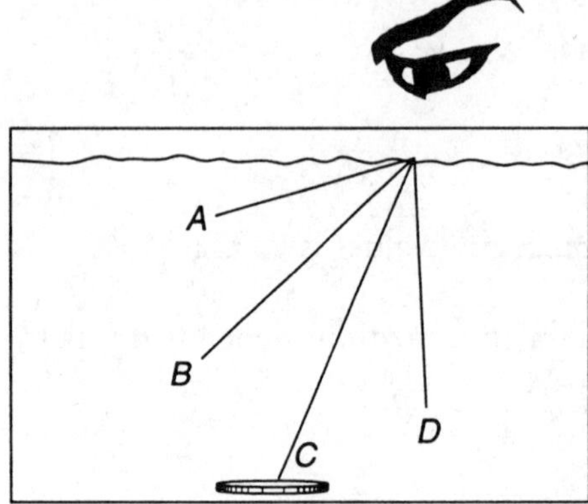

10. If, in question 9, the water is replaced by alcohol ($n = 1.36$), the coin will appear to be

(A) higher
(B) lower
(C) the same
(D) higher or lower, depending on the depth of the alcohol
(E) Not enough information is provided to answer the question.

FREE-RESPONSE

1. (a) Light of wavelength 700 nm is directed onto a diffraction grating with 5000 lines/cm. What are the angular deviations of the first- and second-order maxima from the central maxima?

   (b) Explain why X rays, rather than visible light, are used to study crystal structure?

2. Light is incident on a piece of flint glass ($n = 1.66$) from the air in such a way that the angle of refraction is exactly half the angle of incidence. What are the values of the angles of incidence and the angle of refraction?

3. A ray of light passing through air is incident on a piece of quartz ($n = 1.46$) at an angle of 25°, as shown below. The quartz is 1.5 cm thick. Calculate the deviation d of the ray as it emerges back into the air.

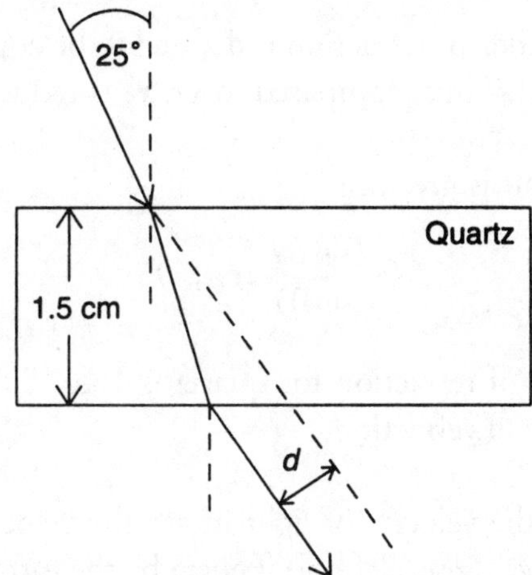

4. Immersion oil is a transparent liquid used in microscopy. It has an absolute index of refraction equal to 1.515. A glass rod attached to the cap of a bottle of immersion oil is practically invisible when viewed at certain angles (under normal lighting conditions). Explain how this might occur.
5. Explain why a diamond sparkles more than a piece of glass of similar size and shape.
6. Explain why total internal reflection occurs at boundaries between transparent media for which the relative index of refraction is less than 1.0.

ANSWERS EXPLAINED

MULTIPLE-CHOICE PROBLEMS

1. **(B)** The velocity of light in air is given by the formula $\mathbf{c} = f\lambda$. The wavelength is 2.2 m, and the velocity of light in air is 3×10^8 m/s. Substituting known values gives us

$$f = 1.36 \times 10^8 \text{ Hz}$$

2. **(D)** The critical angle of incidence is given by the formula

$$\sin \theta_c = \frac{n_2}{n_1}$$

where $n_2 = 1.33$ and $n_1 = 1.52$. Substitution yields a value of 61° for the critical angle.

3. **(B)** The velocity relationship is given by the formula

$$\frac{\mathbf{v}_1}{\mathbf{v}_2} = n_2/n_1$$

Since the relative index of refraction is defined to be equal to the ratio n_2/n_1, we see that $\mathbf{v}_1 = 1.2(\mathbf{v}_2)$. Thus, compared to $\mathbf{v}_1$, $\mathbf{v}_2$ is reduced by 1.2 times.

4. **(A)** Snell's law in air is given by

$$\frac{\sin \theta i}{\sin \theta r} = n_2$$

The absolute index of refraction for quartz is 1.46. Substitution yields a value of 24° for the angle of refraction.

5. **(A)** Compared to the velocity of light in air, the velocity of light in any other transparent substance (here, alcohol) is given by the formula $\mathbf{v} = \mathbf{c}/n$. In this case, $n = 1.36$, and so the velocity of light is equal to 2.2×108 m/s.

6. **(E)** By definition, a medium is said to be "dispersive" if the velocity of light is dependent on its frequency.

7. **(C)** The position and separation of interference maxima are independent of the intensity of the light.

8. **(A)** Since the relative index of refraction is less than 1, the light ray will speed up as it crosses the boundary between the two media and will therefore bend away from the normal, approximately along path A.

9. **(B)** The human eye traces a ray of light back to its apparent source as a straight line. If we follow the line from the eye straight back, we reach point *B*.

10. **(A)** Since alcohol has a higher absolute index of refraction, the light will be bent further away from the normal. Tracing that imaginary line straight back would imply that the image would appear closer to the surface (higher in the alcohol).

FREE-RESPONSE PROBLEMS

1. (a) The general formula for diffraction is

$$n\lambda = d \sin \theta$$

where θ is the angle of deviation from the center. For the first-order maximum, $n = 1$, and d is equal to the reciprocal of the number of lines per meter. Thus, we must convert 5000 lines/cm to 500,000 lines/m. Now,

$$\sin \theta = \frac{n\lambda}{d} = \frac{(1)(7 \times 10^{-7})}{500,000} = 0.35$$

and

$$\theta = 20.5°$$

For the second-order maximum, we have $n = 2$. Thus

$$\sin\theta = \frac{(2)(7 \times 10^{-7})}{500,000} = 0.7$$

and

$$\theta = 44.4°$$

(b) X rays are used to study crystal structure because of their very small wavelengths. These wavelengths are comparable to the spacings between lattices in a crystal and thus make it possible for the X rays to be diffracted from the different layers. Visible light has wavelengths that are much greater than these spacings and so are not affected by them. The use of X rays to probe crystals was one of the first diagnostic applications of these rays in atomic physics at the beginning of the twentieth century.

2. We want the angle of refraction to be equal to half the angle of incidence. This means that $i = 2r$. Now, since the light ray is initially in air ($n = 1.00$), Snell's law can be written:

$$\frac{\sin\theta i}{\sin\theta r} = n \text{ (glass)}$$

Substituting our requirement that $i = 2r$ gives

$$\frac{\sin 2r}{\sin r} = 1.66$$

Now, we recall the following trigonometric identity:

$$\sin 2\theta = 2 \sin\theta \cos\theta$$

Thus

$$\frac{2 \sin r \cos r}{\sin r} = 1.66$$

and

$$2 \cos r = 1.66$$

Therefore

$$\cos r = 0.83$$

and

$$r = 34°$$

which means that

$$i = 68°$$

3. The diagram from the problem has been redrawn as shown below. From our knowledge of refraction, we know that angle θ must be equal to 25°. Angle r is given by Snell's law:

$$\frac{\sin 25}{\sin r} = 1.46$$

$$\sin r = \frac{\sin 25}{1.46} = 0.2894$$

$$r = 16.8°$$

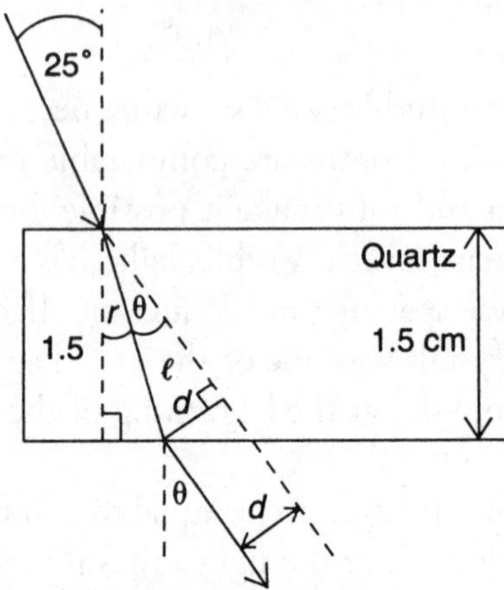

Now, angle θ is equal to the difference between the angle of incidence and the angle of refraction:

$$\theta = 25° - 16.8° = 8.2°$$

Since the quartz is 1.5 cm in thickness, the length of the light ray, in the quartz, at the angle of refraction, can be determined from $\cos r = 1.5/\ell$. This implies that $\ell = 1.57$ cm. Now, since we know the length of the diagonal ℓ, and the angle θ, the deviation d is just part of the right triangle in our diagram. Thus

$$\sin \theta = \frac{d}{\ell}$$

and

$$d = \ell \sin \theta = (1.57) \sin 8.2 = 0.224 \text{ cm}$$

4. The index of refraction for the glass is very nearly equal to that for the immersion oil. When the applicator rod is filled with liquid, the light passes through both media without refracting and makes the glass rod appear invisible.

5. A diamond has a lower critical angle of incidence than glass. Therefore, when turned through various angles, the light passing through a diamond is internally reflected and then dispersed more easily than through glass. This creates the "sparkling" effect.

6. In order to produce total internal reflection, the angle of refraction must exceed 90 degrees. This is possible only when light passes from a high-index material to one with a low index of refraction. Under these conditions, the relative index of refraction for the two media is less than 1.0.

CHAPTER 21

Geometrical Optics

KEY CONCEPTS

- Image Formation in Plane Mirrors
- Image Formation in Curved Mirrors
- Image Formation in Lenses

21.1 IMAGE FORMATION IN PLANE MIRRORS

When you look at yourself in a plane mirror, your image appears to be directly in front of you and on the other side of the mirror. Everything about your image is the same as you, the person, except for a left-right reversal. Since no light can be originating from the other side of the mirror, your image is termed **virtual.**

The formation of a virtual image in a plane mirror is illustrated in Figure 21.1. Using an imaginary point object, we construct two rays of light that merge radially from the object because of ambient light from its environment. Each light ray is incident on the plane mirror at some arbitrary angle and is reflected off at the same angle (relative to the normal). Geometrically, we construct these lines using the law of reflection and a protractor. Since the rays diverge from the object, they continue to diverge after reflection. The human eye, however, perceives the rays as originating from a point on the other side of the mirror and in a direct line with the object.

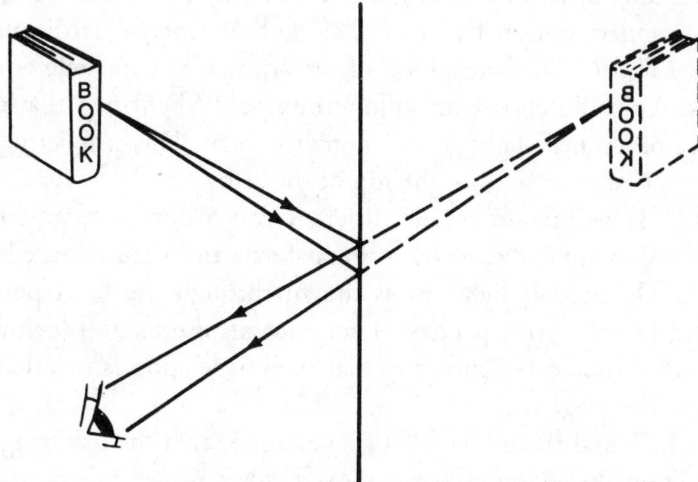

Figure 21.1 Plane Mirror

21.2 IMAGE FORMATION IN CURVED MIRRORS

A. Concave Mirrors

In Chapter 20, we saw that, if we use a concave or convex mirror, the shape will cause parallel rays of light to either converge or diverge. In Figure 21.2 a typical concave mirror is illustrated. The shape of the bisecting axis, called the **principal axis**, should be parabolic in cross section. If the curvature of the mirror is too large, a defect known as **spherical aberration** occurs and distorts the images seen. The real images formed by concave mirrors can be projected onto a screen. The focal length can be determined by using parallel rays of light and observing the point at which they converge.

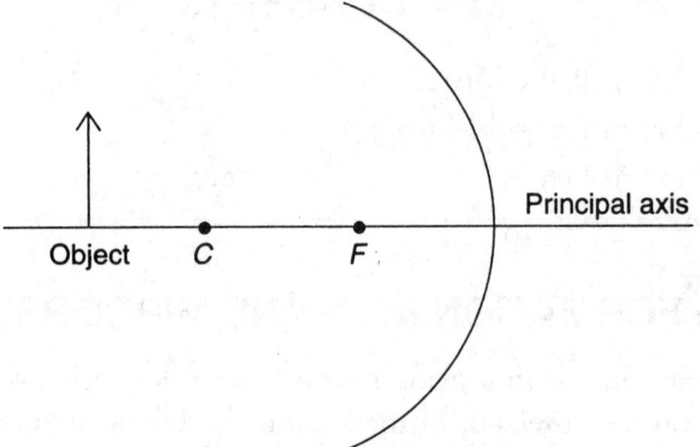

Figure 21.2 Concave Mirror

Another, more interesting method is to aim the mirror out the window at distant objects. Again, the images of those objects can be focused onto a screen. Since the objects are very far away, they are considered to be **at infinity**. Normally, at infinity, an object sends parallel rays of light and appears as a point. Since we are viewing extended objects, we project a smaller image of them, and the distance from the screen to the mirror is the focal length F.

The point along the principal axis labeled C on Figure 21.2 marks the **center of curvature** and is located at a distance $2F$.

To illustrate how to construct an image formed by a concave mirror, we use an arrow as an imaginary object. Its location along the principal axis is measured relative to points F and C. The orientation of the arrow is of course determined by the way it points. We could choose an infinite number of light rays that come off the object because of ambient light from its environment. For simplicity, however, we choose two rays that emerge from the top of the arrow.

In Figure 21.3, we present several different concave-mirror constructions. The first light ray is drawn parallel to the principal axis and then reflected through the focal point F. The second light ray is drawn through the focal point and then reflected parallel to the principal axis. The point of intersection (below the principal axis in cases I, II, and III) indicates that the image appears inverted at the location marked in the diagram.

From cases I, II, and III in Figure 21.3, we can see that the real images are always inverted. Additionally, as the object is moved closer to the mirror, the image gets larger and appears to move further away from the mirror. Notice that, when the object is at point C, as in case II, the image is also at point C and is the same size.

When the object is at the focal point, as in case IV, no image can be seen since the light is reflected parallel from all points on the mirror. If the object is moved even closer, as in case V, we get an enlarged virtual image that is erect. In case IV, a light ray has been drawn toward the center of the mirror; by the law of reflection, it will reflect below the principal axis at the same angle of incidence.

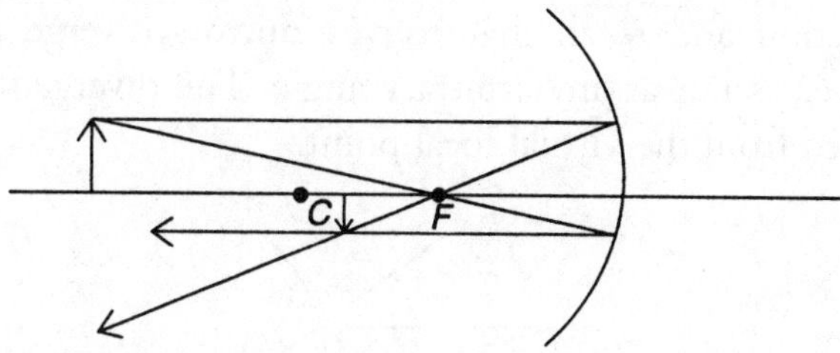

Case I—object beyond *C;* image is real, is between *F* and *C*, is smaller

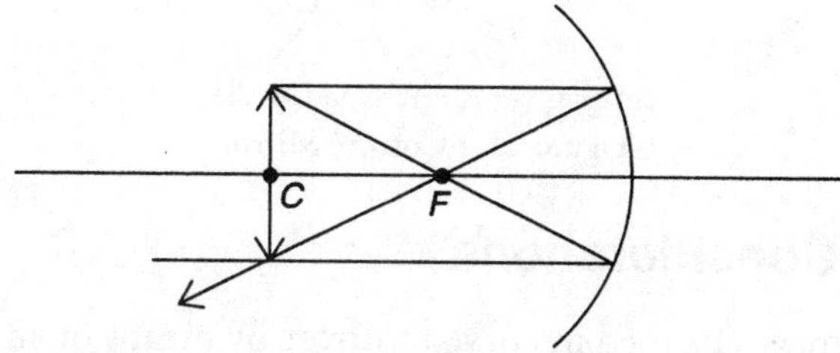

Case II—object at *C*; image is real, is at point *C*, is the same size

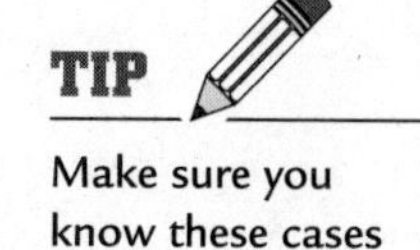

TIP

Make sure you know these cases for both curved mirrors and lenses.

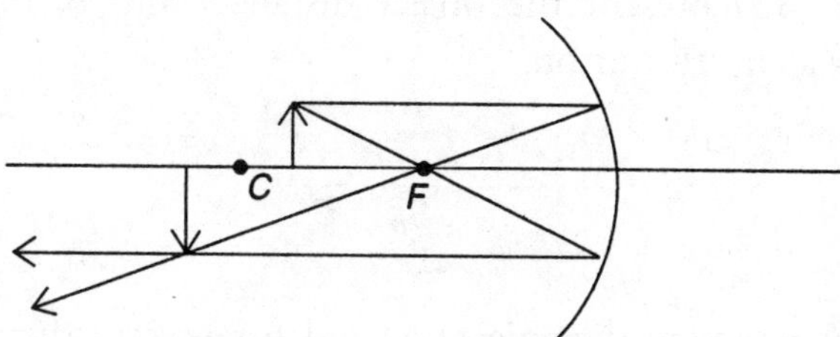

Case III—object between *F* and *C;* image is real, is beyond *C*, is larger

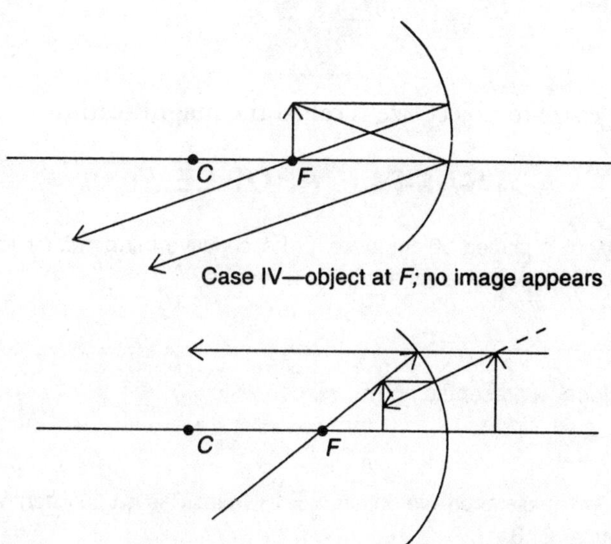

Case IV—object at *F;* no image appears

Case V—object between *F* and the mirror; image is virtual, is larger

Figure 21.3

B. Convex Mirrors

Convex mirrors are used in a variety of situations. In stores or elevators, for example, they have the ability to reveal images (although distorted) from around corners in aisles. An image in a convex mirror is always virtual and always smaller. This fact suggests that only one case construction, as opposed to five for concave mirrors, is necessary to understand image formation in these mirrors. In Figure 21.4, we show a sample construction and recall that convex mirrors diverge parallel rays. This divergence, however, is not at any arbitrary angle. The divergent ray is directed as though it originated from the virtual focal point.

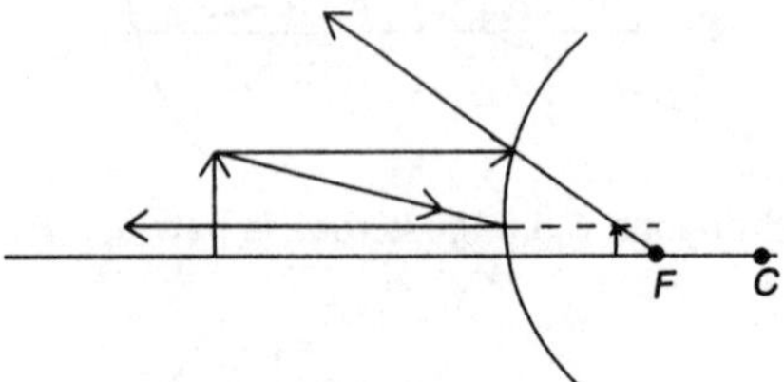

Image is virtual, erect, and smaller.
Figure 21.4 Convex Mirror

C. Algebraic Considerations

We can study the images formed in curved mirrors by means of an algebraic relationship. If we let F represent the focal length (positive in a concave mirror; negative in a convex mirror), S_o represent the object distance, and S_i represent the image distance, then, for a curved mirror.

$$\frac{1}{f} = \frac{1}{S_o} + \frac{1}{S_i}$$

Also, if we let h_o represent the object size and h_i represent the image size, then

$$\frac{h_i}{h_o} = \frac{S_i}{S_o}$$

The ratio of image size to object size is called the **magnification**.

SAMPLE PROBLEM

A 10-cm-tall object is placed 20 cm in front of a concave mirror with a focal length of 8 cm. Where is the image located, and what is its size?

Solution

Using the first formula presented above, we can write

$$\frac{1}{8} = \frac{1}{20} + \frac{1}{S_i}$$

Solving for the image distance, we obtain $S_i = 13.1$ cm. This is consistent with case I as illustrated in Figure 21.3.

The image size can be obtained from the expression

$$h_i = h_o\left(\frac{S_i}{S_o}\right) = 10\left(\frac{13.3}{20}\right) = 6.7 \text{ cm}$$

21.3 IMAGE FORMATION IN LENSES

A. Converging Lenses

When discussing the formation of images in lenses, we usually invoke what is called the **thin lens approximation**; that is, we consider that the light begins to refract from the center of the lens. As a result any curvature effects can be ignored. Since the top and the bottom of the lens are tapered like a prism, however, a defect known as **chromatic aberration** can sometimes occur. This defect causes the different colors of light to disperse in the lens and focus at different places because of their different frequencies and velocities in the lens material.

Figure 21.5 shows a series of cases in which a double convex lens is constructed as a straight line, using the thin lens approximation. The symmetry of the lens creates two real focal points, one on either side, and instead of considering the center of curvature (as in the curved mirror cases), we employ point $2f$ as an analogous location for reference. Our object is again an arrow drawn along the principal axis, which bisects the lens.

From Chapter 20, we know that a light ray parallel to the principal axis will refract through the focal point. Therefore, a light ray originating from the focal point will refract parallel to the principal axis. We will use these two light rays to construct most of our images.

Notice that in case IV no image is produced, and we had to draw a different light ray passing through the optical center of the lens. In case V an enlarged virtual image is produced on the same side as the object. This is an example of a simple magnifying glass.

B. Diverging Lenses

A concave lens will diverge parallel rays of light away from a virtual focal point. Figure 21.6 illustrates a typical ray construction for a diverging lens. Again, if we assume the thin lens approximation, we will draw the lens itself as a line for ease of construction. The context of the problem tells us that the drawing represents a concave, not a convex, lens.

To construct the image, we have drawn a parallel light ray that then diverges away from the focal point. The second light ray is drawn through the optical center of the lens. The resulting virtual image will be erect and smaller and will be in front of the object. This will be true for any location of the object, so only one sample construction is necessary.

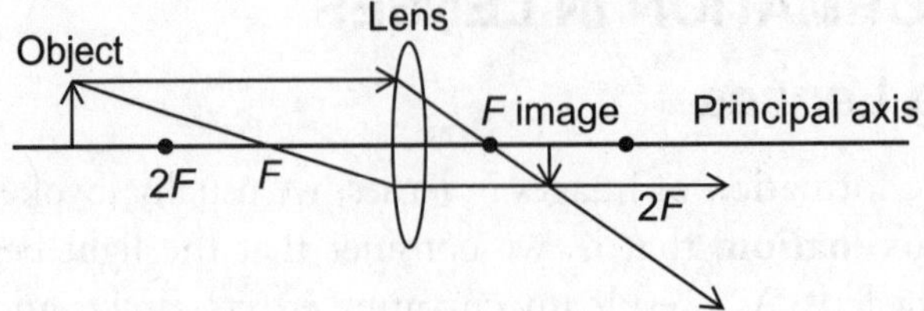

Case I–object beyond 2*F*; image is between *F* and 2*F*, is real, is smaller

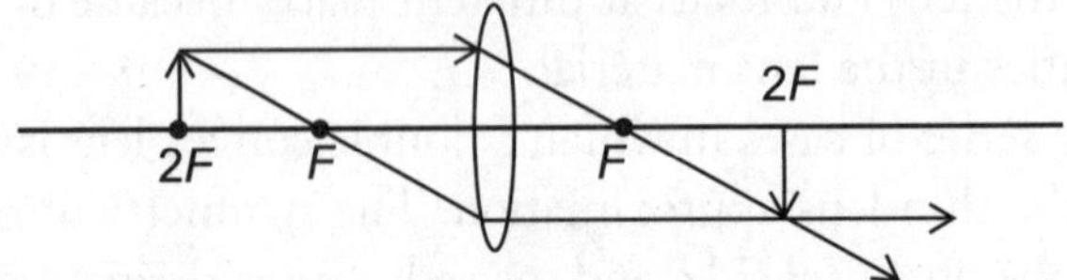

Case II–object at 2*F*; image is at 2*F*, is real, is same size

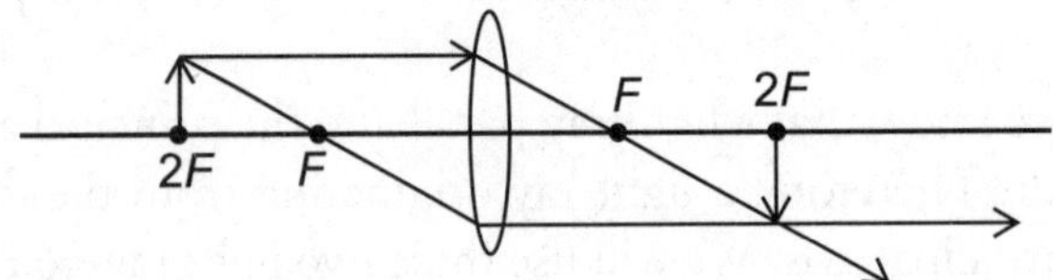

Case III–object between *F* and 2*F*; image is beyond 2*F*; is real, is larger

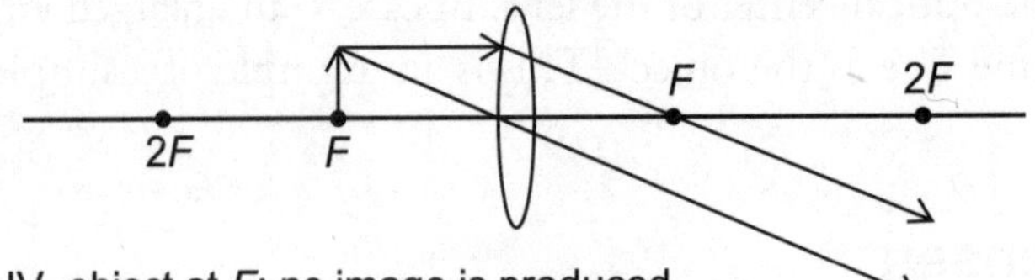

Case IV–object at *F*; no image is produced

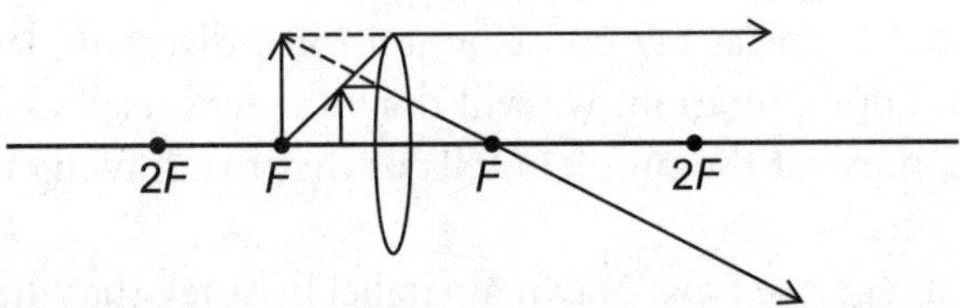

Case V–object between *F* and the lens; image is behind object, is virtual, is larger

Figure 21.5

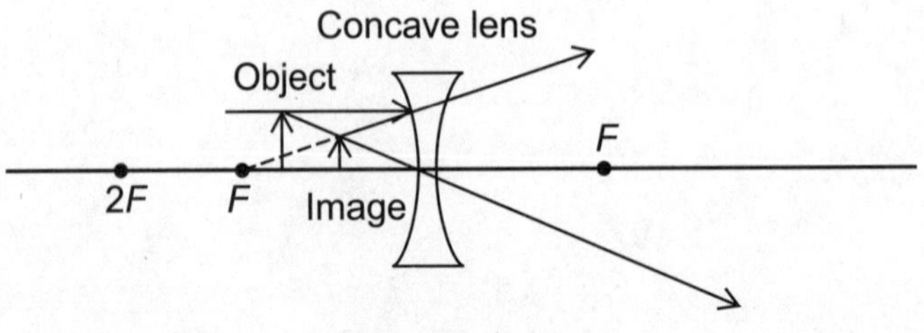

Figure 21.6

C. Algebraic Considerations

The relationship governing the image formation in a thin lens is the same as it was for a curved mirror. Positive image distances occur when the image forms on the other side of the lens. A negative image distance implies a virtual image. A positive focal length implies a convex lens, while a negative focal length implies a concave lens. Therefore, as before, we can write

$$\frac{1}{f} = \frac{1}{S_o} + \frac{1}{S_i}$$

and

$$\frac{h_i}{h_o} = \frac{S_i}{S_o}$$

If two converging lenses are used in combination, separated by some distance x, the combined magnification, m, of the system is given by

$$m = m_1 m_2$$

SAMPLE PROBLEM

A 3-cm-tall object is placed 6 cm in front of a concave lens (f = 3 cm). Calculate the image distance and size.

Solution

We use the modified version of the mirror-lens equation:

$$S_i = \frac{(S_o f)}{(S_o - f)} \qquad \text{Recall that for a concave lens, } f < 0\,!$$

$$S_i = \frac{(6\text{ cm})(-3\text{ cm})}{6\text{ cm} - (-3\text{ cm})} = -2\text{ cm}$$

$$h_i = h_o\left|\frac{S_i}{S_o}\right| = \frac{(3\text{ cm})(2\text{ cm})}{6\text{ cm}} = 2\text{ cm}$$

Chapter Summary

- Images are classified as being real or virtual.
- Real images are formed when light rays converge. Real images can be projected onto a screen and are always inverted in appearance.
- Virtual images are formed as the brain imagines that reflected or refracted light rays converge back from their diverging paths. Virtual images cannot be projected onto a screen and are always erect.
- Plane mirrors, convex mirrors, and concave lenses always produce virtual images. The virtual images produced by convex mirrors and concave lenses are always smaller than the object.

- Concave mirrors and convex lenses produce both real and virtual images. The real images may be larger or smaller than the object, whereas the virtual images are always larger.
- For a curved mirror, the focal length is equal to half the radius of curvature.
- If an object is placed at the focal point of a concave mirror or a convex lens, no image will be formed.

Problem-Solving Strategies for Geometrical Optics

To solve ray-diagram problems, you must remember the types of light rays and understand how they reflect and refract using mirrors and lenses. Numerically, keep in mind that positive focal lengths imply concave mirrors and convex lenses.

Memorize all of the cases illustrated in this chapter, and remember that real images are always inverted. Even if a given problem does not require a construction, draw a sketch. Additionally, remember that the focal length of a lens is dependent on the frequency of light used and the material of which the lens is made.

Practice Exercises

MULTIPLE-CHOICE

1. Which material will produce a converging lens with the longest focal length?

   (A) Lucite
   (B) Crown glass
   (C) Flint glass
   (D) Quartz
   (E) Diamond

2. An object is placed in front of a converging lens in such a way that the image produced is inverted and larger. If the lens were replaced by one with a larger index of refraction, the size of the image would

   (A) increase
   (B) decrease
   (C) increase or decrease, depending on the degree of change
   (D) remain the same
   (E) increase or decrease depending on the wavelength of the light

3. You wish to make an enlarged reproduction of a document using a copying machine. When you push the enlargement button, the lens inside the machine moves to a point

   (A) equal to f
   (B) equal to $2f$
   (C) between f and $2f$
   (D) beyond $2f$
   (F) less than f

4. A negative image distance means that the image formed by a concave mirror will be

   (A) real
   (B) erect
   (C) inverted
   (D) smaller
   (A) both (B) and (D)

5. Real images are always produced by

   (A) plane mirrors
   (B) convex mirrors
   (C) concave lenses
   (D) convex lenses
   (E) both (B) and (D)

6. The focal length of a convex mirror with a radius of curvature of 8 cm is

   (A) 4 cm
   (B) −4 cm
   (C) 8 cm
   (D) −8 cm
   (E) 16 cm

7. An object appears in front of a plane mirror as shown below:

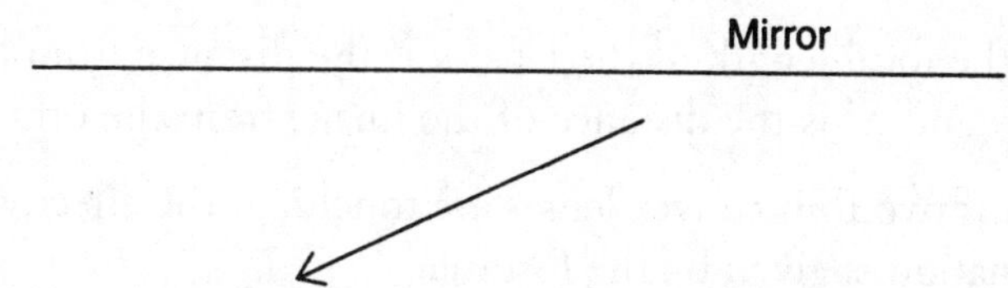

   Which of the following diagrams represents the reflected image of this object?

   (A) 

   (B) 

   (C) 

   (D) 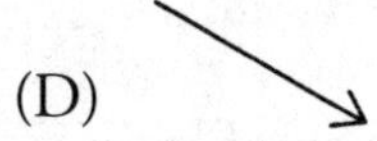

   (E) 

8. An object is located 15 cm in front of a converging lens. An image twice as large as the object appears on the other side of the lens. The image distance must be

(A) 15 cm
(B) 30 cm
(C) 45 cm
(D) 60 cm
(E) 75 cm

9. A 1.6-meter-tall person stands 1.5 m in front of a vertical plane mirror. The height of his image is

(A) 0.8 cm
(B) 2.6 cm
(C) 3.2 cm
(D) 4.5 cm
(E) 1.6 cm

FREE-RESPONSE

CHALLENGE

1. Using a geometric-ray diagram construction, prove Newton's version of the lens equation:

$$f^2 = SS'$$

where f is the focal length of the lens, d is the distance from the object to the focal point, and S' is the distance of the image from the other focal point.

2. Prove that, if two thin convex lenses are touching, the effective focal length of the combination is given by the formula

$$\frac{1}{f} = \frac{1}{f_1} + \frac{1}{f_2}$$

CHALLENGE

3. An object is 25 cm in front of a converging lens with a 5-cm focal length. A second converging lens, with a focal length of 3 cm, is placed 10 cm behind the first one.

(a) Locate the image formed by the first lens.

(b) Locate the image formed by the second lens if the first image is used as an "object" for the second lens.

(c) What is the combined magnification of this combination of lenses?

4. Why do passenger-side mirrors on cars have a warning that states: Objects are closer than they appear?

5. Two lenses have identical sizes and shapes. One is made from quartz (n = 1.46), and the other is made from glass (n = 1.5). Which lens would make a better magnifying glass?

6. Why do lenses produce chromatic aberration whereas spherical mirrors do not?

ANSWERS EXPLAINED

MULTIPLE-CHOICE PROBLEMS

1. **(D)** The longest focal length will be produced by the material that refracts the least, that is, has the smallest index of refraction. Of the five choices, quartz has the lowest index of refraction.

2. **(B)** The lens formula can be rewritten as

$$S_i = \frac{S_o f}{S_o - f}$$

If the object remains in the same position relative to the lens, then using a larger index of refraction will imply a smaller focal length. The image will move closer to the lens and consequently will be smaller.

3. **(C)** To produce an enlarged real image, the object must be located between f and $2f$.

4. **(B)** In a concave mirror, a negative image distance implies a virtual image, which is enlarged and erect.

5. **(D)** Only a convex lens or a concave mirror can produce a real image.

6. **(B)** The radius of curvature of a spherical mirror is twice the focal length. However, in a convex mirror, the focal length is taken to be negative.

7. **(B)** The reflected image must point "away" from the mirror but on the other side, flipped over.

8. **(B)** The equation governing magnification in a converging lens is

$$M = \frac{S_i}{S_o}$$

Since $M = 2$ and $S_o = 15$ cm, the image distance must be $S_i = 30$ cm.

9. **(E)** A plane mirror produces a virtual image that is the same size as the object in all cases.

FREE-RESPONSE PROBLEMS

1. A drawing of the situation is given below:

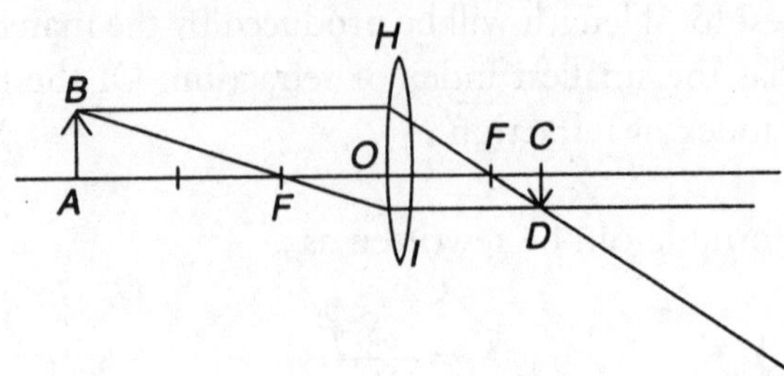

In this construction, we have used the thin lens approximation. On the left side of the lens,

$$AF = S \quad \text{and} \quad FO = f \text{ (focal length)}$$

On the right side,

$$FO = f \quad \text{and} \quad FC = S'$$

Since we have a thin lens,

$$AB = OH \quad \text{and} \quad OI = CD$$

Now, we have ΔFAB similar to ΔFOI and ΔFOH similar to ΔFCD. Thus, the following sides are in proportion:

$$\frac{AB}{OI} = \frac{AF}{FO}$$
$$\frac{OH}{CD} = \frac{FO}{FC}$$

Thus, $f/S = S'/f$, using the relationships defined above. Therefore,

$$f^2 = SS'$$

as desired.

2. If we have two lenses in contact, the image of the first lens becomes a "negative" object for the second lens. For the first lens, we have

$$\frac{1}{f_1} = \frac{1}{S_o} + \frac{1}{S_{i1}}$$

where p is the object distance for the combination. For the second lens, we must now have

$$\frac{1}{f_2} = \frac{1}{S_o} + \frac{1}{S_{i2}}$$

If we combine these equations, we obtain

$$\frac{1}{f_1}+\frac{1}{f_2}=\frac{1}{S_o}+\frac{1}{S_{i2}}$$

Since the two thin lenses are in contact, S_{i2} can be considered the image distance from the first lens also. Therefore, if we set

$$\frac{1}{S_o}+\frac{1}{S_{i2}}=\frac{1}{f}$$

the combination of two thin lenses in contact behaves equivalently as a single lens with a focal length *f*, so that

$$\frac{1}{f}=\frac{1}{f_1}+\frac{1}{f_2}$$

3. (a) For the first lens, we have $f = 5$ cm and $S_o = 25$; thus

$$S_i=\frac{S_o f}{S_o-f}=\frac{(25)(5)}{25-5}=\frac{125}{20}$$

and the location of the image formed by the first lens is given by $S_i = 6.25$ cm.

(b) Since $S = 10$ cm for the second lens (its distance from the first lens) and $S_i = 6.25$ for the first lens, the image q now serves as the new "object" at a distance $S'_o = 10 - 6.25 = 3.75$. The focal length of the second lens is 3 cm; therefore, the distance S'_i of the image formed by the second lens is

$$S'_i=\frac{(3.75)(3)}{3.75-3}=15 \text{ cm}$$

(c) The combined magnification of the system is equal to the product of the separate magnifications. Since $m = S_i/S_o$, in general, we have

$$M_1=\frac{6.25}{25}=0.25$$
$$M_2=\frac{15}{3.75}=4$$

Therefore, the combined magnification of this combination of lenses is given by

$$M=M_1M_2=(0.25)(4)=1$$

4. Passenger side mirrors are convex in shape. This distorts images because of the divergence of the light, making them appear to be smaller and more distant than they actually are.

5. Using the lens construction diagrams, we see that for case V the lens becomes a magnifying glass. Changing the focal length affects the size and location of the image. If the focal length is decreased, the image size and distance will increase. A shorter focal length is obtained by using a material with a higher index of refraction. Hence, glass would make a better magnifying glass.

6. Chromatic aberration is due to the dispersion of white light into the colors of the spectrum. Since a mirror does not disperse light on reflection, this problem does not occur. This observation was the motivation for Sir Isaac Newton to invent the reflecting telescope in 1671.

CHAPTER 22

Quantum Theory

KEY CONCEPTS

- Photoelectric Effect
- Compton Effect and Photon Momentum
- Matter Waves
- Spectral Lines

22.1 PHOTOELECTRIC EFFECT

Light has a very interesting effect on certain metals. Imagine a piece of zinc placed on top of an electroscope and charged negatively by contact (in a vacuum). If the zinc is exposed to various frequencies of light, it is determined that light in the ultraviolet region causes the electroscope to discharge (see Figure 22.1). This phenomenon does not occur, however, when the zinc and electroscope are charged positively.

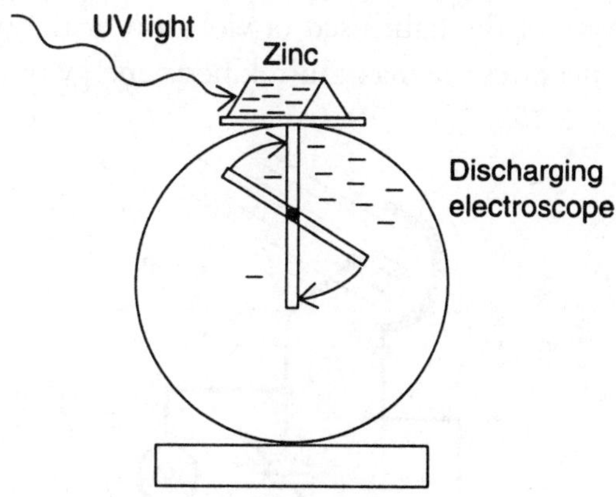

Figure 22.1

The conclusion is that the ultraviolet light causes the zinc to emit electrons. This phenomenon, called the **photoelectric effect**, defines a new view of light that has come to be known as the **quantum theory**. Developed by Albert Einstein, the quantum theory has been extended to particles as well as electromagnetic waves and forms the basis of modern physics.

The photoelectric effect can be studied in a more quantitative way. Figure 22.2 shows an evacuated tube that contains a photoemissive material. Many metals exhibit the photoelectric effect. On the other side of the tube is a collecting plate. The tube is connected to a microammeter, which can be used to measure the current generated by the emitted electrons. When exposed to light of suitable frequency, the ammeter will implicitly measure the number of electrons emitted per second. This number is sensitive only to the intensity of the electromagnetic wave applied. The minimum frequency that emits electrons is called the **threshold frequency** and is designated as f_0.

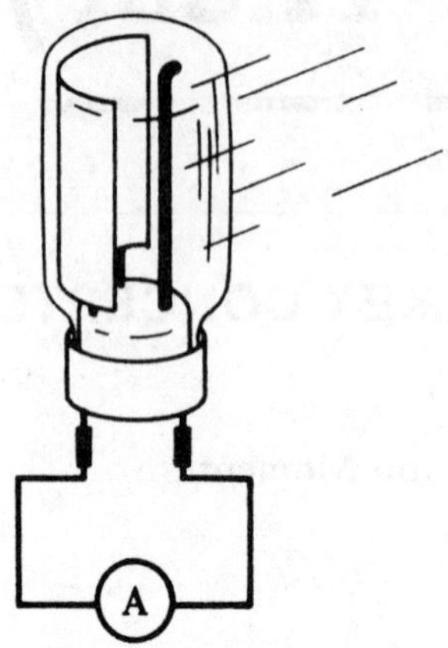

Figure 22.2

To measure the energy of the emitted electrons, we insert in the circuit a source of potential difference (Figure 22.3) such that the negative electrons approach a negatively charged side. This will create a braking force; and if we select the voltage properly, we can stop the current completely. The voltage that performs this feat is called the **stopping voltage** or **stopping potential** and is designated as V_0.

The energy of the electrons stopped by this voltage is equal to the product of the electrons' charge and the stopping voltage, eV_0. Since the stopping voltage is independent of the intensity of the light used (a violation of the wave theory of light), the stopping voltage measures the maximum kinetic energy of the emitted electrons.

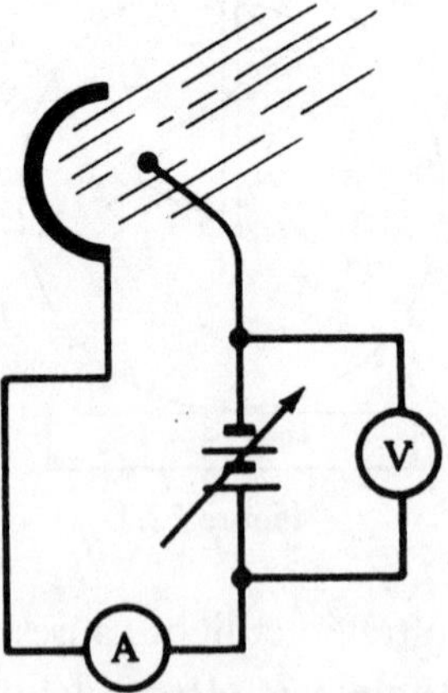

Figure 22.3

As the frequency of light is slowly increased, there is a change in the maximum kinetic energy of the electrons. The fact that this energy is independent of the intensity of the light was used by Einstein to demonstrate that light consists of discrete or **quantized** packets of energy called **photons**. The energy of a given photon is directly proportional to its frequency.

To clarify this relationship, Figure 22.4 illustrates a typical graph of maximum kinetic energy versus frequency.

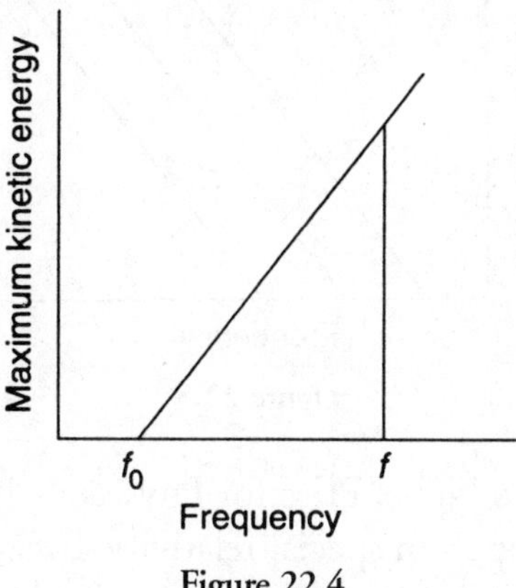

Figure 22.4

Starting with the threshold frequency, the graph is a straight line. For any frequency point, f, the slope of the line is given by

$$\text{Slope} = \frac{KE_{max}}{f - f_0} = h$$

This slope, referred to as **Planck's constant** and symbolized as h, has a value of

$$h = 6.63 \times 10^{-34} \text{ J/Hz} = 6.63 \times 10^{-34} \text{ J} \cdot \text{s}$$

Planck's constant originally appeared in the theoretical work of German physicist Max Planck. His mathematical discovery in 1900 of an elementary quantum of action helped explain the nature of radiation emitted from hot bodies. Planck's constant became a part of Einstein's quantum theory of light and Bohr's quantum theory of the hydrogen atom.

The above equation can be solved for the maximum kinetic energy:

$$KE_{max} = hf - hf_0$$

In this expression, known as the **photoelectric equation**, hf represents the energy of the original photon, while hf_0, called the **work function**, is a property of the metal. The quantity hf_0 is a measure of the minimum amount of energy needed to free an electron. It is designated in some textbooks by the Greek letter phi (ϕ) and in others by W_0. In this book, we use the latter designation for the work function.

Experiments with various metals yield some interesting results. Even though the different metals all have different threshold frequencies, they all obey the same photoelectric equation and have the same slope, h. A typical comparison graph of three metals, A, B, and C, is shown in Figure 22.5.

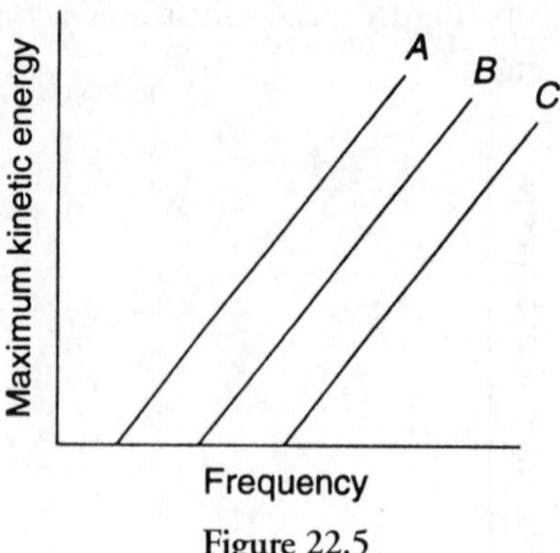

Figure 22.5

Einstein was awarded the Nobel Prize for Physics in 1921, the same year that he published his celebrated paper on special relativity. Because of the controversy over relativity, the Nobel committee awarded the prize on the basis of Einstein's theoretical work in 1905 on the quantum theory of light. His suggestion of the photoelectric effect, whereby light behaves as packets of quantized energy (photons), was in direct conflict with the prevailing wave theory of light, as demonstrated by interference and diffraction. Thus, physicists in the early twentieth century were presented with a **dual nature of light**.

SAMPLE PROBLEM

Light with a frequency of 2×10^{15} Hz is incident on a piece of copper.

(a) What is the energy of the light in joules and in electron-volts?

(b) If the work function for copper is 4.5 eV, what is the maximum kinetic energy, in electron volts, of the emitted electrons?

Solution

(a) The energy in joules is given by $E = hf$. Thus,

$$E = hf = (6.63 \times 10^{-34})(2 \times 10^{15}) = 1.326 \times 10^{-18}\text{J}$$

Since 1 eV = 1.6×10^{-19} J, we have

$$E = 8.28 \text{ eV}$$

(b) To find the maximum kinetic energy, we simply subtract the work function from the photon energy:

$$KE_{max} = 8.28 - 4.5 = 3.79 \text{ eV}$$

22.2 COMPTON EFFECT AND PHOTON MOMENTUM

In the early 1920s, American physicist Arthur Compton conducted experiments with X rays and graphite. He was able to demonstrate that, when an X-ray photon collides with an electron, the collision obeys the law of conservation of momentum. The scattering of photons due to this momentum interaction is known as the **Compton effect** (Figure 22.6). In the Compton effect, the scattered photon has a lower frequency than the incident photon.

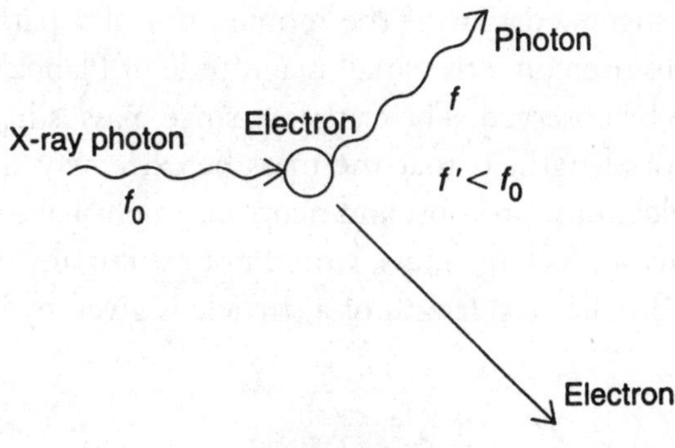

Figure 22.6

The momentum of the photon is given by the relationship $\mathbf{p} = h/\lambda$. Thus X rays, because of their very high frequency, possess a large photon momentum. Even though we state that photons can have momentum, they do not possess what is called rest mass. If light can have momentum as well as wavelength, the question now raised is, can particles have wavelength as well as momentum?

22.3 MATTER WAVES

If a beam of X rays is incident on a crystal of salt, the pattern that emerges is characteristic of the scattering of particles, not waves. If, however, a beam of electrons is incident on a narrow slit (Figure 22.7), the pattern that emerges is a typical wave diffraction pattern. Electron diffraction is one example of the dual nature of matter, which, under certain conditions, exhibits a wave nature. The circumstances are determined by experiment. If you perform an interference experiment with light, you will demonstrate its wave properties. If you do a photoelectric effect experiment, light will reveal its "particle-like" properties.

In the 1920s, French physicist Louis de Broglie used the dual nature of light to suggest that matter can have a "wavelike" nature, as indicated by its de Broglie wavelength. Since a photon's momentum is given by $\mathbf{p} = h/\lambda$, the wavelength can be transposed to give $\lambda = h/\mathbf{p}$.

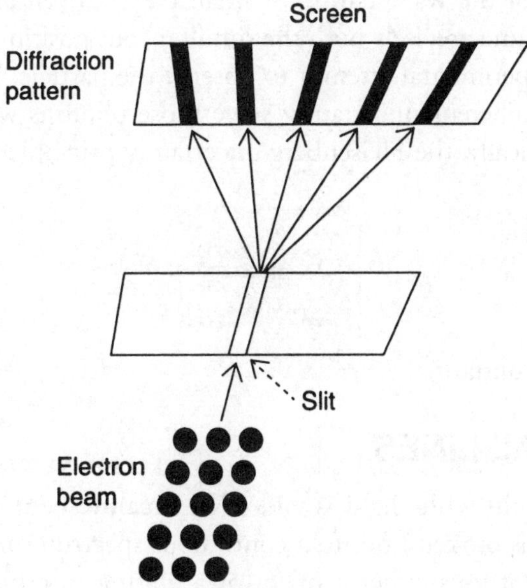

Figure 22.7

Louis de Broglie suggested that, if the momentum of a particle, $\mathbf{p} = m\mathbf{v}$, has a magnitude suitable to overcome the small magnitude of Planck's constant, a significant wavelength can be observed. The criterion, since mass is in the denominator of the expression for wavelength, is that the mass be extremely small (on the atomic scale). Thus, while electrons, protons, and neutrons might have wave characteristics, a falling stone, because of its large mass, would not exhibit any wavelike effects. The equation for the de Broglie wavelength of a particle is given by

$$\lambda = \frac{h}{m\mathbf{v}}$$

SAMPLE PROBLEM

Find the de Broglie wavelength for each of the following:

(a) A 10-g stone moving with a velocity of 20 m/s.
(b) An electron moving with a velocity of 1×10^7 m/s.

Solution

(a) Since $m = 10$ g $= 0.01$ kg, and $\lambda = h/m\mathbf{v}$, we have

$$\lambda = \frac{6.63 \times 10^{-34}}{(0.01)(20)} = 3.315 \times 10^{-33} \text{ m}$$

(b) In this part we have

$$\lambda = \frac{h}{mv} = \frac{6.63 \times 10^{-34}}{(9.1 \times 10^{-31})(1 \times 10^{-7})} = 7.3 \times 10^{-11} \text{ m}$$

This aspect of particle motion, on a small scale, leads to the so-called **Heisenberg uncertainty principle**. Developed by Werner Heisenberg, also in the 1920s, this states that because of the wave nature of small particles (on an atomic scale) it is impossible to determine precisely both the simultaneous position and momentum of the particle. Any experimental attempt to observe the particle, using photon detection, will cause an inherent uncertainty since those photons will interact with the particles. Mathematically, the Heisenberg uncertainty principle states that

$$\Delta x \, \Delta \mathbf{p} \geq \frac{h}{2\pi}$$

where h is Planck's constant.

22.4 SPECTRAL LINES

A hot source that emits white light is said to be **incandescent**. Its light, when analyzed with a prism, is broken up into a continuous spectrum of colors. (A **spectroscope** is a device that uses either a prism or a grating to create a spectrum.) If a colored filter is used over the incandescent source, light of only that one color will be emitted, and the spectrum will appear as a continuous band of that color.

If hydrogen gas contained at low pressure in a tube is electrically sparked, the gas will emit a bluish light. If that light is analyzed with a spectroscope, a discrete series of colored lines ranging from red to violet, rather than a continuous band of color, is revealed. When sparked, other chemical elements also show characteristic **emission line spectra** that are useful for chemical analysis.

Figure 22.8 illustrates a typical hydrogen spectral series in the visible range of the electromagnetic spectrum. This series is often called the **Balmer series** in honor of Jakob Balmer, a Swiss mathematician who empirically studied the hydrogen spectrum in the late nineteenth century. In Figure 22.8, a hydrogen tube is excited and its light is passed through a narrow slit to provide a point source. This light is then passed through a prism or grating. Many emission lines are produced, but only the first four, designated as Hα (red), Hβ, Hλ, and Hδ (violet), are illustrated here. The wavelengths range from about 656 to about 400 nanometers.

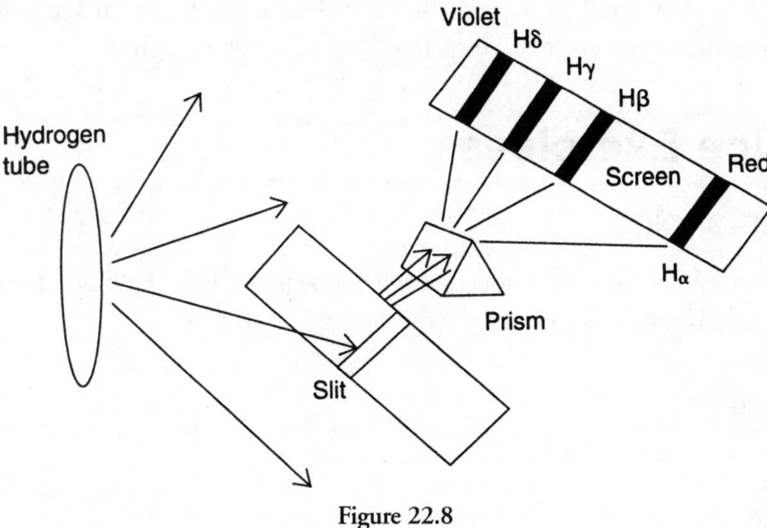

Figure 22.8

Chapter Summary

- The photoelectric effect concludes that light consists of discrete packets of energy called photons.
- Each photon has an energy proportional to its frequency ($E = hf$).
- In the photoelectric effect, the maximum kinetic energy of the emitted electrons is independent of the intensity (number of photons) and depends directly on the frequency of the incident electromagnetic radiation.
- There is a minimum frequency needed to emit electrons called the threshold frequency.
- The maximum kinetic energy of the electrons can be measured by applying a negative electric potential across their paths called the stopping potential.

- The Compton effect demonstrates that photons have a momentum $\mathbf{p} = h/\lambda$.
- Louis de Broglie developed a theory of matter waves in which material particles have a wave-like behavior inversely proportional to their momentum ($\lambda = h/m\mathbf{v}$).
- Spectral lines provide an observational demonstration of discrete energy transfers in an atom.

Problem-Solving Strategies for Quantum Theory

Remember that the photoelectric effect demonstrates the particle nature of light and is independent of the intensity of the electromagnetic waves used. The number of photons is, however, proportional to the intensity of energy.

Note that in some problems energy is expressed in units of electron volts. This is convenient since the maximum kinetic energy is proportional to the stopping potential. Additionally, wavelengths are sometimes expressed in angstroms, and these values must be converted to meters. Be careful with units, and make sure that SI units are used in equations involving energies and other quantities.

Practice Exercises

MULTIPLE-CHOICE

1. How many photons are associated with a beam of light having a frequency of 2×10^{16} Hz and a detectable energy of 6.63×10^{-15} J?

   (A) 25
   (B) 500
   (C) 135
   (D) 8
   (E) 80

2. A photoelectric experiment reveals a maximum kinetic energy of 2.2 eV for a certain metal. The stopping potential for the emitted electrons is

   (A) 1.2 V
   (B) 1.75 V
   (C) 3.5 V
   (D) 4.2 V
   (E) 2.2 V

3. The work function for a certain metal is 3.7 eV. What is the threshold frequency for this metal?

   (A) 9×10^{14} Hz
   (B) 2×10^{15} Hz
   (C) 7×10^{14} Hz
   (D) 5×10^{15} Hz
   (E) 3.5×10^{15} Hz

4. In which kind of waves do the photons have the greatest momentum?

   (A) Radio waves
   (B) Microwaves
   (C) Red light
   (D) Ultraviolet
   (E) X ray

5. What is the momentum of a photon associated with yellow light that has a wavelength of 5500 Å?

   (A) 1.2×10^{-27} kg · m/s
   (B) 1.2×10^{-37} kg · m/s
   (C) 1.2×10^{-17} kg · m/s
   (D) 1.2×10^{-30} kg · m/s
   (E) 5000 kg · m/s

6. What is the de Broglie wavelength for a proton ($m = 1.67 \times 10^{-27}$ kg) with a velocity of 6×10^7 m/s?

   (A) 1.5×10^{14} m
   (B) 1.5×10^{-14} m
   (C) 4.8×10^{-11} m
   (D) 6.6×10^{-15} m
   (E) 3×10^8 m

7. What is the velocity of an electron with a de Broglie wavelength of 3×10^{-10} m?

   (A) 3×10^8 m/s
   (B) 2.4×10^6 m/s
   (C) 7.28×10^4 m/s
   (D) 1.5×10^7 m/s
   (E) 5.7×10^5 m/s

8. Which of the following is a formula for photon momentum?

   (A) $\mathbf{p} = Ec$
   (B) $\mathbf{p} = Ec^2$
   (C) $\mathbf{p} = E/c$
   (D) $\mathbf{p} = mc^2$
   (E) $\mathbf{p} = E + c$

9. Which of the following statements is correct about emission line spectra?

   (A) All of the lines are evenly spaced.
   (B) All elements in the same chemical family have the same spectra.
   (C) Only gases emit emission lines.
   (D) All lines result from discrete energy differences.
   (E) All of the preceding statements are correct.

10. Which of the following is equivalent to the units, joule · seconds, of Planck's constant?

    (A) kg · m/s
    (B) kg · m/s^2
    (C) kg · m^2/s
    (D) kg · m^2/s^2
    (E) kg/s

FREE-RESPONSE

1. (a) Explain the implication of Planck's constant being larger than its present value.

   (b) Explain the basic theory behind electron diffraction.

2. The threshold frequency for calcium is 7.7×10^{14} Hz.

   (a) What is the work function, in joules, for calcium?

   (b) If light of wavelength 2.5×10^{-7} m is incident on calcium, what will be the maximum kinetic energy of the emitted electrons?

   (c) What is the stopping potential for the electrons emitted under the conditions in part (b)?

3. Explain why electrons are diffracted through a crystal.

4. Explain why increasing the intensity of electromagnetic radiation on a photoemissive surface does not affect the kinetic energy of the ejected photoelectrons. Why is this kinetic energy referred to as the "maximum kinetic energy."

5. Why don't we speak of the wavelength nature of large objects such as cars or balls?

ANSWERS EXPLAINED

MULTIPLE-CHOICE PROBLEMS

1. **(B)** If E equals the total detectable energy and E_p equals the energy of a photon, then

$$N = \frac{E}{E_p} = \frac{E}{hf} = \frac{6.63 \times 10^{-15}\text{ J}}{1.326 \times 10^{-17}\text{ J}} = 500$$

2. **(E)** By definition, 1 eV is the amount of energy given to one electron when placed in a potential difference of 1 V. Thus, if $KE_{max} = 2.2$ eV, the stopping potential must be equal to 2.2 V.

3. **(A)** The formula for the work function is $W_0 = hf_0$. The work function must be expressed in units of joules before dividing by Planck's constant:

$$f_0 = \frac{(3.7)(1.6 \times 10^{-19})}{6.63 \times 10^{-34}} = 8.9 \times 10^{14} \text{ Hz}$$

4. **(E)** Since $\mathbf{p} = h/\lambda$, the waves with the smallest wavelength will have the greatest photon momentum. Of the five choices, an X-ray photon has the smallest wavelength and thus the greatest momentum.

5. **(A)** The wavelength must first be converted to meters; then we can use $\mathbf{p} = h/\lambda$:

$$\mathbf{P} = \frac{6.63 \times 10^{-34}}{(5500)(1 \times 10^{-10})} = 1.2 \times 10^{-27} \text{ kg} \cdot \text{m/s}$$

6. **(D)** The formula for the de Broglie wavelength is $\lambda = h/m\mathbf{v}$. Substituting the known values gives

$$\lambda = \frac{6.63 \times 10^{-34}}{(1.67 \times 10^{-27})(6 \times 10^{7})} = 6.6 \times 10^{-15} \text{ m}$$

7. **(B)** Using the de Broglie formula for wavelength gives $\mathbf{v} = h/m\lambda$. Substituting the known values, we obtain

$$\mathbf{v} = \frac{6.63 \times 10^{-34}}{(9.1 \times 10^{-31})(3 \times 10^{-10})} = 2.4 \times 10^{6} \text{ m/s}$$

8. **(C)** We have that $\mathbf{p} = h/\lambda$ and that $E = hf = hc/\lambda$. Thus, $E = \mathbf{p}c$, and so $\mathbf{p} = E/c$.

9. **(D)** Spectral emission lines are different for every atom or molecule, and the lines are not evenly spaced. They do, however, arise from energy difference transitions as discussed in Chapter 23.

10. **(C)** The units of Planck's constant are joule · seconds; and since 1 J = 1 kg · m^2/s^2, the units for h are kilogram · square meters per second (kg · m^2/s). These units are also the units for angular momentum!

FREE-RESPONSE PROBLEMS

1. (a) If Planck's constant were larger than its present value, an implication would be that the uncertainty principle, $\Delta\mathbf{p}\,\Delta x \geq h/2\pi$, would apply to a domain of larger magnitude and we would see quantum effects occur at larger sizes as well.

   (b) Electron diffraction is based on the wave theory of matter. Since an electron has a sufficiently small mass, its de Broglie wavelength is large enough to be diffracted by small openings and thus to exhibit seemingly common wave properties. Scanning electron microscopes, although much more complex in theory, use the same basic principle.

2. (a) The work function is given by

$$w_0 = hf_0 = (6.63 \times 10^{-34})(7.7 \times 10^{14}) = 5.1 \times 10^{-19}\ \text{J}$$

(b) The frequency associated with a wavelength of 2.5×10^{-7} m is given by

$$f = \frac{c}{\lambda} = \frac{3 \times 10^{8}}{2.5 \times 10^{-7}} = 1.2 \times 10^{15}\ \text{Hz}$$

The energy of the associated photon is equal to

$$E = hf = 7.956 \times 10^{-19}\ \text{J}$$

The maximum kinetic energy of the emitted electrons is equal to the difference between this energy and the work function:

$$\text{KE}_{\text{max}} = 7.956 \times 10^{-19} - 5.1\ 10^{-19} = 2.856 \times 10^{-19}\ \text{J}$$

(c) This maximum kinetic energy is equal to the product of the electron charge and the stopping potential. Thus:

$$V_0 = \frac{\text{KE}_{\text{max}}}{e} = \frac{2.856 \times 10^{-19}}{1.6 \times 10^{-19}} = 1.785\ \text{V}$$

3. The de Broglie wavelength of electrons is comparable to the spacing between molecules in the lattice structure of a crystal. Thus, the electrons have a wave-like characteristic, and the crystal acts like a diffraction grating.

4. In the quantum theory of light, each photon interacts with one, and only one, electron. Thus, if the energy of a photon is based on its frequency, the intensity of light is just the number of photons. Thus, each electron emerges with the same kinetic energy. This is true for 1 or 1 billion electrons. Therefore, we refer to this kinetic energy as the "maximum kinetic energy."

5. Using the de Broglie formula (see sample problems), it is easy to see that the wavelength of a moving ball or car is much too small to be detected or have any significant wavelike interactions.

CHAPTER 23

The Atom

KEY CONCEPTS

- Atomic Structure and Rutherford's Model
- The Bohr Model
- Quantum Mechanics and the Electron Cloud Model

23.1 ATOMIC STRUCTURE AND RUTHERFORD'S MODEL

In 1896, French physicist Henri Becquerel was studying the phosphorescent properties of uranium ores. Certain rocks glow after being exposed to sunlight, and many physicists at that time were examining the light emanations from these ores.

One day, while engaged in such a study, Becquerel placed a piece of uranium ore in a drawer on top of some sealed photographic paper. Several days later, he found to his surprise that the paper had been exposed even though no light was incident on it! His conclusion was that radiation emitted by the uranium had penetrated the sealed envelope and exposed the paper.

Subsequent work by Marie and Pierre Curie revealed that this natural radioactivity, as they called it, came from the uranium itself and was a consequence of its instability as an atom. Also, Ernest Rutherford discovered (see Figure 23.1) that the radiation emitted yielded particles that were useful for further atomic studies.

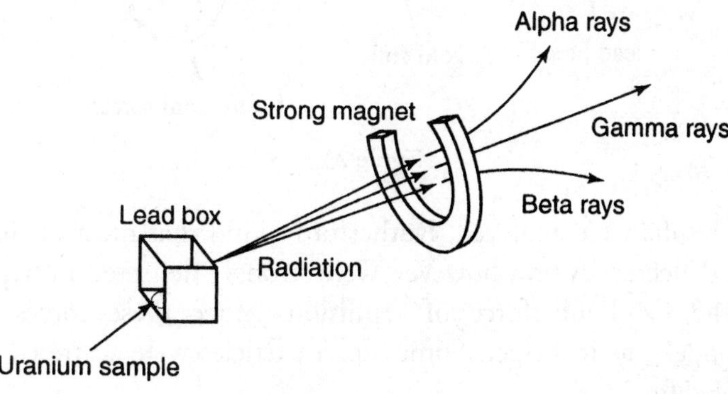

Figure 23.1

When passed through a magnetic field, the emitted radiation split into three separate parts, called **alpha rays**, **beta rays**, and **gamma rays**. Analysis of their trajectories led to the conclusions that alpha particles, later identified as helium nuclei, are positively charged; beta particles, later identified as electrons, are negatively charged; and gamma rays, later identified to be photons, are electrically neutral.

J. J. Thomson, using the data available at the end of the nineteenth century, devised an early model of the atom based on the fact that all atoms are electrically neutral. After concluding that the atom is positively charged and that electrons are also associated with it, Thomson's model proposed that the atom consisted of a relatively large, uniformly distributed, positive mass with negatively charged electrons embedded in it like raisins in a pudding (see Figure 23.2).

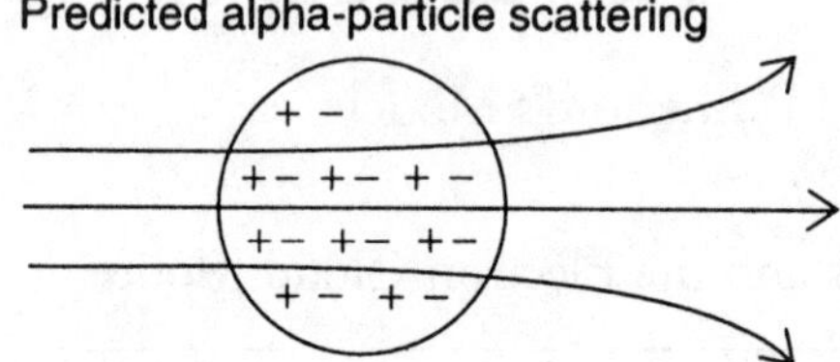

Figure 23.2 Thomson's Model of the Atom

In 1911, Rutherford and his co-workers decided to test Thomson's model using alpha-particle scattering. In the Thomson model, the positive charge was uniformly distributed. Since the negative electrons were embedded inside the nucleus, the Coulomb force of attraction would weaken the deflecting force on a passing positively charged alpha particle.

Rutherford's procedure was to aim alpha particles at a thin metal foil (he used gold) and observe the scattering pattern on a zinc sulfide screen (Figure 23.3).

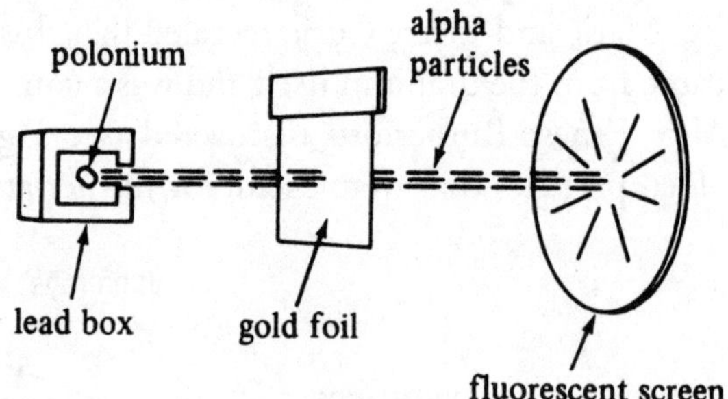

Figure 23.3

When the results were analyzed, Rutherford found that most of the alpha particles were not deflected. A few, however, were strongly deflected in hyperbolic paths because of the Coulomb force of repulsion. Since these forces were strong, Thomson's model was incorrect. Some alpha particles were scattered back close to 180 degrees (Figure 23.4).

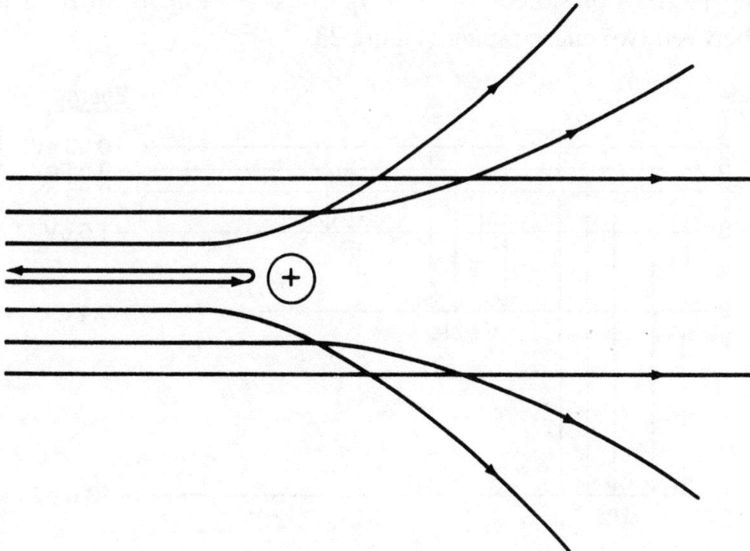

Figure 23.4 Alpha Particle Scattering Trajectories

Rutherford's conclusion was that the atom consisted mostly of empty space and that the nucleus was a very small, densely packed, positive charge. The electrons, he proposed, orbited around the nucleus like planets orbiting the Sun.

Rutherford's model, however, had several major problems, including the fact that it could not account for the appearance of discrete emission line spectra. In Rutherford's model the electrons continuously orbited around the nucleus. This circular, "accelerated" motion should produce a continuous band of electromagnetic radiation, but it did not. Additionally, the predicted orbital loss of energy would cause an atom to disintegrate in a very short time and thus break apart all matter. This phenomenon, too, did not occur.

23.2 THE BOHR MODEL

Niels Bohr, a Danish physicist, decided in 1913 to study the hydrogen spectral lines again. Years earlier, Einstein had proposed a quantum theory of light, developing the concept of a photon and completely explaining the photoelectric effect.

Bohr observed that the Balmer formula for the hydrogen spectrum could be viewed in terms of energy differences. He then realized that these energy differences involved a new set of postulates about atomic structure. The main ideas of his theory are these:

1. Electrons do not emit electromagnetic radiation while in a given orbit (energy state).
2. Electrons cannot remain in any arbitrary energy state. They can remain only in discrete or quantized energy states.
3. Electrons emit electromagnetic energy when they make transitions from higher energy states to lower energy states. The lowest energy state is the **ground state**.

According to the Bohr theory, the energy of a photon is equal to the energy difference between two energy states (Figure 23.5).

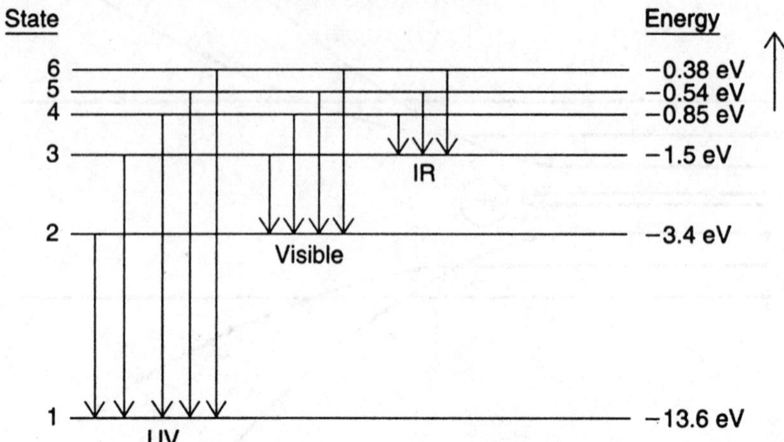

Figure 23.5 Energy Level Diagram for Hydrogen

The different spectral series arise from transitions from higher energy states to specified lower energy states. The ultraviolet Lyman series occurs when electrons fall to level 1; the visible Balmer series, when electrons fall to level 2; the infrared Paschen series, when electrons fall to level 3. Thus, we can write that the energy, in joules, for an emitted photon is given by

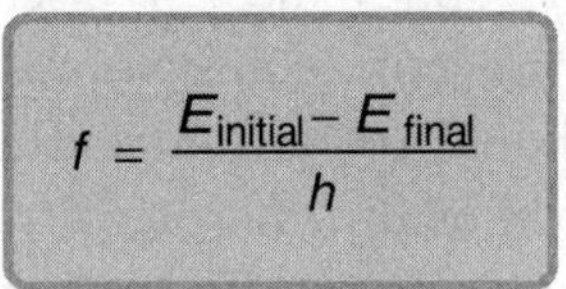

$$f = \frac{E_{\text{initial}} - E_{\text{final}}}{h}$$

The energy level diagram for hydrogen is shown in Figure 23.5. The energy in a given state represents the amount of energy, called the **ionization energy**, needed to free an electron from the atom. A charged atom is termed an **ion**. Electron ionization energies were confirmed by experiments in 1914. The apparatus used by James Franck and Gustav Hertz is shown in Figure 23.6.

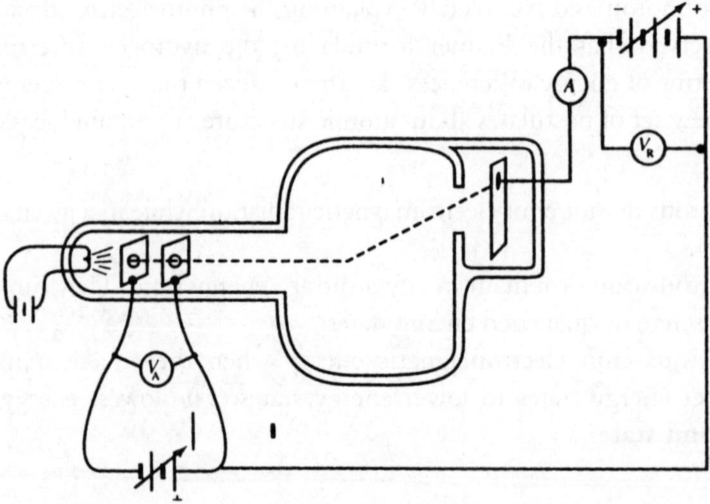

Figure 23.6

Although we designate the levels mathematically as negative so that the ionization level will have a value of zero, the energies are positive quantities. Even though hydrogen has only one electron, the many atoms of hydrogen in a sample of this gas, coupled with the different probabilities of electrons being in any one particular state, produce the multiple lines that appear in the visible spectrum when the gas is "excited."

SAMPLE PROBLEM

What is the wavelength of a photon emitted when an electron makes a transition in a hydrogen atom from level 5 to level 3?

Solution

From level 5 to level 3 involves an energy difference of

$$E = (-0.54\text{ eV}) - (-1.50\text{ eV}) = 0.96\text{ eV} = 1.536 \times 10^{-19}\text{ J}$$

Now, the frequency of the photon will be equal to

$$f = \frac{1.536 \times 10^{-19}}{6.63 \times 10^{-34}} = 2.3 \times 10^{14}\text{ Hz}$$

The wavelength of this photon is given by

$$\lambda = \frac{c}{f} = \frac{3 \times 10^{8}}{2.3 \times 10^{14}} = 1.29 \times 10^{-6}\text{ m}$$

(which is in the infrared part of the electromagnetic spectrum).

It is important to remember that an electron will make a transition from a lower to a higher energy state only when it absorbs a photon equal to a given possible ("allowed") energy difference. This is the nature of the quantization of the atom; the electron can exist only in a specified energy state. To change states, the electron must either absorb or emit a photon whose energy is proportional to Planck's constant and equal to the energy difference between two energy states.

The Heisenberg uncertainty principle, discussed in Chapter 22, introduces the element of "probability" in describing the nature of electron orbits and energy states. This subject is studied in the science of **quantum mechanics**, developed by Erwin Schrödinger and Werner Heisenberg.

23.3 QUANTUM MECHANICS AND THE ELECTRON CLOUD MODEL

The success of the Bohr model was limited to the hydrogen atom and certain hydrogen-like atoms (atoms with one electron in the outermost shell; single-ionized helium is an example). The wave nature of matter led to a problem in predicting the behavior of an electron in a strict Bohr energy state.

In 1926, physicist Erwin Schrödinger developed a new mechanics of the atom, then called **wave mechanics**, but later termed **quantum mechanics**. In the wave mechanics of Schrödinger, the position of an electron was determined by a mathematical function called the **wave function**. This wave function was related to the probability of finding an electron in any one energy state. A **wave equation**, called the **Schrödinger equation**, was derived using ideas about time-dependent motion in classical mechanics.

The energy states of Bohr were now replaced by a probability "cloud." In this **cloud model** the electrons are not necessarily limited to specified orbits. A cloud of uncertainty is produced, with the densest regions corresponding to the highest probability of an electron being in a given state.

The use of probability to explain the structure of the atom led to intense debates between the proponents of the new quantum mechanics and one of its fiercest opponents, Albert Einstein. Einstein, who had helped to develop the quantum theory of light 20 years earlier, was not convinced that the universe could be determined by probability. To paraphrase, Einstein said he could not believe that God would play dice with the Universe. "God is subtle," he said, "but not malicious."

At about the same time that Schrödinger introduced his wave mechanics, Werner Heisenberg developed his **matrix mechanics** for the atom. Instead of using a wave analogy, Heisenberg developed a purely abstract, mathematical theory using matrices. Working with Heisenberg, Wolfgang Pauli advanced the so-called **Pauli exclusion principle**, which states that two electrons having the same spin orientations cannot occupy the same quantum state at the same time.

The new **quantum mechanics** was born when the two rival theories were shown to be equivalent. Electron spin was experimentally verified, and the theory proved to be more successful than the older Bohr theory in explaining atomic and molecular structures and leading to new ideas about solid-state physics and the subsequent development of lasers.

Chapter Summary

- Ernest Rutherford used alpha particles scattering off of gold foil to show that the atom consists of a small positively charged nucleus surrounded by negatively charged electrons.
- Niels Bohr developed a theory to explain spectral line emission as well as the predicted loss of electron energy according to classical radiation theory.
- In an atom, according to Bohr, electrons exist in certain discrete energy states. The lowest energy state is called the ground state. Spectral lines are emitted as electrons fall from higher to lower energy states.
- The visible hydrogen spectral lines (Balmer series) form as electrons fall from higher levels to level two.
- The frequency of emitted photons in a spectral series is proportional to the energy difference between levels ($hf = E_f - E_i$).
- In the quantum mechanical model developed by Erwin Schrödinger, Werner Heisenberg, and others, the atom is surrounded by a probability cloud of electrons due to the uncertainty of locating the exact location and momentum of the electrons (Heisenberg's uncertainty principle).

Problem-Solving Strategies for the Atom

Understanding Bohr's postulates is a key element in solving atomic structure and spectra problems. Semiclassical ideas about Coulomb's law, energy, and uniform circle motion were used by Bohr.

You should either memorize the energy level diagram for hydrogen or else remember that the levels are inversely proportional to the square of the principal quantum number n and directly proportional to the factor −13.6 eV. However, in a frequency or wavelength calculation, you must convert these energies back to joules.

Practice Exercises

MULTIPLE-CHOICE

1. How much energy is needed to ionize a hydrogen atom in the $n = 4$ state?

   (A) 0.85 eV
   (B) 13.6 eV
   (C) 12.75 eV
   (D) 10.2 eV
   (E) 3.4 eV

2. Which electron transition will emit a photon with the greatest frequency?

   (A) $n = 1$ to $n = 4$
   (B) $n = 5$ to $n = 2$
   (C) $n = 3$ to $n = 1$
   (D) $n = 7$ to $n = 3$
   (E) $n = 1$ to $n = 3$

3. An electron in the ground state of a hydrogen atom can absorb a photon with any of the following energies except

   (A) 10.2 eV
   (B) 12.1 eV
   (C) 12.5 eV
   (D) 12.75 eV
   (E) 13.06 eV

4. What is the frequency of a photon emitted in an electron transition from $n = 3$ to $n = 1$ in a hydrogen atom?

   (A) 1.83×10^{34} Hz
   (B) 2.92×10^{15} Hz
   (C) 1.7×10^{15} Hz
   (D) 4.7×10^{14} Hz
   (E) 3.3×10^{15} Hz

5. Which of the following statements about the atom is correct?

(A) Orbiting electrons can sharply deflect passing alpha particles.
(B) The nucleus of the atom is electrically neutral.
(C) The nucleus of the atom deflects alpha particles into parabolic trajectories.
(D) The nucleus of the atom contains most of the atomic mass.
(E) None of the preceding statements is correct.

6. Which of the following statements about the Bohr theory reflect how it differs from classical predictions about the atom?

I. An electron can orbit without a net force acting.
II. An electron can orbit about a nucleus.
III. An electron can be accelerated without radiating energy.
IV. An orbiting electron has a quantized angular momentum that is proportional to Planck's constant.

(A) I and II
(B) I and III
(C) III and IV
(D) II and IV
(E) I and IV

7. What is the maximum angle by which an alpha particle can be deflected by a nucleus?

(A) 30°
(B) 60°
(C) 90°
(D) 135°
(E) 180°

8. An electron makes a transition from a higher energy state to the ground state in a Bohr atom. As a result of this transition

(A) the total energy of the atom is increased
(B) the force of attraction on the electron is increased
(C) the energy of the ground state is increased
(D) the charge on the electron is increased
(E) the electron's energy remains the same

9. How many different photon frequencies can be emitted if an electron is in excited state $n = 4$ in a hydrogen atom?

(A) 1
(B) 3
(C) 5
(D) 6
(E) 8

10. A radioactive atom emits a gamma ray photon. As a result
    (A) the energy of the nucleus is decreased
    (B) the charge in the nucleus is decreased
    (C) the ground-state energy is decreased
    (D) the force on an orbiting electron is decreased
    (E) None of the preceding statements is correct.

FREE-RESPONSE

1. In Rutherford's model of the atom, what happens to the frequency of the emitted electron radiation as the electron spirals in its orbit?

2. If a source of light from excited hydrogen gas is moving toward or away from an observer, what changes, if any, are observed in the emission spectral lines?

3. A hydrogen atom consists of only one electron. Explain why the emission spectrum of excited hydrogen gas shows the presence of many different spectral lines.

ANSWERS EXPLAINED

MULTIPLE-CHOICE PROBLEMS

1. **(A)** In the $n = 4$ state, the energy level is -0.85 eV. Thus, to ionize the electron in this state, $+0.85$ eV must be supplied.

2. **(C)** Transitions to level 1 produce ultraviolet photons that have frequencies greater than those produced by transitions to other levels.

3. **(C)** An electron in the ground state can absorb a photon that has an energy equal to the energy difference between the ground state and any other level. Only 12.5 eV is not such an energy difference.

4. **(B)** The energy difference from level 3 to level 1 is 12.1 eV. The emitted frequency is

$$f = \frac{E}{h} = \frac{(12.1)(1.6 \times 10^{-19})}{6.63 \times 10^{-34}} = 2.92 \times 10^{15} \text{ Hz}$$

5. **(D)** "The nucleus of the atom contains most of the atomic mass" is a correct statement.

6. **(C)** The Bohr theory challenged classical theories by stating that an electron can be accelerated without radiating energy (statement III) and that its angular momentum is quantized and proportional to Planck's constant (statement IV).

7. **(E)** Rutherford's experiments revealed that some alpha particles are backscattered through angles as large as 180°.

8. **(B)** As an electron gets closer to the nucleus, the force of attraction on the electron increases.

9. **(D)** There are six possible photon frequencies emitted from state $n = 4$: (1) $n = 4$ to $n = 1$, (2) $n = 4$ to $n = 3$, (3) $n = 4$ to $n = 2$, (4) $n = 3$ to $n = 2$, (5) $n = 3$ to $n = 2$, (6) $n = 2$ to $n = 1$.

10. **(A)** Since photons do not have any rest mass, their emission releases nuclear energy only.

FREE-RESPONSE PROBLEMS

1. According to the Rutherford model, electrons emit continuous radiation while orbiting the nucleus. The appearance of emission spectral lines refutes this. Additionally, the loss of energy would make the electrons spiral into the nucleus, which would be accompanied by a shift in frequency. As electrons spiral in, the frequency would increase and move toward the violet part of the spectrum. This does not occur in nature.

2. If the source of the radiation is moving toward or away from an observer, the spectral lines are shifted according to the Doppler effect. In stars, we see redshifts as the stars move away from Earth, and blueshifts as they move toward us.

3. The different spectral lines we observe from excited hydrogen gas are due to the many atoms of hydrogen in different excited states simultaneously.

CHAPTER 24

The Nucleus

KEY CONCEPTS

- Nuclear Structure and Stability
- Binding Energy
- Radioactive Decay
- Fission
- Fusion

24.1 NUCLEAR STRUCTURE AND STABILITY

The experiments of Hans Geiger and Ernest Marsden (under Rutherford's supervision) in 1911 confirmed that the nucleus of an atom is compact and positively charged. Experiments on nuclear masses using a **mass spectrograph**, which deflects the positively charged nuclei into curved paths with different radii, confirmed that the estimate of nuclear mass was too low. The nucleus could not, therefore, be made entirely of positively charged protons (see Figure 24.1).

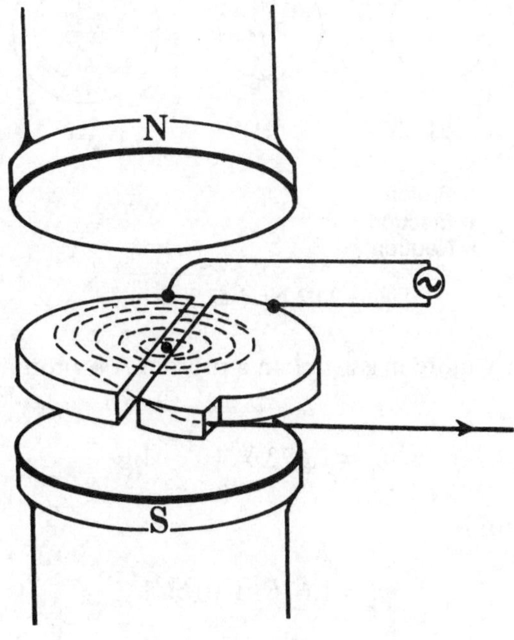

Figure 24.1

Physicists theorized that another electrically neutral particle, called the **neutron**, must be present since (a) all of the charge on the nucleus could be accounted for by the presence of the protons, (b) more neutral mass was needed, and (c) a method for countering the repulsive Coulomb forces created by the protons was required.

Thus, the question of nuclear stability depended on the particular structure of the nucleus. The work on radioactive decay by Rutherford and Marie and Pierre Curie confirmed that the radioactive disintegration of an atom is a nuclear phenomenon and is tied to stability as well.

The basic structure of the atom, using a Bohr-Rutherford planetary model, consists of a nucleus with a certain number of protons and neutrons (together called **nucleons**). The number of protons characterizes an element since no two elements have the same number of protons. This **atomic number**, Z, also measures the relative charge on the nucleus. The **mass number**, A, is a measure of the total number of nucleons (protons plus neutrons) in the nucleus. Both of these numbers are whole numbers, but the actual nuclear mass is not a whole number.

Since an atom is electrically neutral, the initial numbers of protons and electrons are the same. Schematically, an atom is designated by a capital letter, possibly followed by one lowercase letter, indicating the element it represents, with the two nuclear numbers A and Z written as a superscript and a subscript, respectively, on the left-hand side. For an unknown element X we write ${}^{A}_{Z}\mathrm{X}$. The designation ${}^{1}_{1}\mathrm{H}$ represents the element hydrogen, with one nucleon (a proton) and an atomic number of 1. Some elements have **isotopes**, that is, nuclei with the same number of protons but different numbers of neutrons. For example, the element helium is designated as ${}^{4}_{2}\mathrm{He}$, where the 4 indicates 2 protons and 2 neutrons. An isotope of this atom is helium-3, designated as ${}^{2}_{3}\mathrm{He}$. Some nuclear structures are illustrated in Figure 24.2.

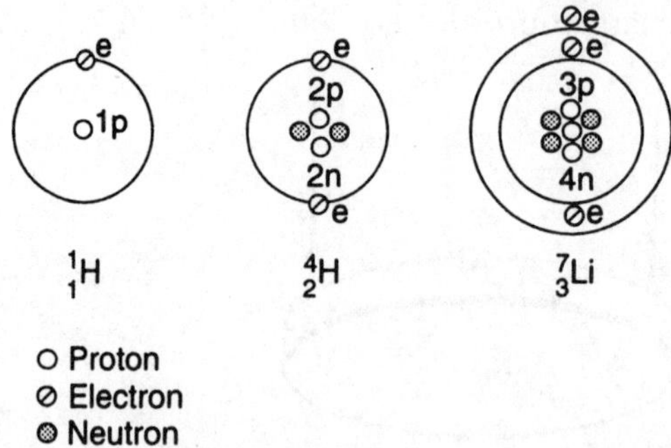

Figure 24.2 Nuclear Structures

A neutron is slightly more massive than a proton. A proton mass is approximately equal to

$$m_\mathrm{p} = 1.673 \times 10^{-27} \text{ kg}$$

and a neutron mass to

$$m_\mathrm{n} = 1.675 \times 10^{-27} \text{ kg}$$

Nuclear masses are usually expressed in **atomic mass units** (u). One atomic mass unit is defined to be equal to one-twelfth the mass of a carbon-12 isotope:

$$1\ \text{u} = 1.66 \times 10^{-27}\ \text{kg}$$

On this scale, the proton's mass is given as 1.0078 atomic mass units, and the neutron's mass is given as 1.0087 atomic mass units.

If one looks at all stable isotopes of known nuclei (see Figure 24.3), something qualitatively interesting emerges. Experiments show that lighter nuclei have about the same number of protons as neutrons. However, heavier stable nuclei have more neutrons than protons. This suggests that the neutrons must somehow help overcome the repulsive forces in a tightly packed positive nucleus. Also, this **strong nuclear force** must operate within the short distances in an atomic nucleus. Unstable, radioactive nuclei lie off this stability curve (see Figure 24.2); and as they decay, the alpha, beta and other particle emissions **transmute** the original nucleus into a more stable one.

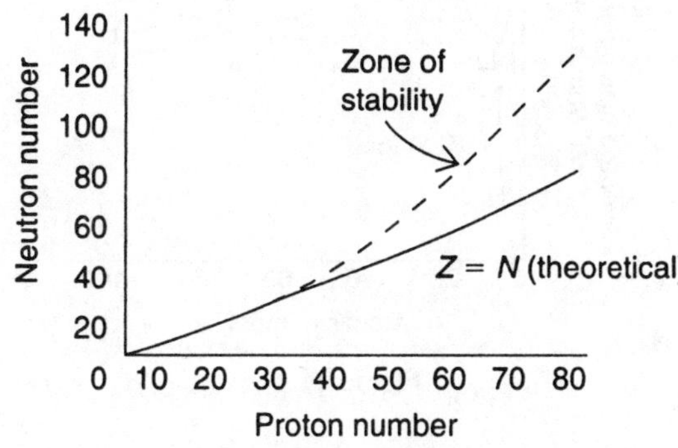

Figure 24.3 Proton-Neutron Plot

24.2 BINDING ENERGY

If we were to determine the total mass of a particular isotope, say $^{4}_{2}\text{He}$, based on the number of nucleons, we would arrive at a nucleus that was more massive than in reality. For example, the helium-4 nucleus contains two protons (each with a mass of 1.0078 u) and two neutrons (each with a mass of 1.0087 u). The total predicted mass would be 4.0330 u. Experimentally, however, the mass of the helium-4 nucleus is found to be 4.0026 u. The difference of 0.0304 u, called the **mass defect**, is proportional to the **binding energy** of the nucleus.

Using Einstein's special theory of relativity, we know that mass can be converted to energy. The energy equivalent of 1 atomic mass unit is equal to 931.5 million electron volts (MeV). The binding energy of any nucleus is therefore determined by using the relationship

$$\text{BE (MeV)} = \text{Mass defect} \times 931.5\ \text{MeV/u}$$

In our example, the binding energy of the helium-4 nucleus is

$$\text{BE} = 0.0304 \times 931.5 = 28.3\ \text{MeV}$$

It is useful to think about the average binding energy per nucleon, which is equal to BE/A, where A = the mass number of the element. Thus, for helium-4, where $A = 4$,

$$\text{Avg BE} = \frac{28.3}{4} = 7.08 \text{ MeV per nucleon}$$

In Figure 24.4, a graph of average binding energy per nucleon is plotted against atomic number Z. The graph shows that a maximum point is reached at about $Z = 26$ (iron). The greater the average binding energy, the greater the stability of the nucleus. Thus, since hydrogen isotopes have less stability than helium isotopes, the **fusion** of hydrogen into helium increases the stability of the nucleus through the release of energy. Likewise, the unstable uranium isotopes yield more stable nuclei if they are split **(fission)** into lighter nuclei (also releasing energy).

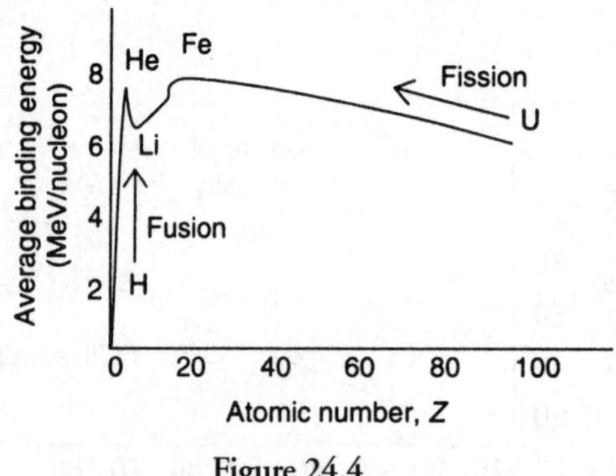

Figure 24.4

24.3 RADIOACTIVE DECAY

Radioactivity was discovered in 1895 by Henri Becquerel, and the relationship between radioactivity and nuclear stability was studied by Pierre and Marie Curie. Ernest Rutherford pioneered the experiments of nuclear bombardment and introduced the word **transmutation** into the nuclear physics vocabulary. The discovery of alpha- and beta-particle emission from nuclei led to the use of these particles to probe the structure of the nucleus in the first decades of the twentieth century. Subsequent research showed the existence of positively charged electrons called **positrons**, and even negatively charged protons!

The question of how these particles could originate from a nucleus was answered using Einstein's theory of mass-energy equivalence ($E = mc^2$). Schematically, we have the following designations for some of these particles:

Proton (hydrogen nucleus): ${}^{1}_{1}\text{H}$
Alpha particle (helium nucleus): ${}^{4}_{2}\text{He}$
Beta particle (electron): ${}^{0}_{-1}\text{e}$
Neutron: ${}^{1}_{0}\text{n}$
Positron (positive electron): ${}^{0}_{1}\text{e}$
Photon: ${}^{0}_{0}\gamma$

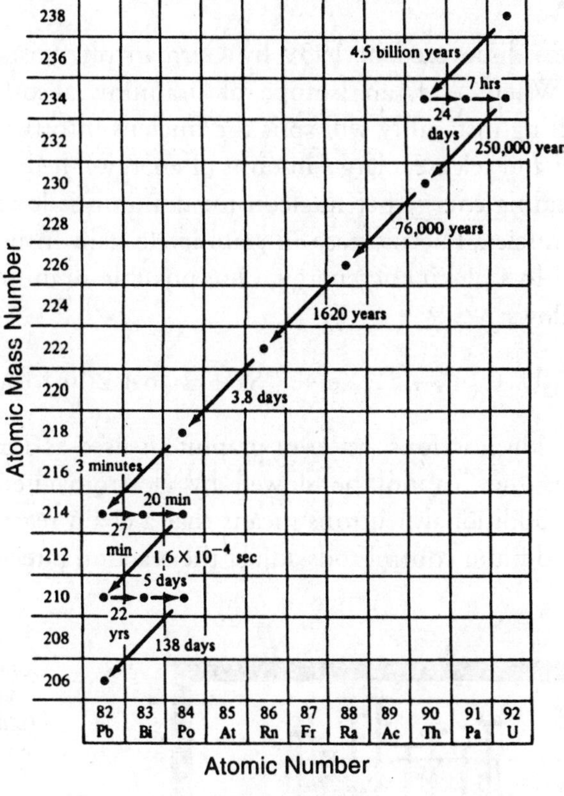

Figure 24.5

When $^{238}_{92}U$, an isotope of uranium, decays, it emits an alpha particle according to the following equation:

$$^{238}_{92}U \rightarrow {}^{234}_{90}Th + {}^{4}_{2}He$$

When, in turn, the radioactive thorium isotope decays, it emits a beta particle (see Figure 24.5).

$$^{234}_{90}Th \rightarrow {}^{234}_{91}Pa + {}^{0}_{-1}e$$

In the first example, both the atomic number and the mass number changed as the uranium transmuted into thorium. In the decay of the thorium nucleus, however, only the atomic number changed. The reason is that in beta decay a neutron in the nucleus is converted into a proton, an electron, energy, and a particle called an **antineutrino**.

In addition to alpha and beta decay, there are several other scenarios for nuclear disintegration. In **electron capture**, an isotope of copper captures an electron into its nucleus and transmutes into an isotope of nickel:

$$^{64}_{29}Cu + {}^{0}_{-1}e \rightarrow {}^{64}_{28}Ni$$

In **positron emission**, the same isotope of copper decays by emitting a positron and transmutes into the same isotope of nickel:

$$^{64}_{29}Cu \rightarrow {}^{0}_{-1}e + {}^{64}_{28}Ni$$

24.4 FISSION

Nuclear fission was discovered in 1939 by German physicists Otto Hahn and Fritz Strassman. When $^{235}_{92}U$, an isotope of uranium, absorbs a slow-moving neutron, the resulting instability will split the nucleus into two smaller (but still radioactive) nuclei and release a large amount of energy. The curve in Figure 24.4 shows that the binding energy per nucleon for uranium is lower than the values for other, lighter nuclei. The increase in stability (a gain in potential energy) is accompanied by a loss of kinetic energy. One possible uranium fission reaction could occur as follows:

$$^{235}_{92}U + ^{1}_{0}n \rightarrow ^{140}_{54}Xe + ^{94}_{38}Sr + 2\ ^{1}_{0}n + 200\ \text{MeV}$$

In controlled fission reactions, water or graphite is used as a **moderator** to slow the neutrons since they cannot be slowed by electromagnetic processes. The production of two additional neutrons means that a chain reaction is possible. In a fission reactor, cadmium control rods adjust the reaction rates to desirable levels. See Figure 24.6.

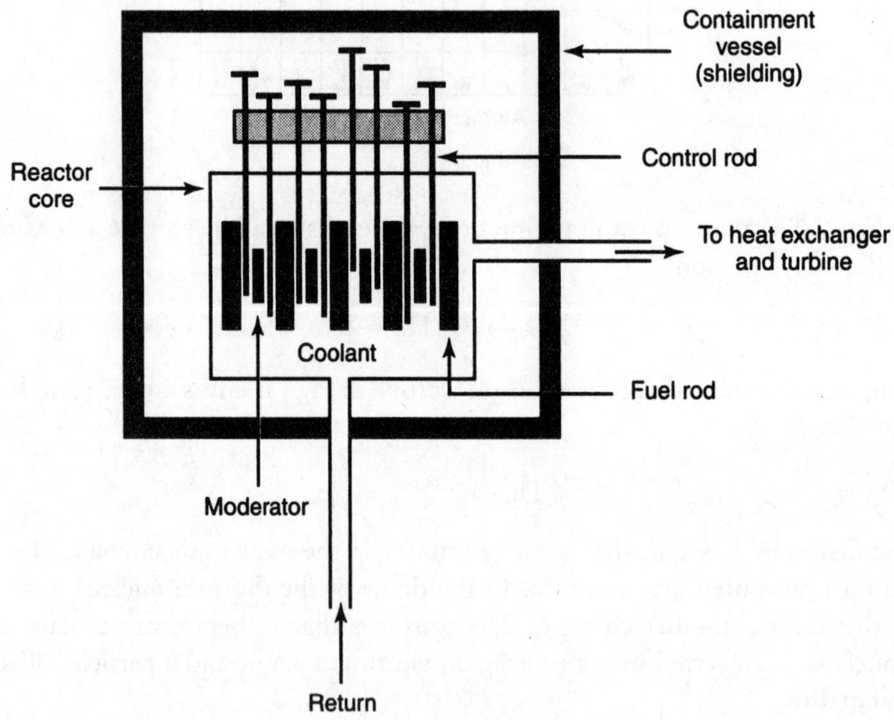

Figure 24.6

In the production of plutonium as a fissionable material, another isotope of uranium, $^{238}_{92}U$, is used. First, the uranium absorbs a neutron to form neptunium, a rare element. The neptunium then decays to form plutonium:

$$^{238}_{92}U + ^{1}_{0}n \rightarrow ^{239}_{92}U \rightarrow ^{239}_{93}Np + ^{0}_{-1}e$$

$$^{239}_{93}Np \rightarrow ^{239}_{94}Pu + ^{0}_{-1}e$$

24.5 FUSION

The binding energy curve (Figure 24.4) indicates that, if hydrogen can be fused into helium, the resulting nucleus will be more stable and energy can be released. In a fusion reaction, isotopes of hydrogen known as deuterium ($^2_1$H) and tritium ($^3_1$H) are used. The following equations represent one kind of hydrogen fusion reaction:

$$^1_1\text{H} + {}^1_1\text{H} \rightarrow {}^2_1\text{H} + {}^0_1\text{e} + 0.4 \text{ MeV}$$
$$^1_1\text{H} + {}^2_1\text{H} \rightarrow {}^3_2\text{He} + 5.5 \text{ MeV}$$
$$^3_2\text{He} + {}^3_2\text{He} \rightarrow {}^4_2\text{He} + 2\ {}^1_1\text{H} + 12.9 \text{ MeV}$$

In this reaction, hydrogen is first fused into deuterium. The deuterium is then fused with hydrogen to form the isotope helium-3. Finally, the helium-3 isotope is fused into the stable element helium.

In a different reaction, the isotope tritium is fused with deuterium to form the stable element helium in one step:

$$^2_1\text{H} + {}^3_1\text{H} \rightarrow {}^4_2\text{He} + {}^1_0\text{n} + 17.6 \text{ MeV}$$

Chapter Summary

- The nucleus of an atom consists of neutrons and protons.
- The number of protons (its charge) is called the atomic number.
- The number of protons and neutrons is called the atomic mass.
- Protons and neutrons are collectively referred to as nucleons.
- Isotopes of nuclei contain the same number of protons, but a different number of neutrons.
- For light nuclei, the ratio of the number of protons to the number of neutrons is almost one-to-one. As atomic numbers increase, there tend to be more neutrons than protons in stable isotopes.
- The binding energy of the nucleus is found from Einstein's equation $E = mc^2$. It is the energy needed to bind the nucleus together and is observed because the actual mass of a nucleus is less than the additive sum of its constituent nucleons.
- The average binding energy per nucleon is the binding energy divided by the total number of nucleons in a given isotope.
- Iron has the largest average binding energy per nucleon.
- Radioactive decay is the name given to unstable isotopes (such as uranium), which emit particles and photons.
- Alpha decay is the process by which unstable nuclei emit helium nuclei.
- Beta decay is the process by which unstable nuclei emit electrons.
- Gamma decay is the process by which unstable nuclei emit photons.

- Decay processes will continue until a stable nucleus (like lead-206) is produced.
- Fission is the process by which a heavy unstable nucleus (like uranium) is split into two smaller daughter nuclei by the absorption of a slow-moving neutron.
- The speed of the neutron can be controlled by a moderator such as graphite or water.
- A fission reaction can by controlled using rods made of cadmium.
- Fusion is the process by which two light nuclei (such as deuterium) are fused into a more stable isotope (such as helium).
- Fusion does not produce radioactive by-products as fission does.
- The Sun and other stars are powered by the fusion of hydrogen into helium. Older stars may fuse helium into other, heavier elements.

Problem-Solving Strategies for the Nucleus

Solving nuclear equations requires that you remember two rules. First, make sure that you have conservation of charge. That means that all of the atomic numbers, which measure the number of protons (and hence the charge) should add up on both sides. However, to get the actual charge, you must multiply the atomic number by the proton charge in units of coulombs.

The second rule to consider is the conservation of mass. The mass numbers that appear as superscripts should add up. Any missing mass should be accounted for by the appearance of energy (sometimes designated by the letter Q). The mass energy equivalence is that 1 u is equal to 931.5 MeV. This second rule applies to problems involving binding energy and mass defect. Remember that an actual nucleus has less mass than the sum of its nucleons!

Practice Exercises

MULTIPLE-CHOICE

1. How many neutrons are in a nucleus of $^{213}_{84}Po$?

   (A) 84
   (B) 129
   (C) 213
   (D) 297
   (E) 546

2. The nitrogen isotope $^{13}_{7}N$ emits a beta particle as it decays. The new isotope formed is

   (A) $^{14}_{7}N$
   (B) $^{13}_{6}C$
   (C) $^{13}_{8}O$
   (D) $^{14}_{8}O$
   (E) Not enough information is provided.

3. Which radiation has the greatest ability to penetrate matter?

   (A) X ray
   (B) Alpha particle
   (C) Gamma ray
   (D) Beta particle
   (E) Positron

4. As a radioactive material undergoes nuclear disintegration,

   (A) its atomic number always remains the same
   (B) its mass number never increases
   (C) its mass number never decreases
   (D) it always emits an alpha particle
   (E) None of the preceding statements is correct.

5. If all of the following particles had the same velocity, which one would have the shortest wavelength?

   (A) Alpha particle
   (B) Beta particle
   (C) Neutron
   (D) Proton
   (E) Positron

6. Which of the following statements is correct about nuclear isotopes?

   (A) They all have the same binding energy.
   (B) They all have the same mass number.
   (C) They all have the same number of nucleons.
   (D) They all have the same number of protons.
   (E) All of these statements are correct.

7. In the following nuclear reaction, an isotope of aluminum is bombarded by alpha particles:

$$^{27}_{13}\text{Al} + ^{4}_{2}\text{He} \rightarrow ^{30}_{15}\text{P} + \text{Y}$$

   Quantity Y must be

   (A) a neutron
   (B) an electron
   (C) a positron
   (D) a photon
   (E) a proton

8. When a gamma-ray photon is emitted from a radioactive nucleus, which of the following occurs?

   (A) The nucleus goes to an excited state.
   (B) The nucleus goes to a more stable state.
   (C) An electron goes to an excited state.
   (D) The atomic number of the nucleus decreases.
   (E) The number of nucleons decreases.

FREE-RESPONSE

1. Given the isotope $^{56}_{26}Fe$, which has an actual mass of 55.934939 u:
   (a) Determine the mass defect of the nucleus in atomic mass units.
   (b) Determine the average binding energy per nucleon in units of million electron volts.

2. Why do heavier stable nuclei have more neutrons than protons?

3. Why are the conditions for the fusion of hydrogen into helium favorable inside the core of a star?

4. Explain the radioactive disintegration series of uranium-238 into stable lead-206 in terms of the neutron-proton plot.

5. Albert Einstein, in his special theory of relativity, stated that energy and mass were related by the expression $E = mc^2$. Explain how the concept of binding energy confirms this claim.

ANSWERS EXPLAINED

MULTIPLE-CHOICE PROBLEMS

1. **(B)** The number of neutrons is given by

$$N = A - Z = 213 - 84 = 129$$

2. **(C)** The emission of a beta particle involves the transformation of a neutron into a proton and an electron. Thus, the total number of nucleons (Z) remains the same, but the number of protons increases by one. The new isotope is therefore oxygen-13.

3. **(C)** Gamma rays have the greatest energy with the greatest penetration power.

4. **(B)** In nuclear *disintegration*, the nucleus always either loses mass or releases energy in the form of photons.

5. **(A)** From Chapter 22, we know that, at the same velocity, the particle with the largest mass will have the shortest wavelength. Of the four choices, the alpha particle, being a helium nucleus, has the largest mass and hence the shortest wavelength.

6. **(D)** Isotopes, by definition, are of the same chemical element. Hence they must have the same number of protons (but different numbers of neutrons and different binding energies).

7. **(A)** We must have conservation of charge and mass. The subscripts already add up to 15 on each side, so we are left with a subscript of 0 for Y; the particle is therefore neutral. The superscripts on the right must add up to 31 to balance. Since the missing particle needs a superscript of 1, particle Y is a neutron.

8. **(B)** When radioactive nuclei emit photons, they settle down to a more stable state since no mass has been lost.

FREE-RESPONSE PROBLEMS

1. (a) The actual mass of iron-56 is 55.934939 u. The total additive mass is determined by noting that this isotope of iron has 26 protons and 30 neutrons. Each proton has a mass of 1.0078 u, and each neutron has a mass of 1.0087 u. Thus the total constituent mass is

$$\begin{aligned} 26(1.0078) &= 26.2028 \text{ u} \\ 30(1.0087) &= 30.2610 \text{ u} \\ \hline \text{Total mass} &= 56.4638 \text{ u} \end{aligned}$$

The mass defect of the nucleus is therefore the difference between this mass and the actual mass:

$$\text{Mass defect} = 56.463800 - 55.934939 = 0.528861 \text{ u}$$

(b) The binding energy is given by

$$\text{BE} = 0.528861 \times 931.5 = 492.63402 \text{ MeV}$$

The average binding energy per nucleon is

$$\text{Avg. BE} = \frac{492.63402}{56} = 8.797 \text{ MeV}$$

2. Heavier nuclei have more protons, which increases the Coulomb force of repulsion between them. To maintain stability, extra neutrons are present to provide more particles for the strong nuclear force to act on and overcome this repulsive force (which is indifferent to charge).

3. The density, pressure, and temperature inside the core of a star are sufficient to overcome the natural repulsive tendency of hydrogen atoms to fuse into helium.

4. Radioactive uranium-238 is unstable and off the neutron-proton plot. Through the process of alpha and beta decay, the emission of neutrons and protons eventually brings the isotopes formed closer to that line of stable nuclei.

5. The release of energy and the missing mass in binding energy are accounted for completely using Einstein's formula. This is true for radioactivity decay, fission, and fusion, as well as matter-antimatter annihilation.

Practice Tests

The following two practice tests were designed to simulate the actual AP Physics B examination. As mentioned earlier, ETS copyrights its examination, so that the level of difficulty, percentage of topics covered, and wording of questions many differ from the actual examination.

Do not use a formula sheet for Section I since you will not be provided one during the actual examination.

Please note the following:

1. A formula sheet is provided for Section II.
2. Calculators are not allowed for use in Section I.

Finally, doing these practice problems, even if they are different from the actual examination problems, can only help your understanding of physics and problem solving and, of course, help prepare you for the real thing.

After each exam, use the self-assessment guide and score improvement sheet to help you do better. Good luck!

Answer Sheet

PRACTICE TEST 1

1 Ⓐ Ⓑ Ⓒ Ⓓ Ⓔ	25 Ⓐ Ⓑ Ⓒ Ⓓ Ⓔ	49 Ⓐ Ⓑ Ⓒ Ⓓ Ⓔ
2 Ⓐ Ⓑ Ⓒ Ⓓ Ⓔ	26 Ⓐ Ⓑ Ⓒ Ⓓ Ⓔ	50 Ⓐ Ⓑ Ⓒ Ⓓ Ⓔ
3 Ⓐ Ⓑ Ⓒ Ⓓ Ⓔ	27 Ⓐ Ⓑ Ⓒ Ⓓ Ⓔ	51 Ⓐ Ⓑ Ⓒ Ⓓ Ⓔ
4 Ⓐ Ⓑ Ⓒ Ⓓ Ⓔ	28 Ⓐ Ⓑ Ⓒ Ⓓ Ⓔ	52 Ⓐ Ⓑ Ⓒ Ⓓ Ⓔ
5 Ⓐ Ⓑ Ⓒ Ⓓ Ⓔ	29 Ⓐ Ⓑ Ⓒ Ⓓ Ⓔ	53 Ⓐ Ⓑ Ⓒ Ⓓ Ⓔ
6 Ⓐ Ⓑ Ⓒ Ⓓ Ⓔ	30 Ⓐ Ⓑ Ⓒ Ⓓ Ⓔ	54 Ⓐ Ⓑ Ⓒ Ⓓ Ⓔ
7 Ⓐ Ⓑ Ⓒ Ⓓ Ⓔ	31 Ⓐ Ⓑ Ⓒ Ⓓ Ⓔ	55 Ⓐ Ⓑ Ⓒ Ⓓ Ⓔ
8 Ⓐ Ⓑ Ⓒ Ⓓ Ⓔ	32 Ⓐ Ⓑ Ⓒ Ⓓ Ⓔ	56 Ⓐ Ⓑ Ⓒ Ⓓ Ⓔ
9 Ⓐ Ⓑ Ⓒ Ⓓ Ⓔ	33 Ⓐ Ⓑ Ⓒ Ⓓ Ⓔ	57 Ⓐ Ⓑ Ⓒ Ⓓ Ⓔ
10 Ⓐ Ⓑ Ⓒ Ⓓ Ⓔ	34 Ⓐ Ⓑ Ⓒ Ⓓ Ⓔ	58 Ⓐ Ⓑ Ⓒ Ⓓ Ⓔ
11 Ⓐ Ⓑ Ⓒ Ⓓ Ⓔ	35 Ⓐ Ⓑ Ⓒ Ⓓ Ⓔ	59 Ⓐ Ⓑ Ⓒ Ⓓ Ⓔ
12 Ⓐ Ⓑ Ⓒ Ⓓ Ⓔ	36 Ⓐ Ⓑ Ⓒ Ⓓ Ⓔ	60 Ⓐ Ⓑ Ⓒ Ⓓ Ⓔ
13 Ⓐ Ⓑ Ⓒ Ⓓ Ⓔ	37 Ⓐ Ⓑ Ⓒ Ⓓ Ⓔ	61 Ⓐ Ⓑ Ⓒ Ⓓ Ⓔ
14 Ⓐ Ⓑ Ⓒ Ⓓ Ⓔ	38 Ⓐ Ⓑ Ⓒ Ⓓ Ⓔ	62 Ⓐ Ⓑ Ⓒ Ⓓ Ⓔ
15 Ⓐ Ⓑ Ⓒ Ⓓ Ⓔ	39 Ⓐ Ⓑ Ⓒ Ⓓ Ⓔ	63 Ⓐ Ⓑ Ⓒ Ⓓ Ⓔ
16 Ⓐ Ⓑ Ⓒ Ⓓ Ⓔ	40 Ⓐ Ⓑ Ⓒ Ⓓ Ⓔ	64 Ⓐ Ⓑ Ⓒ Ⓓ Ⓔ
17 Ⓐ Ⓑ Ⓒ Ⓓ Ⓔ	41 Ⓐ Ⓑ Ⓒ Ⓓ Ⓔ	65 Ⓐ Ⓑ Ⓒ Ⓓ Ⓔ
18 Ⓐ Ⓑ Ⓒ Ⓓ Ⓔ	42 Ⓐ Ⓑ Ⓒ Ⓓ Ⓔ	66 Ⓐ Ⓑ Ⓒ Ⓓ Ⓔ
19 Ⓐ Ⓑ Ⓒ Ⓓ Ⓔ	43 Ⓐ Ⓑ Ⓒ Ⓓ Ⓔ	67 Ⓐ Ⓑ Ⓒ Ⓓ Ⓔ
20 Ⓐ Ⓑ Ⓒ Ⓓ Ⓔ	44 Ⓐ Ⓑ Ⓒ Ⓓ Ⓔ	68 Ⓐ Ⓑ Ⓒ Ⓓ Ⓔ
21 Ⓐ Ⓑ Ⓒ Ⓓ Ⓔ	45 Ⓐ Ⓑ Ⓒ Ⓓ Ⓔ	69 Ⓐ Ⓑ Ⓒ Ⓓ Ⓔ
22 Ⓐ Ⓑ Ⓒ Ⓓ Ⓔ	46 Ⓐ Ⓑ Ⓒ Ⓓ Ⓔ	70 Ⓐ Ⓑ Ⓒ Ⓓ Ⓔ
23 Ⓐ Ⓑ Ⓒ Ⓓ Ⓔ	47 Ⓐ Ⓑ Ⓒ Ⓓ Ⓔ	
24 Ⓐ Ⓑ Ⓒ Ⓓ Ⓔ	48 Ⓐ Ⓑ Ⓒ Ⓓ Ⓔ	

Table of Information

Useful Constants

1 atomic mass unit	$1\ u = 1.66 \times 10^{-27}$ kg
Rest mass of the proton	$m_p = 1.67 \times 10^{-27}$ kg
Rest mass of the neutron	$m_n = 1.67 \times 10^{-27}$ kg
Rest mass of the electron	$m_e = 9.11 \times 10^{-31}$ kg
Magnitude of the electron charge	$e = 1.60 \times 10^{-19}$ C
Avogadro's number	$N_0 = 6.02 \times 10^{23}$ per mol
Universal gas constant	$R = 8.32$ J/(mol · K)
Boltzmann's constant	$k_B = 1.38 \times 10^{-23}$ J/K
Speed of light	$c = 3 \times 10^8$ m/s
Planck's constant	$h = 6.63 \times 10^{-34}$ J · s $= 4.14 \times 10^{-15}$ eV · s
1 electron volt	$1\ eV = 1.6 \times 10^{-19}$ J
Vacuum permittivity	$\epsilon_0 = 8.85 \times 10^{-12}\ C^2/N \cdot m^2$
Coulomb's law constant	$k = (1/4)\pi\epsilon_0 = 9 \times 10^9\ N \cdot m^2/C^2$
Vacuum permeability	$\mu_0 = 4\pi \times 10^{-7}$ Wb/(A · m)
Magnetic constant	$k' = k/c^2 = \mu_0/4\pi = 10^{-7}$ Wb/(A · m)
Acceleration due to gravity at Earth's surface	$\mathbf{g} = 9.8\ m/s^2$
Universal gravitational constant	$G = 6.67 \times 10^{-11}\ m^3/(kg \cdot s^2)$
1 atmosphere pressure	$1\ atm = 1.0 \times 10^5\ N/m^2 = 1.0 \times 10^5$ Pa
1 nanometer	$1\ nm = 1.0 \times 10^{-9}$ m

Unit Symbols

Meter, m	Mole, mol	Watt, W	Farad, F
Kilogram, kg	Hertz, Hz	Coulomb, C	Tesla, T
Second, s	Newton, N	Volt, V	Degree Celsius, °C
Ampere, A	Pascal, Pa	Ohm, Ω	Electron volt, eV
Kelvin, K	Joule, J	Henry, H	

Prefixes

Factor	**Prefix**	**Symbol**
10^9	Giga-	G
10^6	Mega-	M
10^3	Kilo-	k
10^{-2}	Centi-	c
10^{-3}	Milli-	m
10^{-6}	Micro-	μ
10^{-9}	Nano-	n
10^{-12}	Pico-	p

Values of Trigonometric Functions for Common Angles

θ	**0°**	**30°**	**37°**	**45°**	**53°**	**60°**	**90°**
sin θ	0	$\frac{1}{2}$	$\frac{3}{5}$	$\frac{\sqrt{2}}{2}$	$\frac{4}{5}$	$\frac{\sqrt{3}}{2}$	1
cos θ	1	$\frac{\sqrt{3}}{2}$	$\frac{4}{5}$	$\frac{\sqrt{2}}{2}$	$\frac{3}{5}$	$\frac{1}{2}$	0
tan θ	0	$\frac{\sqrt{3}}{3}$	$\frac{3}{4}$	1	$\frac{4}{3}$	$\sqrt{3}$	∞

Practice Test 1

Section I

MULTIPLE-CHOICE QUESTIONS

70 QUESTIONS

90 MINUTES

50 PERCENT OF TOTAL GRADE

Directions: For each of the questions or incomplete statements below there are five choices. In each case select the best answer or completion and fill in the corresponding oval on the answer sheet. You may not use a calculator for Section I.

I. All frames of reference are assumed to be inertial.
II. Electrical current will follow the direction of a positive charge (conventional current).
III. For any isolated charge, the potential at infinity is taken to be equal to zero.
IV. The work done by a thermodynamic system is defined as a positive quantity.

1. Two objects are thrown vertically upward. One object has twice the initial velocity of the other. Neglecting any air resistance, the object with the greater initial velocity will rise to a maximum height that is

(A) the same as the other object
(B) $\sqrt{2}$ times that of the other object
(C) twice that of the other object
(D) eight times that of the other object
(E) four times that of the other object

2. A 2-kilogram cart has a velocity of 4 meters per second to the right. It collides with a 5-kilogram cart moving to the left at 1 meter per second. After the collision, the two carts stick together. What are the magnitude and the direction of the velocity of the two carts after the collision?

(A) 3/7 m/s left
(B) 7/3 m/s left
(C) 3/7 m/s right
(D) 7/3 m/s right
(E) 15/7 m/s right

Practice Test 1

3. A projectile is launched at a 30° angle to the ground with a velocity of 200 m/s. Approximately how long will it take for the projectile to reach its maximum height?
   (A) 5 s
   (B) 10 s
   (C) 15 s
   (D) 20 s
   (E) 25 s

4. A 5 kg mass is sitting at rest on a horizontal surface. A horizontal force of 10 N is needed to just get the mass started. What is the value of the maximum coefficient of static friction between the mass and the surface?
   (A) 0.2
   (B) 0.30
   (C) 0.10
   (D) 0.25
   (E) 0.15

5. Which of the following graphs represents an object moving with no net force acting on it?

(A)

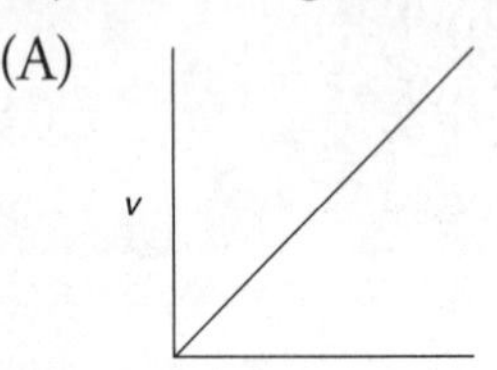

(B)

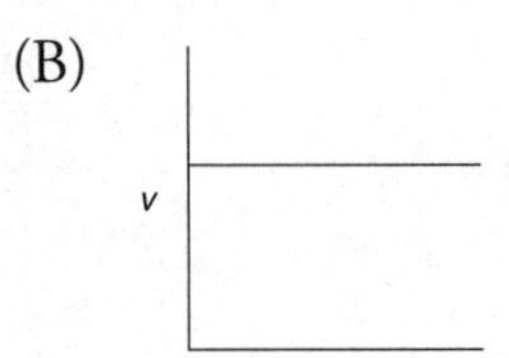

(C)

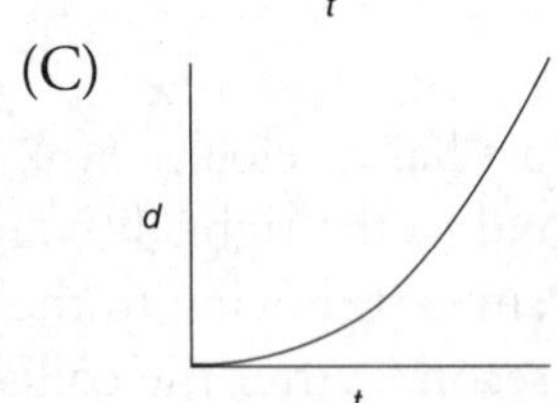

(D)

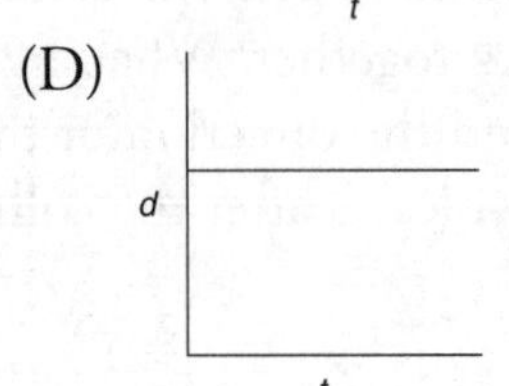

(E)

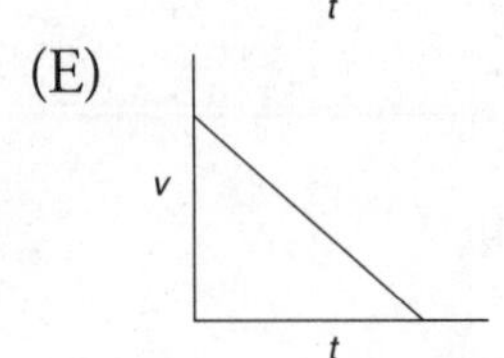

6. A projectile is launched horizontally with an initial velocity $\mathbf{v}_0$ from a height h. If it is assumed that there is no air resistance, which of the following expressions represents the vertical trajectory of the projectile?
   (A) $-\mathbf{g}\mathbf{v}_0^2/2x^2$
   (B) $-\mathbf{g}\mathbf{v}_0^2x^2$
   (C) $-\mathbf{g}\mathbf{v}/2\mathbf{v}_0^2$
   (D) $-2\mathbf{g}x^2/\mathbf{v}_0^2$
   (E) $-\mathbf{g}x^2/\mathbf{v}_0^2$

Questions 7 and 8 are based on the information and diagram below:

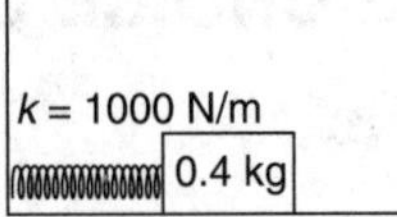

A 0.4-kilogram mass is attached to a spring that has a force constant of k = 1,000 newtons per meter.

7. What is the approximate maximum velocity attained by the mass if the amplitude of oscillations is 0.2 meter?
   (A) 2 m/s
   (B) 60 m/s
   (C) 15 m/s
   (D) 10 m/s
   (E) 30 m/s

8. Which of the following statements concerning the oscillatory motion described above is correct?
   (A) The maximum velocity and maximum acceleration occur at the same time.
   (B) The maximum velocity occurs when the acceleration is a minimum.
   (C) The velocity is always directly proportional to the displacement.
   (D) The maximum velocity occurs when the displacement is a maximum.
   (E) None of the preceding statements is correct.

9. A pendulum of length ℓ is swinging in an arc *AB* as shown below. It has a velocity **v** at point *O*. What is the tension *T* in the string at point *O*?

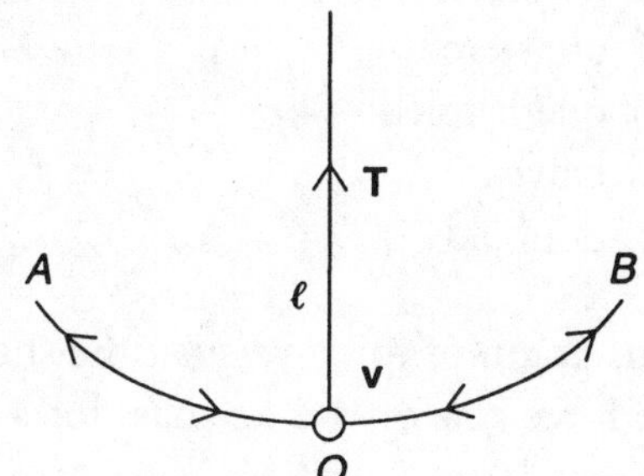

(A) $m\mathbf{g}$
(B) $m\mathbf{v}^2/\ell$
(C) $m\mathbf{v}^2/\ell + m\mathbf{g}$
(D) $m\mathbf{g} - m\mathbf{v}^2/\ell$
(E) $m\mathbf{v}^2/\ell - m\mathbf{g}$

10. A conical pendulum consists of a mass *M* attached to a light string of length ℓ. The mass swings around in a horizontal circle, making an angle θ with the vertical, as shown below. What is the tension **T** in the string?

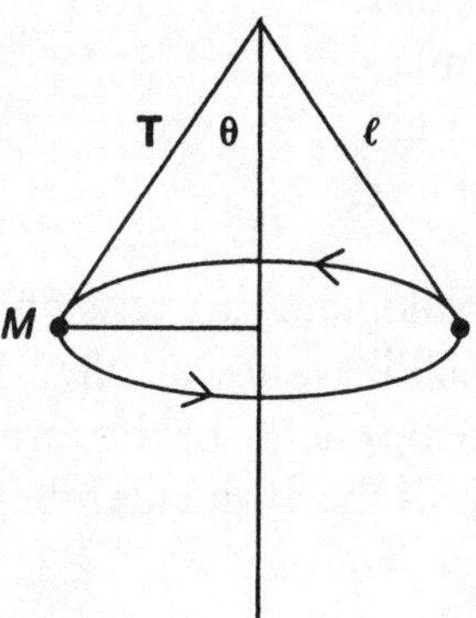

(A) $m\mathbf{g}/\cos\theta$
(B) $m\mathbf{g}\cos\theta$
(C) $m\mathbf{g}/\sin\theta$
(D) $m\mathbf{g}\sin\theta$
(E) $m\mathbf{g}/\tan\theta$

11. A 10 kg mass is being pulled horizontally by a force of 500 N along a rough surface at constant velocity. Which of the following is the value of the coefficient of kinetic friction?

(A) 0.02
(B) 0.50
(C) 1.0
(D) 0.10
(E) 0.20

12. A 10-newton force is applied to two masses, 4 kilograms and 1 kilogram, respectively, in contact as shown below. The horizontal motion is along a frictionless plane. What is the magnitude of the contact force, **P**, between the two masses?

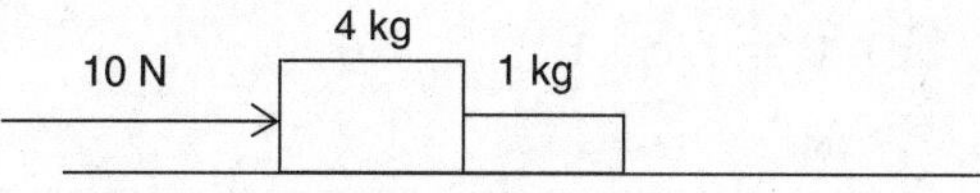

(A) 10 N
(B) 2 N
(C) 4 N
(D) 5 N
(E) 6 N

13. An object with mass *m* is dropped from height *h* above the ground. Neglecting air resistance, which formula best describes the power generated if the object takes time *t* to fall?

(A) $m\mathbf{g}h$
(B) $m\mathbf{g}ht$
(C) $\mathbf{g}h/t$
(D) $m\mathbf{g}/t$
(E) $m\mathbf{g}h/t$

14. A 1500-kilogram car has a velocity of 25 meters per second. If it is brought to a stop in 15 seconds, what was the magnitude of the braking force applied?

(A) 22,500 N
(B) 2500 N
(C) 14,700 N
(D) 16,000 N
(E) 1500 N

15. A block of mass M rests on a rough incline, as shown below. The angle of elevation of the incline is increased until an angle θ is reached. At that angle, the mass begins to slide down the incline. Which of the following is an expression for the coefficient of static friction μ?

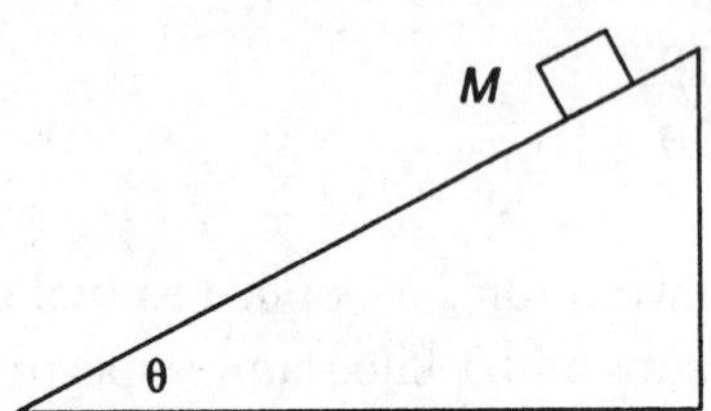

(A) $\mu = \tan\theta$
(B) $\mu = \sin\theta$
(C) $\mu = \cos\theta$
(D) $\mu = 2\sin\theta$
(E) $\mu = 1/\cos\theta$

16. A child is on a swing that varies in height from 75 cm at its lowest height above the ground to a maximum height of 225 cm above the ground. At its lowest point, the velocity of the swing is approximately equal to

(A) 5.5 m/s
(B) 30 m/s
(C) 4 m/s
(D) 8.5 m/s
(E) Not enough information is given

17. The gravitational force of attraction between two masses is 36 N when the masses are separated by a distance of 3 m. If the distance between them is reduced to 1 m, then the new force of attraction will be equal to

(A) 18 N
(B) 108 N
(C) 72 N
(D) 324 N
(E) 9 N

18. A pendulum of length ℓ has a period T. If the period is to be doubled, the length of the pendulum should be

(A) increased by $\sqrt{2}$
(B) quartered
(C) quadrupled
(D) halved
(E) doubled

19. This graph of force versus time shows how the force acts on a 2 kg mass for a total time of 10 s.

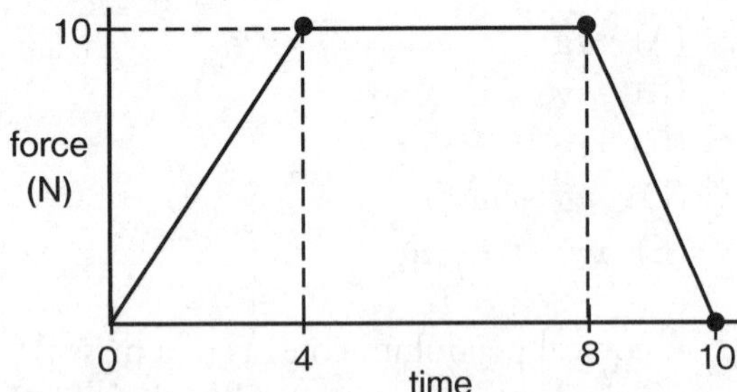

What is the magnitude of the change in velocity of the mass during that time?

(A) 35 m/s
(B) 70 m/s
(C) 7 m/s
(D) 3.5 m/s
(E) 70 m/s

20. A 200-kilogram cart rests on top of a frictionless hill as shown below. What will be the kinetic energy of the cart when it is at the top of the 10-meter hill?

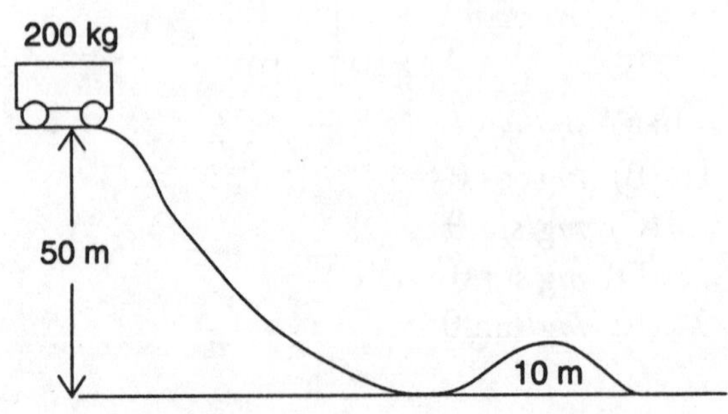

(A) 98,000 J
(B) 8,000 J
(C) 78,400 J
(D) 10,000 J
(E) 19,000 J

Practice Test 1

Questions 21 and 22 are based on the following information:

The magnitude of the one-dimensional momentum of a 2-kilogram particle obeys the relationship $\mathbf{p} = 2t + 3$.

21. What was the velocity of the particle at $t = 1$ second?

(A) 5 m/s
(B) 2 m/s
(C) 4 m/s
(D) 1.5 m/s
(E) 2.5 m/s

22. What was the average force applied to the particle from $t = 1$ second to $t = 3$ seconds?

(A) 2 N
(B) 5 N
(C) 9 N
(D) 10 N
(E) 14 N

23. A 50-kilogram box is pushed up a 1.5-meter incline with an effort of 200 newtons. The top of the incline is 0.5 meter above the ground. How much work was done to overcome friction?

(A) 55 J
(B) 75 J
(C) 245 J
(D) 300 J
(E) 490 J

24. A 0.2 kg mass is pushed down onto an elastic spring ($k = 20$ N/m). The spring is compressed by 20 cm and then released. Assuming no air resistance, to what height above the spring will the mass rise?

(A) 2 m
(B) 0.4 m
(C) 4 m
(D) 0.8 m
(E) 0.2 m

25. A cart with a mass of 1.5 kilograms needs to complete a 1.6-meter radius loop-the-loop, as shown below. What is the approximate minumum velocity required to achieve this goal?

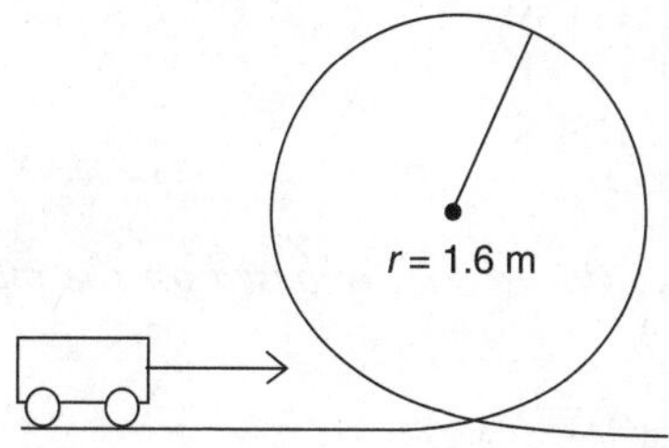

(A) 4 m/s
(B) 12 m/s
(C) 16 m/s
(D) 10 m/s
(E) 18 m/s

26. Which materials have the highest heat conductivity?

(A) Gases
(B) Liquids
(C) Papers
(D) Metals
(E) Woods

27. Isobaric changes in an ideal gas imply that there is no change in

(A) volume
(B) temperature
(C) pressure
(D) internal energy
(E) potential energy

28. Which of the following is equivalent to 1 pascal of gas pressure?

(A) 1 kg · m^2/s^2
(B) 1 kg/m · s^2
(C) 1 kg · m^3/s^2
(D) 1 kg · m/s
(E) 1 kg · m^2/s^3

29. A change in temperature of 75° Celsius is equivalent to a change in absolute temperature of

(A) 348 K
(B) 75 K
(C) 273 K
(D) 198 K
(E) 175 K

Questions 30 and 31 are based on the following graphs.

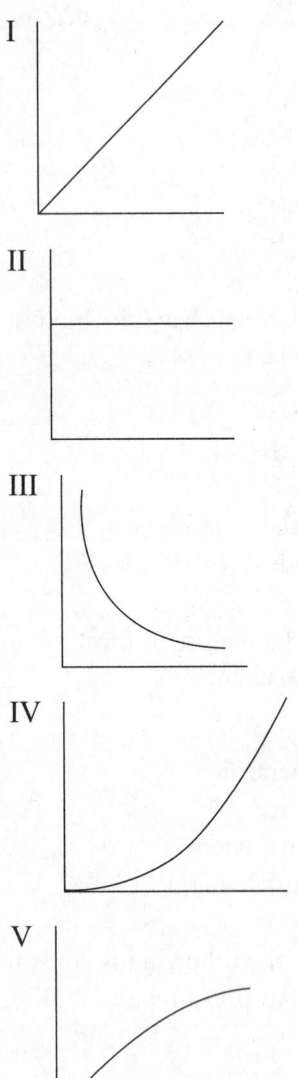

30. Which graph best represents the relationship between pressure and volume for an ideal confined gas at constant temperature?

(A) I
(B) II
(C) III
(D) IV
(E) V

31. Which graph best represents the relationship between the average kinetic energy of the molecules in an ideal gas and its absolute temperature?

(A) I
(B) II
(C) III
(D) IV
(E) V

Questions 32 and 33 are based on the following circuit:

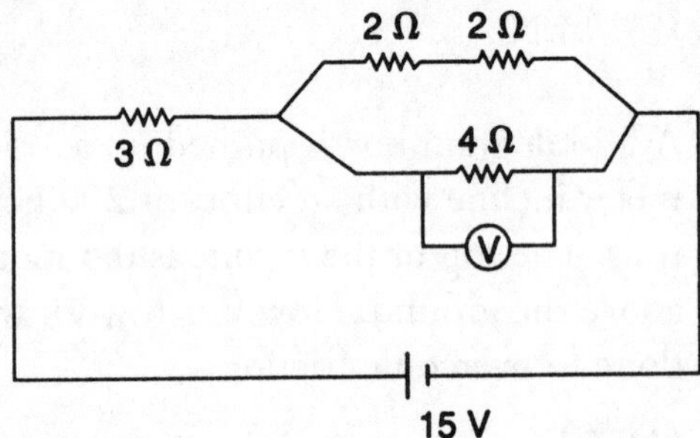

32. What is the equivalent resistance of the circuit?

(A) 6 Ω
(B) 8 Ω
(C) 5 Ω
(D) 15 Ω
(E) 3 Ω

33. What is the reading of the voltmeter across the 4-ohm resistor?

(A) 9 V
(B) 8 V
(C) 15 V
(D) 12 V
(E) 6 V

34. The diagram below shows a leaf electroscope that has been charged positively by a negatively charged rod. Which of the following statements is correct?

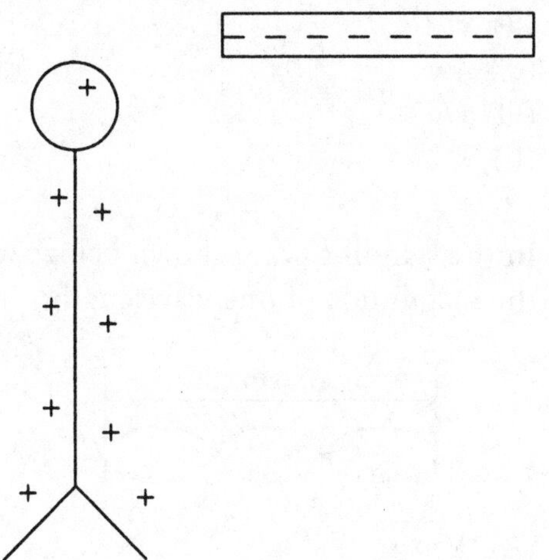

(A) The electroscope was charged by conduction.
(B) The electroscope was charged by contact.
(C) If the rod is brought closer, protons will be attracted to the top of the electroscope.
(D) If the rod is brought closer, protons will be repelled from the top of the electroscope.
(E) None of the preceding statements is correct.

35. What is the equivalent capacitance of the circuit shown below?

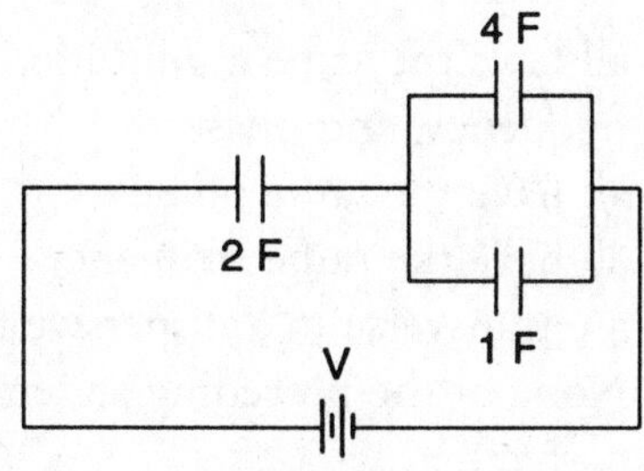

(A) 7/10 F
(B) 10/7 F
(C) 7 F
(D) 14/5 F
(E) 13/4 F

36. A wire segment in a circuit with a cross-sectional area *A* and length ℓ is replaced by one made of the same material that has twice the area but half the length. The resistance of the new segment, compared to the original segment, will

(A) be reduced by half
(B) double
(C) be reduced by one-quarter
(D) quadruple
(E) remain the same

Questions 37 and 38 are based on the circuit shown below:

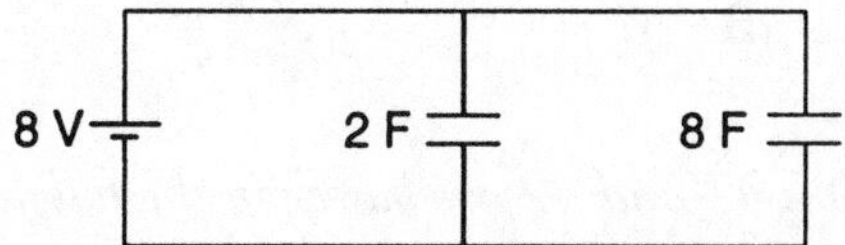

37. What is the charge stored in the 2-farad capacitor?

(A) 4 C
(B) 16 C
(C) 10 C
(D) 6 C
(E) 64 C

38. What is the energy stored in the 8-farad capacitor?

(A) 64 J
(B) 24 J
(C) 32 J
(D) 128 J
(E) 256 J

39. An electron is moving through a pair of crossed electric and magnetic fields in such a way that its path remains straight. Which of the following is an expression for the kinetic energy of the electron?

(A) $(1/2)M\mathbf{E}\mathbf{B}$
(B) $M/2\mathbf{E}\mathbf{B}$
(C) $M\mathbf{E}^2\mathbf{B}/2$
(D) $1/2\ M(\mathbf{E}/\mathbf{B})$
(E) $m\mathbf{E}^2/2\mathbf{B}^2$

40. An electron is trapped in a circular path because of a uniform perpendicular magnetic field **B**. The velocity of the electron is **v** and the radius of the path is *r*. Which of the following expressions represents the angular velocity **ω**?

(A) $\frac{\mathbf{B}er}{M}$

(B) $\mathbf{B}erM$

(C) $\frac{\mathbf{B}M}{er}$

(D) $\frac{er}{\mathbf{B}M}$

(E) $\mathbf{B}e^2M$

Questions 41 and 42 are based on the following information and diagram:

A square wire frame is pulled to the right with a velocity of 4 meters per second across an inward uniform magnetic field of strength 10 teslas. The length of each side of the frame is 0.2 meter.

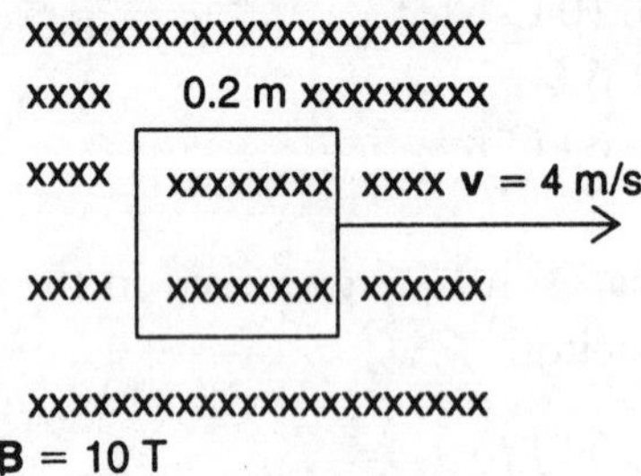

41. What is the magnitude of the induced motional electromotive force in the wire?

(A) 40 V
(B) 20 V
(C) 8 V
(D) 16 V
(E) 2 V

42. As the wire is moved to the right, a force appears to oppose it. This force is explained by

(A) Lenz's law
(B) Faraday's law
(C) Ohm's law
(D) Coulomb's law
(E) Henry's law

43. If the tension in a taut string is doubled, the velocity of propagation of transverse waves in the string will change from **v** to

(A) 2 **v**
(B) **v**/2
(C) 4 **v**
(D) **v**/4
(E) $\sqrt{2}\mathbf{v}$

44. In the standing wave shown below, what is the magnitude of one wavelength?

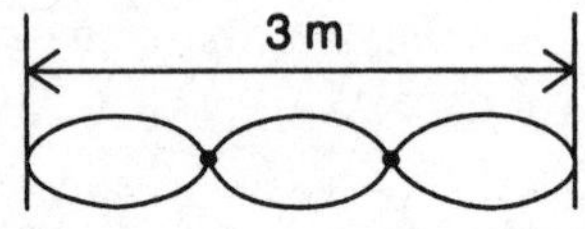

(A) 1 m
(B) 1.5 m
(C) 2.0 m
(D) 2.5 m
(E) 3.0 m

45. It takes 3 seconds to detect a sound produced while the air temperature at sea level is 25° Celsius. What is the distance to the sound source?

(A) 331 m
(B) 1038 m
(C) 993 m
(D) 632 m
(E) 852 m

46. A set of waves is coherent if the waves

(A) all have the same amplitude, frequency, and phase
(B) all have the same phase
(C) all have the same frequency
(D) are transverse in a dispersive medium
(E) None of the preceding statements is correct.

Practice Test 1

47. A 340-hertz tuning fork sets an air column vibrating in fundamental resonance as shown below. A hollow tube is inserted into a column of water, and the height of the tube is adjusted until strong resonance is heard. In this situation, the air column is 25 centimeters in length. What is the velocity of the sound?

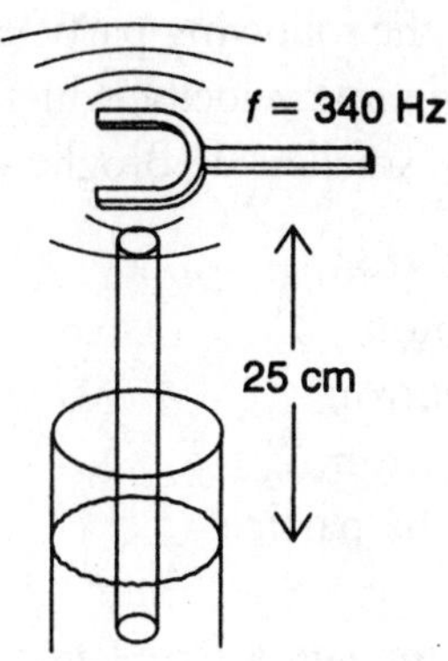

(A) 340 m/s
(B) 331 m/s
(C) 342 m/s
(D) 350 m/s
(E) 360 m/s

48. A shadow is formed by a point source of light. Upon closer inspection, the edges of the shadow seem to be diffuse and fuzzy. This phenomenon is probably caused by

(A) reflection
(B) refraction
(C) diffraction
(D) polarization
(E) dispersion

49. A light ray is incident on a glass-air interface as shown below. Which path will the light ray follow after it enters the air?

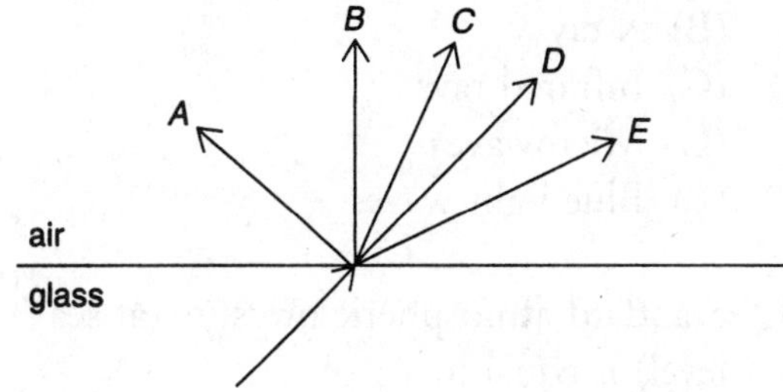

(A) *A*
(B) *B*
(C) *C*
(D) *D*
(E) *E*

50. As the angle of incidence for a ray of light passing from glass to air increases, the critical angle of incidence for the glass

(A) increases
(B) decreases
(C) increases and then decreases
(D) decreases and then increases
(E) remains the same

51. A 5-centimeter-tall object is located 10 centimeters in front of a converging lens that has a 6-centimeter focal length. What is the size of the image produced?

(A) 5 cm
(B) 7.5 cm
(C) 10 cm
(D) 15 cm
(E) 30 cm

52. Which of the following statements about a diverging mirror is correct?

(A) The mirror must be concave in shape.
(B) The images are sometimes larger than the actual objects.
(C) The images are always erect.
(D) The images are sometimes real.
(E) Algebraically, the focal length is designated as $+f$.

53. Which of the following waves cannot be polarized?

(A) Sound waves
(B) X rays
(C) Infrared rays
(D) Microwaves
(E) Blue light waves

54. Standard atmospheric pressure (at sea level) is equal to

(A) 1240 kPa.
(B) 101 kPa.
(C) 250 kPa.
(D) 157 kPa.
(E) 57 kPa.

55. The "flapping" of a flag in the wind is best explained using

(A) Archimedes' principle.
(B) Bernoulli's principle.
(C) Newton's principle.
(D) Pascal's principle.
(E) Torricelli's principle.

56. Fluid pressure can be measured in all of the following units except

(A) Pa
(B) Bar
(C) N/m^2
(D) $Kg \cdot s^2/m$
(E) $Kg/m \cdot s^2$

57. In a photoelectric effect experiment, the emitted electrons could be stopped with a retarding potential of 12 volts. What was the maximum kinetic energy of these electrons?

(A) 12 eV
(B) 24 eV
(C) 36 eV
(D) 1.92×10^{-18} eV
(E) 1.6×10^{-19} eV

58. What units are associated with the quantity h/mc^2?

(A) N
(B) J/s
(C) J
(D) $kg \cdot m/s$
(E) s

59. If all of the following particles were moving with the same velocity, which one would have the smallest de Broglie wavelength?

(A) Electron
(B) Proton
(C) Neutron
(D) Positron
(E) Alpha particle

Questions 60 through 62 are based on the following simulated energy level diagram for a mythical hydrogen-like atom:

Level	Energy
5	−4.2 eV
4	−5.7 eV
3	−6.3 eV
2	−8.2 eV
1	−15.2 eV

60. What energy is associated with a photon emitted during a transition from level 4 to level 2?

(A) 9.2 eV
(B) 5.7 eV
(C) 3.5 eV
(D) 6.3 eV
(E) 14.9 eV

61. How much energy is required to ionize an electron in level 3?

(A) 8.1 eV
(B) 6.3 eV
(C) 15.2 eV
(D) 2.5 eV
(E) 4.2 eV

62. Which of the following level transitions will result in the emission of a photon with the largest energy?

    (A) 1 to 3
    (B) 5 to 2
    (C) 2 to 1
    (D) 5 to 3
    (E) 4 to 2

63. One atomic mass unit is defined to be equal to

    (A) 1/2 the mass of a hydrogen molecule
    (B) 1/12 the mass of the carbon atom
    (C) 1/16 the mass of the oxygen atom
    (D) 1/4 the mass of the helium atom
    (E) 1/56 the mass of the iron atom

64. How many neutrons are contained in the isotope ${}^{238}_{92}U$?

    (A) 92
    (B) 100
    (C) 146
    (D) 330
    (E) 238

65. Which of the following represents the units of Planck's constant in terms of kilograms, meters, and seconds?

    (A) $kg \cdot m^2/s^2$
    (B) $kg \cdot m^2/s$
    (C) $kg \cdot m/s^2$
    (D) $kg \cdot m/s$
    (E) $kg \cdot m^2/s^3$

66. In the reaction

    $${}^{27}_{13}AI + {}^{4}_{2}He \rightarrow {}^{30}_{15}P + X$$

    what is the mass number for particle X?

    (A) 1
    (B) 2
    (C) 0
    (D) −1
    (E) 4

67. Radon gas (${}^{222}_{86}Rn$) is radioactive with a half-life of 4 days. A sample is sealed in an evacuated tube for more than 1 week. At that time, the presence of a second gas is detected. This gas is most probably

    (A) hydrogen
    (B) helium
    (C) oxygen
    (D) lithium
    (E) argon

68. According to the scale of binding energy per nucleon, which atom has the most stable nuclear isotope?

    (A) Hydrogen
    (B) Helium
    (C) Iron
    (D) Lithium
    (E) Uranium

69. A boat that is anchored at a dock is impacted by wave crests that are 50 m apart and traveling at 10 m/s. The waves will reach the boat once every

    (A) 0.2 s
    (B) 5 s
    (C) 10 s
    (D) 50 s
    (E) 500 s

70. An example of a wave that has both longitudinal and transverse properties is a
    (A) light wave
    (B) water wave
    (C) sound wave
    (D) wave in plucked string
    (E) microwave

This is the end of section I. You may use any remaining time to check your work in this section.

Formula Sheet for Section II

Newtonian Mechanics		Electricity and Magnetism	
$v = v_0 + at$	$P = Fv\cos\theta$	$F = \frac{1}{4\pi\epsilon_0}\frac{q_1q_2}{r^2}$	$V = IR$
$x = x_0 + v_0t + \frac{1}{2}at^2$	$\mathbf{F}_s = -k\mathbf{x}$	$\mathbf{E} = \frac{\mathbf{F}}{q}$	$P = IV$
$v^2 = v_0^2 + 2a(x - x_0)$	$U_s = \frac{1}{2}kx^2$	$U_E = qV = \frac{1}{4\pi\epsilon_0}\frac{q_1q_2}{r}$	$C_p = \sum_i C_i$
$\Sigma\mathbf{F} = \mathbf{F}_{net} = m\mathbf{a}$	$T_s = 2\pi\sqrt{\frac{m}{k}}$	$E_{avg} = -\frac{V}{d}$	$\frac{1}{C_s} = \sum_i \frac{1}{C_i}$
$F_{fric} \leq \mu N$	$T_p = 2\pi\sqrt{\frac{\ell}{g}}$	$V = \frac{1}{4\pi\epsilon_0}\sum_i \frac{q_i}{r_i}$	$R_s = \sum_i R_i$
$a_c = \frac{v^2}{r}$	$T = \frac{1}{f}$	$C = \frac{Q}{V}$	$\frac{1}{R_p} = \sum_i \frac{1}{R_i}$
$\tau = rF\sin\theta$	$F_G = -\frac{Gm_1m_2}{r^2}$	$C = \frac{\epsilon_0 A}{d}$	$F_B = qvB\sin\theta$
$\mathbf{p} = m\mathbf{v}$	$U_G = -\frac{Gm_1m_2}{r}$	$U_c = \frac{1}{2}QV = \frac{1}{2}CV^2$	$F_B = BI\ell\sin\theta$
$\mathbf{J} = \mathbf{F}\Delta t = \Delta\mathbf{p}$		$I_{avg} = \frac{\Delta Q}{\Delta t}$	$B = \frac{\mu_0}{2\pi}\frac{I}{r}$
$K = \frac{1}{2}mv^2$		$R = \frac{\rho\ell}{A}$	$\phi_m = BA\cos\theta$
$\Delta U_g = mgh$			$\epsilon_{avg} = -\frac{\Delta\phi_m}{\Delta t}$
$W = F\Delta r\cos\theta$			$\epsilon = B\ell v$
$P_{avg} = \frac{W}{\Delta t}$			
a = acceleration	r = radius or distance	A = area	R = resistance
F = force	T = period	B = magnetic field	r = distance
f = frequency	t = time	C = capacitance	t = time
h = height	U = potential energy	d = distance	U = potential (stored) energy
J = impulse	v = velocity or speed	E = electric field	V = electrical potential or potential difference
K = kinetic energy	W = work done on a system	ϵ = emf	v = velocity or speed
k = spring constant	x = position	F = force	ρ = resistivity
ℓ = length	μ = coefficient of friction	I = current	θ = angle
m = mass	θ = angle	ℓ = length	ϕ_m = magnetic flux
N = normal force	τ = torque	P = power	
P = power		Q = charge	
p = momentum		q = point charge	

Fluid Mechanics and Thermal Physics

$P = P_0 + \rho g h$

$F_{buoy} = \rho V g$

$A_1 \upsilon_1 = A_2 \upsilon_2$

$P + \rho g y + \frac{1}{2}\rho v^2 = \text{const.}$

$\Delta \ell = \alpha \ell_0 \Delta T$

$H = \frac{kA\Delta T}{L}$

$P = \frac{F}{A}$

$PV = nRT = Nk_B T$

$K_{avg} = \frac{3}{2} k_B T$

$v_{rms} = \sqrt{\frac{3RT}{M}} = \sqrt{\frac{3k_B T}{\mu}}$

$W = -P\Delta V$

$\Delta U = Q + W$

$e = \left| \frac{W}{Q_H} \right|$

$e_c = \frac{T_H - T_C}{T_H}$

A = area

e = efficiency

F = force

h = depth

H = rate of heat transfer

k = thermal conductivity

K_{avg} = average molecular kinetic energy

ℓ = length

L = thickness

M = molar mass

n = number of moles

N = number of molecules

P = pressure

Q = heat transferred to a system

T = temperature

U = internal energy

V = volume

υ = velocity or speed

υ_{rms} = root-mean-square velocity

W = work done on a system

y = height

α = coefficient of linear expansion

μ = mass of molecule

ρ = density

Atomic and Nuclear Physics

$E = hf = pc$

$K_{max} = hf - \phi$

$\lambda = \frac{h}{p}$

$\Delta E = (\Delta m)c^2$

E = energy

f = frequency

K = kinetic energy

m = mass

p = momentum

λ = wavelength

ϕ = work function

Waves and Optics

$\upsilon = f\lambda$

$n = \frac{c}{v}$

$n_1 \sin\theta_1 = n_2 \sin\theta_2$

$\sin\theta_c = \frac{n_2}{n_1}$

$\frac{1}{s_i} + \frac{1}{s_0} = \frac{1}{f}$

$M = \frac{h_i}{h_0} = -\frac{s_i}{s_0}$

$f = \frac{R}{2}$

$x_m \approx \frac{m\lambda L}{d}$

d = separation

f = frequency or focal length

h = height

L = distance

M = magnification

m = an integer

n = index of refraction

R = radius of curvature

s = distance

υ = speed

x = position

λ = wavelength

θ = angle

Geometry and Trigonometry

Rectangle

$A = bh$

Triangle

$A = \frac{1}{2}bh$

Circle

$A = \pi r^2$

$C = 2\pi r$

Parallelepiped

$V = \ell wh$

Cylinder

$V = \pi r^2 \ell$

$S = 2\pi r\ell + 2\pi r^2$

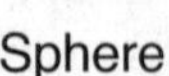

Sphere

$V = \frac{4}{3}\pi r^3$

$S = 4\pi r^3$

Right Triangle

$a^2 + b^2 = c^2$

$\sin\theta = \frac{a}{c}$

$\cos\theta = \frac{b}{c}$

$\tan\theta = \frac{a}{b}$

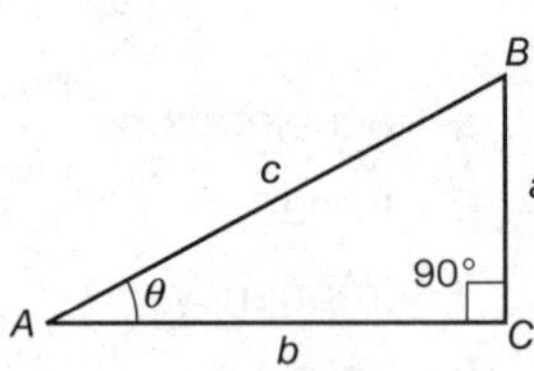

A = area

C = circumference

V = volume

S = surface area

b = base

h = height

ℓ = length

w = width

r = radius

Section II

FREE-RESPONSE QUESTIONS

6 QUESTIONS

90 MINUTES

50 PERCENT OF TOTAL GRADE

Directions: Solve the following problems. Be sure to show all work, including substitutions with units, and to explain clearly. You may use a calculator for Section II.

1. (15 points) A 2-kilogram mass is twirled in a vertical circle as shown. It is attached to a 2-meter rope. As the mass just clears the ground (point *A*), its velocity is 10 m/s to the right. Neglect any air resistance. When the mass reaches point *B*, it makes a 30° angle to the horizontal as shown.

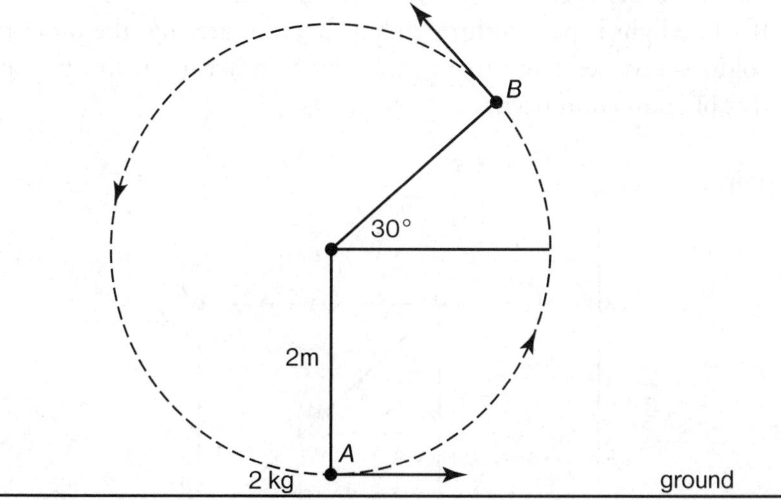

   (a) Determine the magnitude of the tension in the rope when the mass is at point *A*.
   (b) Determine the speed of the mass at point *B*.
   (c) At point *B*, the rope is released and the mass becomes a projectile.
      i. Determine the magnitudes of the vertical and horizontal components of the velocity at point *B*.
      ii. Determine the total height above the ground that the projectile reaches.

2. (15 points) A mass m is resting at a height h above the ground. When released, the mass can slide down a frictionless track to a loop-the-loop of radius R, as shown below:

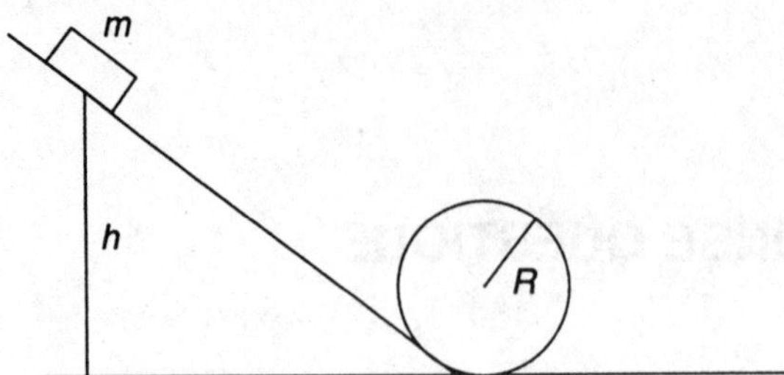

(a) Draw free-body diagrams for the mass when it is (i) at the top and (ii) at the bottom of the loop.
(b) Derive expressions for the reaction force of the track on the mass when the mass is (i) at the top and the (ii) at the bottom.
(c) At what height h must the mass be released so that the reaction force of the track on the mass when it is at the top of the loop is exactly equal to its weight?

3. (15 points) The 501.5-nanometer line of helium is observed at an angle of 45 degrees in the second-order spectrum of a diffraction grating.

(a) How many lines per millimeter does this diffraction grating have?
(b) Calculate the angular deviation for the 667.8-nanometer helium line in the first-order spectrum for the same grating.
(c) If white light is passed through a diffraction grating, the order of the colors is reversed from that formed by a refracting prism. Explain why this phenomenon occurs.

4. (15 points)

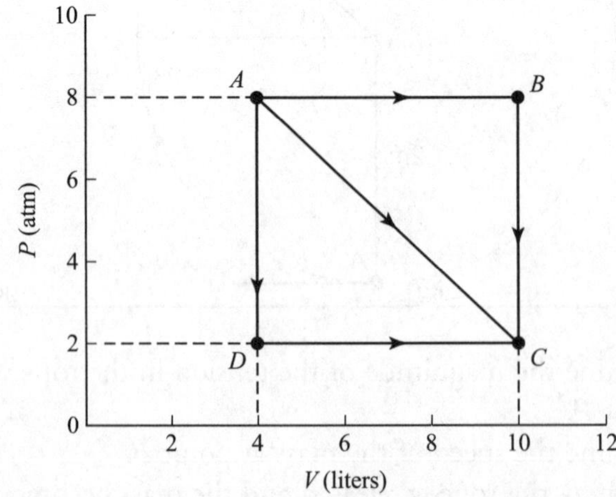

A gas expands from points A to C along three possible paths. Calculate the work done along path:
i. ABC
ii. AC
iii. ADC

5. (10 points) Given the following information:

$$\text{Proton mass} = 1.0078\ \text{u}$$
$$\text{Neutron mass} = 1.0087\ \text{u}$$
$$\text{Mass of } ^{226}_{88}\text{Ra} = 226.0244\ \text{u}$$

(a) Determine the mass defect for this isotope.
(b) Determine the binding energy per nucleon for this isotope.
(c) Radium-88 undergoes alpha decay via the following equation:

$$^{226}_{88}\text{Ra} \rightarrow\ ^{222}_{86}\text{Rn} +\ ^{4}_{2}\text{He} + Q$$

If the mass of the radon-86 isotope is 222.0165 u and the mass of the helium-4 isotope is 4.0026 u, how much energy, in million electron volts, is associated with quantity Q?

6. (10 points) You are given three 2-Ω resistors, some wire, a variable dc voltage supply, a voltmeter, and an ammeter.

(a) Draw a schematic diagram of a circuit that will produce an equivalent resistance of 3 Ω, as well as measure the circuit current and source voltage.
(b) State the complete law that relates the voltage and current in this circuit.
(c) A student wishes to measure the voltage and current in a simple circuit using small light bulbs instead of commercially manufactured resistors. She finds that after a few minutes the current in the ammeter is decreasing. How can she account for this?

This is the end of section II. You may use any remaining time to check your work in this section.

Answer Key

SECTION I

1. E	19. A	37. B	55. B
2. C	20. C	38. E	56. D
3. B	21. E	39. E	57. A
4. A	22. A	40. A	58. E
5. B	23. A	41. C	59. E
6. C	24. E	42. A	60. C
7. D	25. A	43. E	61. B
8. B	26. D	44. C	62. C
9. C	27. C	45. B	63. B
10. A	28. B	46. A	64. C
11. E	29. B	47. A	65. B
12. B	30. C	48. C	66. A
13. E	31. A	49. E	67. B
14. B	32. C	50. E	68. C
15. A	33. E	51. B	69. B
16. A	34. E	52. C	70. B
17. D	35. B	53. A	
18. C	36. C	54. B	

ANSWER EXPLANATIONS

SECTION I

1. **(E)** The maximum height reached by a projectile thrown vertically upward is given by

$$\mathbf{y}_{max} = \frac{(\mathbf{v}_y)^2}{2\mathbf{g}}$$

Therefore, if the initial vertical velocity is doubled, the projectile with the greater initial velocity will go four times higher.

2. **(C)** Conservation of linear momentum implies that the total momentum before the collision of the two carts must equal the total momentum after the collision. This collision is inelastic since the two carts stick together. Thus, taking the "left" as negative, we have

$$(2 \text{ kg})(4 \text{ m/s}) - (5 \text{ kg})(1 \text{ m/s}) = (2 \text{ kg} + 5 \text{ kg})\mathbf{v}_f$$

and $\mathbf{v}_f = 3/7$ m/s to the right (since the answer is positive).

3. **(B)** We need the vertical component of the launch velocity:

$$\mathbf{v}_{oy} = \mathbf{v}_o \sin\theta$$

$$\mathbf{v}_{oy} = (200 \text{ m/s}) \sin(30°) = 100 \text{ m/s}$$

Now, the time to maximum height is the time it takes gravity to decelerate this initial vertical velocity to zero (assume $\mathbf{g} = 10 \text{ m/s}^2$, so no calculator is needed!):

$$\mathbf{a} = \frac{\Delta\mathbf{v}}{\Delta t}$$

$$t = \frac{\mathbf{v}_{oy}}{\mathbf{g}} = 10 \text{ s}$$

4. **(A)** We know that at maximum static friction

$$\mathbf{F}_f = \mu_s\mathbf{F_N}$$

In this case, $\mathbf{F}_N = m\mathbf{g} = 50$ N (approximately). Thus,

$$\mu_s = \frac{10 \text{ N}}{50 \text{ N}} = 0.20$$

5. **(B)** Graph B shows a constant velocity. This means that there is no acceleration and thus no net force acting on the object.

6. **(C)** The vertical trajectory is given by $y = -(1/2)\mathbf{g}t^2$. The horizontal velocity is constant, and so the time t is given by the initial velocity $\mathbf{v}_0$ and the horizontal displacement x: $x/\mathbf{v}_0$. Thus we get

$$y = \frac{-\mathbf{g}x^2}{2\mathbf{v}_0^2}$$

7. **(D)** We can use

$$\frac{1}{2}\,m\mathbf{v}^2_{max} = \frac{1}{2}\,kA^2$$

Thus $\mathbf{v}_{max} = 10$ m/s.

8. **(B)** The maximum velocity occurs when the acceleration is a minimum in simple harmonic motion.

9. **(C)** At the bottom of the swing

$$\mathbf{F} = m\mathbf{a} = \frac{m\mathbf{v}^2}{r} = \mathbf{T} - m\mathbf{g}$$

Thus,

$$\mathbf{T} = \frac{m\mathbf{v}^2}{r} + m\mathbf{g} = \frac{m\mathbf{v}^2}{\ell} + m\mathbf{g}$$

10. **(A)** From the diagram we can see that the upward component of the tension, in magnitude, is given by

$$\mathbf{T}\cos\theta = m\mathbf{g}$$

Thus,

$$\mathbf{T} = \frac{m\mathbf{g}}{\cos\theta}$$

11. **(E)** If the mass is moving at a constant velocity, then the net force acting on it is equal to zero. Therefore, the force of friction must be equal (but opposite to) the applied force. In this case, since $\mathbf{F}_f = \mu\mathbf{F}_N$, we see that the normal force is equal to the weight:

$$\mathbf{F}_g = \mathbf{F}_N = m\mathbf{g} = 100\text{ N (approximately)}$$

Therefore, we see that $\mu = 100\text{ N}/500\text{ N} = 0.20$.

12. **(B)** Since the two masses are in contact, force $\mathbf{F} = 10$ N gives them the same acceleration:

$$\mathbf{a} = 10\text{ N}/5\text{ kg} = 2.0\text{ m/s}^2$$

If we isolate the 1-kg mass in a free-body diagram, the only force acting on it is **P**, the contact force exerted by the 4-kg mass. However, in magnitude, $\mathbf{P} = m\mathbf{a}$, where $m = 1$ kg and **a** is the acceleration of the system. Thus we find that $\mathbf{P} = 2$ N.

13. **(E)** The power generated is equal to the work done divided by the time. For a mass falling without air resistance, the work done is equal to the loss of potential energy:

$$W = \Delta U_g = m\mathbf{g}h$$

The power is given by $m\mathbf{g}h/t$.

14. **(B)** Since $\mathbf{F} = m\mathbf{a}$, we have

$$\mathbf{F} = \frac{m\,\Delta\mathbf{v}}{\Delta t} = 2500 \text{ N}$$

upon substitution.

15. **(A)** On the incline, the force of friction must just balance the component of weight down the incline. The normal force on an incline is given by $m\mathbf{g}\cos\theta$, and the force down the incline (in magnitude) is given by $m\mathbf{g}\sin\theta$. Thus, we write

$$m\mathbf{g}\sin\theta = \mu m\mathbf{g}\cos\theta$$

and

$$\mu = \tan\theta$$

This angle θ is sometimes called the "angle of repose."

16. **(A)** We can use conservation of mechanical energy to solve this problem. Since the lowest position of the swing can be considered to have zero potential energy, the change in kinetic energy is equal to the change in potential energy:

$$\Delta\text{KE} = \Delta U_g$$

The change in potential energy depends on the change in height, which is 150 cm (or 1.5 meters) in this case. The swing also stops at its highest point. Let the magnitude of $\mathbf{g} = 10 \text{ m/s}^2$:

$$\tfrac{1}{2}mv^2 = m\mathbf{g}(\Delta h)$$
$$v^2 = 2\mathbf{g}(\Delta h)$$
$$v^2 = 2(10 \text{ m/s}^2)(1.5 \text{ m}) = 30 \text{ m}^2/\text{s}^2$$

A calculator is not needed to find the square root, and v is approximately equal to 5.5 m/s.

17. **(D)** Gravitation is an inverse square law. Thus, if the distance is reduced by one-third, then the force increases by nine times. Thus, 36 N × 9 = 324 N.

18. **(C)** The period of a pendulum is proportional to the square root of its length. To double the period, the length must therefore be quadrupled.

19. **(A)** The change in velocity is found by recalling that the area under a graph of force versus time is equal to the impulse applied to the mass. The impulse is equal to the change in momentum ($\mathbf{F}t = m\Delta\mathbf{v}$). If we divide the total area by the mass, we will obtain the change in velocity. The total area is easily found by dividing the shape into simple rectangles and triangles. The total area is equal to 70 kg · m/s. Divide this by 2 kg to obtain $\Delta\mathbf{v} = 35$ m/s.

20. **(C)** Using the law of conservation of energy, we see that the total energy at the start is $E = m\mathbf{g}h$ which is (200)(9.8)(50) = 98,000 J. At the top of the second (10-m) hill, the kinetic energy is equal to the difference in potential energies:

$$98{,}000 \text{ J} - 19{,}600 \text{ J} = 78{,}400 \text{ J}$$

21. **(E)** At $t = 1$ s, the momentum of the particle is given by $\mathbf{p} = 5$ kg · m/s. With a mass of 2 kg, that implies an initial velocity of 2.5 m/s.

22. **(A)** The average force is given by $\mathbf{F} = \Delta\mathbf{p}/\Delta t$. In this situation, $\Delta t = 2$ s and $\Delta\mathbf{p} = 4$ kg · m/s. Thus, $\mathbf{F} = 2$ N.

23. **(A)** The work done by friction is the difference between the work input and the work output. In this problem

$$W_{\text{input}} = (200 \text{ N})(1.5 \text{ m}) = 300 \text{ J}$$

We can take $g = 9.8$ N/kg here. Then

$$W_{\text{output}} = (50 \text{ kg})(9.8 \text{ N/kg})(0.5 \text{ m}) = 245 \text{ J}$$

Thus

$$W_{\text{friction}} = 55 \text{ J}$$

24. **(E)** We can use conservation of mechanical energy to solve this problem. The work done by the compressed spring will be transferred into potential energy. Let $\mathbf{g} = 10$ m/s^2. Note that $x = 20$ cm = 0.2 m in this case:

$$\Delta U_s = \Delta U_g$$

$$\tfrac{1}{2}kx^2 = m\mathbf{g}(\Delta h)$$

$$\tfrac{1}{2}(20 \text{ N/m})(0.2 \text{ m})^2 = (0.2 \text{ kg})(10 \text{ m/s}^2)\Delta h$$

$$\Delta h = 0.2 \text{ m}$$

The mass will rise 0.2 m above the spring.

25. **(A)** The minimum velocity is given by the formula

$$\mathbf{v} = \sqrt{\mathbf{g}r} = \sqrt{10 \text{ m/s}^2(1.6 \text{ m})} \approx 4.00 \text{ m/s}$$

26. **(D)** Metals have the highest values of heat conductivity.

27. **(C)** Isobaric changes in an ideal gas imply that the pressure in the gas remains constant.

28. **(B)** One pascal of gas pressure is equivalent to

$$1\ \text{N/m}^2 = 1\ \text{kg/m} \cdot \text{s}^2$$

29. **(B)** A change of 75°C is equal to a change in absolute temperature of 75 K. The magnitudes are the same.

30. **(C)** This relationship describes Boyle's law and is represented by a hyperbola.

31. **(A)** In an ideal gas the absolute temperature is directly proportional to the average kinetic energy of the molecules. Graph A best describes this relationship.

32. **(C)** First we reduce the series portion in the parallel branch:

$$R = 2\ \Omega + 2\ \Omega = 4\ \Omega$$

Now, this 4-Ω resistance is in parallel with the other 4-Ω resistor. Recalling that

$$\frac{1}{R} = \frac{1}{R_1} + \frac{1}{R_2}$$

we find that the next equivalent resistance is equal to 2 Ω. Finally, this resistance is in series with the 3-Ω resistance, giving a final equivalent resistance of $3\ \Omega + 2\ \Omega = 5\ \Omega$.

33. **(E)** With a 5-Ω equivalent resistance, Ohm's law states that the circuit current will be equal to

$$I = \frac{15\ \text{V}}{5\ \Omega} = 3\ \text{A}$$

In a series circuit, the potential drop across each resistor is given by $V = IR$. Now, the 4-Ω resistor is part of a branch that has an equivalent resistance of 2 Ω. Thus, the potential drop across that entire branch is equal to 6 V. The voltmeter recording the potential difference across the 4-Ω resistor will register 6 V also.

34. **(E)** None of the statements is correct about the electroscope described in the problem.

35. **(B)** Capacitors add up directly in parallel and add up reciprocally in series. This is the opposite of what resistors do. Thus, in the parallel branch, the equivalent capacitance is 5 F.
For the two series capacitors, we have that

$$\frac{1}{C} = \frac{1}{2} + \frac{1}{5}$$

which implies $C = 10/7$ F.

36. **(C)** For a resistor at constant temperature we have $R = \rho L/A$, where ρ is the resistivity. When the area is made twice as large and the length one-half as long, the result is a reduction in resistance of one-quarter.

37. **(B)** Both capacitors have the same potential difference, 8 V. Thus, $Q = CV = 16$ C.

38. **(E)** The energy stored in the 8-F capacitor is given by

$$E = \left(\frac{1}{2}\right)QV = 256 \text{ J}$$

39. **(E)** In crossed electric and magnetic fields, the velocity of the electron is given by $\mathbf{v} = \mathbf{E}/\mathbf{B}$. Thus we have

$$\text{KE} = \left(\frac{1}{2}\right)M\mathbf{v}^2 = \left(\frac{1}{2}\right)\frac{M\mathbf{E}^2}{\mathbf{B}^2} = \frac{M\mathbf{E}^2}{2\mathbf{B}^2}$$

40. **(A)** In a uniform perpendicular magnetic field, the centripetal force on the electron is balanced by the magnetic force:

$$\frac{m\mathbf{v}^2}{r} = \mathbf{B}e\mathbf{v}$$

Thus

$$\mathbf{v} = \frac{\mathbf{B}er}{M}$$

41. **(C)** The motional emf = $\mathbf{B}\ell\mathbf{v}$ in this case. Substituting yields

$$\text{emf} = (10 \text{ T})(0.2 \text{ m})(4 \text{ m/s}) = 8 \text{ V}$$

42. **(A)** Lenz's law will provide an opposing force by inducing a magnetic field to oppose the initial change in B.

43. **(E)** In a taut string, the velocity of propagation of transverse waves is proportional to the square root of the tension. Thus, if the tension is doubled, the velocity will increase from **v** to $\sqrt{2}\,\mathbf{v}$.

44. **(C)** In a standing wave, the nodes occur at one-half wavelength intervals. Since the entire segment is 3 m long, each nodal segment is 1.0 m long. Thus, $\lambda = 2.0$ m.

45. **(B)** The velocity of sound increases by approximately 0.6 m/s for each 1°C rise in the temperature of the air. At 25°C, the velocity of sound in air is 346 m/s. At 3 s of travel, the distance to the source is 1,038 m.

46. **(A)** A set of waves is coherent if all the waves have the same amplitude, frequency, and phase.

47. **(A)** In a closed-end tube, the standing wave resonance point occurs at a length, ℓ, equal to $\lambda/4$. Thus, $\lambda = 1.0$ m; and since $\mathbf{v} = f\lambda$, $\mathbf{v} = 340$ m/s in this case.

48. **(C)** In the seventeenth century, the Italian scientist Grimaldi discovered that the fuzzy appearance of the edges of shadows was caused by the diffraction of light near the edges of the object.

49. **(E)** Light will bend away from the normal as it enters air from glass. This is best represented by path *E*.

50. **(E)** The critical angle of incidence for a transparent material is not affected by the described change.

51. **(B)** To find the size of the image we must first locate its position:

$$\frac{1}{f} = \frac{1}{S_o} + \frac{1}{S_i}$$
$$\frac{1}{6} = \frac{1}{10} + \frac{1}{S_i}$$
$$S_i = 15 \text{ cm}$$

The magnification of the image is given by the ratio $\frac{S_i}{S_o} = 1.5$, and so

$$h_i = 5 \times 1.5 = 7.5 \text{ cm}$$

52. **(C)** Diverging mirrors always produce erect virtual images.

53. **(A)** A sound wave is a longitudinal wave that cannot be polarized.

54. **(B)** Standard atmospheric pressure at sea level is equal to 101 kPa.

55. **(B)** The flapping of a flag is due to the reduced pressure as air rushes over one side. This is best explained by Bernoulli's principle which states that the pressure exerted by a moving fluid is reduced with an increase in fluid velocity.

56. **(D)** The SI units of pressure are N/m^2, which can alternately be expressed in units of pascals, bars, or $\text{kg/m} \cdot \text{s}^2$. Thus, choice D is not an equivalent unit.

57. **(A)** The stopping potential of 12 V will imply a maximum kinetic energy of 12 eV for each electron since $KE_{max} = eV$, and for one electron

$$KE_{max} = 12 \text{ V} \times 1 \text{ electron}$$

which equals 12 eV.

58. **(E)** Planck's constant is expressed in units of joule · seconds, and the quantity mc^2 is in units of joules. Thus the ratio of these two quantities leaves seconds (s) as the units.

59. **(E)** The de Broglie wavelength varies inversely with momentum. Since momentum varies directly with mass and all the particles are moving with the same velocity, the smallest de Broglie wavelength would correspond to the most massive particle listed. This particle is the alpha particle, which is equivalent to a helium nucleus.

60. **(C)** The energy of the emitted photon is equal to the difference between energy levels 4 and 2. Thus:

$$-5.7 \text{ eV} - (-8.2 \text{ eV}) = 3.5 \text{ eV}$$

61. **(B)** Since the third level has an energy of −6.3 eV, an electron in this state would need a minimum of 6.3 eV to become ionized.

62. **(C)** In any hydrogen-like atom, the greatest energy emitted is a transition to the lowest energy state, level 1. In this case a transition from level 2 to level 1 emits the most energy.

63. **(B)** By definition, 1 u is equal to 1/12 the mass of the carbon atom.

64. **(C)** The number of neutrons is equal to the difference

$$A - Z = 238 - 92 = 146$$

65. **(B)** Planck's constant $h = 6.6 \times 10^{-34}$ J·s. Since 1 J = 1 N·m = 1 kg·m/s^2, J·s = kg·m^2/s.

66. **(A)** Balancing the superscripts on the two sides of the arrow, we see that the mass number for X must be 31 − 30 = 1.

67. **(B)** Radon decays by alpha-particle emission. Alpha particles are helium nuclei.

68. **(C)** Iron has the most stable nuclear isotope, based on the graph of average binding energy per nucleon.

69. **(B)** If the wave crests are separated by 50 m and traveling at 10 m/s, the time period for the waves is given by $\mathbf{d} = \mathbf{v}t$. This means that $t = 5$ s.

70. **(B)** A water wave has both transverse and longitudinal characteristics.

SECTION II

1. (a) (5 points) A free-body diagram shows that the net force acting on the mass when at the bottom is given by a combination of the tension in the string (directed upward) and the weight of the mass (directed downward):

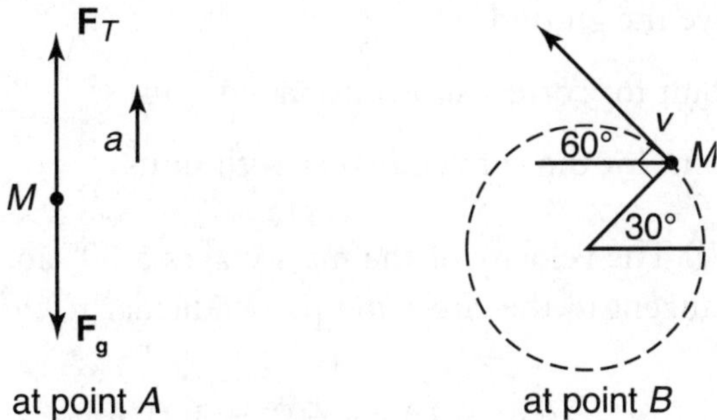

The net force at point A is the centripetal force. Let acceleration upward be in the positive direction:

$$\Sigma \mathbf{F} = m\mathbf{a}$$

$$\mathbf{F}_T - m\mathbf{g} = \frac{m\mathbf{v}^2}{r}$$

$$\mathbf{F}_T = (2 \text{ kg})(9.8 \text{ m/s}^2) + \frac{(2 \text{ kg})(10 \text{ m/s})^2}{(2 \text{ m})} = 119.6 \text{ N}$$

Award 1 point for correctly stating Newton's second law.

Award 2 points for writing the correct equation.

Award 1 point for correct substitutions with units.

Award 1 point for the correct answer with units.

(b) (5 points) When the rope was horizontal, the mass was 2 m above the ground. At point B, given the 30° angle relative to the horizontal, the additional vertical height is equal to:

$$y_B = (2 \text{ m})(\sin 30°) = 1 \text{ m}$$

This means that the height of the mass at point B is equal to 3 m. We can now use the conservation of mechanical energy to obtain the velocity of the mass at point B (assuming that there is effectively zero height at point A):

Total mechanical energy at point A = Total mechanical energy at point B

$$\tfrac{1}{2}\, m(\mathbf{v}_A)^2 = \tfrac{1}{2}\, m(\mathbf{v}_B)^2 + m\mathbf{g}(\Delta h)$$

This expression is independent of mass.

$$\tfrac{1}{2}\,(2 \text{ kg})(10 \text{ m/s})^2 = \tfrac{1}{2}\,(2 \text{ kg})(\mathbf{v}_B)^2 + (2 \text{ kg})(9.8 \text{ m/s}^2)(3\text{m})$$

$$100 \text{ m}^2/\text{s}^2 = \mathbf{v}_B^2 + 58.8 \text{ m}^2/\text{s}^2$$

$$\mathbf{v}_B = 6.42 \text{ m/s}$$

Answer Explanations

Award 1 point for recognizing the conservation of mechanical energy.

Award 1 point for writing the correct expression for the conservation of mechanical energy.

Award 1 point for recognizing that the height of the mass at point *B* is equal to 3 m above the ground.

Award 1 point for correct substitution with units.

Award 1 point for the correct answer with units.

(c) (5 points) (i) The velocity of the mass makes a 60° angle to the horizontal since it is tangent to the circle and perpendicular to the radius of the rope. Therefore:

$$\mathbf{v}_x = (6.42 \text{ m/s})(\cos 60°) = 3.21 \text{ m/s}$$
$$\mathbf{v}_y = (6.42 \text{ m/s})(\sin 60°) = 5.56 \text{ m/s}$$

(ii) We can calculate the maximum height above the release point using:

$$y_{max} = \frac{(\mathbf{v}_y)^2}{2\mathrm{g}} = \frac{(5.56 \text{ m/s})^2}{(19.6 \text{ m/s}^2)} = 1.58 \text{ m}$$

The total height above the ground is then equal to $y_{max} + 3 \text{ m} = 4.58 \text{ m}$.

Award 1 point for recognizing that the velocity makes a 60° angle to the horizontal.

Award 1 point for each correct velocity component with units (for a total of 2 points).

Award 2 points for correctly solving for the total height above the ground with correct units.

2. (a) (4 points) The free-body diagrams for the top (i) and bottom (ii) situations are given below:

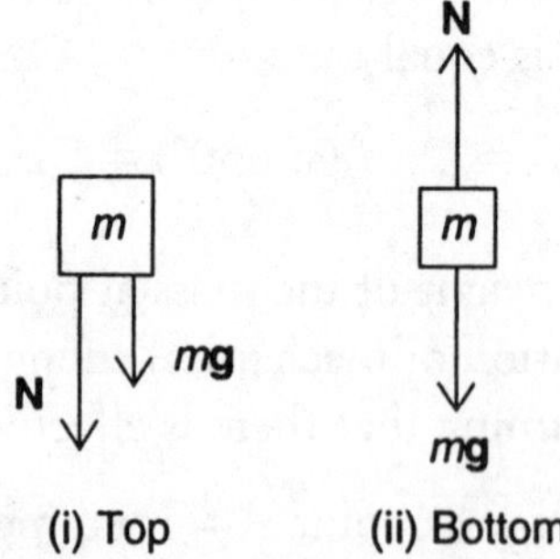

Award 1 point each for the correct four forces drawn and labeled for the top and the bottom.

(b) (6 points) (i) For the top:

$$\Sigma\mathbf{F} = -m\mathbf{a} = -\frac{m\mathbf{v}^2}{R} = -\mathbf{N} - m\mathbf{g}$$

This implies that

$$\mathbf{N} = \frac{m\mathbf{v}^2}{R} - m\mathbf{g}$$

(ii) For the bottom:

$$\Sigma\mathbf{F} = m\mathbf{a} = \frac{m\mathbf{v}^2}{R} = \mathbf{N} - m\mathbf{g}$$

This implies that

$$\mathbf{N} = \frac{m\mathbf{v}^2}{R} + m\mathbf{g}$$

Award 1 point for showing at the top that $\mathbf{F} = m\mathbf{a}$ and two points for the correct expression.

Award 1 point for again showing that at the bottom, $\mathbf{F} = m\mathbf{a}$ and two points for the correct expression.

(c) We want the reaction force **N** at the top to be equal to $m\mathbf{g}$. Thus,

$$\mathbf{N} = \frac{m\mathbf{v}^2}{R} - m\mathbf{g}$$

$$m\mathbf{g} = \frac{m\mathbf{v}^2}{R} - m\mathbf{g}$$

$$2\mathbf{g}R = \mathbf{v}^2$$

We now need to find the square of the velocity, using conservation of energy:

$$m\mathbf{g}h = m\mathbf{g}(2R) + \left(\frac{1}{2}\right)m\mathbf{v}^2$$

$$\mathbf{v}^2 = 2\mathbf{g}h - 4\mathbf{g}R$$

Substituting, we get

$$2\mathbf{g}R = 2\mathbf{g}h - 4\mathbf{g}R$$

which implies that $h = 3R$.

Award 1 point for recognizing that energy is conserved.

Award 1 point for solving for the square of the velocity.

Award 3 points for the remaining correct solving for $h = 3R$.

3. (a) (6 points) For diffraction our relationship is

$$n\lambda = d \sin\theta$$

The information given is that $\lambda = 501.5$ nm, $n = 2$, and $\theta = 45°$. Solving for d, we get $d = 1.433 \times 10^{-6}$ m. The reciprocal of d is the number of lines per meter. Thus

$$\frac{1}{d} = 69{,}706.28 \text{ lines/m} = 698 \text{ lines/mm}$$

Award 2 points for the correct equation for the diffraction pattern.

Award 2 points for a correct solution for *d*, showing all substitutions with units.

Award 2 points for the correct separation of the gradient slits.

(b) (3 points) Along with the value for *d* obtained in part (a), we use $n = 1$ and $\lambda = 667.8$ nm to solve for θ. Direct substitution yields $\theta = 27.8°$.

Award 3 points for the correct answer for the angle showing all substitutions with units.

(c) (6 points) The diffraction displacement of a color depends directly on its wavelength. Since violet light has a shorter wavelength than red light, it appears closer (is diffracted least) to the center. A prism refracts and disperses colors based on their frequency. Since the frequency of violet light is greater than that of red light, it is refracted more and appears after the red light (ROYGBIV).

Award 6 points for a satisfactory explanation about the positioning of the various colors based on frequency and wavelength in the electromagnetic spectrum.

4. The work done is going to be equal to the area under each segment of the P-V graph. Also, recall that 1 atm = 101 kPa = 101,000 N/m^2 and 1 L = 0.001 m^3.

   i. Along path *ABC*, there is no work done from $B \rightarrow C$ because there is no change in volume.

   Thus, the area is just the area under $AB = (8 \text{ atm})(6 \text{ L}) = 4{,}848$ J.

   ii. Along path *AC*, the total area is equal to:

   $$\frac{1}{2}(6 \text{ atm})(6 \text{ L}) + (2 \text{ atm})(6 \text{ L}) = 3030 \text{ J}.$$

   iii. Along path *ADC*, there is no work done from $A \rightarrow D$ because there is no change in volume.

   Thus, the area is just the area under $DC = (2 \text{ atm})(6 \text{ L}) = 1212$ J.

   Award 5 points for each correct subsection (i, ii, and iii).

5. (a) (4 points) To find the mass defect we first find the total constituent mass:

$$\begin{aligned} 88 \text{ protons: } m &= (88)(1.0078) = 88.6864 \text{ u} \\ 138 \text{ neutrons: } m &= (138)(1.0087) = 139.2006 \text{ u} \\ M(\text{total}) &= 227.8870 \text{ u} \end{aligned}$$

Then we have

$$\text{Mass defect} = 227.8870 \text{ u} - 226.0244 \text{ u} = 1.8626 \text{ u}$$

Award 1 point for the correct mass for the total number of protons.

Award 1 point for the correct mass for the total number of neutrons.

Award 2 points for the correct value of the mass defect.

Answer Explanations

(b) (4 points) The total binding energy is

$$\text{BE} = (931.5\ \text{MeV/u})(1.8626\ \text{u}) = 1735.0119\ \text{MeV}$$

The binding energy per nucleon is

$$\text{BE/nucleon} = 1735.0119/226 = 7.677\ \text{Mev}$$

Award 2 points for the correct value for the binding energy showing all substitutions and units.

Award 2 points for the correct value for the average binding energy per nucleon.

(c) (2 points) The equivalent mass of quantity Q is given by

$$Q = 226.0244 - 222.0165 - 4.0026 = 0.0053\ \text{u}$$

Then the energy associated with Q is

$$E = (931.5\ \text{MeV/u})(0.0053\ \text{u}) = 4.94\ \text{MeV}$$

Award 2 points for the correct value of the energy associated with the equivalent mass Q.

6. (a) (6 points)

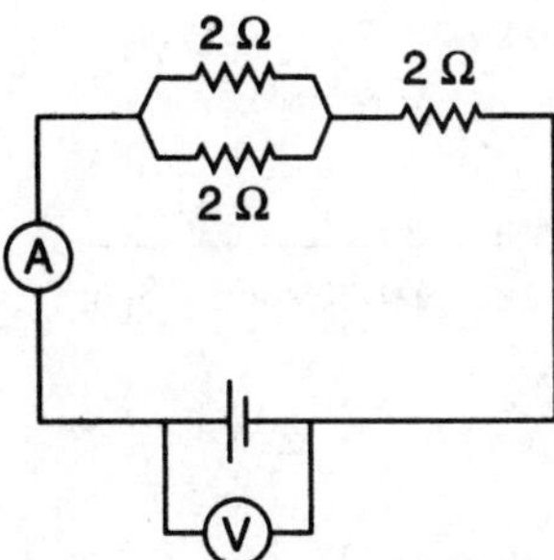

Award a total of 6 points. Award 1 point for each correctly drawn item.

(b) (2 points) Ohm's law relates the voltage and current in a circuit. It states that for solid conductors, with no heat losses, the change in voltage is directly proportional to the change in current.

Award 2 points for the correct statement of Ohm's law.

(c) (2 points) The light bulbs get hot very quickly, which increases their electrical resistance. This, in turn, reduces the current measured by the ammeter.

Award 2 points for a correct explanation.

Test Analysis
Practice Test 1

Section I: Multiple-Choice

Number correct (out of 70) = ____________

Raw Score = number correct × .714 = ____________
Multiple-Choice Score

Section II: Free-Response

Question 1 = ____________
(out of 15)

Question 2 = ____________
(out of 15)

Question 3 = ____________
(out of 15)

Question 4 = ____________
(out of 15)

Question 5 = ____________
(out of 10)

Question 6 = ____________
(out of 10)

Raw Score: ____________ × .625 = ____________
Free-Response Score

Final Score

____________ + ____________ = ____________

Multiple-Choice Score + Free-Response Score = Final Score (rounded to the nearest whole number)

Final Score Range	AP Score*
81–100	5
61–80	4
51–60	3
41–50	2
0–40	1

*The score range corresponding to each grade varies from exam to exam and is approximate.

SELF-ASSESSMENT GUIDE

How well did you do? Remember that on the actual AP Physics B exam, grading and scoring are based on the year and guidelines set up by the College Board for the readers who will be grading your exam. Use the results of this assessment only as a guide to further your studying and not as an absolute predictor of an AP grade.

Use the table below to help you locate the topics in the book for which you need further study.

Topic	Multiple-Choice Question Number
Motion	1, 3, 6, 8, 18
Forces and Momentum	2, 4, 5, 7, 9, 10, 11, 12, 14, 15, 17, 19, 21, 22, 25, 55, 56
Work and Energy	13, 16, 20, 23, 24
Heat and Gases	26, 27, 28, 29, 30, 31, 54
Waves and Sound	43, 44, 45, 46, 47, 69, 70
Light and Optics	48, 49, 50, 51, 52, 53
Electricity and Magnetism	32, 33, 34, 35, 36, 37, 38, 39, 40, 41, 42
Modern Physics	57, 58, 59, 60, 61, 62, 63, 64, 65, 66, 67, 68

SCORE IMPROVEMENT

Not satisfied with your score? Don't worry. Here are some tips for improvement.

For the multiple-choice section:

1. Write the numbers of the questions you left blank or answered incorrectly in the first column below.
2. Go over the answers to the problems in the Answers Explained section and write the main ideas and concepts behind the problems in the second column.
3. Go back and reread the sections covering the material in the second column.
4. Retake the skipped and incorrect questions.
5. Recalculate your score to see how much you improved.

Questions	Main Idea or Concept

For the free-response section:

1. Go over the answers to the problems in the Answers Explained section, and circle any mistakes in your answers.
2. Go back and reread the sections covering the material.
3. Retake the missed free-response questions.
4. Recalculate your score to see how much you improved.

Answer Sheet

PRACTICE TEST 2

1 Ⓐ Ⓑ Ⓒ Ⓓ Ⓔ	25 Ⓐ Ⓑ Ⓒ Ⓓ Ⓔ	49 Ⓐ Ⓑ Ⓒ Ⓓ Ⓔ
2 Ⓐ Ⓑ Ⓒ Ⓓ Ⓔ	26 Ⓐ Ⓑ Ⓒ Ⓓ Ⓔ	50 Ⓐ Ⓑ Ⓒ Ⓓ Ⓔ
3 Ⓐ Ⓑ Ⓒ Ⓓ Ⓔ	27 Ⓐ Ⓑ Ⓒ Ⓓ Ⓔ	51 Ⓐ Ⓑ Ⓒ Ⓓ Ⓔ
4 Ⓐ Ⓑ Ⓒ Ⓓ Ⓔ	28 Ⓐ Ⓑ Ⓒ Ⓓ Ⓔ	52 Ⓐ Ⓑ Ⓒ Ⓓ Ⓔ
5 Ⓐ Ⓑ Ⓒ Ⓓ Ⓔ	29 Ⓐ Ⓑ Ⓒ Ⓓ Ⓔ	53 Ⓐ Ⓑ Ⓒ Ⓓ Ⓔ
6 Ⓐ Ⓑ Ⓒ Ⓓ Ⓔ	30 Ⓐ Ⓑ Ⓒ Ⓓ Ⓔ	54 Ⓐ Ⓑ Ⓒ Ⓓ Ⓔ
7 Ⓐ Ⓑ Ⓒ Ⓓ Ⓔ	31 Ⓐ Ⓑ Ⓒ Ⓓ Ⓔ	55 Ⓐ Ⓑ Ⓒ Ⓓ Ⓔ
8 Ⓐ Ⓑ Ⓒ Ⓓ Ⓔ	32 Ⓐ Ⓑ Ⓒ Ⓓ Ⓔ	56 Ⓐ Ⓑ Ⓒ Ⓓ Ⓔ
9 Ⓐ Ⓑ Ⓒ Ⓓ Ⓔ	33 Ⓐ Ⓑ Ⓒ Ⓓ Ⓔ	57 Ⓐ Ⓑ Ⓒ Ⓓ Ⓔ
10 Ⓐ Ⓑ Ⓒ Ⓓ Ⓔ	34 Ⓐ Ⓑ Ⓒ Ⓓ Ⓔ	58 Ⓐ Ⓑ Ⓒ Ⓓ Ⓔ
11 Ⓐ Ⓑ Ⓒ Ⓓ Ⓔ	35 Ⓐ Ⓑ Ⓒ Ⓓ Ⓔ	59 Ⓐ Ⓑ Ⓒ Ⓓ Ⓔ
12 Ⓐ Ⓑ Ⓒ Ⓓ Ⓔ	36 Ⓐ Ⓑ Ⓒ Ⓓ Ⓔ	60 Ⓐ Ⓑ Ⓒ Ⓓ Ⓔ
13 Ⓐ Ⓑ Ⓒ Ⓓ Ⓔ	37 Ⓐ Ⓑ Ⓒ Ⓓ Ⓔ	61 Ⓐ Ⓑ Ⓒ Ⓓ Ⓔ
14 Ⓐ Ⓑ Ⓒ Ⓓ Ⓔ	38 Ⓐ Ⓑ Ⓒ Ⓓ Ⓔ	62 Ⓐ Ⓑ Ⓒ Ⓓ Ⓔ
15 Ⓐ Ⓑ Ⓒ Ⓓ Ⓔ	39 Ⓐ Ⓑ Ⓒ Ⓓ Ⓔ	63 Ⓐ Ⓑ Ⓒ Ⓓ Ⓔ
16 Ⓐ Ⓑ Ⓒ Ⓓ Ⓔ	40 Ⓐ Ⓑ Ⓒ Ⓓ Ⓔ	64 Ⓐ Ⓑ Ⓒ Ⓓ Ⓔ
17 Ⓐ Ⓑ Ⓒ Ⓓ Ⓔ	41 Ⓐ Ⓑ Ⓒ Ⓓ Ⓔ	65 Ⓐ Ⓑ Ⓒ Ⓓ Ⓔ
18 Ⓐ Ⓑ Ⓒ Ⓓ Ⓔ	42 Ⓐ Ⓑ Ⓒ Ⓓ Ⓔ	66 Ⓐ Ⓑ Ⓒ Ⓓ Ⓔ
19 Ⓐ Ⓑ Ⓒ Ⓓ Ⓔ	43 Ⓐ Ⓑ Ⓒ Ⓓ Ⓔ	67 Ⓐ Ⓑ Ⓒ Ⓓ Ⓔ
20 Ⓐ Ⓑ Ⓒ Ⓓ Ⓔ	44 Ⓐ Ⓑ Ⓒ Ⓓ Ⓔ	68 Ⓐ Ⓑ Ⓒ Ⓓ Ⓔ
21 Ⓐ Ⓑ Ⓒ Ⓓ Ⓔ	45 Ⓐ Ⓑ Ⓒ Ⓓ Ⓔ	69 Ⓐ Ⓑ Ⓒ Ⓓ Ⓔ
22 Ⓐ Ⓑ Ⓒ Ⓓ Ⓔ	46 Ⓐ Ⓑ Ⓒ Ⓓ Ⓔ	70 Ⓐ Ⓑ Ⓒ Ⓓ Ⓔ
23 Ⓐ Ⓑ Ⓒ Ⓓ Ⓔ	47 Ⓐ Ⓑ Ⓒ Ⓓ Ⓔ	
24 Ⓐ Ⓑ Ⓒ Ⓓ Ⓔ	48 Ⓐ Ⓑ Ⓒ Ⓓ Ⓔ	

Table of Information

Useful Constants

1 atomic mass unit	$1\ u = 1.66 \times 10^{-27}$ kg
Rest mass of the proton	$m_p = 1.67 \times 10^{-27}$ kg
Rest mass of the neutron	$m_n = 1.67 \times 10^{-27}$ kg
Rest mass of the electron	$m_e = 9.11 \times 10^{-31}$ kg
Magnitude of the electron charge	$e = 1.60 \times 10^{-19}$ C
Avogadro's number	$N_0 = 6.02 \times 10^{23}$ per mol
Universal gas constant	$R = 8.32$ J/(mol · K)
Boltzmann's constant	$k_B = 1.38 \times 10^{-23}$ J/K
Speed of light	$c = 3 \times 10^8$ m/s
Planck's constant	$h = 6.63 \times 10^{-34}$ J · s $= 4.14 \times 10^{-15}$ eV · s
1 electron volt	1 eV $= 1.6 \times 10^{-19}$ J
Vacuum permittivity	$\varepsilon_0 = 8.85 \times 10^{-12}$ C^2/N · m^2
Coulomb's law constant	$k = (1/4)\pi\varepsilon_0 = 9 \times 10^9$ N · m^2/C^2
Vacuum permeability	$\mu_0 = 4\pi \times 10^{-7}$ Wb/(A · m)
Magnetic constant	$k' = k/c^2 = \mu_0/4\pi = 10^{-7}$ Wb/(A · m)
Acceleration due to gravity at Earth's surface	$\mathbf{g} = 9.8$ m/s^2
Universal gravitational constant	$G = 6.67 \times 10^{-11}$ m^3/(kg · s^2)
1 atmosphere pressure	1 atm $= 1.0 \times 10^5$ N/m^2 $= 1.0 \times 10^5$ Pa
1 nanometer	1 nm $= 1.0 \times 10^{-9}$ m

Unit Symbols

Meter, m	Mole, mol	Watt, W	Farad, F
Kilogram, kg	Hertz, Hz	Coulomb, C	Tesla, T
Second, s	Newton, N	Volt, V	Degree Celsius, °C
Ampere, A	Pascal, Pa	Ohm, Ω	Electron volt, eV
Kelvin, K	Joule, J	Henry, H	

Prefixes

Factor	**Prefix**	**Symbol**
10^9	Giga-	G
10^6	Mega-	M
10^3	Kilo-	k
10^{-2}	Centi-	c
10^{-3}	Milli-	m
10^{-6}	Micro-	μ
10^{-9}	Nano-	n
10^{-12}	Pico-	p

Values of Trigonometric Functions for Common Angles

θ	**0°**	**30°**	**37°**	**45°**	**53°**	**60°**	**90°**
sin θ	0	$\frac{1}{2}$	$\frac{3}{5}$	$\frac{\sqrt{2}}{2}$	$\frac{4}{5}$	$\frac{\sqrt{3}}{2}$	1
cos θ	1	$\frac{\sqrt{3}}{2}$	$\frac{4}{5}$	$\frac{\sqrt{2}}{2}$	$\frac{3}{5}$	$\frac{1}{2}$	0
tan θ	0	$\frac{\sqrt{3}}{3}$	$\frac{3}{4}$	1	$\frac{4}{3}$	$\sqrt{3}$	∞

Practice Test 2

Section I

MULTIPLE-CHOICE QUESTIONS

70 QUESTIONS

90 MINUTES

50 PERCENT OF TOTAL GRADE

Directions: For each of the questions or incomplete statements below there are five choices. In each case select the best answer or completion and fill in the corresponding oval on the answer sheet. You may not use a calculator for Section I.

I. All frames of reference are assumed to be inertial.
II. Electrical current will follow the direction of a positive charge (conventional current).
III. For any isolated charge, the potential at infinity is taken to be equal to zero.
IV. The work done by a thermodynamic system is defined as a positive quantity.

1. A vector has the following components: $\mathbf{a}_x = 3$ units and $\mathbf{a}_y = 5$ units. Which of the following statements is correct concerning the angle that this vector makes with the positive *x*-axis?

   (A) It is less than 45°.
   (B) It is equal to 45°.
   (C) It is greater than 45° but less than 90°.
   (D) It is equal to 90°.
   (E) It is greater than 90°.

2. Two vectors, **A** and **B**, have components $\mathbf{a}_x = -2$, $\mathbf{a}_y = 3$, $\mathbf{b}_x = 1$, and $\mathbf{b}_y = 4$. What is the approximate magnitude of the vector **A** + **B**?

   (A) 3
   (B) 6
   (C) 8
   (D) 5
   (E) 7

3. As the angle between two vectors increases from 0° to 180°, the magnitude of their resultant

(A) increases, only
(B) increases and then decreases
(C) decreases, only
(D) decreases and then increases
(E) None of the above

4. A car with a 500-newton driver goes over a hill that has a radius of 50 meters, as shown below. The velocity of the car is 20 meters per second. What are the approximate force and the direction that the car exerts on the driver?

(A) 900 N, up
(B) 100 N, down
(C) 100 N, up
(D) 900 N, down
(E) 500 N, up

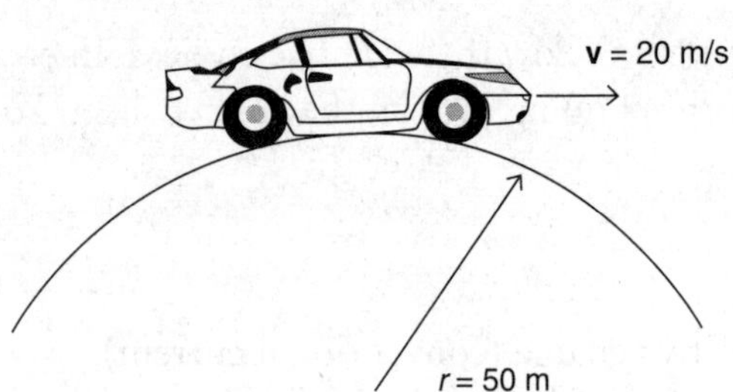

5. At what angle should a projectile be launched in order to achieve the maximum range for a given initial velocity?

(A) 0°
(B) 30°
(C) 45°
(D) 60°
(E) 90°

6. An object is dropped from a height of 45 m. Neglecting air resistance, what is the approximate velocity of the object as it hits the ground?

(A) 10 m/s
(B) 15 m/s
(C) 20 m/s
(D) 25 m/s
(E) 30 m/s

7. A boat moving due north crosses a 190-meter-wide river with a velocity of 8 meters per second relative to the water. The river flows east with a velocity of 4 meters per second. How long will the boat take to cross the river?

(A) 27 s
(B) 24 s
(C) 21 s
(D) 29 s
(E) 26 s

Questions 8 and 9 are based on the following information:

A variable force acts on a 2-kilogram mass according to the graph below.

8. How much work was done to displace the mass 10 meters?

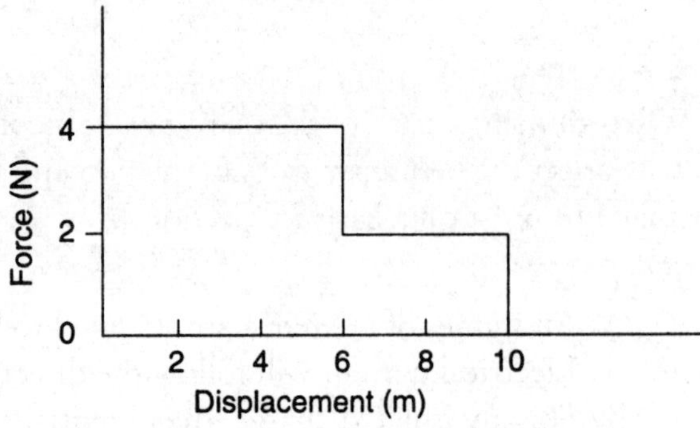

(A) 40 J
(B) 38 J
(C) 32 J
(D) 30 J
(E) 26 J

9. What was the average force supplied to the mass for the entire 10-meter displacement?

(A) 3.2 N
(B) 1.2 N
(C) 4.4 N
(D) 4 N
(E) 2.6 N

10. A block is pushed along a frictionless surface for a distance of 2.5 meters, as shown below. How much work has been done if a force of 10 newtons makes an angle of 60 degrees with the horizontal?

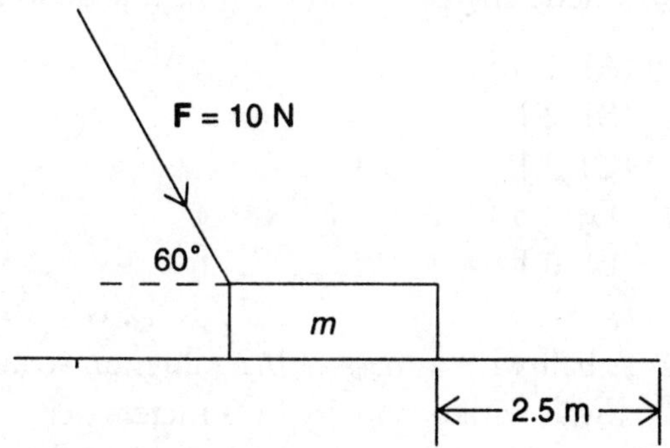

(A) 10.5 J
(B) 8.5 J
(C) 6.5 J
(D) 12.5 J
(E) 25 J

11. What is the instantaneous power due to the gravitational force acting on a 3-kilogram projectile the instant the projectile is traveling with a velocity of 10 meters per second at an angle of 30 degrees to the horizontal?

(A) 15 W
(B) 147 W
(C) −29 W
(D) −15 W
(E) −147 W

12. A person weighing 490 newtons is standing on a bathroom scale in an elevator that is accelerating upward at a rate of 0.2 meter per second squared. What does the bathroom scale indicate the person's weight is?

(A) 480 N
(B) 490 N
(C) 588 N
(D) 500 N
(E) 392 N

13. Two tuning forks are struck at the same time. A beat frequency of 12 beats per second is observed. If one tuning fork has a frequency of 384 Hz, the frequency of the second tuning fork could be equal to

(A) 256 Hz
(B) 372 Hz
(C) 360 Hz
(D) 408 Hz
(E) 390 Hz

14. A hockey puck with a mass of 0.2 kilogram is sliding along ice that can be considered frictionless with a velocity of 10 meters per second. The puck then crosses over onto a rough floor that has a coefficient of kinetic friction equal to 0.2. How far will the puck travel before friction stops it?

(A) 17 m
(B) 42 m
(C) 25 m
(D) 60 m
(E) 65 m

15. A spring with a force constant k is stretched a distance x. By what factor must the spring's elongation be changed so that the elastic potential energy in the spring is doubled?

(A) 1/4
(B) 1/2
(C) 2
(D) 4
(E) $\sqrt{2}$

Questions 16 and 17 are based on the following information:

A 10-kg projectile is launched at a 60° angle to the ground with a velocity of 200 m/s. Neglect air resistance.

16. What is the value of the initial horizontal velocity?

(A) 100 m/s
(B) 200 m/s
(C) 141 m/s
(D) 172 m/s
(E) 50 m/s

17. As the launch angle is lowered to 45°, the maximum horizontal distance traveled by the projectile will

(A) decrease, only
(B) increase only
(C) increase, then decrease
(D) decrease, then increase
(E) remain the same

18. Which of the following is an expression for power?

(A) $\mathbf{F}t/m$
(B) $\mathbf{F}m/t$
(C) $\mathbf{F}^2m/t$
(D) $\mathbf{F}m^2/t$
(E) $\mathbf{F}^2t/m$

19. A 0.2 kilogram mass rests on top of a spring that has been compressed by 0.4 meter. The force constant for this spring is 20 newtons per meter. When released, how high will the mass rise?

(A) 1.24 m
(B) 0.75 m
(C) 0.80 m
(D) 1.04 m
(E) 1.34 m

20. Which of the following is an equivalent expression of the units for a spring's force constant *k*?

(A) $\text{kg} \cdot \text{m}^2/\text{s}^2$
(B) $\text{kg} \cdot \text{m/s}$
(C) $\text{kg} \cdot \text{s}^2$
(D) kg/s^2
(E) $\text{kg} \cdot \text{s/m}$

21. What braking force is applied to a 2500-kilogram car having a velocity of 30 meters per second if the car is brought to a stop in 15 seconds?

(A) 5000 N
(B) 6000 N
(C) 8000 N
(D) 10,000 N
(E) 12,000 N

22. A 1-kilogram object is moving to the right with a velocity of 6 meters per second. It collides with, and sticks to, a 2-kilogram mass, also moving to the right, with a velocity of 3 meters per second. How much kinetic energy was lost in this interaction?

(A) 1.5 J
(B) 2 J
(C) 3 J
(D) 3.5 J
(E) 0 J

23. A ball with a mass of 0.2 kilogram strikes a wall with a velocity of 3 meters per second. It bounces straight back with a velocity of 1 meter per second. What was the magnitude of the change in momentum for this ball?

(A) 0.1 kg · m/s
(B) 0.8 kg · m/s
(C) 0.4 kg · m/s
(D) 0.6 kg · m/s
(E) 0.7 kg · m/s

24. A 50-kilogram person is sitting on a seesaw 1.2 meters from the balance point. On the other side, a 70-kilogram person is balanced. How far from the balance point is the second person sitting?

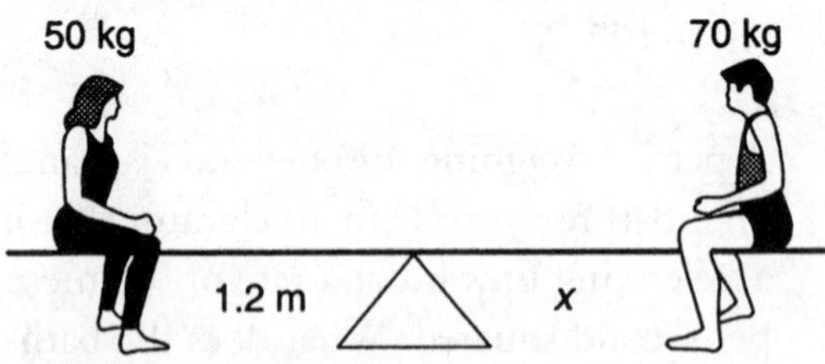

(A) 0.57 m
(B) 0.75 m
(C) 0.63 m
(D) 0.86 m
(E) 1.2 m

Practice Test 2

25. What is the net torque acting on the pivot supporting a 2-meter-long, 10-kilogram beam, as shown below?

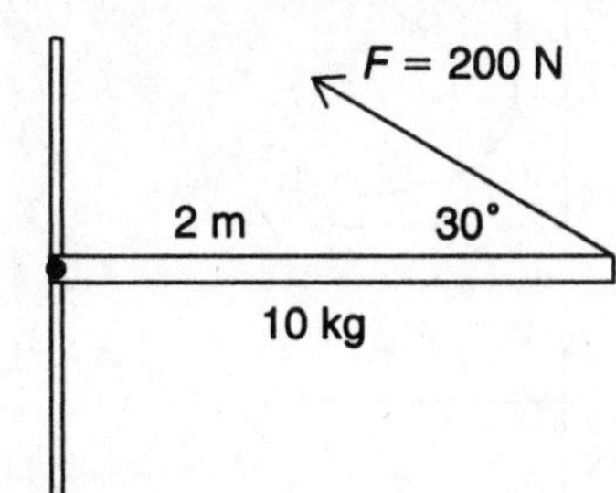

(A) 198 N · m
(B) −198 N · m
(C) −102 N · m
(D) 102 N · m
(E) −120 N · m

26. Which of the following is an equivalent expression for the maximum velocity attained by a mass oscillating horizontally along a frictionless surface? The mass is attached to a spring with a force constant *k* and has an amplitude of oscillations of *A* meters.

(A) $\sqrt{Ak/m}$

(B) $\sqrt{A^2mk}$

(C) $\sqrt{A^2m/k}$

(D) $\sqrt{A^2k/m}$

(E) Akm

27. A 0.5-kilogram mass is attached to a spring with a force constant of 50 newtons per meter. What is the total energy stored in the mass-spring system if the amplitude of oscillations is 2 centimeters?

(A) 0.01 J
(B) 0.1 J
(C) 0.5 J
(D) 0.3 J
(E) 0.2 J

28. What is the value of **g** at a position above Earth's surface that is equal to Earth's radius?

(A) 9.8 m/s^2
(B) 4.9 m/s^2
(C) 3.35 m/s^2
(D) 2.45 m/s^2
(E) 1.6 m/s^2

29. Which of the following expressions is equivalent to the magnitude of the escape velocity in terms of the magnitude of the orbital velocity $\mathbf{v}_0$ for a spacecraft?

(A) $\sqrt{2\mathbf{v}_0}$
(B) $\mathbf{v}_0$
(C) $\mathbf{v}_0^{\,2}$
(D) $2\mathbf{v}_0$
(E) $\sqrt{\mathbf{v}_0}$

30. Which of the following graphs correctly shows the relationship between gravitational force and distance between two masses?

(A)

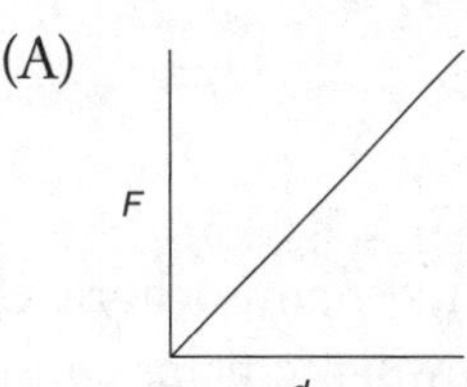

(B)

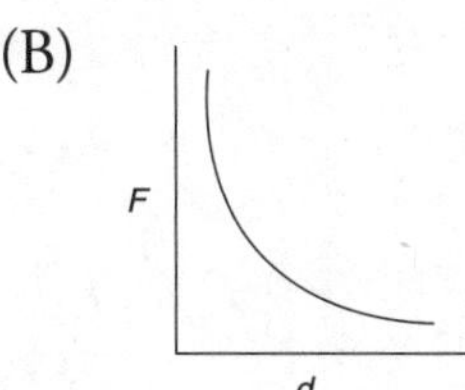

(C)

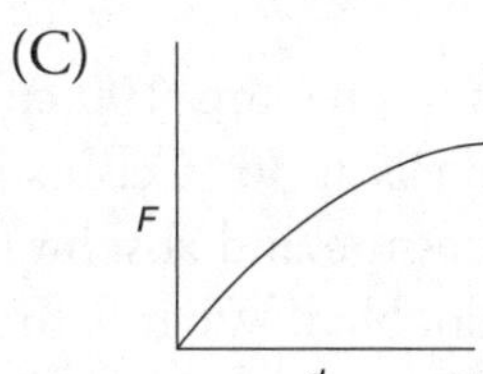

(D)

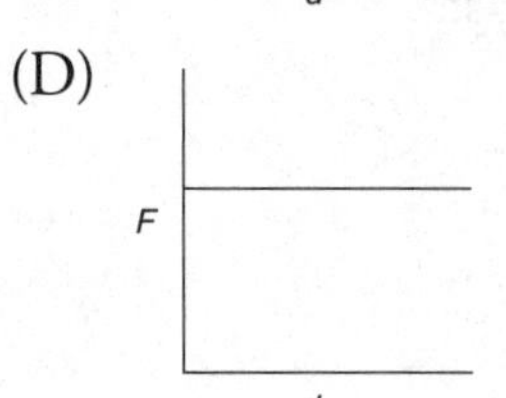

(E)

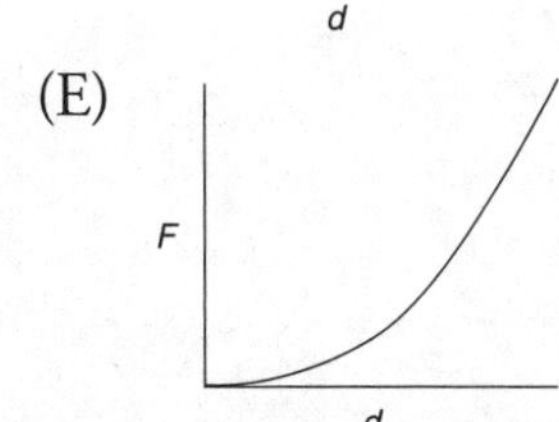

31. The function of a refrigerator is to

(A) add cold to an area
(B) remove heat from an area
(C) convert heat into cold
(D) increase entropy
(E) compress a gas

32. Heat transfer may be performed by

(A) radiation only
(B) conduction, only
(C) convection, only
(D) conduction and convection
(E) conduction, convection, and radiation

33. Standing wave nodes in a string occur every 20 cm. If the velocity of the incident waves is equal to 5 m/s, the frequency of the waves is equal to

(A) 25 Hz
(B) 5 Hz
(C) 2 Hz
(D) 12.5 Hz
(E) 1 Hz

34. An ideal gas undergoes an isochoric change. Which of the following remains constant?

(A) Phase
(B) Temperature
(C) Internal energy
(D) Pressure
(E) Volume

35. At 2 atmospheres of pressure, 100 cubic meters of an ideal gas at 50° Celsius is heated until its pressure and absolute temperature are doubled. What is the new volume of the gas?

(A) 62 m^3
(B) 100 m^3
(C) 58 m^3
(D) 50 m^3
(E) 112 m^3

36. Which of the following graphs represents an isothermal expansion of an ideal gas?

(A) 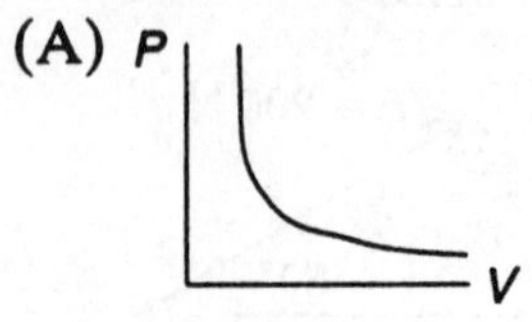

(B) 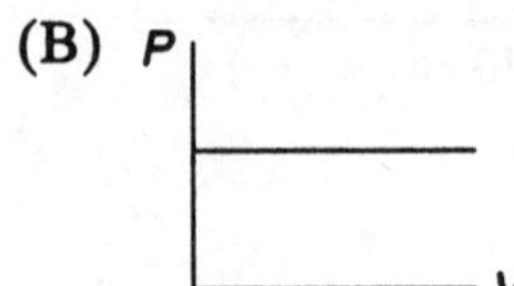

(C) 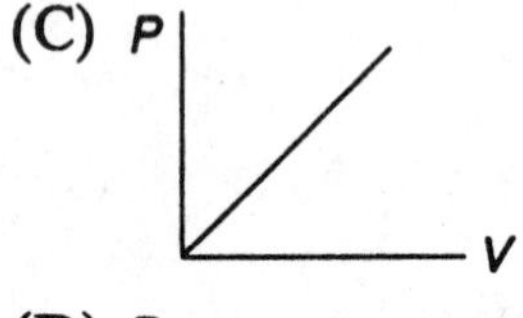

(D) 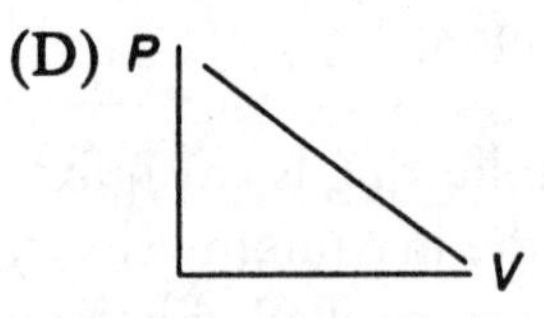

(E) 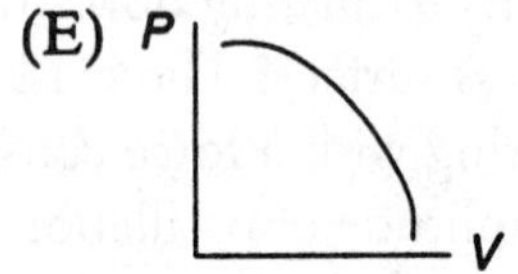

37. At a temperature of 30° Celsius, 1 mole of hydrogen gas and 1 mole of nitrogen gas must have the same

(A) density
(B) number of molecules
(C) root mean square velocity
(D) volume
(E) pressure

38. Which of the following graphs best represents the relationship between absolute temperature and volume for an ideal confined gas at constant pressure?

(A)
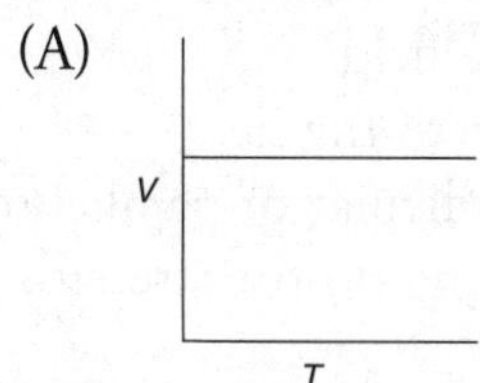

(B)
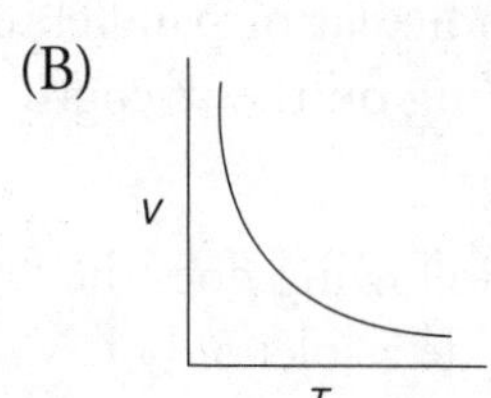

(C)
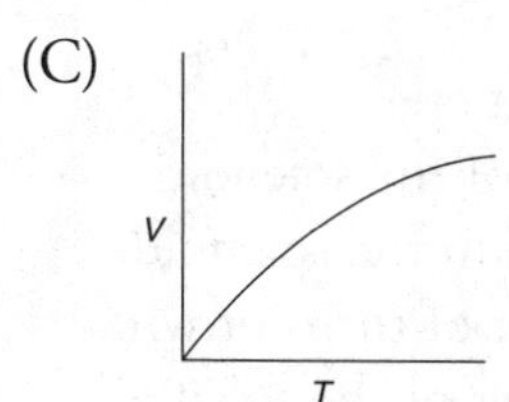

(D)
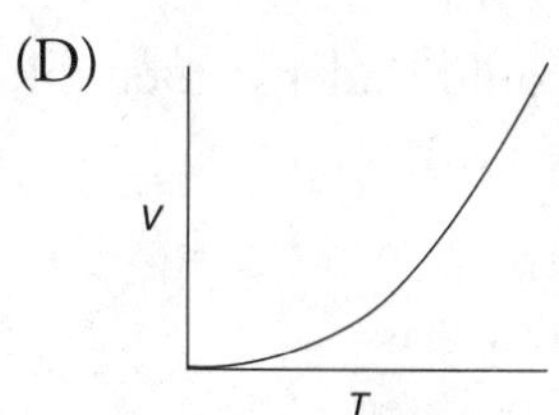

(E)
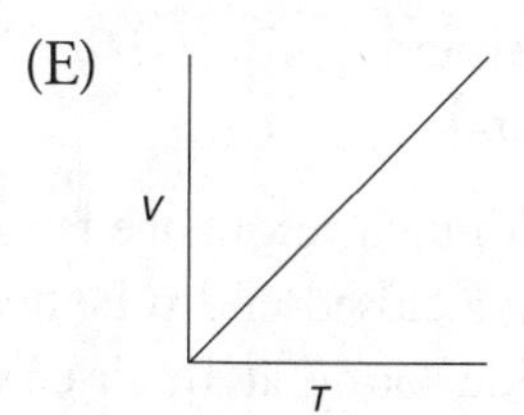

39. Which of the following statements about the adiabatic expansion of an ideal gas is correct?

(A) The temperature may change during the expansion.
(B) The temperature must remain constant during the expansion.
(C) There will be no change in the internal energy.
(D) The gas cannot do any work during the expansion.
(E) The entropy of the gas will decrease.

40. Which of the following processes is not involved in an ideal Carnot cycle?

(A) Isothermal expansion
(B) Isobaric expansion
(C) Adiabatic expansion
(D) Adiabatic compression
(E) Isothermal compression

41. A charged rod attracts a suspended pith ball. Which of the following statements is correct?

(A) The pith ball has the same sign charge as the rod.
(B) The rod must be negatively charged.
(C) The pith ball might be neutral.
(D) The pith ball could be grounded.
(E) The pith ball could not have a metal coating.

42. Which of the following units is equivalent to the unit of electric field strength N/C?

(A) V/m
(B) C/s
(C) V · m
(D) N/V
(E) J/C

43. The ratio of Coulomb's constant k to the magnetic constant k' is equal, where **c** is the speed of light, to

(A) $\mathbf{c}^2$
(B) $\sqrt{\mathbf{c}}$
(C) **c**
(D) 2**c**
(E) **c**/2

44. One farad is equal to one unit of electrical capacitance. Which of the following units is equivalent to one farad?

(A) J/C
(B) V/C
(C) C/V
(D) N/C
(E) C/m

45. A circuit consists of two 10-microfarad capacitors in series, which are then connected in parallel to a 5-microfarad capacitor, as shown below. What is the equivalent capacitance of this circuit?

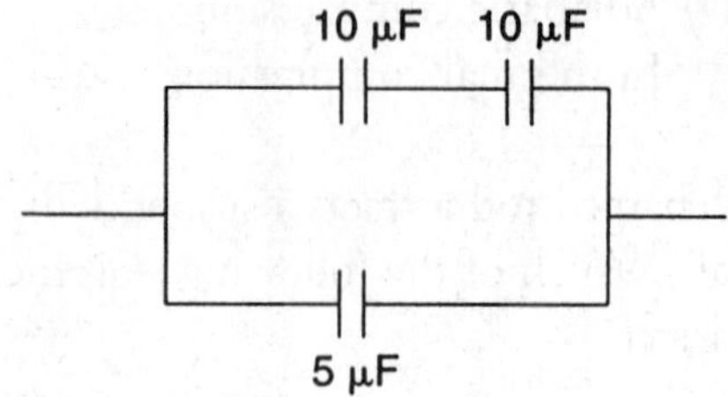

(A) 10 μF
(B) 5 μF
(C) 2 μF
(D) 4 μF
(E) 3 μF

46. Which of the following units is equivalent to 1 ohm of resistance?

(A) kg · m^2/C^2
(B) kg · m/C
(C) kg · m^2/sC^2
(D) kg · C^2/m · s^2
(E) kg · m · C/s

47. An electric motor operates from a 120-volt source, using 4 amperes of current. What is the power rating of this motor?

(A) 30 W
(B) 48 W
(C) 120 W
(D) 480 W
(E) 124 W

48. Which of the following correctly describes the magnetic field near a long, straight wire?

(A) The field consists of straight lines perpendicular to the wire.
(B) The field consists of straight lines parallel to the wire.
(C) The field consists of radial lines originating from the wire.
(D) The field consists of spirals wrapping around the wire.
(E) The field consists of concentric circles centered on the wire.

49. A current-carrying wire is rotating by itself in a uniform magnetic field. The rotation will continue until the plane of the loop is

(A) perpendicular to the field
(B) parallel to the field
(C) at a 45° angle to the field
(D) either perpendicular or parallel to the field, depending on the strength of the current
(E) either perpendicular or parallel to the field, depending on the strength of the field

50. On which of the following does the magnetic field inside a solenoid of *N* turns not depend?

(A) The core material
(B) The length of the solenoid
(C) The current in the solenoid
(D) The number of turns of wire
(E) The diameter of the solenoid

51. A ray of light could undergo total internal reflection if the ray goes from

(A) air to water
(B) water to flint glass
(C) flint glass to water
(D) water to diamond
(E) air to diamond

52. A student performing a resonance lab experiment using a closed-end tube hears a fundamental mode sound at an air column length of 14 cm. What is the approximate wavelength of the sound?

(A) 56 cm
(B) 28 cm
(C) 42 cm
(D) 14 cm
(E) 7 cm

53. Which of the following represents the units for sound wave intensity?

(A) J/s
(B) W/m^2
(C) W · m^2
(D) J/m^2
(E) W/s

54. A ray of light is incident from air to a glass plate at an angle of 30 degrees. If the absolute index of refraction of the glass is equal to 1.6, the sine of the angle of refraction is

(A) 0.500
(B) 0.742
(C) 0.425
(D) 0.750
(E) 0.333

55. The colors observed in thin films are caused by interference and

(A) reflection
(B) refraction
(C) diffraction
(D) polarization
(E) luminosity

56. Which of the following *cannot* produce polarized light?

(A) A polaroid filter
(B) Reflection
(C) Scattering
(D) Selective absorption
(E) Diffraction

57. As the number of lines on a diffraction grating are increased,

(A) the spacing between the spectral lines increases.
(B) the spacing between the spectral lines decreases.
(C) the intensity of the spectral lines increases.
(D) the intensity of the spectral lines decreases.
(E) everything about the spectral lines remains the same.

58. If all of the following particles were traveling with the same velocity, which particle would have the longest de Broglie wavelength?

(A) electron
(B) proton
(C) alpha particle
(D) neutron
(E) deuteron

59. What is the function of a "moderator" in a fission reactor?

(A) control the speed of neutrons
(B) act as a source of fissionable material
(C) control the costs of running the reactor
(D) produce a new source of neutrons
(E) control the half-life of the radioactive material

60. What energy is associated with a 300-nanometer photon?

(A) 350 J
(B) 3.5×10^{-9} J
(C) 6.6×10^{-19} J
(D) 2.32×10^{-14} J
(E) 1.37×10^{-19} J

61. What is the de Broglie wavelength for a 3000-kilogram car moving with a velocity of 11 meters per second?

(A) 2.7×10^{-19} m
(B) 2×10^{-38} m
(C) 7.2×10^{-36} m
(D) 6.63×10^{-34} m
(E) 8.7×10^{-33} m

62. In a photoelectric effect experiment, increasing the intensity of the incident electromagnetic radiation will

(A) increase only the number of emitted electrons
(B) have no effect on any aspect of the experiment
(C) increase the maximum kinetic energy of the emitted electrons
(D) increase the de Broglie wavelength of the emitted electrons
(E) increase the stopping potential of the experiment

63. In a vacuum, all photons have the same

(A) frequency
(B) wavelength
(C) velocity
(D) amplitude
(E) period

64. Which of the following photons would have the largest momentum?

(A) X ray
(B) Ultraviolet
(C) Infrared
(D) Microwave
(E) Radio

65. Which element has the highest binding energy per nucleon?

(A) Hydrogen
(B) Iron
(C) Helium
(D) Carbon
(E) Uranium

66. Photons can scatter electrons, and the trajectories of the electrons will obey the law of conservation of momentum. These facts are observed in the

(A) Rutherford scattering experiment
(B) Michelson-Morley experiment
(C) Compton effect
(D) Doppler effect
(E) Lorentz-Fitzgerald effect

67. As hydrogen is fused into helium, the stability of the new nuclei, compared to that of hydrogen, will

(A) be greater
(B) be less
(C) be the same
(D) depend on how much hydrogen is used
(E) depend on which atoms are being fused

68. How many neutrons are contained in the isotope $^{211}_{84}Po$?

(A) 84
(B) 211
(C) 295
(D) 127
(E) 156

69. In the reaction:

$$^{6}_{3}Li + X \rightarrow ^{7}_{4}Be + ^{1}_{0}n$$

element X is given by

(A) $^{1}_{1}H$
(B) $^{3}_{1}H$
(C) $^{4}_{2}He$
(D) $^{2}_{1}H$
(E) $^{3}_{2}He$

70. Which of the following particles could not be deflected by an electromagnetic field?

(A) Neutron
(B) Electron
(C) Proton
(D) Positron
(E) Alpha particle

STOP

This is the end of section I. You may use any remaining time to check your work in this section

Formula Sheet for Section II

Newtonian Mechanics		Electricity and Magnetism	
$v = v_0 + at$	$P = Fv\cos\theta$	$F = \frac{1}{4\pi\epsilon_0}\frac{q_1 q_2}{r^2}$	$V = IR$
$x = x_0 + v_0 t + \frac{1}{2}at^2$	$\mathbf{F}_s = -k\mathbf{x}$	$\mathbf{E} = \frac{\mathbf{F}}{q}$	$P = IV$
$v^2 = v_0^2 + 2a(x - x_0)$	$U_s = \frac{1}{2}kx^2$	$U_E = qV = \frac{1}{4\pi\epsilon_0}\frac{q_1 q_2}{r}$	$C_p = \sum_i C_i$
$\Sigma\mathbf{F} = \mathbf{F}_{net} = m\mathbf{a}$	$T_s = 2\pi\sqrt{\frac{m}{k}}$	$E_{avg} = -\frac{V}{d}$	$\frac{1}{C_s} = \sum_i \frac{1}{C_i}$
$F_{fric} \le \mu N$	$T_p = 2\pi\sqrt{\frac{\ell}{g}}$	$V = \frac{1}{4\pi\epsilon_0}\sum_i \frac{q_i}{r_i}$	$R_s = \sum_i R_i$
$a_c = \frac{v^2}{r}$	$T = \frac{1}{f}$	$C = \frac{Q}{V}$	$\frac{1}{R_p} = \sum_i \frac{1}{R_i}$
$\tau = rF\sin\theta$	$F_G = -\frac{Gm_1 m_2}{r^2}$	$C = \frac{\epsilon_0 A}{d}$	$F_B = qvB\sin\theta$
$\mathbf{p} = m\mathbf{v}$	$U_G = -\frac{Gm_1 m_2}{r}$	$U_c = \frac{1}{2}QV = \frac{1}{2}CV^2$	$F_B = BI\ell\sin\theta$
$\mathbf{J} = \mathbf{F}\Delta t = \Delta\mathbf{p}$		$I_{avg} = \frac{\Delta Q}{\Delta t}$	$B = \frac{\mu_0}{2\pi}\frac{I}{r}$
$K = \frac{1}{2}mv^2$		$R = \frac{\rho\ell}{A}$	$\phi_m = BA\cos\theta$
$\Delta U_g = mgh$			$\epsilon_{avg} = -\frac{\Delta\phi_m}{\Delta t}$
$W = F\Delta r\cos\theta$			$\epsilon = B\ell v$
$P_{avg} = \frac{W}{\Delta t}$			

a = acceleration	r = radius or distance	A = area	R = resistance
F = force	T = period	B = magnetic field	r = distance
f = frequency	t = time	C = capacitance	t = time
h = height	U = potential energy	d = distance	U = potential (stored) energy
J = impulse	v = velocity or speed	E = electric field	V = electrical potential or potential difference
K = kinetic energy	W = work done on a system	ϵ = emf	v = velocity or speed
k = spring constant	x = position	F = force	ρ = resistivity
ℓ = length	μ = coefficient of friction	I = current	θ = angle
m = mass	θ = angle	ℓ = length	ϕ_m = magnetic flux
N = normal force	τ = torque	P = power	
P = power		Q = charge	
p = momentum		q = point charge	

Fluid Mechanics and Thermal Physics

$P = P_0 + \rho gh$

$F_{buoy} = \rho Vg$

$A_1 \upsilon_1 = A_2 \upsilon_2$

$P + \rho gy + \frac{1}{2}\rho v^2 = \text{const.}$

$\Delta \ell = \alpha \ell_0 \Delta T$

$H = \frac{kA\Delta T}{L}$

$P = \frac{F}{A}$

$PV = nRT = Nk_BT$

$K_{avg} = \frac{3}{2}k_BT$

$v_{rms} = \sqrt{\frac{3RT}{M}} = \sqrt{\frac{3k_BT}{\mu}}$

$W = -P\Delta V$

$\Delta U = Q + W$

$e = \left|\frac{W}{Q_H}\right|$

$e_c = \frac{T_H - T_C}{T_H}$

A = area
e = efficiency
F = force
h = depth
H = rate of heat transfer
k = thermal conductivity
K_{avg} = average molecular kinetic energy
ℓ = length
L = thickness
M = molar mass
n = number of moles
N = number of molecules
P = pressure
Q = heat transferred to a system
T = temperature
U = internal energy
V = volume
υ = velocity or speed
υ_{rms} = root-mean-square velocity
W = work done on a system
y = height
α = coefficient of linear expansion
μ = mass of molecule
ρ = density

Atomic and Nuclear Physics

$E = hf = pc$

$K_{max} = hf - \phi$

$\lambda = \frac{h}{p}$

$\Delta E = (\Delta m)c^2$

E = energy
f = frequency
K = kinetic energy
m = mass
p = momentum
λ = wavelength
ϕ = work function

Waves and Optics

$\upsilon = f\lambda$

$n = \frac{c}{v}$

$n_1 \sin\theta_1 = n_2 \sin\theta_2$

$\sin\theta_c = \frac{n_2}{n_1}$

$\frac{1}{s_i} + \frac{1}{s_0} = \frac{1}{f}$

$M = \frac{h_i}{h_0} = -\frac{s_i}{s_0}$

$f = \frac{R}{2}$

$x_m \approx \frac{m\lambda L}{d}$

d = separation
f = frequency or focal length
h = height
L = distance
M = magnification
m = an integer
n = index of refraction
R = radius of curvature
s = distance
υ = speed
x = position
λ = wavelength
θ = angle

Geometry and Trigonometry

Rectangle
$A = bh$

Triangle
$A = \frac{1}{2}bh$

Circle
$A = \pi r^2$
$C = 2\pi r$

Parallelepiped
$V = \ell wh$

Cylinder
$V = \pi r^2 \ell$
$S = 2\pi r\ell + 2\pi r^2$

Sphere
$V = \frac{4}{3}\pi r^3$
$S = 4\pi r^3$

Right Triangle
$a^2 + b^2 = c^2$
$\sin\theta = \frac{a}{c}$
$\cos\theta = \frac{b}{c}$
$\tan\theta = \frac{a}{b}$

B
c
a
90°
θ
A
b
C

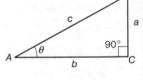

A = area
C = circumference
V = volume
S = surface area
b = base
h = height
ℓ = length
w = width
r = radius

Practice Test 2

Section II

FREE-RESPONSE QUESTIONS

7 QUESTIONS

90 MINUTES

50 PERCENT OF TOTAL GRADE

Directions: Solve the following problems. Be sure to show all work, including substitutions with units, and to explain clearly. You may use a calculator for Section II.

1. (15 points)

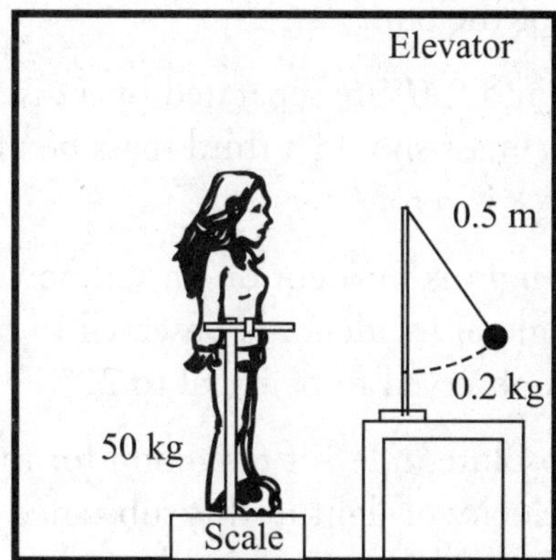

A 50-kg girl is standing on a bathroom scale in an elevator, which is at rest. Next to her, on a table, is a pendulum consisting of a string 0.5 m long and a 0.2-kg mass.

(a) What is the weight of the girl while the elevator is at rest?

(b) What is the period of the pendulum if it is undergoing simple harmonic motion while the elevator is at rest?

(c) The elevator now has an upward acceleration of $+3\ m/s^2$. What will be the apparent weight of the girl as indicated on the scale?

(d) The pendulum is swinging while the elevator has the upward acceleration of $+3\ m/s^2$.

Compared to the period of the pendulum when the elevator was at rest, the period of the pendulum now will be

_______ more _______ less _______ the same

Check your choice for the best answer, and explain your reasoning.

2. (15 points) Two cars are separated by 25 meters. Both are initially at rest. Then the car in front begins to accelerate uniformly at 2 meters per second squared. The second car, behind, begins to accelerate at 3 meters per second squared.

   (a) How long does it take the faster car to catch up with the slower car?
   (b) How far does the faster car go during that time?
   (c) What is the velocity of each car at the moment the second car catches up with the first car?

3. (15 points) (a) Explain why, when you blow across a sheet of paper, it begins to rise.

   (b) Explain why aircraft carriers need to turn into the wind before launching their planes.
   (c) Don't try this yourself, but some people can actually sleep on a bed of nails! Explain how this is possible.
   (d) How does Archimedes' principle explain why a boat floats?

4. (10 points) (a) Given that the average distance from Earth to the Sun is 1.5×10^{11} m and that the average orbital velocity of Earth is 3×10^4 m/s, determine the mass of the Sun.

   (b) Two masses, M and $9M$, are separated by a distance d. At what distance from the smaller mass should a third mass be placed such that the net gravitational force is zero?

5. (10 points) A ray of light is incident on an unknown transparent substance from the air. The angle of incidence is observed to be equal to 40°, and the angle of refraction is observed to be equal to 22°.

   (a) Calculate the absolute index of refraction for this substance.
   (b) Calculate the velocity of light in this substance.
   (c) The substance is now submerged in glycerol ($n = 1.47$). Calculate the critical angle of incidence for light going from this substance into glycerol.
   (d) The substance is now shaped into a convex lens. How does the focal length of this lens compare with the focal length of a similar shaped lens made out of crown glass ($n = 1.52$)?

6. (10 points) You are given a wooden block, a wooden board, some standard laboratory masses, and a spring scale calibrated in newtons.

   (a) Explain how to use this equipment to measure the force of friction between the block and the board.
   (b) Describe the relationship between the load and the force of friction in this situation.
   (c) If all sides of the block are identical in smoothness, how does the force of friction vary with a change in the contact surface area?

7. (10 points) (a) A brick wall is 5 meters long, 3 meters high, and 30 cm thick. The thermal conductivity of brick is 0.6 W/m·°C. If the outside air temperature is 0°C while the inside air temperature is 20°C, what is the rate of conductive heat flow?

   (b) How many joules are transferred in 1 hour?

   (c) Explain why brick is an efficient way to keep rooms warm in the winter.

This is the end of section II. You may use any remaining time to check your work in this section.

Answer Key

SECTION I

1. C
2. E
3. C
4. C
5. C
6. E
7. B
8. C
9. A
10. D
11. E
12. D
13. B
14. C
15. E
16. A
17. B
18. E
19. C
20. D
21. A
22. C
23. B
24. D
25. C
26. D
27. A
28. D
29. A
30. B
31. B
32. E
33. D
34. E
35. B
36. A
37. B
38. E
39. A
40. B
41. C
42. A
43. A
44. C
45. A
46. C
47. D
48. E
49. A
50. E
51. C
52. A
53. B
54. E
55. A
56. E
57. A
58. A
59. A
60. C
61. B
62. A
63. C
64. A
65. B
66. D
67. A
68. D
69. D
70. A

ANSWER EXPLANATIONS

SECTION I

1. **(C)** The tangent of the angle that the vector makes with the positive *x*-axis is equal to 5/3, which means that the angle must be greater than 45°.

2. **(E)** The sum of two vectors can be found by adding their components:

$$\mathbf{c}_x = \mathbf{a}_x + \mathbf{b}_x = -2 + 1 = -1 \quad \text{and} \quad \mathbf{c}_y = \mathbf{a}_y + \mathbf{b}_y = 3 + 4 = 7$$

The magnitude of the sum is now given by the Pythagorean theorem:

$$|\mathbf{A} + \mathbf{B}| \sqrt{(-1)^2 + (7)^2} = \sqrt{50} = 7.1$$

3. **(C)** The magnitude of the resultant between two vectors decreases as the angle between them increases.

4. **(C)** The force that the car exerts on the driver is equal to the normal force **N**. If we write Newton's second law for the car at the top of the hill, we get

$$\Sigma\mathbf{F}_y = m\mathbf{a} = \frac{-m\mathbf{v}^2}{r} = \mathbf{N} - m\mathbf{g}$$

Since $\mathbf{W} = m\mathbf{g} = 500$ N, we know that the mass is approximately equal to 50 kg. Making the necessary substitutions for velocity and radius, we obtain $\mathbf{N} = +100$ N, up.

5. **(C)** The maximum range possible for a given initial velocity is achieved at an angle of 45°.

6. **(E)** We use

$$\mathbf{v}_f^2 = 2\mathbf{g}\mathbf{y}$$
$$\mathbf{v}_f = 30 \text{ m/s}$$

7. **(B)** The two velocities are independent of each other. Thus, to find the time needed to cross the river, we consider only the velocity component involved in heading across the river:

$$t = \frac{190 \text{ m}}{8 \text{ m/s}} = 24 \text{ s (actually 23.75 s)}$$

8. **(C)** The work done by the variable force is equal to the total area under the graph. Using rectangles, we find this total area to be equivalent to 32 J of work.

9. **(A)** The average force applied is equal to the total work done, divided by the displacement:

$$\mathbf{F}_{\text{avg}} = \frac{32 \text{ J}}{10 \text{ m}} = 3.2 \text{ N}$$

10. **(D)** The work done is equal to the product of the magnitude of the force component parallel to the direction of motion and the displacement. In this case, the force component is equal to

$$(10\text{ N})\cos 60 = 50\text{ N}$$

Thus,

$$W = \mathbf{Fd} = (5\text{ N})(2.5\text{ m}) = 12.5\text{J}$$

11. **(E)** The force of gravity acting on the projectile (assuming $\mathbf{g} = -9.8\text{ m/s}^2$) is equal to −29.4 N. The rate at which work is done is equal to the power, which is found by multiplying the force and the instantaneous velocity along the same line. We need the vertical component of the velocity:

$$\mathbf{v}_y = (10\text{ m/s})\sin 30 = 5\text{ m/s}$$

Thus, $P = -147$ W.

12. **(D)** The upward acceleration of the elevator will increase the force that the scale exerts on the person, who will observe an increase in his weight, determined as follows: Let **F** be the force that the scale exerts on the person (his perceived weight).

$$\mathbf{F} - m\mathbf{g} = m\mathbf{a}$$

The person weighs 490 N, implying a mass of 50 kg. The acceleration is 0.2 m/s^2. Making the necessary substitutions yields **F** = 500 N.

13. **(B)** The beat frequency is equal to the difference between the two frequencies. Since one frequency is equal to 384 Hz, the other frequency could be either 372 Hz or 396 Hz. Since 396 Hz is not a choice, the correct answer is 372 Hz.

14. **(C)** The force of friction is given by $\mathbf{f} = \mu m\mathbf{g}$ in this case. Since this is a net force acting on the puck,

$$\mathbf{f} = m\mathbf{a} = -\mu m\mathbf{g}$$

Thus, $\mathbf{a} = -1.96\text{ m/s}^2$. We know that the puck is going to stop, so $\mathbf{v}_0^2 = 2\mathbf{ax}$. Given that the initial velocity is 10 m/s, and using the acceleration obtained above, we get $x \approx 25$ m.

15. **(E)** The elastic potential energy stored in a stretched spring is proportional to the square root of the spring's elongation. To double this quantity, the elongation must be increased by a factor equal to $\sqrt{2}$.

16. **(A)** We use

$$\mathbf{v}_{ox} = \mathbf{v}_o \cos\theta = (200\text{ m/s})\cos(60°) = 100\text{ m/s}$$

17. **(B)** We know that the maximum horizontal range for a projectile occurs at 45°. So if we change the launch angle from 60° to 45°, the maximum horizontal range will increase.

18. **(E)** We know that $\mathbf{P} = \mathbf{Fv}$ where $\mathbf{v} = \mathbf{at}$ (assuming we start from rest). Also, $\mathbf{a} = \mathbf{F}/m$; thus $P = \mathbf{F}^2 t/m$.

19. **(C)** According to the law of conservation of energy, the potential energy stored in the spring will be transferred into kinetic energy as the mass rises. This energy, in turn, will be converted into gravitational potential energy. Thus the starting elastic potential energy is equal to the final gravitational potential energy:

$$\left(\frac{1}{2}\right)kx^2 = m\mathbf{g}h$$

$$\left(\frac{1}{2}\right)(20 \text{ N/m})(0.4 \text{ m})^2 = (0.2 \text{ kg})(9.8 \text{ m/s}^2)h$$

$$h = 0.82 \text{ m}$$

20. **(D)** The units of k are newtons per meter. Newtons can be expressed as kilogram · meters per second squared (kg · m/s²). Thus, the units for k can be expressed as kilograms per second squared (kg/s²).

21. **(A)** We use

$$\mathbf{F} = m\mathbf{a} = m\frac{\Delta\mathbf{v}}{\Delta\mathbf{t}} = \frac{(2500 \text{ kg})(30 \text{ m/s})}{15 \text{ s}} = 5000 \text{ N}$$

22. **(C)** We first find the velocity of the two masses after they collide and stick together. Using conservation of momentum, we know that

$$(1 \text{ kg})(6 \text{ m/s}) + (2 \text{ kg})(3 \text{ m/s}) = (1 + 2)\mathbf{v}$$

and $\mathbf{v} = 4$ m/s. The total initial kinetic energy is equal to 27 J, and the final kinetic energy is 24 J. Thus 3 J of kinetic energy was lost.

23. **(B)** The change in momentum is given by $\Delta\mathbf{p} = m\,\Delta\mathbf{v}$. In this case, there is a change in direction, so

$$\Delta\mathbf{v} = \mathbf{v_f} - \mathbf{v_i} = 3 \text{ m/s} - (-1 \text{ m/s}) = 4 \text{ m/s}$$

Thus,

$$\Delta\mathbf{p} = (0.2 \text{ kg})(4 \text{ m/s}) = 0.8 \text{ kg} \cdot \text{m/s}$$

24. **(D)** We need to balance the torques on the two sides of the pivot:

$$(50 \text{ kg})(\mathbf{g})(1.2 \text{ m}) = (70 \text{ kg})(\mathbf{g})(x)$$

implies that $x = 0.86$ m.

25. **(C)** We take counterclockwise torques as positive. Acting on the pivot, we have two forces contributing to a net torque.

First, the component of force **F** acting at right angles to the beam contributes a torque:

$$\tau_1 = (200 \text{ N})(\sin 30)(2 \text{ m}) = 200 \text{ N} \cdot \text{m}$$

Second, the weight of the beam acts from its center of mass ($x = 1$ m, since the beam is uniform). This torque will be clockwise and negative:

$$\tau_2 = -(98 \text{ N})(1 \text{ m}) = -98 \text{ N}$$

Thus, the net torque is -102 N · m.

26. **(D)** The maximum velocity attained by an oscillating mass on a spring is found by $\frac{1}{2}m\mathbf{v}_{max}^2 = \frac{1}{2}kA^2$, so

$$\mathbf{v}_{max} = \sqrt{\frac{A^2k}{m}}$$

27. **(A)** The total energy is given by $E = (1/2)kA^2$, where A is in meters. Here $A = -0.02$ m, so $E = 0.01$ J.

28. **(D)** Since **g** varies inversely as the square of the distance from Earth's center, if we start at the surface and go to a height equal to Earth's radius, we have effectively doubled the radius. Thus **g** is decreased by one-fourth to 2.45 m/s^2.

29. **(A)** The orbital velocity is given by

$$\mathbf{v}_0 = \sqrt{\frac{GM}{R}}$$

The escape velocity is given by

$$\mathbf{v}_{esc} = \sqrt{\frac{2GM}{R}} = \sqrt{2}\mathbf{v}_0$$

30. **(B)** The gravitational force is inversely proportional to the square of the distance between the masses. This relationship is best described by graph B.

31. **(B)** A refrigerator removes heat from an enclosed area.

32. **(E)** Heat transfer can be performed by conduction, convection, and radiation.

33. **(D)** Standing wave nodes occur every half wavelength. In this case, the nodes occur every 20 cm (0.2 m), which means that the wavelength is equal to 0.4 m. Since $v = f\lambda$, we have 5 m/s = f(0.4 m). Thus, f = 12.5 Hz.

34. **(E)** In isochoric change the volume of an ideal gas remains the same.

35. **(B)** Using the ideal gas law, we first convert 50°C to 323 K. We know that for an ideal gas PV/T = constant. Thus, if both the pressure and the absolute temperature are doubled, the volume of gas will remain the same (100 m^3).

36. **(A)** In an isothermal expansion, the temperature remains the same and the pressure varies inversely with the volume (Boyle's law). Graph A displays this case.

37. **(B)** At the same temperature, 1 mole of each gas contains Avogadro's number of molecules; that is, the two gases have the same number of molecules.

38. **(E)** The description in the question describes Charles's law, which describes a direct relationship between volume and absolute temperature for a confined ideal gas at constant pressure. The correct answer is graph E.

39. **(A)** Adiabatic means that no heat enters or leaves the system. During an adiabatic process the temperature, however, may change.

40. **(B)** An isobaric (constant-pressure) expansion is *not* part of the Carnot cycle.

41. **(C)** If the pith ball is initially attracted to the charged rod, it is either neutral or is charged with the opposite sign (not one of the choices).

42. **(A)** The units of electric field strength can be expressed as newtons per coulomb or volts per meter.

43. **(A)** The ratio $k/k' = c^2$, as can be verified from the Useful Constants table.

44. **(C)** The capacitance of a capacitor is defined from the equation $C = Q/V$. Thus, one farad is equal to one coulomb per volt.

45. **(A)** The two series capacitors add up reciprocally:

$$\frac{1}{C} = \frac{1}{C_1} + \frac{1}{C_2} = \frac{1}{10\ \mu\text{F}} + \frac{1}{10\ \mu\text{F}}$$

implies that $C = 5\ \mu\text{F}$. This capacitor is now in parallel with another 5-μF capacitor. The equivalent capacitance is equal to the sum of these two. Thus, $C_{eq} = 10\ \mu\text{F}$.

46. **(C)** One ohm is equal to 1 V/A; 1 V is equal to 1 J/C; and 1 A is equal to 1 C/s. Thus, $1\ \Omega = 1\ \text{kg} \cdot \text{m}^2/\text{s} \cdot \text{C}^2$.

47. **(D)** Power is equal to potential difference times current. Thus,

$$P = VI = 480\ \text{W}$$

48. **(E)** The magnetic field around a long, straight-current carrying wire is a set of concentric circles centered on the wire.

49. **(A)** The rotation will continue until the plane of the loop is perpendicular to the field. Induced force will be zero at that point. The use of commutator rings makes it possible for the rotation to continue by alternating the direction of the current.

50. **(E)** In calculating the strength of the magnetic field inside a solenoid, one does *not* consider the diameter of the solenoid.

51. **(C)** A light ray can undergo total internal reflection if it exceeds the critical angle of incidence during refraction from a higher-index material to a lower-index material. The only combination that goes from a high index to a low index is flint glass to water.

52. **(A)** You should remember that in a closed-end resonance tube, the fundamental mode occurs when the air column length is equal to one-quarter of the wavelength. Therefore, if the air column length is 14 cm, the wavelength must be equal to 14 cm × 4 = 56 cm.

53. **(B)** The units for sound wave intensity are watts per square meter, which indicate the amount of energy falling on a unit area each second.

54. **(E)** Snell's law states that

$$\frac{\sin \theta\, i}{\sin \theta\, r} = \frac{n_2}{n_1}$$

In this case $n_1 = 1.0$ and $n_2 = 1.5$. Given the angle of incidence, we get

$$\frac{\sin 30}{\sin \theta\, r} = 1.5$$

and $\sin \theta r = 0.383$.

55. **(A)** The colors observed in thin films are caused by the reflection and interference of light.

56. **(E)** Diffraction is a process that cannot produce polarized light.

57. **(A)** Increasing the number of lines on a diffraction grating decreases the spacing between each two lines. This has the effect of increasing the spacing between the spectral lines. The formula $\lambda/d = x/L$, where λ is the wavelength of light, d is the spacing between the grating lines, L is the distance from the grating to the screen, and x is the spectral line separation. If λ and L remain constant, then, as d decreases, x increases.

58. **(A)** The de Broglie wavelength varies inversely with momentum. Since all the listed particles are moving with the same velocity, the particle with the smallest mass will have the longest wavelength. This particle is the electron.

59. **(A)** A moderator such as water or graphite controls the speed of the neutrons in a fission reactor.

60. **(C)** The frequency of the 300-nm photon is equal to 1.0×10^{15} s^{-1}. Now,

$$E = hf = (6.63 \times 10^{-34} \text{ J} \cdot \text{s})(1.0 \times 10^{15} \text{ s}^{-1}) = 6.6 \times 10^{-19} \text{ J}$$

61. **(B)** The de Broglie wavelength is given by the formula $\lambda = h/m\mathbf{v}$. Using the given values, we have

$$\lambda = \frac{h}{m\mathbf{v}} = \frac{(6.63 \times 10^{-34} \text{ J} \cdot \text{s})}{(3000 \text{ kg})(11 \text{ m/s})} = 2.0 \times 10^{-38} \text{ m}$$

62. **(A)** Increasing the intensity of the incident electromagnetic radiation onto a photoelectric metal will increase only the number (current) of emitted electrons. No change will occur in the kinetic energy of the electrons. This was the contradiction with the wave theory of light!

63. **(C)** In a vacuum, all photons have the same velocity.

64. **(A)** We know from the Compton effect that $\mathbf{p} = h/\lambda$. Of the choices, X ray photons have the shortest wavelength and thus have the largest momentum.

65. **(B)** Iron has the highest binding energy per nucleon.

66. **(D)** The scattering of electrons by photons and the fact that both electrons and photons obey the law of conservation of momentum are detected in the Compton effect.

67. **(A)** The fusion process results in a higher binding energy per nucleon. This increases the stability of the new nucleus (and releases energy).

68. **(D)** The number of neutrons is equal to the difference between the mass number and atomic number. For this isotope

$$211 - 84 = 127 \text{ neutrons}$$

69. **(D)** By matching the numbers on the top and the bottom, we see that element X is ${}^{2}_{1}\text{H}$.

70. **(A)** Since it is electrically neutral, a neutron could not be accelerated by an electromagnetic field.

SECTION II

1. (a) (3 points) The weight of the girl is given by $\mathbf{F}_g = m\mathbf{g}$ = (50 kg)(9.8 m/s^2) = 490 N.

   Award 1 point for the correct equation for the weight.

   Award 2 points for the correct answer with substitution and units.

   (b) (3 points) The period of the pendulum is given by

$$T = \frac{2\pi\sqrt{L}}{\mathbf{g}} = 1.42 \text{ s}$$

   Award 1 point for the correct equation.

   Award 2 points for the correct answer with substitution and units.

   (c) (4 points) There are two forces acting on the girl; the force of gravity and the upward (normal force) of the scale. These combine to produce the net force ($m\mathbf{a}$):

$$\mathbf{F}_N - m\mathbf{g} = m\mathbf{a}$$

   The scale reading is the normal force:

$$\mathbf{F}_N = m\mathbf{g} + m\mathbf{a} = (50 \text{ kg})(9.8 \text{ m/s}^2 + 3 \text{ m/s}^2) = 640 \text{ N}$$

   Award 2 points for the correct expression for the net force.

   Award 2 points for the correct answer with substitution and units.

   (d) (5 points) The pendulum will have a shorter period when the elevator has positive acceleration. The reason for this is that when the elevator has positive acceleration, the system acts as if there is a greater acceleration due to gravity. If the girl looked at the scale reading of 640 N, and knew her mass was 50 kg, she might conclude that

$$\mathbf{g}' = \frac{640 \text{ N}}{50 \text{ kg}} = 12.8 \text{ m/s}^2$$

   Thus,

$$T' = \frac{2\pi\sqrt{L}}{\mathbf{g}'}$$

   And the period would be less because the effective acceleration due to gravity is higher (in this case). Additionally, if the elevator had negative acceleration, then the period would be longer.

   Award 2 points for recognizing that the pendulum will have a shorter period when the elevator has positive acceleration.

   Award 3 points for recognizing that the "effective gravitational acceleration" in that frame of references is greater than **g**.

2. (a) (7 points) At the beginning, car 1 is 25 m ahead of car 2. The inital velocity of each car is zero. When they meet, the position of each car, relative to the origin, must be the same. Since each car undergoes uniformly accelerated motion from rest, we can write

$$\left(\frac{1}{2}\right)\mathbf{a}_2 t^2 = \left(\frac{1}{2}\right)\mathbf{a}_1 t^2 + 25\text{ m}$$
$$\left(\frac{1}{2}\right)(3\text{ m/s}^2)t^2 = \left(\frac{1}{2}\right)(2\text{ m/s}^2)t^2 + 25\text{ m}$$
$$t^2 = 50\text{ s}^2$$
$$t = 7.07\text{ s}$$

Award 2 points for recognizing that there is uniform acceleration for both cars.

Award 2 points for the correct expression equating the distances for both cars.

Award 3 points for the correct solution for the time showing substitutions and units.

(b) (3 points) Car 2, the faster car, in 7.07 s goes a distance of

$$\left(\frac{1}{2}\right)\mathbf{a}_2 t^2 = (0.5)(3\text{ m/s}^2)(7.07\text{ s})^2 = 75\text{ m}$$

Award 3 points for the correct calculation of the time showing substitutions and units.

(c) (5 points) Since each car has zero initial velocity, for any time t, $\mathbf{v} = \mathbf{a}t$:

$$\text{car 1: } \mathbf{v} = (2\text{ m/s}^2)(7.07\text{ s}) = 14.14\text{ m/s}$$
$$\text{car 2: } \mathbf{v} = (3\text{ m/s}^2)(7.07\text{ s}) = 21.21\text{ m/s}$$

Award 3 points for correctly recognizing that $\mathbf{v} = \mathbf{a}t$ for both cars.

Award 2 points for the correct answers for the velocities showing substitutions and units.

3. (a) (4 points) The sheet rises because of Bernoulli's principle, which states that the pressure exerted by a moving fluid varies inversely with its velocity. When air is blown across the top, the pressure is reduced, causing the sheet to rise as a net force is created from the area of higher to lower pressure.

Award 4 points for a correct explanation.

(b) (4 points) Aircraft carriers turn into the wind to have extra wind velocity passing over the upper surface of the wings. The lift created to make an airplane fly is similar to the procedure outlined in part (a).

Award 4 points for a correct explanation.

(c) (4 points) The bed of nails distributes the weight of the person over a large area, reducing the pressure of any one individual nail.

Award 4 points for a correct explanation.

(d) (3 points) When the wind blows across a chimney, the reduced pressure causes the smoke from a fireplace to rise more quickly because of the reduced pressure across the top.

Award 3 points for a correct explanation.

4. (a) (5 points) We can assume that Earth orbits the Sun in a nearly circular orbit. Since the net force due to gravity sets up a centripetal force, we can write:

$$\Sigma \mathrm{F} = m\mathbf{a}$$

$$\frac{Gm_S m_E}{r^2} = \frac{m_E \mathbf{v}^2}{r}$$

$$m_S = \frac{rv^2}{G}$$

$$M_S = \frac{(1.5 \times 10^{11}\ \mathrm{m})(3 \times 10^4\ \mathrm{m/s})^2}{(6.67 \times 10^{-11}\ \mathrm{N \cdot m^2/kg^2})}$$

$$M_S = 2.02 \times 10^{30}\ \mathrm{kg}$$

Award 1 point for setting up Newton's second law.

Award 2 points for setting up the correct expressions for gravitation and centripetal force.

Award 1 point for correct substitution with units.

Award 1 point for the correct answer with units.

(b) (5 points) If we let the masses be equal to M and $9M$ and let the test mass be equal to m, we can write the balanced equation relative to the small mass M as

$$\frac{GmM}{x^2} = \frac{Gm(9M)}{(d - x)^2}$$

where d is the distance between the masses M and $9M$ and x is the distance between the third mass and small mass M.

Thus:

$$\frac{1}{x^2} = \frac{9}{(d - x)^2}$$

Cross-multiplying we get

$$9x^2 = (d - x)^2$$

Take the square root of both sides and we get

$$3x = d - x$$

Which means

$$x = d/4$$

Award 2 points for the correct setup of the gravitational forces.

Award 3 points for showing the correct solution for the distance x.

5. (a) (3 points) We use

$$n_1 \sin \theta_1 = n_2 \sin \theta_2$$

Thus, since air has the index $n_1 = 1.00$, we have

$$n_2 = \frac{\sin \theta_1}{\sin \theta_2}$$
$$n_2 = \frac{\sin 40°}{\sin 22°} = 1.72$$

Award 1 point for the correct equation for Snell's law.

Award 2 points for the correct answer showing substitutions and units.

(b) (3 points) We use

$$\mathbf{v} = \frac{\mathbf{c}}{n} = \frac{3 \times 10^8 \text{ m/s}}{1.72} = 1.74 \times 10^8 \text{ m/s}$$

Award 1 point for the correct equation.

Award 2 points for the correct answer showing substitutions and units.

(c) (3 points) For the critical angle, we know that the angle of refraction will be 90°:

$$\sin \theta_c = \frac{n_2}{n_1} = \frac{1.47}{1.74} = 0.8448$$
$$\theta_c = 57°.65$$

Award 1 point for the correct equation.

Award 2 points for the correct answer showing substitutions and units.

(d) (1 point) The focal length of the lens made from the unknown substance will be shorter than the focal length of the lens made from crown glass. Since the index of refraction of the substance is greater than that of crown glass, light will refract more, and thus, the converging point for the focal length will be closer to the lens.

Award 1 point for a correct explanation.

6. (a) (4 points) The force of friction between the block and the board can be measured by pulling the block at a constant velocity. The spring scale should read a steady force which is the magnitude of the force of friction (since the net force is zero).

Award 4 points for a correct explanation.

(b) (3 points) Friction varies directly with the load of normal force. This can be tested by adding the weights on top of the block and recording the changes in the magnitude of the force of friction.

Award 3 points for a correct explanation.

(c) (3 points) Friction is independent of the surface area of contact (if the surfaces are the same smoothness).

Award 3 points for a correct explanation.

7. (a) (5 points) The thickness of the brick is 0.30 m and the temperature difference is $\Delta T = 20°C$.

The area of the wall is $A = (5 \text{ m})(3 \text{ m}) = 15 \text{ m}^2$. The formula for conductive heat transfer is

$$H = \frac{kA\Delta T}{L} = \frac{(0.6 \text{ W/m} \cdot °\text{C})(15 \text{ m}^2)(20°\text{C})}{(0.30 \text{ m})} = 600 \text{ W} = 600 \text{ J/s}$$

Award 1 point for correctly changing the units of the thickness to meters.

Award 1 point for the correct formula.

Award 1 point for calculating the correct area with units.

Award 1 point for the correct substitutions with units.

Award 1 point for the correct answer with units.

(b) (3 points) Remember that 600 W = 600 J/s and 1 hour = 3600 seconds. Since the heat transfer units are of J/s, we need to multiply the answer to part (a) by 3600 s:

$$\text{Energy} = (600 \text{ J/s})(3600 \text{ s}) = 2{,}160{,}000 \text{ J}$$

Award 1 point for correctly changing from 1 hour to 3600 s.

Award 1 point for correctly knowing that the heat transfer rate needs to be multiplied by the time in seconds.

Award 1 point for the correct final answer with units.

(c) (2 points) Brick walls are efficient for holding in heat since they have a low coefficient of thermal conductivity.

Award 2 points for a correct explanation consistent with the question.

Test Analysis
Practice Test 2

Section I: Multiple-Choice

Number correct (out of 70) = ________

Raw Score = number correct × .714 = ________
Multiple-Choice Score

Section II: Free-Response

Question 1 = ________
(out of 15)

Question 2 = ________
(out of 15)

Question 3 = ________
(out of 15)

Question 4 = ________
(out of 10)

Question 5 = ________
(out of 10)

Question 6 = ________
(out of 10)

Question 7 = ________
(out of 10)

Raw Score: ________ × .588 = ________
Free-Response Score

Final Score

________ + ________ = ________

Multiple-Choice Score + Free-Response Score = Final Score (rounded to the nearest whole number)

Final Score Range	AP Score*
81–100	5
61–80	4
51–60	3
41–50	2
0–40	1

*The score range corresponding to each grade varies from exam to exam and is approximate.

SELF-ASSESSMENT GUIDE

How well did you do? Remember that on the actual AP Physics B exam, grading and scoring are based on the year and guidelines set up by the College Board for the readers who will be grading your exam. Use the results of this assessment only as a guide to further your studying and not as an absolute predictor of an AP grade.

Use the table below to help you locate the topics in the book for which you need further study.

Topic	Multiple-Choice Question Number
Motion	1, 2, 3, 5, 6, 7, 16, 17
Forces and Momentum	4, 8, 9, 10, 11, 12, 14, 15, 18, 19, 20, 21, 23, 24, 25, 28, 29, 30
Work and Energy	22, 26, 27
Heat and Gases	31, 32, 34, 35, 36, 37, 38, 39, 40
Waves and Sound	13, 33, 52, 53
Light and Optics	51, 54, 55, 56, 57
Electricity and Magnetism	41, 42, 43, 44, 45, 46, 47, 48, 49, 50
Modern Physics	58, 59, 60, 61, 62, 63, 64, 65, 66, 67, 68, 69, 70

SCORE IMPROVEMENT

Not satisfied with your score? Don't worry. Here are some tips for improvement.

For the multiple-choice section:

1. Write the numbers of the questions you left blank or answered incorrectly in the first column below.
2. Go over the answers to the problems in the Answers Explained section and write the main ideas and concepts behind the problems in the second column.
3. Go back and reread the sections covering the material in the second column.
4. Retake the skipped and incorrect questions.
5. Recalculate your score to see how much you improved.

Questions	Main Idea or Concept

For the free-response section:

1. Go over the answers to the problems in the Answers Explained section, and circle any mistakes in your answers.
2. Go back and reread the sections covering the material.
3. Retake the missed free-response questions.
4. Recalculate your score to see how much you improved.

SCORE IMPROVEMENT

Multiple-choice sections

Free-response sections

Glossary

A

absolute index of refraction For a transparent material, a number that represents the ratio of the speed of light in a vacuum to the speed of light in the material.

absolute temperature A measure of the average kinetic energy of the molecules in an object; on the Kelvin scale, equal to the Celsius temperature of an object plus 273.

absolute zero The theoretical lowest possible temperature, designated as 0 K.

absorption spectrum A continuous spectrum crossed by dark lines representing the absorption of particular wavelengths of radiation by a cooler medium.

acceleration A vector quantity representing the time rate of change of velocity.

action A force applied to an object that leads to an equal but opposite reaction; the product of total energy and time in units of joules times seconds; the product of momentum and position, especially in the case of the Heisenberg uncertainty principle.

activity The number of decays per second of a radioactive atom.

adiabatic In thermodynamics, referring to the process that occurs when no heat is added or subtracted from the system.

alpha decay The spontaneous emission of a helium nucleus from certain radioactive atoms.

alpha particle A helium nucleus that is ejected from a radioactive atom.

alternating current An electric current that changes its direction and magnitude according to a regular frequency.

ammeter A device that, when placed in series, measures the current in an electric circuit; a galvanometer with a low-resistance coil placed across it.

ampere (A) The SI unit of electric current, equal to 1 C/s.

amplitude The maximum displacement of an oscillating particle, medium, or field relative to its rest position.

angle of incidence The angle between a ray of light and the normal to a reflecting or transparent surface at the location where the ray intercepts the surface.

angle of reflection The angle between a reflected light ray and the normal to a mirror or other reflecting surface at the location where the ray intercepts the surface.

angle of refraction The angle between an emerging light ray in a transparent material and the normal to the surface at the location where the ray first enters the material.

angstrom (Å) A unit of wavelength measurement equal to 1×10^{-10} m.

antinodal lines A region of maximum displacement in a medium where waves are interacting with each other.

Archimedes' principle A body wholly or partially immersed in a fluid will be buoyed up by a force equal to the weight of the fluid it displaces.

armature The rotating coil of wire in an electric motor.

atomic mass unit (u) A unit of mass equal to one-twelfth the mass of a carbon-12 nucleus. Older texts use the abbreviation "amu."

atomic number The number of protons in an atom's nucleus.

atmosphere (atm) A unit of pressure, equal to 101 kPa at sea level.

average speed A scalar quantity equal to the ratio of the total distance to the total elapsed time.

Avogadro's number The number of molecules in 1 mole of an ideal gas, equal to 6.02×10^{23}. Named for Italian chemist/physicist Amedeo Avogadro.

B

Balmer series In a hydrogen atom, the visible spectral emission lines that correspond to electron transitions from higher excited states to lower level 2.

battery A combination of two or more electric cells.

beats The interference caused by two sets of sound waves with only a slight difference in frequency.

becquerel (Bq) A unit of radioactive decay, equal to one decay event per second.

Bernoulli's principle If the speed of a fluid particle increases as it travels along a streamline, the pressure of the fluid must decrease.

beta decay The spontaneous ejection of an electron from the nuclei of certain radioactive atoms.

beta particle An electron that is spontaneously ejected by a radioactive nucleus.

binding energy The energy required to break apart an atomic nucleus; the energy equivalent of the mass defect.

Boltzmann's constant In thermodynamics, a constant equal to the ratio of the universal gas constant, R, to Avogadro's number, N_A, equal to 1.38×10^{-23} J/K. Named for German physicist Ludwig Boltzmann.

Boyle's law At constant temperature, the pressure in an ideal gas varies inversely with the volume of the gas. Named for British physicist/chemist Robert Boyle.

bright-line spectrum The display of brightly colored lines on a screen or photograph indicating the discrete emission of radiation by a heated gas at low pressure.

C

capacitance The ratio of the total charge to the potential difference in a capacitor.

capacitor A pair of conducting plates, with either a vacuum or an insulator between them, used in an electric circuit to store current.

Carnot cycle In thermodynamics, a sequence of four steps in an ideal gas confined in a cylinder with a movable piston (a Carnot engine). The cycle includes an isothermal expansion, an adiabatic expansion, an isothermal compression, and an adiabatic compression. Named for French physicist Sadi Carnot.

Cartesian coordinate system A set of two or three mutually perpendicular reference lines, called axes and usually designated as x, y, and z, that are used to define the location of an object in a frame of reference; a coordinate system named for French scientist Rene Descartes.

cathode ray tube An evacuated gas tube into which a beam of electrons is projected. Their energy produces an image on a fluorescent screen when deflected by external electric or magnetic fields.

Celsius temperature scale A metric temperature scale in which, at sea level, water freezes at 0°C and boils at 100°C. Named for Swedish astronomer Anders Celsius.

center of curvature A point that is equidistant from all other points on the surface of a spherical mirror; a point equal to twice the focal length of a spherical mirror.

center of mass The weighted mean distribution point where all the mass of an object can be considered to be located; the point at which, if a single force is applied, translational motion will result.

centripetal acceleration The acceleration of mass moving uniformly in a circle at a constant speed directed radially inward toward the center of the circular path.

centripetal force The deflecting force, directed radially inward toward a given point, that causes an object to follow a circular path.

chain reaction In nuclear fission, the uncontrolled reaction of neutrons splitting uranium nuclei and creating more neutrons that continue the process on a self-sustained basis.

Charles's law At constant pressure, the volume of an ideal gas varies directly with the absolute temperature of the gas. Named for J. A. C. Charles.

chromatic aberration In optics, the defect in a converging lens that causes the dispersion of white light into a continuous spectrum, with the result that the lens refracts the colors to different focal points.

coefficient of friction The ratio of the force of friction to the normal force when one surface is sliding (or attempting to slide) over another surface.

coherent Referring to a set of waves that have the same wavelength, frequency, and phase.

component One of two mutually perpendicular vectors that lie along the principal axes in a coordinate system and can be combined to form a given resultant vector.

concave lens A diverging lens that causes parallel rays of light to emerge in such a way that they appear to diverge away from a focal point behind the lens.

concave mirror A converging spherical mirror that causes parallel rays of light to converge to a focal point in front of the mirror.

concurrent forces Two or more forces that act at the same point and at the same time.

conductor A substance, usually metallic, that allows the relatively easy flow of electric charges.

conservation of energy A principle of physics that states that the total energy of an isolated system remains the same during all interactions within the system.

conservation of mass-energy A principle of physics that states that, in the conversion of mass into energy or energy into mass, the total mass-energy of the system remains the same.

conservation of momentum A principle of physics that states that, in the absence of any external forces, the total momentum of an isolated system remains the same.

conservative force A force such that any work done by this force can be recovered without any loss; a force whose work is independent of the path taken.

constructive interference The additive result of two or more waves interacting with the same phase relationship as they move through a medium.

continuous spectrum A continuous band of colors, consisting of red, orange, yellow, green, blue, and violet, formed by the dispersion or diffraction of white light.

control rod A device, usually made of cadmium, that is inserted in a nuclear reactor to control the rate of fission.

converging lens A lens that will cause parallel rays of light incident on its surface to refract and converge to a focal point; a convex lens.

converging mirror A spherical mirror that will cause parallel rays of light incident on its surface to reflect and converge to a focal point; a concave mirror.

convex lens See **converging lens.**

convex mirror See **diverging mirror**.

coordinate system A set of reference lines, not necessarily perpendicular, used to locate the position of an object within a frame of reference by applying the rules of analytic geometry.

core The interior of a solenoidal electromagnet, usually made of a ferromagnetic material; the part of a nuclear reactor where the fission reaction occurs.

coulomb (C) The SI unit of electrical charge, defined as the amount of charge 1 A of current contains each second.

Coulomb's law The electrostatic force between two point charges is directly proportional to the product of the charges and inversely proportional to the square of the distance separating them. Named for French physicist Charles-Augustin de Coulomb.

critical angle of incidence The angle of incidence to a transparent substance such that the angle of refraction equals 90° relative to the normal drawn to the surface.

current A scalar quantity that measures the amount of charge passing a given point in an electric circuit each second.

current length A relative measure of the magnetic field strength produced by a length of wire carrying current, equal to the product of the current and the length.

cycle One complete sequence of periodic events or oscillations.

D

damping The continuous decrease in the amplitude of mechanical oscillations due to a dissipative force.

deflecting force Any force that acts to change the direction of motion of an object.

derived unit Any combination of fundamental physical units.

destructive interference The result produced by the interaction of two or more waves with opposite phase relationships as they move through a medium.

deuterium An isotope of hydrogen containing one proton and one neutron in its nucleus; heavy hydrogen (a component of heavy water).

dielectric An electric insulator placed between the plates of a capacitor to alter its capacitance.

diffraction The ability of waves to pass around obstacles or squeeze through small openings.

diffraction grating A reflecting or transparent surface with many thousands of lines ruled on it, used to diffract light into a spectrum.
direct current Electric current that is moving in one direction only around an electric circuit.
dispersion The separation of light into its component colors or spectrum.
dispersive medium Any medium that produces the dispersion of light; any medium in which the velocity of a wave depends on its frequency.
displacement A vector quantity that determines the change in position of an object by measuring the straight-line distance and direction from the starting point to the ending point.
dissipative force Any force, such as friction, that removes kinetic energy from a moving object; a nonconservative force.
distance A scalar quantity that measures the total length of the path taken by a moving object.
diverging lens A lens that causes parallel rays of light incident on its surface to refract and diverge away from a focal point on the other side of the lens; a concave lens.
diverging mirror A spherical mirror that causes parallel rays of light incident on its surface to reflect and diverge away from a focal point on the other side of the mirror.
Doppler effect The apparent change in the wavelength or frequency of a wave as the source of the wave moves relative to an observer. Named for Austrian physicist Christian Doppler.
dynamics The branch of mechanics that studies the effects of forces on objects.

E

elastic collision A collision between two objects in which there is a rebounding and no loss of kinetic energy occurs.
elastic potential energy The energy stored in a spring when work is done to stretch or compress it.
electrical ground The passing of charges to or from Earth to establish a potential difference between two points.
electric cell A chemical device for generating electricity.
electric circuit A closed conducting loop consisting of a source of potential difference, conducting wires, and other devices that operate on electricity.
electric field The region where an electric force is exerted on a charged object.
electric field intensity A vector quantity that measures the ratio of the magnitude of the force to the magnitude of the charge on an object.
electromagnet A coil of wire, wrapped around a ferromagnetic core (usually made of iron), that generates a magnetic field when current is passed through it.
electromagnetic field The field produced by an electromagnet or moving electric charges.
electromagnetic induction The production of a potential difference in a conductor due to the relative motion between the conductor and an external magnetic field, or due to the change in an external magnetic flux near the conductor.
electromagnetic spectrum The range of frequencies covering the discrete emission of energy from oscillating electromagnetic fields; included are radio waves, microwaves, infrared waves, visible light, ultraviolet light, X rays, and gamma rays.
electromagnetic wave A wave generated by the oscillation of electric charges producing interacting electric and magnetic fields that oscillate in space and travel at the speed of light in a vacuum.
electromotive force (emf) The potential difference caused by the conversion of different forms of energy into electrical energy; the energy per unit charge.
electron A negatively charged particle that orbits a nucleus in an atom; the fundamental carrier of negative electric charge.
electron capture The process in which an orbiting electron is captured by a nucleus possessing too many neutrons with respect to protons; also called K-capture.
electron cloud A theoretical probability distribution of electrons around the nucleus due to the Heisenberg uncertainty principle. The most probable location for an electron is in the densest regions of the cloud.
electron volt (eV) A unit of energy related to the kinetic energy of a moving charge and equal to 1.6×10^{-19} J.
electroscope A device for detecting the presence of static charges on an object.
elementary charge The fundamental amount of charge of an electron.
emf See **electromotive force**.

emission spectrum The discrete set of colored lines representing the electromagnetic energy produced when atomic compounds are excited into emitting light because of heat, sparks, or atomic collisions.

energy A scalar quantity representing the capacity to work.

energy level One of several regions around a nucleus where electrons are considered to reside.

entropy The degree of randomness or disorder in a thermodynamic system.

equilibrant The force equal in magnitude and opposite in direction to the resultant of two or more forces that brings a system into equilibrium.

equilibrium The balancing of all external forces acting on a mass; the result of a zero vector sum of all forces acting on an object.

escape velocity The velocity attained by an object such that, if coasting, the object would not be pulled back toward the planet from which it came.

excitation The process by which an atom absorbs energy and causes its orbiting electrons to move to higher energy levels.

excited state In an atom, the situation in which its orbiting electrons are residing in higher energy levels.

F

farad (F) A unit of capacitance equal to 1 C/V.

Faraday's law of electromagnetic induction The magnitude of the induced emf in a conductor is equal to the rate of change of the magnetic flux. Named for British chemist/physicist Michael Faraday.

ferromagnetic substances A metal or a compound made of iron, cobalt, or nickel that produces very strong magnetic fields.

field A region characterized by the presence of a force on a test body like a unit mass in a gravitational field or a unit charge in an electric field.

field intensity A measure of the force exerted on a unit test body; the force per unit mass; the force per unit charge.

first law of thermodynamics A statement of the conservation of energy as applied to thermodynamic systems: The change in energy of a system is equal to the change in the internal energy plus any work done by the system.

fission The splitting of a uranium nucleus into two smaller, more stable nuclei by means of a slow-moving neutron, with the release of a large amount of energy.

flux A measure, in webers, of the product of the perpendicular component of a magnetic field crossing an area and the magnitude of the area.

flux density A measure, in webers per square meter, of the field intensity per unit area.

focal length The distance along the principal axis from a lens or spherical mirror to the principal focus.

focus The point of convergence of light rays caused by a converging mirror or lens; either of two fixed points in an ellipse that determines its shape.

force A vector quantity that corresponds to any push or pull due to an interaction of matter that changes the motion of an object.

force constant See **spring constant.**

forced vibration A vibration caused by the application of an external force.

frame of reference A point of view consisting of a coordinate system in which observations are made.

free-body diagram A diagram that illustrates all of the forces acting on a mass at any given time.

frequency The number of completed periodic cycles per second in an oscillation or wave motion.

friction A force that opposes the motion of an object as it slides over another surface.

fuel rods Rods packed with fissionable material that are inserted into the core of a nuclear reactor.

fundamental unit An arbitrary scale of measurement assigned to certain physical quantities, such as length, time, mass, and charge, that are considered to be the basis for all other measurements. In the SI system, the fundamental units used in physics are the meter, kilogram, second, ampere, kelvin, and mole.

fusion The combination of two or more light nuclei to produce a more stable, heavier nucleus, with the release of energy.

G

galvanometer A device used to detect the presence of small electric currents when connected in series in a circuit.
gamma radiation High-energy photons emitted by certain radioactive substances.
gravitation The mutual force of attraction between two uncharged masses.
gravitational field strength A measure of the gravitational force per unit mass in a gravitational field.
gravity Another name for gravitation or the gravitational force; the tendency of objects to fall to Earth.
ground state The lowest energy level of an atom.

H

heat The energy observed due to particle collisions in matter; the energy produced when matter interacts with particles that are colliding randomly with each other, as in a gas.
hertz (Hz) The SI unit of frequency, equal to 1 s^{-1}.
Hooke's law The stress applied to an elastic material is directly proportional to the strain produced. Named for English scientist Robert Hooke.

I

ideal gas A gas for which the assumptions of the kinetic theory are valid.
image An optical reproduction of an object by means of a lens or mirror.
impulse A vector quantity equal to the product of the average force applied to a mass and the time interval in which the force acts; the area under a force versus time graph.
induced potential difference A potential difference created in a conductor because of its motion relative to an external magnetic field.
induction coil A transformer in which a variable potential difference is produced in a secondary coil when a direct current, applied to the primary, is turned on and off.
inelastic collision A collision in which two masses interact and stick together, leading to an apparent loss of kinetic energy.
inertia The property of matter that resists the action of applied force trying to change the motion of an object.
inertial frame of reference A frame of reference in which the law of inertia holds; a frame of reference moving with constant velocity relative to Earth.
instantaneous velocity The slope of a tangent line to a point in a displacement versus time graph.
insulator A substance that is a poor conductor of electricity because of the absence of free electrons.
interference The interaction of two or more waves, producing an enhanced or a diminished amplitude at the point of interaction; the superposition of one wave on another.
interference pattern The pattern produced by the constructive and destructive interference of waves generated by two point sources.
isobaric In thermodynamics, referring to a process in which the pressure of a gas remains the same.
isochoric In thermodynamics, referring to a process in which the volume of a gas remains the same.
isolated system A combination of two or more interacting objects that are not being acted upon by external force.
isotope An atom with the same number of protons as a particular element but a different number of neutrons.

J

joule (J) The SI unit of work, equal to 1 N · m; the SI unit of mechanical energy, equal to 1 kg · m^2/s^2.
junction The point in an electric circuit where a parallel connection branches off.

K

K-capture See **electron capture.**
kelvin (K) The SI unit of absolute temperature, defined in such a way that 0 K equals −273°C.

Kepler's first law The orbital paths of all planets are elliptical. Named for German astronomer Johannes Kepler.

Kepler's second law A line from the Sun to a planet sweeps out equal areas in equal time.

Kepler's third law The ratio of the cube of the mean radius to the Sun to the square of the period is a constant for all planets orbiting the Sun.

kilogram (kg) The SI unit of mass.

kilojoule (kJ) A unit representing 1000 J.

kilopascal (kPa) A unit representing 1000 Pa.

kinematics In mechanics, the study of how objects move.

kinetic energy The energy possessed by a mass because of its motion relative to a frame of reference.

kinetic friction The friction induced by sliding one surface over another.

kinetic theory of gases The theory that all matter consists of molecules that are in a constant state of motion.

Kirchhoff's first law The algebraic sum of all currents at a junction in a circuit equals zero. Named for German physicist Gustav Kirchhoff.

Kirchhoff's second law The algebraic sum of all potential drops around any closed loop in a circuit equals zero.

L

laser A device for producing an intense coherent beam of monochromatic light; an acronym for *l*ight *a*mplification by the *s*timulated *e*mission of *r*adiation.

law of inertia See **Newton's first law of motion.**

lens A transparent substance, with one or two curved surfaces, used to direct rays of light by refraction.

Lenz's law An induced current in a conductor is always in a direction such that its magnetic field opposes the magnetic field that induced it. Named for German physicist Emil Lenz.

line of force An imaginary line drawn in a gravitational, electrical, or magnetic field that indicates the direction a test particle would take while experiencing a force in that field.

longitudinal wave A wave in which the oscillating particles vibrate in a direction parallel to the direction of propagation.

M

magnet An object that exerts a force on ferrous materials.

magnetic field A region surrounding a magnet in which ferrous materials or charged particles experience a force.

magnetic field strength The force on a unit current length in a magnetic field.

magnetic flux density The total magnetic flux per unit area in a magnetic field.

magnetic pole A region on a magnet where magnetic lines of force are most concentrated.

mass The property of matter used to represent the inertia of an object; the ratio of the net force applied to an object and its subsequent acceleration as specified by Newton's second law of motion.

mass defect The difference between the actual mass of a nucleus and the sum of the masses of the protons and neutrons it contains.

mass number The total number of protons and neutrons in an atomic nucleus.

mass spectrometer A device that uses magnetic fields to cause nuclear ions to assume circular trajectories and then determines the masses of the ions based on their charges, the radius of the path, and the external field strength.

mean radius For a planet, the average distance to the Sun.

meter (m) The SI unit of length.

moderator A material, such as water or praphite, that is used to slow neutrons in a fission reaction.

mole (mol) The SI unit for the amount of substance containing Avogadro's number of molecules.

moment arm distance A line drawn from a pivot point.

momentum A vector quantity equal to the product of the mass and the velocity of a moving object.

monochromatic light Light consisting of only one frequency.

N

natural frequency The frequency with which an elastic body will vibrate if disturbed.
net force The resultant force acting on a mass.
neutron A subatomic particle, residing in the nucleus, that has a mass comparable to that of a proton but is electrically neutral.
newton (N) The SI unit of force, equal to 1 kg · m/s^2.
Newton's first law of motion An object at rest tends to remain at rest, and an object in motion tends to remain in motion at a constant velocity, unless acted upon by an external force; the law of inertia. Named for British mathematician/physicist Sir Isaac Newton.
Newton's second law of motion The acceleration of a body is directly proportional to the applied net force; $\mathbf{F} = m\mathbf{a}$.
Newton's third law of motion For every action, there is an equal, but opposite, reaction.
nodal line A line of minimum displacement when two or more waves interfere.
nonconservative force A force, such as friction, that decreases the amount of kinetic energy after work is done by or against the force.
normal A line perpendicular to a surface.
normal force A force that is directed perpendicularly to a surface when two objects are in contact.
nuclear force The force between nucleons in a nucleus that opposes the Coulomb force repulsion of the protons; the strong force.
nucleon Either a proton or a neutron as it exists in a nucleus.

O

ohm (Ω) The SI unit of electrical resistance, equal to 1 V/A.
Ohm's law In a circuit at constant temperature, the ratio of the potential difference to the current is a constant (called the resistance). Named for German physicist Georg Ohm.
optical center The geometric center of a converging or diverging lens.

P

parallel circuit A circuit in which two or more devices are connected across the same potential difference and provide an alternative path for charge flow.
pascal (Pa) The SI unit of pressure, equal to 1 N/m^2.
Pascal's principle A change in the pressure applied to an enclosed fluid is transmitted undiminished to every portion of the fluid and to the walls of the containing vessel.
period The time, in seconds, to complete one cycle of repetitive oscillations or uniform circular motion.
permeability The property of matter that affects an external magnetic field in which the material is placed, causing it to align free electrons within the substance and hence magnetizing it; a measure of a material's ability to become magnetized.
phase The relative position of a point on a wave with respect to another point on the same wave.
photoelectric effect The process by which surface electrons in a metal are freed through the incidence of electromagnetic radiation above a certain minimum frequency.
photon A single packet of electromagnetic energy.
Planck's constant A universal constant representing the ratio of the energy of a photon to its frequency; the fundamental quantum of action, having a value of 6.63×10^{-34} kg · m^2/s. Named for German physicist Max Planck.
polar coordinate system A coordinate system in which the location of a point is determined by a vector from the origin of a Cartesian coordinate system and the acute angle it makes with the positive horizontal axis.
polarization The process by which the vibrations of a transverse wave are selected to lie in a preferred plane perpendicular to the direction of propagation.
polarized light Light that has been polarized by passing it through a suitable filter.
positron A subatomic particle having the same mass as an electron but a positive electrical charge.
potential difference The difference in work per unit charge between any two points in an electric field.

potential energy The energy possessed by a body because of its vertical position relative to the surface of the Earth or any other arbitrarily chosen base level.

power The ratio of the work done to the time needed to complete the work.

pressure The force per unit area.

primary coil In a transformer, the coil connected to an alternating potential difference source.

principal axis An imaginary line passing through the center of curvature of a spherical mirror or the optical center of a lens.

prism A transparent triangular shape that disperses white light into a continuous spectrum.

pulse A single vibratory disturbance in an elastic medium.

Q

quantum A discrete packet of electromagnetic energy; a photon.

quantum theory The theory that light consists of discrete packets of energy that are absorbed or emitted in units.

R

radioactive decay The spontaneous disintegration of a nucleus due to its instability.

radius of curvature The distance from the center of curvature to the surface of a spherical mirror.

ray A straight line used to indicate the direction of travel of a light wave.

real image An image formed by a converging mirror or lens that can be focused onto a screen.

refraction The bending of a light ray when it passes obliquely from one transparent substance to another in which its velocity is different.

resistance The ratio of the potential difference across a conductor to the current in the conductor.

resistivity A property of matter that measures the ability of that substance to act as a resistor.

resolution of forces The process by which a given force is decomposed into a pair of perpendicular forces.

resonance The production of sympathetic vibrations in a body, at its natural vibrating frequency, caused by another vibrating body.

rest mass The mass of an object when the object is not moving.

rotational equilibrium The situation when the vector sum of all torques acting on a rotating mass equals zero.

rotational inertia The ability of a substance to resist the action of a torque; moment of inertia.

S

scalar A physical quantity, such as mass or speed, that is characterized by magnitude only.

second (s) The SI unit of time.

secondary coil In a transformer, a coil in which an alternating potential difference is induced.

second law of thermodynamics Heat flows only from a hot body to a cold one; the entropy of the universe is always increasing.

series circuit A circuit in which electrical devices are connected sequentially in a single conducting loop, allowing only one path for charge flow.

sliding friction A force that resists the sliding motion of one surface over another; kinetic friction.

Snell's law For light passing obliquely from one transparent substance to another, the ratio of the sine of the angle of incidence to the sine of the angle of refraction is equal to the relative index of refraction for the two media. Named for Willebrord Snell.

solenoid A coil of wire used for electromagnetic induction.

specific heat The amount of energy, in kilojoules, needed to change the temperature of 1 kg of a substance by 1°C.

speed A scalar quantity measuring the time rate of change of distance.

spherical aberration In a converging spherical mirror or a converging lens, the inability to properly focus parallel rays of light because of the shape of the mirror or lens.

spring constant The ratio of the applied force and resultant displacement of a spring.

standard pressure At sea level, 1 atm, which is equal to 101.3 kPa.

standing wave A stationary wave pattern formed in a medium when two sets of waves with equal wavelength and amplitude pass through each other, usually after a reflection.

starting friction The force of friction overcome when one of two objects begins to slide over the other.
static electricity Stationary electric charges.
static friction The force that prevents one object from sliding over another.
statics The study of the forces acting on an object that is at rest relative to a frame of reference.
superposition The ability of waves to pass through each other, interfere, and then continue on their way unimpeded.

T

temperature The relative measure of warmth or cold relative to a standard; see also **absolute temperature**.
tesla (T) The SI unit of magnetic field strength, equal to 1 W/m^2.
test charge A small, positively charged mass used to detect the presence of a force in an electric field.
thermionic emission The emission of electrons from a hot filament.
thermometer An instrument that makes a quantitative measurement of temperature based on an accepted scale.
third law of thermodynamics The temperature of any system can never be reduced to absolute zero.
threshold frequency The minimum frequency of electromagnetic radiation necessary to induce the photoelectric effect in a metal.
torque The application, from a pivot, of a force at right angles to a designated line that tends to produce circular motion; the product of a force and a perpendicular moment arm distance.
total internal reflection The process by which light is incident on a medium in which its velocity would increase at an angle greater than the critical angle of incidence.
total mechanical energy In a mechanical system, the sum of the kinetic and potential energies.
transmutation The process of changing one atomic nucleus into another by radioactive decay or nuclear bombardment.
transverse wave A wave in which the vibrations of the medium or field are at right angles to the direction of propagation.

U

uniform circular motion Motion around a circle at a constant speed.
uniform motion Motion at a constant speed in a straight line.
unit An arbitrary scale assigned to a physical quantity for measurement and comparison.
universal law of gravitation The gravitational force between any two masses is directly proportional to the product of their masses, and inversely proportional to the square of the distance between them.

V

vector A physical quantity that is characterized by both magnitude and direction; a directed arrow drawn in a Cartesian coordinate system used to represent a quantity such as force, velocity, or displacement.
velocity A vector quantity representing the time rate of change of displacement.
virtual focus The point at which the rays from a diverging lens would meet if they were traced back with straight lines.
virtual image An image, formed by a mirror or lens, that cannot be focused onto a screen.
visible light The portion of the electromagnetic spectrum that can be detected by the human eye.
volt (V) The SI unit of potential difference, equal to 1 J/C.
voltmeter A device used to measure the potential difference between two points in a circuit when connected in parallel; a galvanometer with a high-resistance coil placed in series with it.
volt per meter (V/m) The SI unit of electric field intensity, equal to 1 N/C.

W

watt (W) The SI unit of power, equal to 1 J/s.

wave A series of periodic disturbances in an elastic medium or field.

wavelength The distance between any two successive points in phase on a wave.

weber (Wb) The SI unit of magnetic flux, equal to $1\ \text{T} \times 1\ \text{m}^2$.

weight The force of gravity exerted on a mass at the surface of a planet.

work A scalar measure of the relative amount of change in mechanical energy; the product of the magnitude of the displacement of an object and the component of applied force in the same direction as the displacement; the area under a graph of force versus displacement.

work function The minimum amount of energy needed to free an electron from the surface of a metal using the photoelectric effect.

Appendix

A. Review of Mathematics

I. ALGEBRA

Physical relationships are often expressed as mathematical equations, and the techniques of algebra are often needed to solve these equations as part of the process of analyzing the physical world. In general, the letters *u*, *v*, *w*, *x*, *y*, and *z* are used as variable unknown quantities, with *x*, *y*, and *z* the most popular choices. The letters *a*, *b*, *c*, *d*, *e*, . . . often represent numbers or coefficients, with *a*, *b*, and *c* the most popular.

An equation of the form

$$y = ax + b$$

is called a **linear** equation, and its graph is a straight line. The coefficient *a* is called the **slope** of the line, and *b* is termed the ***y*-intercept**, that is, the point at which the line crosses the *y*-axis (using standard Cartesian coordinates). An equation of the form

$$y = ax^2 + bx + c$$

is called a **quadratic** equation, and a graph of this relationship is a parabola. The **order** of the equation is the highest power of *x*, which is the **independent variable**.

The variables *x* and *y* can represent any physical quantities being studied. For example, the following quadratic equation represents the displacement of a particle undergoing one-dimensional uniformly accelerated motion (see Chapter 5):

$$x = x_0 + v_0 t + \left(\frac{1}{2}\right)\mathbf{a}t^2$$

In this equation x_0, v_0, and *a* are all constants, while *x* and *t* are the variables (*x* representing displacement and *t*, time).

While there is only one solution to a linear equation, a quadratic equation has two solutions. However, in physics, it is possible that only one solution is physically reasonable; for example, there is no "negative time" in physics.

A linear equation represents a **direct relationship** (see Figure A.1.a) between two quantities. We would say that *y* varies directly with *x*. A quadratic equation also expresses a direct relationship, but not a linear one. An **inverse relationship** (Figure

A.1.b) is one in which, as the independent variable increases, the dependent variable decreases. This relationship is expressed as

$$y = \frac{a}{x}$$

and its graph is a hyperbola. It is possible that the power of x is larger than 1 (like the inverse square law of gravity; see Chapter 12) in the denominator, but the graph remains a hyperbola, although a steeper one (Figure A.1.c).

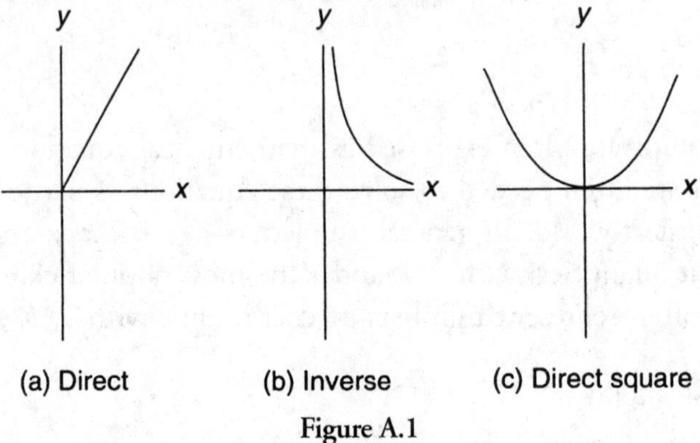

Figure A.1

Solutions of Algebraic Equations

A set of points (x, y) is a solution of an algebraic equation if the number of points for each variable is the same and if, for each x, there is one and only one y value (this is also the definition of a **function**) when the value of x is substituted into the equation and the subsequent arithmetical operations are performed. In other words, the equation $3x + 2 = -4$ has only one solution since there is only one variable, x (the solution being $x = -2$). The equation $y = 3x + 2$, however, requires a pair of numbers (x, y) since there are two variables. We could try an imprecise method of trial and error, but an easier way is to graph the equation using test values for the independent variable x. A table of sample numbers for x and the subsequent y values is presented below.

x	y
0	2
1	5
2	8
−1	−1
−2	−4

In Figure A.2, a graph of the equation $y = 3x + 2$ is given. The "solution" is somewhat arbitrary in the sense that the value of y for any given x can be interpolated (or extrapolated) from the graph.

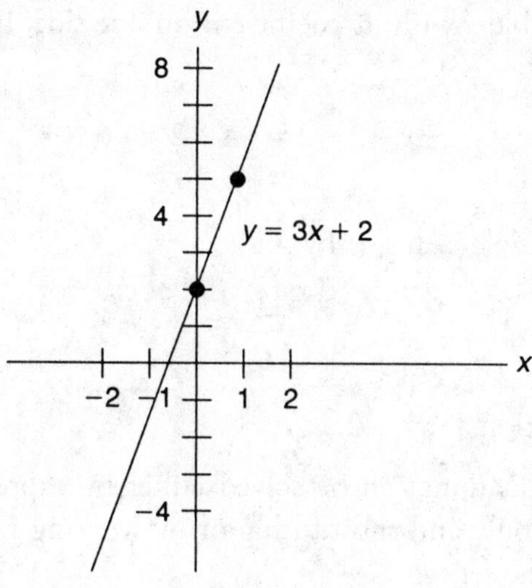

Figure A.2

Technically, there is no one solution to $y = 3x + 2$. Since both x and y are variables, we need another equation to make a coupled system of two equations and two unknowns. Suppose such a set consists of

$$y = 3x + 2$$
$$x - y = -1$$

The solution to this system is a unique set of numbers (x, y) that satisfies both equations. Graphically, we can solve this system by plotting both equations (with sample data) and then looking for the points of intersection (see Figure A.3).

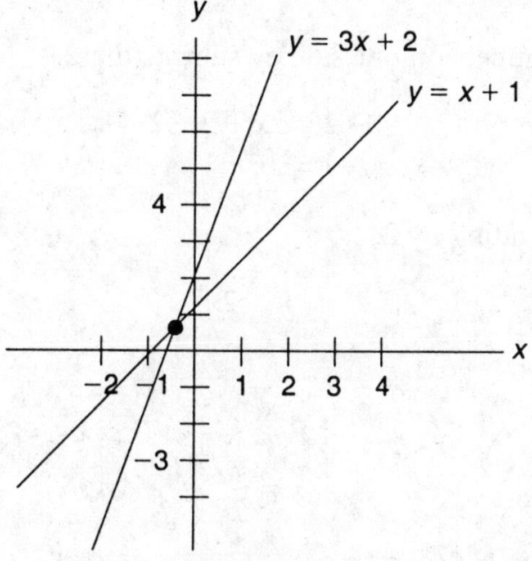

Figure A.3

The intersection points are $x = -0.5$ and $y = 0.5$. Substitution of these values into both equations will yield similar results.

The same results can be achieved algebraically. The equation $3x + 2 = -4$ can be solved using the rules of algebra.

Step 1: Place the variable x with its coefficient on one side, by itself, by subtracting 2 on each side:

$$3x + 2 - 2 = -4 - 2 = -6$$
$$3x = -6$$

Step 2: Find x by dividing each side by 3:

$$\frac{3x}{3} = \frac{-6}{3} = -2$$

Step 3: Write the solution as $x = -2$.

A system of two equations can be solved either by expressing one equation in terms of one variable only and substituting or by working with both equations. In our example we had the set

$$y = 3x + 2$$
$$x - y = -1$$

Step 1: Write the second equation as $y = x + 1$.

Step 2: In the first equation substitute for y: $x + 1 = 3x + 2$.

Step 3: Following the preceding example, get all the x variables on one side through addition or subtraction. In this case subtract x from both sides of the equation obtained in step 2:

$$x + 1 - x = 3x + 2 - x$$
$$1 = 2x + 2$$

Step 4: Get all the numbers on one side by subtracting 2:

$$1 - 2 = 2x + 2 - 2$$
$$-1 = 2x$$

Step 5: Isolate x by dividing by 2:

$$-\frac{1}{2} = \frac{2x}{2}$$

which implies that

$$x = \frac{1}{2}$$

Step 6: Substitute the value for x into any equation to solve for y:

$$y = \frac{1}{2} + 1 = \frac{1}{2}$$

Step 7: Write the final solution as $x = -1/2$ and $y = 1/2$.

Quadratic equations can be solved using the **quadratic formula**. Given the quadratic equation

$$ax^2 + bx + c = 0$$

we can write

$$x = \frac{-b \pm \sqrt{b^2 - 4ac}}{2a}$$

The quantity under the radical sign, called the **discriminant**, can be positive, negative, or zero. If the discriminant is negative, the square root is an **imaginary number**, which usually has no physical meaning for our study.

For example, suppose we wish to solve $x^2 + 3x + 2 = 0$ using the quadratic formula. In this equation $a = 1$, $b = 3$, and $c = 2$. By direct substitution and taking the necessary square roots, we find that $x = -1$ or $x = -2$.

Another way to solve this quadratic equation is by **factoring**. Some quadratic equations can be written as the product of two linear factors. In this case, we can write

$$x^2 + 3x + 2 = (x + 2)(x + 1) = 0$$

The solution is found by recognizing that, for this equation to be zero, either $x + 2 = 0$ or $x + 1 = 0$. Thus, either $x = -2$ or $x = -1$ is a solution (as before).

Often, equations given in one form need to be expressed in an alternative form. The rules of algebra allow us to manipulate the form of an equation. For example, we might have this problem: Given $x = (1/2)at^2$, solve for t. First, we clear the fraction by multiplying by 2, obtaining $2x = at^2$. Now, we divide both sides by a to get $2x/a = t^2$. Finally, to solve for t, we take the square root of each side. Mathematically, there are two solutions. However, in physics, we must allow for the physical reality of a solution. Since this equation represents uniformly accelerated motion from rest, the concept of "negative time" is not realistic. Hence, we discard the negative square root solution and state simply that

$$t = \sqrt{\frac{2x}{a}}$$

Exponents and Scientific Notation

Any number, n, can be written in the form of some base number, B, raised to a power a. The number a is called the **exponent** of the base number, B. In other words, we can write

$$n = B^a$$

One common base number is 10. The use of products of numbers with powers of 10 is called **scientific notation**. Some examples of powers of 10 are $10 = 10^1$, $100 = 10 \times 10 = 10^2$, and $1000 = 10 \times 10 \times 10 = 10^3$.

By definition, any number raised to the zero power is equal to 1; that is, $10^0 = 1$.

Decimal numbers less than 1 have negative exponents, since they are fractions of the powers of 10 previously discussed. Some examples are $0.1 = 10^{-1}$, $0.01 = 10^{-2}$, and $0.001 = 10^{-3}$. The use of negative exponents for small numbers comes from the law of division of exponents, which is defined as follows:

$$\frac{10^a}{10^b} = 10^{a-b}$$

Since a reciprocal means "1 over . . . ," and 10 raised to the power of zero is equal to 1, then, if, in the division example above, $a = 0$ and $b =$ any number, we have negative exponents for small decimal numbers.

The law of multiplication of exponents is as follows:

$$10^a 10^b = 10^{a+b} \quad \text{and} \quad (10^a)^b = 10^{ab}$$

Scientific notation involves the use of numbers and powers of 10 as products. For example, the number 200 can be expressed in scientific notation as 2.0×10^2, and the number 3450 as 3.45×10^3. Finally, the number 0.045 is expressed in scientific notation as 4.5×10^{-2}.

On most scientific calculators, scientific notation can be activated using the "exp" button. It is not necessary to push "×10" or "times 10" on the calculator. The "exp" button automatically implies "times 10 to the . . ." in the notation. Since the use of a scientific calculator is permitted on the AP Physics Examination, we will not spend time reviewing the addition, subtraction, multiplication, and division of numbers in scientific notation. Consult a mathematics textbook for more examples and details.

The following prefixes designating exponentiation are often used in physics:

10^{18}	*exa-*	10^{-1}	*deci-*
10^{15}	*peta-*	10^{-2}	*centi-*
10^{12}	*tera-*	10^{-3}	*milli-*
10^{9}	*giga-*	10^{-6}	*micro-*
10^{6}	*mega-*	10^{-9}	*nano-*
10^{3}	*kilo-*	10^{-12}	*pico-*
10^{2}	*hecto-*	10^{-15}	*femto-*
10^{1}	*deca-*	10^{-18}	*atto-*

II. GEOMETRY

Some common formulas from geometry are of use in physics. We first review some of the more common geometric shapes and their equations:

Straight line with slope m and y-intercept b: $y = mx + b$

Circle of radius R centered at the origin: $x^2 + y^2 = R^2$

Parabola whose vertex is at $y = b$: $y = ax^2 + b$

Hyperbola: $xy =$ constant

Ellipse with semimajor axis a and semiminor axis b: $\dfrac{x^2}{a^2} + \dfrac{y^2}{b^2} = 1$

We now review the equations for some of the physical characteristics of selected geometric shapes.

Circle of radius R: Area $= \pi R^2$; Circumference $= 2\pi R$

Sphere of radius R: Volume $= \dfrac{4}{3}\pi R^3$; Surface area $= 4\pi R^2$

Cylinder of radius R and length ℓ: Volume $= \pi R^2 \ell$

Right circular cone of base radius R and height h: Volume $= \frac{1}{3}\pi R^2 h$

Ellipse with semimajor axis a and semiminor axis b: Area $= \pi ab$

Rectangle of length ℓ and width w: Area $= \ell w$

Triangle with base b and altitude h: Area $= \frac{1}{2}bh$

III. TRIGONOMETRY

Trigonometry is the branch of mathematics that deals with the algebraic relationships between angles in triangles. If we have a right triangle, with hypotenuse c and sides a and b, the relationship between these sides is given by the Pythagorean theorem:

$$a^2 + b^2 = c^2$$

The ratios of the sides of a right triangle to its hypotenuse define the **trigonometric functions** of sine, cosine, and tangent (see Figure A.4.a) and set 1 below.

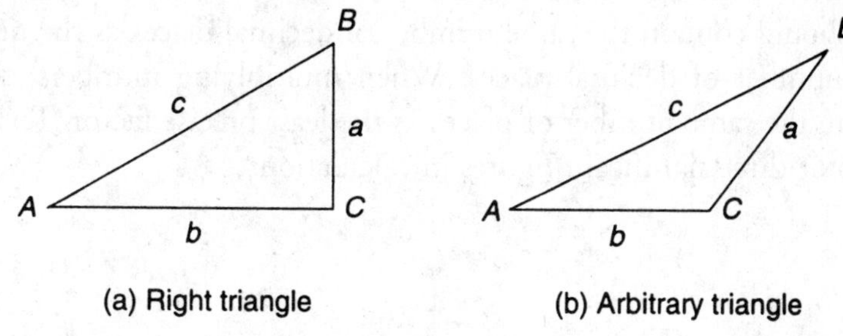

Figure A.4

1. $\sin A = \frac{a}{c}$ $\qquad \cos A = \frac{b}{c}$ $\qquad \tan A = \frac{a}{b}$ $\qquad \tan A = \frac{\sin A}{\cos A}$

2. $\sec A = \frac{c}{b}$ $\qquad \csc A = \frac{c}{a}$ $\qquad \cot A = \frac{b}{a}$

The functions in set 2 are the reciprocals of set 1. A table of values for the standard trigonometric functions is provided in Appendix B. For any angle θ, in a right triangle,

$$\sin\theta = \cos(90 - \theta) \quad \text{and} \quad \cos\theta = \sin(90 - \theta)$$

In the arbitrary general triangle shown in Figure A.4.b, the sides are related through the law of cosines and the law of sines. These formulas are extremely useful for the study of vectors (see Chapter 4):

Law of cosines: $\quad c^2 = a^2 + b^2 - 2ab\cos C$

Law of sines: $\quad \frac{a}{\sin A} = \frac{b}{\sin B} = \frac{c}{\sin C}$

For angles between 0 and 90 degrees, all of the trigonometric functions have positive values. For angles between 90 and 180 degrees, only the sine function is positive. For angles between 180 and 270 degrees, only the tangent function is positive. Finally, between 270 and 360 (0) degrees, only the cosine function is positive.

Some useful identities (relationships) between the various trigonometric functions often encountered in the study of physics are listed below without proof:

$$\sin^2 \theta + \cos^2 \theta = 1$$
$$\sin 2\theta = 2 \sin \theta \cos \theta$$
$$\cos 2\theta = \cos^2 \theta - \sin^2 \theta$$

IV. SIGNIFICANT FIGURES FOR LABORATORY SKILLS

Here is a brief reminder about measurements and significant figures. In physics, measurements are judged on the basis of their accuracy and precision. **Accuracy** refers to the extent to which a given measurement is exact relative to an accepted value. **Precision** refers to the agreement among several similar measurements.

The use of significant figures informs the reader how accurate and precise given measurements are. For example, calculations may produce numerical answers that extend out to six places past the decimal point. If the given information is not accurate to that many places, then you should round off to the nearest significant figure.

A good rule of thumb is that when adding or subtracting numbers, the final answer should contain the same number of decimal places as the number with the smallest number of decimal places. When multiplying numbers, the product should contain the same number of places as the least precise factor. Textbooks typically use two or three significant figures in calculations.

B. Table of Trigonometric Functions

Angle	Sine	Cosine	Tangent	Angle	Sine	Cosine	Tangent
0	0.000	1.000	0.000				
1	0.017	1.000	0.017	46	0.719	0.695	1.036
2	0.035	0.999	0.035	47	0.731	0.682	1.072
3	0.052	0.999	0.052	48	0.743	0.669	1.111
4	0.070	0.998	0.070	49	0.755	0.656	1.150
5	0.087	0.996	0.087	50	0.766	0.643	1.192
6	0.105	0.995	0.105	51	0.777	0.629	1.235
7	0.122	0.993	0.123	52	0.788	0.616	1.280
8	0.139	0.990	0.141	53	0.799	0.602	1.327
9	0.156	0.988	0.158	54	0.809	0.588	1.376
10	0.174	0.985	0.176	55	0.819	0.574	1.428
11	0.191	0.982	0.194	56	0.829	0.559	1.483
12	0.208	0.978	0.213	57	0.839	0.545	1.540
13	0.225	0.974	0.231	58	0.848	0.530	1.600
14	0.242	0.970	0.249	59	0.857	0.515	1.664
15	0.259	0.966	0.268	60	0.866	0.500	1.732
16	0.276	0.961	0.287	61	0.875	0.485	1.804
17	0.292	0.956	0.306	62	0.883	0.469	1.881
18	0.309	0.951	0.325	63	0.891	0.454	1.963
19	0.326	0.946	0.344	64	0.899	0.438	2.050
20	0.342	0.940	0.364	65	0.906	0.423	2.145
21	0.358	0.934	0.384	66	0.914	0.407	2.246
22	0.375	0.927	0.404	67	0.921	0.391	2.356
23	0.391	0.921	0.424	68	0.927	0.375	2.475
24	0.407	0.914	0.445	69	0.934	0.358	2.605
25	0.423	0.906	0.466	70	0.940	0.342	2.748
26	0.438	0.899	0.488	71	0.946	0.326	2.904
27	0.454	0.891	0.510	72	0.951	0.309	3.078
28	0.469	0.883	0.532	73	0.956	0.292	3.271
29	0.485	0.875	0.554	74	0.961	0.276	3.487
30	0.500	0.866	0.577	75	0.966	0.259	3.732
31	0.515	0.857	0.601	76	0.970	0.242	4.011
32	0.530	0.848	0.625	77	0.974	0.225	4.332
33	0.545	0.839	0.649	78	0.978	0.208	4.705
34	0.599	0.829	0.675	79	0.982	0.191	5.145
35	0.574	0.819	0.700	80	0.985	0.174	5.671
36	0.588	0.809	0.727	81	0.988	0.156	6.314
37	0.602	0.799	0.754	82	0.990	0.139	7.115
38	0.616	0.788	0.781	83	0.993	0.122	8.144
39	0.629	0.777	0.810	84	0.995	0.105	9.514
40	0.643	0.766	0.839	85	0.996	0.087	11.430
41	0.656	0.755	0.869	86	0.998	0.070	14.300
42	0.669	0.743	0.900	87	0.999	0.052	19.080
43	0.682	0.731	0.933	88	0.999	0.035	28.640
44	0.695	0.719	0.966	89	1.000	0.017	57.290
45	0.707	0.707	1.000	90	1.000	0.000	

C. Values of Some Fundamental Physical Constants

Quantity	Symbol	Value
Speed of light (vacuum)	c	2.99792458×10^{8} m/s
Permeablility of vacuum	μ_0	$4\pi \times 10^{-7}$ H/m
Permittivity of the vacuum	ϵ_0	$8.854187818 \times 10^{-12}$ F/m
Planck's constant	h	6.626176×10^{-34} J/Hz
Elementary charge	e	$1.6021892 \times 10^{-19}$ C
Avogadro's number	N_A	6.022045×10^{23}/mol
Atomic mass unit	u	$1.6605655 \times 10^{-27}$ kg $= 931.5016 \times 10^{6}$ eV
Gravitational constant	G	6.6720×10^{-11} kg $\cdot$ m^2/kg^2
Bohr radius	a_0	$5.2917706 \times 10^{-11}$ m
Charge-to-mass ratio of electron	e/m_e	1.7588047×10 C/kg
Compton wavelength of electron	λ_C	$2.4263089 \times 10^{-12}$ m
Rest mass of the electron	m_e	9.109534×10^{-31} kg $= 5.485803 \times 10^{-4}$ u
Rest mass of proton	m_p	$1.6726485 \times 10^{-27}$ kg = 1.0072764 u
Rest mass of neutron	m_n	$1.6749543 \times 10^{-27}$ kg = 1.008665 u
Rydberg constant	R_∞	$1.097373177 \times 10^{7}$/m
Boltzmann's constant	k	1.380662×10^{-23} J/K
Molar gas constant	R	8.31441 J/mol $\cdot$ K
Coulomb constant	k	9×10^{9} N $\cdot$ m^2/C^2
Faraday constant	F	9.65×10^{4} C/mol

D. The Greek Alphabet

Alpha	Α	α	Nu	Ν	ν
Beta	Β	β	Xi	Ξ	ξ
Gamma	Γ	γ	Omicron	Ο	ο
Delta	Δ	δ	Pi	Π	π
Epsilon	Ε	ϵ	Rho	Ρ	ρ
Zeta	Ζ	ζ	Sigma	Σ	σ
Eta	Η	η	Tau	Τ	τ
Theta	Θ	θ	Upsilon	Υ	υ
Iota	Ι	ι	Phi	Φ	ϕ
Kappa	Κ	κ	Chi	Χ	χ
Lambda	Λ	λ	Psi	Ψ	ψ
Mu	Μ	μ	Omega	Ω	ω

E. Values of Some Often Used Physical Quantities

Quantity	Value
Acceleration due to gravity (g)	9.8 m/s^2
Density of air (20°C, 1 atm)	1.2 kg/m^3
Density of water (20°C, 1 atm)	1.00×10^3 kg/m^3
Standard atmospheric pressure (1 atm)	1.013×10^5 Pa
Mass of Earth	5.98×10^{24} kg
Mass of the Moon	7.36×10^{22} kg
Mass of the Sun	1.99×10^{30} kg
Mean radius of Earth	6.37×10^6 m
Mean radius of the Moon	1.74×10^6 m
Mean Earth-Moon distance	3.84×10^8 m
Mean Earth-Sun distance	1.496×10^{11} m

Index

THE FOLLOWING DOCUMENTATION APPLIES IF YOU PURCHASED
AP Physics B, 5th Edition **with CD-ROM.**
Please disregard this information if your edition does not contain the CD-ROM.

SYSTEM REQUIREMENTS

The program will run on a PC with:
Windows® Intel® Pentium II 450 MHz
or faster, 128MB of RAM
1024 X 768 display resolution
Windows 2000, XP, Vista
CD-ROM Player

The program will run on a Macintosh® with:
PowerPC® G3 500 MHz
or faster, 128MB of RAM
1024 X 768 display resolution
Mac OS X v.10.1 through 10.4
CD-ROM Player

Installation Instructions

The software is not installed on your computer; it runs directly from the CD-ROM. Barron's CD-ROM includes an "autorun" feature that automatically launches the application when the CD is inserted into the CD-ROM drive. In the unlikely event that the autorun feature is disabled, follow the manual launching instructions below.

Windows®
1. Click on the Start button and choose "My Computer."
2. Double-click on the CD-ROM drive, which will be named **AP_Physics.exe**.
3. Double-click **AP_Physics.exe** application to launch the program.

Macintosh®
1. Insert the CD-ROM.
2. Double-click the CD-ROM icon.
3. Double-click the **AP_Physics_B** icon to start the program.